NanoScience and Technology

NanoScience and Technology

Series Editors:
P. Avouris B. Bhushan D. Bimberg K. von Klitzing H. Sakaki R. Wiesendanger

The series NanoScience and Technology is focused on the fascinating nano-world, mesoscopic physics, analysis with atomic resolution, nano and quantum-effect devices, nanomechanics and atomic-scale processes. All the basic aspects and technology-oriented developments in this emerging discipline are covered by comprehensive and timely books. The series constitutes a survey of the relevant special topics, which are presented by leading experts in the field. These books will appeal to researchers, engineers, and advanced students.

**Magnetic Microscopy
of Nanostructures**
Editors: H. Hopster and H.P. Oepen

Applied Scanning Probe Methods I
Editors: B. Bhushan, H. Fuchs,
S. Hosaka

The Physics of Nanotubes
Fundamentals of Theory, Optics
and Transport Devices
Editors: S.V. Rotkin and S. Subramoney

**Single Molecule Chemistry
and Physics**
An Introduction
By C. Wang, C. Bai

**Atomic Force Microscopy, Scanning
Nearfield Optical Microscopy
and Nanoscratching**
Application to Rough
and Natural Surfaces
By G. Kaupp

Applied Scanning Probe Methods II
Scanning Probe Microscopy
Techniques
Editors: B. Bhushan, H. Fuchs

Applied Scanning Probe Methods III
Characterization
Editors: B. Bhushan, H. Fuchs

Applied Scanning Probe Methods IV
Industrial Application
Editors: B. Bhushan, H. Fuchs

Nanocatalysis
Editors: U. Heiz, U. Landman

**Roadmap
of Scanning Probe Microscopy**
Editors: S. Morita

**Nanostructures –
Fabrication and Analysis**
Editor: H. Nejo

Applied Scanning Probe Methods V
Scanning Probe Microscopy Techniques
Editors: B. Bhushan, H. Fuchs,
S. Kawata

Applied Scanning Probe Methods VI
Characterization
Editors: B. Bhushan, S. Kawata

Applied Scanning Probe Methods VII
Biomimetics and Industrial Applications
Editors: B. Bhushan, H. Fuchs

Applied Scanning Probe Methods VIII
Scanning Probe Microscopy Techniques
Editors: B. Bhushan, H. Fuchs,
M. Tomitori

Applied Scanning Probe Methods IX
Characterization
Editors: B. Bhushan, H. Fuchs,
M. Tomitori

Applied Scanning Probe Methods X
Biomimetics and Industrial Applications
Editors: B. Bhushan, H. Fuchs,
M. Tomitori

Bharat Bhushan
Harald Fuchs
Masahiko Tomitori

Applied Scanning Probe Methods VIII

Scanning Probe Microscopy Techniques

With 289 Figures and 9 Tables
Including 44 Color Figures

Springer

Editors:

Professor Bharat Bhushan
Nanotribology Laboratory for Information
Storage and MEMS/NEMS (NLIM)
W 390 Scott Laboratory, 201 W. 19th Avenue
The Ohio State University, Columbus
Ohio 43210-1142, USA
e-mail: Bhushan.2@osu.edu

Professor Dr. Harald Fuchs
Institute of Physics, FB 16
University of Münster
Wilhelm-Klemm-Str. 10
48149 Münster, Germany
e-mail: fuchsh@uni-muenster.de

Professor Dr. Masahiko Tomitori
Advanced Institute of Science & Technology
School of Materials Science
Asahidai, Nomi 1-1
923-1292 Ishikawa, Japan
e-mail: tomitori@jaist.ac.jp

Series Editors:

Professor Dr. Phaedon Avouris
IBM Research Division
Nanometer Scale Science & Technology
Thomas J. Watson Research Center, P.O. Box 218
Yorktown Heights, NY 10598, USA

Professor Bharat Bhushan
Nanotribology Laboratory for Information
Storage and MEMS/NEMS (NLIM)
W 390 Scott Laboratory, 201 W. 19th Avenue
The Ohio State University, Columbus
Ohio 43210-1142, USA

Professor Dr. Dieter Bimberg
TU Berlin, Fakutät Mathematik,
Naturwissenschaften,
Institut für Festkörperphysik
Hardenbergstr. 36, 10623 Berlin, Germany

Professor Dr., Dres. h. c. Klaus von Klitzing
Max-Planck-Institut für Festkörperforschung
Heisenbergstrasse 1, 70569 Stuttgart, Germany

Professor Hiroyuki Sakaki
University of Tokyo
Institute of Industrial Science,
4-6-1 Komaba, Meguro-ku, Tokyo 153-8505, Japan

Professor Dr. Roland Wiesendanger
Institut für Angewandte Physik
Universität Hamburg
Jungiusstrasse 11, 20355 Hamburg, Germany

ISBN 978-3-540-74079-7   e-ISBN 978-3-540-74080-3

DOI 10.1007/978-3-540-74080-3

NanoScience and Technology ISSN 1434-4904

Library of Congress Control Number: 2007937292

© 2008 Springer-Verlag Berlin Heidelberg

Typesetting: LE-TEX Jelonek, Schmidt & Vöckler GbR, Leipzig
Production: LE-TEX Jelonek, Schmidt & Vöckler GbR, Leipzig
Cover design: WMXDesign GmbH, Heidelberg

Printed on acid-free paper

9 8 7 6 5 4 3 2 1

springer.com

Preface

The success of the Springer Series Applied Scanning Probe Methods I–VII and the rapidly expanding activities in scanning probe development and applications worldwide made it a natural step to collect further specific results in the fields of development of scanning probe microscopy techniques (Vol. VIII), characterization (Vol. IX), and biomimetics and industrial applications (Vol. X). These three volumes complement the previous set of volumes under the subject topics and give insight into the recent work of leading specialists in their respective fields. Following the tradition of the series, the chapters are arranged around techniques, characterization and biomimetics and industrial applications.

Volume VIII focuses on novel scanning probe techniques and the understanding of tip/sample interactions. Topics include near field imaging, advanced AFM, specialized scanning probe methods in life sciences including new self sensing cantilever systems, combinations of AFM sensors and scanning electron and ion microscopes, calibration methods, frequency modulation AFM for application in liquids, Kelvin probe force microscopy, scanning capacitance microscopy, and the measurement of electrical transport properties at the nanometer scale.

Vol. IX focuses on characterization of material surfaces including structural as well as local mechanical characterization, and molecular systems. The volume covers a broad spectrum of STM/AFM investigations including fullerene layers, force spectroscopy for probing material properties in general, biological films .and cells, epithelial and endothelial layers, medical related systems such as amyloidal aggregates, phospholipid monolayers, inorganic films on aluminium and copper oxides, tribological characterization, mechanical properties of polymer nanostructures, technical polymers, and nearfield optics.

Volume X focuses on biomimetics and industrial applications such as investigation of structure of gecko feet, semiconductors and their transport phenomena, charge distribution in memory technology, the investigation of surfaces treated by chemical-mechanical planarization, polymeric solar cells, nanoscale contacts, cell adhesion to substrates, nanopatterning, indentation application, new printing techniques, the application of scanning probes in biology, and automatic AFM for manufacturing.

As a result, Volumes VIII to X of Applied Scanning Probes microscopies cover a broad and impressive spectrum of recent SPM development and application in many fields of technology, biology and medicine, and introduce many technical concepts and improvements of existing scanning probe techniques.

We are very grateful to all our colleagues who took the efforts to prepare manuscripts and provided them in timely manner. Their activity will help both students and established scientists in research and development fields to be informed about the latest achievements in scanning probe methods. We would like to cordially thank Dr. Marion Hertel, Senior Editor Chemistry and Mrs. Beate Siek of Springer for their continuous professional support and advice which made it possible to get this volume to the market on time.

Bharat Bhushan
Harald Fuchs
Masahiko Tomitori

Contents – Volume VIII

Contents – Volume IX

Contents – Volume X

Contents – Volume I

Part III   Industrial Applications

Contents – Volume II

Contents – Volume III

Contents – Volume IV

Contents – Volume V

Contents – Volume VI

Contents – Volume VII

List of Contributors – Volume VIII

Jean-Pierre Aimé
CPMOH, Université Bordeaux 1
351 Cours de la Libération, 33405 Talence cedex, France
e-mail: jp.aime@cpmoh.u-bordeaux1.fr

Matteo Albani
Dipartimento di Ingegneria dell'Informazione, Università di Siena
Via Roma 56, 53100 Siena, Italy
e-mail: matteo.albani@ing.unisi.it

Maria Allegrini
Dipartimento di Fisica "Enrico Fermi", Università di Pisa
Largo Bruno Pontecorvo 3, 56127 Pisa, Italy
e-mail: maria.allegrini@df.unipi.it

Guillaume Bachelier
Laboratoire de Spectrométrie Ionique et Moléculaire (LASIM)
Université Claude Bernard – Lyon 1, Bât. A. Kastler
43 Boulevard du 11 Novembre 1918, 69622 Villeurbanne Cedex, France
e-mail: guillaume.bachelier@lasim.univ-lyon1.fr

Charlotte Bernard
CPMOH, Université Bordeaux 1
351 Cours de la Libération, 33405 Talence cedex, France
e-mail: c.bernard@cpmoh.u-bordeaux1.fr

Bharat Bhushan
Nanotribology Lab for Information Storage and MEMS/NEMS (NLIM)
The Ohio State University
201 W. 19th Avenue, W390 Scott Laboratory, Columbus, Ohio 43210-1142
USA
e-mail: bhushan.2@osu.edu

Amir Boag
School of Electrical Engineering, Faculty of Engineering, Tel Aviv University
Ramat-Aviv, 69978, Israel
e-mail: boag@eng.tau.ac.il

Rodolphe Boisgard
CPMOH, Université Bordeaux 1
351 Cours de la Libération, 33405 Talence cedex, France
e-mail: r.boisgard@cpmoh.u-bordeaux1.fr

Anne Marie Bonnot
LEPES
25 av. des Martyrs, 38042 Grenoble cedex, France
e-mail: bonnot@grenoble.cnrs.fr

P. Carl
Institute of Physiology II, University of Münster
Robert-Koch-Straße 27b, 48149 Münster, Germany
e-mail: pcarl@uni-muenster.de

Ignacio Casuso
Laboratory of Nanobioengineering, Barcelona Science Park
and Department of Electronics, University of Barcelona
C/ Marti i Franque 1, 08028-Barcelona, Spain
e-mail: icasuso@pcb.ub.es

Eugenio Cefalì
Dipartimento di Fisica della Materia e Tecnologie Fisiche Avanzate
Università di Messina
Salita Sperone n. 31, 98166 Messina, Italy
e-mail: eugcef@libero.it

Charles Clifford
Analytical Science Team, Quality of Life Division, National Physical Laboratory
Teddington, Middlesex TW11 0LW, UK
e-mail: charles.clifford@npl.co.uk

Peter J. Cumpson
Analytical Science Team, Quality of Life Division, National Physical Laboratory
Teddington, Middlesex TW11 0LW, UK
e-mail: peter.cumpson@npl.co.uk

Gregory A. Dahlen
Research Scientist
112 Robin Hill Road, Santa Barbara, CA 93117, USA
e-mail: gdahlen@veeco.com

Dirk Dietzel
Physikalisches Institut, Universität Münster
Wilhelm-Klemm-Straße 10, 48149 Münster, Germany
e-mail: dirk.dietzel@uni-muenster.de

Robert Eibl
Plainburgstraße 8, 83457 Bayerisch Gmain, Germany
e-mail: robert_eibl@yahoo.com

Horacio D. Espinosa
Department of Mechanical Engineering, Northwestern University
2145 Sheridan Rd., Evanston, IL 60208-3111, USA
e-mail: espinosa@northwestern.edu

Giorgio Ferrari
Dipartimento di Elettronica e Informazione, Politecnico di Milano
P.za L. da Vinci 32, 20133 Milano, Italy
e-mail: ferrari@elet.polimi.it

Vinzenz Friedli
EMPA – Materials Science and Technology
Feuerwerker Strasse 39, CH-3602 Thun, Switzerland
e-mail: vinzenz.friedli@empa.ch

Takeshi Fukuma
Department of Physics, Kanazawa University
Kakuma-machi, Kanazawa, 920-1192, Japan
email: takeshi.fukuma@tcd.ie

Laura Fumagalli
Laboratory of Nanobioengineering Barcelona Science Park
and Department of Electronics, University of Barcelona
C/ Marti i Franque 1, 08028-Barcelona, Spain
e-mail: Laura.fumagalli@polimi.it

Renato Gardelli
Dipartimento di Fisica della Materia e Tecnologie Fisiche Avanzate
Università di Messina
Salita Sperone n. 31, 98166 Messina, Italy
e-mail: rgardelli@unime.it

G. Gomila
Laboratory of Nanobioengineering, Barcelona Science Park
and Department of Electronics, University of Barcelona
C/ Marti i Franque 1, 08028-Barcelona, Spain
e-mail: ggomila@el.ub.es

Pietro Giuseppe Gucciardi
CNR – Istituto per i Processi Chimico-Fisici, Sezione Messina
Via La Farina 237, 98123 Messina, Italy
e-mail: gucciardi@me.cnr.it

Andrea Ho
Department of Mechanical Engineering, Northwestern University
2145 Sheridan Road, Evanston, IL 60208-3111, USA

Samuel Hoffmann
EMPA – Materials Science and Technology
Feuerwerker Strasse 39, CH-3602 Thun, Switzerland
e-mail: samuel.hoffmann@empa.ch

Sumio Hosaka
Department of Production Science and Technology, Gunma University
1-5-1 Tenjin-cho, Kiryu, Gunma, 376-8515, Japan

Suzanne P. Jarvis
Conway Institute of Biomolecular and Biomedical Research
University College Dublin, Belfield, Dublin 4, Ireland
e-mail: suzi.jarvis@ucd.ie

James Johnstone
DEPC, National Physical Laboratory
Hampton Road, Teddington TW11 0LW, UK
e-mail: james.johnstone@npl.co.uk

P.M. Koenraad
Semiconductor Physics Group
Department of Applied Physics, Eindhoven University of Technology
P.O. Box 513, 5600 MB Eindhoven, The Netherlands
e-mail: P.M.Koenraad@tue.nl

Stefan Lanyi
Institute of Physics, Slovakian Academy o Sciences
Dubravska cesta 9, SK-845 11 Bratislava, Slovakia
e-mail: lanyi@savba.sk

Eli Lepkifker
School of Electrical Engineering, Faculty of Engineering, Tel Aviv University
Ramat-Aviv, 69978, Israel
e-mail: lepkifke@post.tau.ac.il

Hao-Chih Liu
SPM Probe Scientist
112 Robin Hill Road, Santa Barbara, CA 93117, USA
e-mail: hcliu@veeco.com

Sophie Marsaudon
CPMOH, Université Bordeaux 1
351 Cours de la Libération, 33405 Talence cedex, France
e-mail: s.marsaudon@cpmoh.u-bordeaux1.fr

Johann Michler
EMPA – Materials Science and Technology
Feuerwerker Strasse 39, CH-3602 Thun, Switzerland
e-mail: johann.michler@empa.ch

Martin Munz
Analytical Science Team, Quality of Life Division, National Physical Laboratory
Teddington, Middlesex TW11 0LW, UK
e-mail: martin.munz@npl.co.uk

Cattien V. Nguyen
Research Scientist, Eloret Corp./NASA Ames Research Center
MS 229-1, Moffett Field, CA 94035-1000, USA
e-mail: cvnguyen@mail.arc.nasa.gov

Jason R. Osborne
Eng. Manager, Advanced Development Group
112 Robin Hill Road, Santa Barbara, CA 93117, USA
e-mail: josborne@veeco.com

Salvatore Patanè
Dipartimento di Fisica della Materia e Tecnologie Fisiche Avanzate
Università di Messina
Salita Sperone n. 31, 98166 Messina, Italy
e-mail: salvatore.patane@unime.it

Jose Portoles
Laboratory of Biophysics and Surface Analysis, University of Nottingham
Nottingham, NG72RD, UK
e-mail: paxjp1@nottingham.ac.uk

Yossi Rosenwaks
School of Electrical Engineering, Faculty of Engineering, Tel Aviv University
Ramat-Aviv, 69978, Israel
e-mail: yossir@eng.tau.ac.il

John E. Sader
Department of Mathematics and Statistics, The University of Melbourne
Victoria 3010, Australia
e-mail: jsader@unimelb.edu.au

Shimon Saraf
School of Electrical Engineering, Faculty of Engineering, Tel Aviv University
Ramat-Aviv, 69978, Israel
e-mail: saraf@eng.tau.ac.il

Alex Schwarzman
School of Electrical Engineering, Faculty of Engineering, Tel Aviv University
Ramat-Aviv, 69978, Israel

Hayato Sone
Department of Production Science and Technology, Gunma University
1-5-1 Tenjin-cho, Kiryu, Gunma, 376-8515, Japan
e-mail: sone@el.gunma-u.ac.jp

Salvatore Spadaro
Dipartimento di Fisica della Materia e Tecnologie Fisiche Avanzate
Università di Messina
Salita Sperone n. 31, 98166 Messina, Italy
e-mail: salvatore.spadaro@ortica.unime.it

Stephan J. Stranick
National Institute of Standards and Technology Surface
and Microanalysis Science Division (837) 100 Bureau Drive
Gaithersburg, MD 20899-8372, USA
e-mail: stephan.stranick@nist.gov

Oren Tal
School of Electrical Engineering, Faculty of Engineering, Tel Aviv University
Ramat-Aviv, 69978, Israel
e-mail: tal@Physics.LeidenUniv.nl

Ivo Utke
EMPA – Materials Science and Technology
Feuerwerker Strasse 39, CH-3602 Thun, Switzerland
e-mail: ivo.utke@empa.ch

List of Contributors – Volume IX

Rehana Afrin
Department of Life Science, Tokyo Institute of Technology
Nagatsuta, Midori-ku, Yokohama 226-8501, Japan
e-mail: rafrin@bio.titech.ac.jp

David Alsteens
Unité de Chimie des Interfaces, Université catholique de Louvain
Croix du Sud 2/18, 1348 Louvain-la-Neuve, Belgium
e-mail: alsteens@cifa.ucl.ac.be

Matthias Amrein
Department of Cell Biology and Anatomy, Faculty of Medicine
University of Calgary, Canada
e-mail: mamrein@ucalgary.ca

Guillaume Andre
Unité de Chimie des Interfaces, Université catholique de Louvain
Croix du Sud 2/18, 1348 Louvain-la-Neuve, Belgium
e-mail: andre@cifa.ucl.ac.be

Hideo Arakawa
Bio Nanomaterials Group, National Institute for Materials Science
1-1, Namiki, Tsukuba 305-0044, Japan
e-mail : ARAKAWA.Hideo@nims.go.jp

Bharat Bhushan
Nanotribology Lab for Information Storage and MEMS/NEMS (NLIM)
The Ohio State University
201 W. 19th Avenue, W390 Scott Laboratory, Columbus, Ohio 43210-1142, USA
e-mail: bhushan.2@osu.edu

Claudio Canale
Physics Department, University of Genoa
Via Dodecaneso 33, I-16146 Genoa, Italy
e-mail: canale@fisica.unige.it

Ornella Cavalleri
Physics Department, University of Genoa
Via Dodecaneso 33, I-16146 Genoa, Italy
e-mail: cavalleri@fisica.unige.it

Cinzia Cepek
Laboratorio Nazionale TASC-INFM
Strada Statale 14, km 163.5 Basovizza, I-34012 Trieste, Italy
e-mail: cepek@tasc.infm.it

David T. Cramb
Department of Chemistry, University of Calgary, Canada
e-mail: dcramb@ucalgary.ca

Etienne Dague
Unité de Chimie des Interfaces, Université catholique de Louvain
Croix du Sud 2/18, 1348 Louvain-la-Neuve, Belgium
e-mail: dague@cifa.ucl.ac.be

James A. DeRose
EMPA
Ueberlandstrasse 129, CH-8600 Duebendorf, Switzerland
e-mail: james.derose@empa.ch

Y. F. Dufrene
Unité de Chimie des Interfaces, Université catholique de Louvain
Croix du Sud 2/18, 1348 Louvain-la-Neuve, Belgium
e-mail: dufrene@cifa.ucl.ac.be

Eric Finot
Institut Carnot de Bourgogne, UMR 5209 CNRS
Université de Bourgogne, France
e-mail: efinot@u-bourgogne.fr

Luca Gavioli
Dipartimento di Matematica e Fisica, Università Cattolica del Sacro Cuore
Via dei Musei 41, I-25121 Brescia, Italy
e-mail: l.gavioli@dmf.bs.unicatt.it

Alexander Gigler
Sektion Kristallographie, Ludwig-Maximilians-Universität München
Theresienstraße 41/II, 80333 München, Germany
e-mail: gigler@lrz.uni-muenchen.de

Yann Gilbert
Unité de Chimie des Interfaces, Université catholique de Louvain
Croix du Sud 2/18, 1348 Louvain-la-Neuve, Belgium
e-mail: gilbert@cifa.ucl.ac.be

Alessandra Gliozzi
Physics Department, University of Genoa
Via Dodecaneso 33, I-16146 Genoa, Italy
e-mail: gliozzi@fisica.unige.it

Xavier Haulot
Unité de Chimie des Interfaces, Université catholique de Louvain
Croix du Sud 2/18, 1348 Louvain-la-Neuve, Belgium
e-mail: haulot@cifa.ucl.ac.be

Enamul Hoque
Department of Chemistry and Materials Science, Washington State University
Pullman, WA 99164-4630, USA
e-mail: ehoque@ad.wsu.edu

Atsushi Ikai
Department of Life Science, Tokyo Institute of Technology
Nagatsuta, Midori-ku, Yokohama 226-8501, Japan
e-mail: aikai@bio.titech.ac.jp

K. S. Kanaga Karuppiah
Department of Mechanical Engineering, 2025 H. M. Black Engineering Building
Iowa State University, Ames, IA 50011-2161, USA
e-mail: kskarup@iastate.edu

Hyonchol Kim
Division of Biosystems, Institute of Biomaterials and Bioengineering
Tokyo Medical and Dental University
2-3-10, Kanda-surugadai, Chiyoda-ku, Tokyo, 101-0062, Japan
e-mail: kim.bmi@tmd.ac.jp

Sung-Kyoung Kim
Department of Chemistry, Hanyang University
17 Haengdangdong, Seoul 133-791, Korea
e-mail: s_ksk92@hanmail.net

Tiziana Svaldo-Lanero
Physics Department, University of Genoa
Via Dodecaneso 33, I-16146 Genoa, Italy
e-mail: svaldo@fisica.unige.it

Haiwon Lee
Department of Chemistry, Hanyang University
17 Haengdangdong, Seoul 133-791, Korea
e-mail: haiwon@hanyang.ac.kr

Zoya Leonenko
Department of Cell Biology and Anatomy, Faculty of Medicine
University of Calgary, Canada
e-mail: zleonenk@ucalgary.ca

Othmar Marti
Experimental Physics, Ulm University
89069 Ulm, Germany
e-mail: othmar.marti@uni-ulm.de

Hans Jörg Mathieu
EPFL – IMX, Station 12, 1015 Lausanne, Switzerland
e-mail: HansJoerg.Mathieu@EPFL.ch

Vincent T. Moy
Department of Physiology & Biophysics, University of Miami
1600 NW 12th Ave., RMSB 5077, Miami, FL 33136, USA
e-mail: vmoy@miami.edu

H. Oberleithner
University of Münster, Institute of Physiology II
Robert-Koch-Straße 27b, 48149 Münster, Germany
e-mail: oberlei@uni-muenster.de

Toshiya Osada
Department of Life Science, Tokyo Institute of Technology
Nagatsuta, Midori-ku, Yokohama 226-8501, Japan
e-mail: tosada@bio.titech.ac.jp

Bruno Pignataro
Dipartimento di Chimica Fisica "F. Accascina", Università di Palermo
V.le delle Scienze – Parco d'Orleans II, 90128 Palermo, Italy
e-mail: bruno.pignataro@unipa.it

Annalisa Relini
Physics Department, University of Genoa
Via Dodecaneso 33, I-16146 Genoa, Italy
e-mail: relini@fisica.unige.it

Félix Rico
Department of Physiology & Biophysics, University of Miami
1600 NW 12th Ave, RMSB 5077, Miami, FL 33136, USA
e-mail: frico@med.miami.edu

C. Riethmüller
Institute of Physiology II, University of Münster
Robert-Koch-Straße 27b, 48149 Münster, Germany
e-mail: chrth@uni-muenster.de

Ranieri Rolandi
Physics Department, University of Genoa
Via Dodecaneso 33, I-16146 Genoa, Italy
e-mail: rolandi@fisica.unige.it

Toshiharu Saiki
Department Electronics and Electrical Engineering, Keio University
3-14-1 Hiyoshi, Kohoku-ku, Yokohama 223-8522, Japan
e-mail: saiki@elec.keio.ac.jp

Hiroshi Sekiguchi
Department of Life Science, Tokyo Institute of Technology
Nagatsuta, Midori-ku, Yokohama 226-8501, Japan
e-mail: hsekiguc@bio.titech.ac.jp

Sriram Sundararajan
Department of Mechanical Engineering, 2025 H. M. Black Engineering Building
Iowa State University, Ames IA 50011-2161, USA
e-mail: srirams@iastate.edu

Nikhil S. Tambe
Material Systems Technologies, GE Global Research, #122, EPIP, Phase 2
Hoodi Village, Whitefield Road, Bangalore 560066, India
e-mail: nikhil.tambe@ge.com

Hironori Uehara
Department of Life Science, Tokyo Institute of Technology
Nagatsuta, Midori-ku, Yokohama 226-8501, Japan
e-mail: huehara@bio.titech.ac.jp

Claire Verbelen
Unité de Chimie des Interfaces, Université catholique de Louvain
Croix du Sud 2/18, 1348 Louvain-la-Neuve, Belgium
e-mail: verbelen@cifa.ucl.ac.be

Ewa P. Wojocikiewicz
Department of Physiology & Biophysics, University of Miami
1600 NW 12th Ave, RMSB 5077, Miami, FL 33136, USA
e-mail: e.wojcikiewicz@miami.edu

List of Contributors – Volume X

Alexander Alexeev
NT-MDT Co., Post Box 158, Building 317, Zelenograd, Moscow 124482, Russia
e-mail: alexander@ntmdt.ru

Tianming Bao
Veeco Instruments Inc, 14990 Blakehill Dr, Frisco, TX 75035, USA
e-mail: tbao@veeco.com

Bharat Bhushan
Nanotribology Lab for Information Storage and MEMS/NEMS (NLIM)
The Ohio State University
201 W. 19th Avenue, W390 Scott Laboratory, Columbus, Ohio 43210-1142, USA
e-mail: bhushan.2@osu.edu

Yasuo Cho
Research Institute of Electrical Communication, Tohoku University
2-1-1 Katahira Aoba-ku Sendai, 980-8577, Japan
e-mail: cho@riec.tohoku.ac.jp

Gianluca Ciardelli
Department of Mechanics, Politecnico di Torino
Corso Duca degli Abruzzi 24, 10129 Torino, Italy
e-mail: gianluca.ciardelli@polito.it

Mario D'Acunto
Department of Chemical Engineering and Materials Science, University of Pisa
via Diotisalvi 2, I-56126 Pisa, Italy
e-mail: m.dacunto@ing.unipi.it

Patrick Fiorenza
CNR-IMM, Stradale Primosole 50, I-95121 Catania, Italy
e-mail: patrick.fiorenza@imm.cnr.it

David Fong
Veeco Instruments Inc, 112 Robin Hill Rd, Santa Barbara, CA 93117, USA
e-mail: david.fong@veeco.com

Filippo Giannazzo
CNR-IMM, Stradale Primosole 50, I-95121 Catania, Italy
e-mail: filippo.giannazzo@imm.cnr.it

Paolo Giusti
Department of Chemical Engineering and Materials Science, University of Pisa
via Diotisalvi 2, I-56126 Pisa, Italy
e-mail: giusti@ing.unipi.it

Sean Hand
Veeco Instruments Inc, 112 Robin Hill Rd, Santa Barbara, CA 93117, USA

Yasushi Kadota
Ricoh, 1-3-6 Nakamagame, Ohta-ku, Tokyo 143-8555, Japan
e-mail: yasushi.kadota@nts.ricoh.co.jp

Toshi Kasai
Cabot Microelectronics
870 N. Commons Drive, Aurora, IL 60504, USA
e-mail: toshi_kasai@cabotcmp.com

Kazushige Kawabata
Division of Biological Sciences, Graduate School of Science, Hokkaido University
North 10 West 8, Kita-ku, Sapporo 060-0810, Japan
e-mail: kaw@sci.hokudai.ac.jp

Joachim Loos
Department of Chemical Engineering and Chemistry
Eindhoven University of Technology
P.O. Box 513, 5600 MB Eindhoven, The Netherlands
e-mail: j.loos@tue.nl

Franco Montevecchi
Department of Mechanics, Politecnico di Torino
Corso Duca degli Abruzzi 24, I-10129 Torino, Italy
e-mail: franco.montevecchi@polito.it

Vito Raineri
CNR-IMM, Stradale Primosole 50, I-95121 Catania, Italy
e-mail: vito.raineri@imm.cnr.it

Robert A. Sayer
Nanotribology Laboratory for Information Storage and MEMS/NEMS (NLIM)
The Ohio State University
201 West 19th Avenue, Columbus, OH 43210-1142, USA

Ken-ichi Shinohara
School of Materials Science, Japan Advanced Institute of Science and Technology
Asahi-dai, Nomi, Ishikawa 923-1292, Japan
e-mail: shinoken@jaist.ac.jp

Hiroyuki Sugimura
Kyoto University, Sakyo, Kyoto 606-8501, Japan
e-mail: hiroyuki-sugimura@mtl.kyoto-u.ac.jp

Tatsuo Ushiki
Division of Microscopic Anatomy
Graduate School of Medical and Dental Sciences, Niigata University
757 Asahimachi-dori, Chuo-ku, Niigata 951-8510, Japan
e-mail: t-ushiki@med.niigata-u.ac.jp

Jiping Ye
Nano Analysis Section, Research Department, Nissan ARC
1 Natsushima-cho, Yokosuka 237-0061, Japan
e-mail: ye@nissan-arc.co.jp

1 Background-Free Apertureless Near-Field Optical Imaging

Pietro Giuseppe Gucciardi · Guillaume Bachelier · Stephan J. Stranick · Maria Allegrini

Abstract. The goal of this chapter is to review the theoretical background and the experimental techniques of apertureless scanning near-field optical microscopy (SNOM) artifact-free imaging. We describe the principles of apertureless SNOM, detailing the different detection schemes for artifact-free imaging: homodyne and heterodyne detection. Additionally, we detail the physical origin of the measured signals, describe optical artifacts, and discuss experimental techniques capable of artifact-free imaging. Finally, we provide an overview of the potential application of these techniques in materials science, nanophotonics, nanoplasmonics and soft-matter science.

Key words: Near-field, Scanning near-field optical microscopy, Nano-optics, Nanophotonics, Apertureless, Proximal probe microscopy

Abbreviations

AFM	Atomic force microscopy
AOM	Acousto-optical modulator
CNT	Carbon nanotube
NA	Numerical aperture
PMMA	Poly(methyl methacrylate)
SNOM	Scanning near-field optical microscopy
STM	Scanning tunneling microscopy

1.1
Introduction

Nanotechnology is one of the leading research themes in engineering, physics and materials science. The main role in the characterization of nanostructured systems is played by scanning electron microscopy, transmission electron microscopy, atomic force microscopy (AFM), and scanning tunneling microscopy (STM). Such techniques are based on the concept of raster scanning a probe (an electron beam or a sharp tip) over the sample and recovering surface and near-surface properties, e.g., topography. While optical microscopy and spectroscopy can provide valuable structural, chemical and electronic information on an extremely wide class of materials, their major drawback for applications in nanotechnology is, however, that they are micrometer-scale techniques. Conventional optical microscopes with high numerical aperture (NA) objectives (NA = 1.4) allow one to push the resolution to the submicron scale, but ultimately these microscopes reach the physical barrier imposed by diffraction, which limits the lateral resolution to

$$\Delta x \approx \frac{\lambda}{2\,\mathrm{NA}}\,,$$ (1.1)

typically 250–300 nm for wavelengths λ in the visible range. This is an order of magnitude too large for applications in nanotechnology and modern biotechnology. Although recent developments have pushed the spatial resolution of fluorescence confocal microscopy down to the 50-nm range [1, 2], these techniques do not have a straightforward application to other imaging and spectroscopy modes.

Nano-optics [3] addresses the issues of confining light to the 10–100-nm scale, far beyond the classical diffraction limit. With the development of scanning near-field optical microscopy (SNOM) we are able to visualize, characterize and manipulate single nanostructures with nanometer-scale precision. The first implementations of SNOM were developed in the early 1980s [4, 5], based on a concept proposed by Synge [6] in 1928. Synge recognized that the resolution limit of conventional optics could be circumvented by illuminating the sample through a nearby aperture that was significantly smaller than the wavelength of light, while keeping the sample-to-aperture separation fixed at a distance that was also much smaller than the wavelength of light: in the optical near-field. In this way, the image's resolution is a function of the aperture size and its distance from the sample and is no longer limited by diffraction. An important breakthrough in near-field microscopy came with the introduction of tapered, metal-coated optical-fiber sensors [7]. Such probes, ending with an aperture with a diameter in the 50–200-nm range, could be implemented in a scanning probe microscopy framework, exploiting the scanning technology and control developed for AFM and STM. The success of aperture SNOM has triggered a wide range of applications in chemistry, physics and materials science (reviewed in [8–10]). However, as the complexity of systems under study grew, so did the potential for artifacts. This is especially true in elastic scattering images [11, 12] which cannot be easily interpreted unless specific experimental configurations are used or a complete model of the system can be developed [13].

Interest in near-field optics received a boost with the introduction of the field-enhancing, nanoantenna properties of sharp metallic tips: apertureless probes. Apertureless SNOM, also named "scattering SNOM", was introduced in the mid-1990s [14–16] by several groups who demonstrated elastic scattering images with sub-10-nm resolution. Similar spatial resolution has been attained with spectroscopic imaging applications as well (reviewed in [17]). Apertureless SNOM is based on Babinet's reciprocity principle for which a sharp tip or nanoparticle behaves similarly to a small hole (its reciprocal). The tip plays a dual role, confining the field on length scales comparable to its radius of curvature (5–50 nm), and acting, at the same time, as a nanoantenna capable of collecting the near-fields scattered by an irradiated sample surface and radiating them into the far-field.

One key to the practical realization of this technique has been the development of methods to extract the tiny near-field scattering signal from the huge far-field background. Contributions from the extended structure of the macroscopic tip and non-localized scattering from the sample complicate any straightforward subtraction of this background signal. Thus, a more complicated experimental protocol is required.

This has been accomplished through interferometric techniques [18–22], and, when applicable, by exploiting material-dependent phonon-polariton resonances [23, 24]. These techniques, when combined with near-field probe modulation and detected field demodulation at the higher harmonics of the tip oscillation, allow one to achieve artifact free imaging with 5–10-nm spatial resolution.

The goal of this chapter is to review the theoretical background and the experimental techniques of apertureless SNOM artifact-free imaging. The chapter is organized as follows. In Sect. 1.2, we describe the principles of apertureless SNOM, detailing the different detection schemes for artifact-free imaging: homodyne and heterodyne detection. In Sect. 1.3, we analyze the physical origin of the measured signals, describe optical artifacts, and discuss experimental techniques capable of artifact-free imaging. Finally, in Sect. 1.4 we give an overview of the potential application of these techniques in materials science, nanophotonics, nanoplasmonics and soft-matter science.

1.2
Principles of Apertureless SNOM

Near-field optical microscopy is based on the exploitation of the evanescent components of the electromagnetic field to overcome the diffraction limit. Owing to the uncertainty principle, the typical wave vectors involved in the optical properties of the sample fulfil the well-known relation $\Delta k_\parallel \Delta r_\parallel \sim 2\pi$. Hence, small particles of lateral size $\Delta r_\parallel$ will spread the spectrum of the scattered wavevectors (k-spectrum) to higher and higher values $\Delta k_\parallel \sim 2\pi/\Delta r_\parallel$. The different components of the wavevector (parallel, $k_\parallel$, or orthogonal, $k_\perp$, to the sample surface) are, however, linked by the usual dispersion relation $\omega^2/c^2 = k_\parallel^2 + k_\perp^2$, where ω is the electric field frequency and c the speed of light. Therefore, the k-spectrum of an object of dimensions $L_\parallel$ much smaller than the wavelength of light λ, ($L_\parallel \ll \lambda$) will extend up to spatial frequencies $k_\parallel \sim 2\pi/L_\parallel$ that largely exceed $2\pi/\lambda = \omega/c$. The associated values of $k_\perp$ will be therefore purely imaginary, $k_\perp^2$ being negative. Such waves are intrinsically evanescent, explaining why the details on a nanometer scale cannot be seen in a far-field experiment. One way to access the optical properties on a nanometer scale is to couple the evanescent components of the near-field scattered light to propagating far-field radiation. In apertureless SNOM (Fig. 1.1) such a task is accomplished by focusing light on a metallic or dielec-

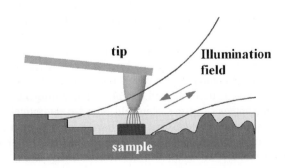

Fig. 1.1. Schematic diagram of an apertureless scanning near-field optical microscopy (SNOM)

tric tip, namely, the probe of a scanning tunneling microscope [25], an atomic force microscope [26] or a shear-force microscope [27–29]. The resulting localized fields near the tip apex are used to probe the sample's optical properties. Near-field interactions, e. g., scattering, take place over a distance proportional to the apex diameter: typically 10–30 nm. The strength of the interaction is governed by the sample's local dielectric constant, the probe's electromagnetic properties and the probe–sample separation. Recently, interest has focused on exploring these interactions as they are found to be at the origin of surprisingly high optical response in metallic nanostructures. For example, in Raman scattering experiments, enhancement effects due to metallic, nanometer-scale structured surfaces are utilized to enhance the scattering cross sections to allow for single-molecule detection. In these applications, networks of metal clusters a few nanometers in diameter have strongly confined and intense near-field distributions at the nanometer gaps between the particles [30, 31], similar to what occurs between the apertureless scanning near-field optical microscope tip and the sample surface. The nature of these "gap modes" is still not fully understood. Experiments based on precise fabrication of engineered gaps, namely, nanodipoles, are at the forefront of research in nano-optics that seeks to fully understand these phenomena. One generalization that can be made is that metallic nanometer-scale particles generate strong near-fields at specific resonance wavelengths. The wavelength dependence of this effect is related to plasmon resonance modes and is sensitive to the particle's size, geometry and metal composition. One interesting system in the visible range consists of nanometer-sized Ag or Au spheres which show strong plasmon resonances. A concept that exploits the enhanced fields present around these metal nanoparticles was proposed by Wessel [32] in 1984. Wessel proposed a concept where a laser-irradiated, elongated metal particle is used to establish an enhanced, localized light field that would enable high-resolution optical imaging. One way to implement this approach would be to attach a resonant nanoparticle to the end of a scanning probe microscope tip. However, using the sharp apex of a metallic or metallized tip as a nanoscale scatterer has turned out to be the easiest experimental configuration to implement. In this technique the sample area under investigation is illuminated using diffractive far-field optics. The sharp apex of the tip is brought to the center of the illuminated area. The sample is then raster-scanned under the tip while the scattered light is collected. A tapping-mode-based feedback maintains a constant tip–sample gap. This allows one to acquire the topography of the surface while simultaneously recording its optical response. A lens is used to collect the scattered light, which is composed of locally scattered light as well as background scattering from the extended structures of the tip and sample. The local near-field scattering is discriminated from the background by means of interferometric amplification and demodulation at the tip's tapping frequency, and its higher harmonics. The spatial resolution achievable by this method can be better than 10 nm [33], and is limited mainly by the radius of the tip apex. Because an interferometric detection scheme is used, the full complex optical information can be obtained, i. e., not only the intensity of the near-field but also its optical phase. Different interferometric apertureless SNOM concepts have been proposed (Fig. 1.2; reviewed in [34]), and can be classified into homodyne, heterodyne and pseudo-heterodyne apertureless SNOM.

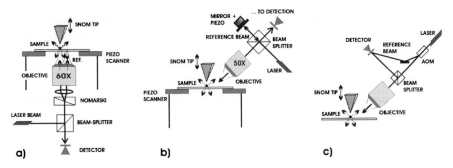

Fig. 1.2. Sketches of the experimental setups for interferometric apertureless scanning near-field optical microscopy using Normarski illumination (**a**), conventional homodyne amplification (**b**), heterodyne amplification (**c**) and pseudo-heterodyne amplification

1.2.1
The Homodyne Apertureless SNOM Concept

One of the first examples of apertureless SNOM with homodyne detection was shown by the Wickramasinghe group [14] at IBM. As can be seen in (Fig. 1.2a), a linearly polarized laser beam is focused by a Nomarski objective into two diffraction-limited spots on the sample surface. An atomic force microscope tip approaches one of the spots and locally enhances the light scattering. The reflection from the second spot is used as the reference beam. The two backscattered fields are collected and recombined on the detector, whose output allows one to extract information concerning both the amplitude and the phase of the field. Alternatively [35, 36], as shown in Fig. 1.2b, a conventional microscope objective can be used to focus the light on the tip, and detect the tip–sample backscattered field (amplitude and phase) interferometrically, after combining it with an external reference beam. This configuration has the advantage of being applicable to opaque samples. Although this technique demonstrated spatial resolutions better than 10 nm, the possible occurrence of topography artifacts was soon envisaged [37, 38].

More recently, several groups have developed new experimental techniques that strongly enhance the near-field scattering signal over the far-field background which contributes to topography artifacts. Our group [20, 39] working in the visible–near IR range and several other the groups working in the mid-IR region [21–23, 40, 41] have shown that by oscillating the probe vertically and using demodulation at higher harmonics of this oscillation frequency, one can extract the pure "near-field" optical interaction and artifact-free imaging.

In the double-modulation setup of Fig. 1.2a [20, 39] the beam of a laser diode ($\lambda = 780$ nm) is expanded and linearly polarized by means of a polarizing beam splitter cube. The beam crosses a Nomarski prism with the optical axis orientated at 45° with respect to the beam polarization. This creates two slightly divergent, cross-linearly polarized beams with adjustable relative phase. The beams are focused on two diffraction-limited spots on the outer surface of the sample using a ×60, 0.85 NA objective. The backscattered light coming from the two spots is collected by the same objective, and recombined by the Nomarski prism and polarizing cube. The latter lets the vertical polarization component of each beam cross the cube, and deviates the horizontal component towards the detector. Light is detected by a photodiode and is filtered at 780 nm to reduce the effects of ambient light. Atomic force microscope tips (n$^+$-doped silicon) are used as scattering sources, with a typical curvature radius between 5 and 10 nm. The tip approaches one of the spots and is maintained in the near-field region by tapping-mode AFM feedback. The probe is glued to a quartz tuning fork, and the system is operated at the resonance frequency of the tuning fork ($\omega \sim 32$ kHz) with oscillation amplitudes up to 100 nm peak-to-peak. The tip position is controlled in x, y and z in a standalone configuration. For the optical detection, an additional modulation is given to the tip in the vertical direction, at a frequency ($\Omega \sim 1$ kHz) far below the fork resonance. The optical signal is therefore demodulated by using a lock-in amplifier referenced to one of the harmonics of the vertical dithering $n\Omega$. Both the amplitude and the phase signals are acquired for imaging purposes. The resulting background suppression of this scheme is evidenced in Fig. 1.3, which shows line scan maps along the x-direction, parallel to the sample surface, while the tip is approaching the sample (z-direction, top to bottom) for the topography, the DC optical signal and the AC optical signal demodulated at the second harmonic. The domain where the scan is performed at constant height is represented by the upper zone in the topographic section of Fig. 1.3a where the line scan is entirely black. The DC optical signal in Fig. 1.3b is the replica with inverted contrast, clearly suggesting the presence of the background-induced z-motion artifact [12]. The second-harmonic signal in Fig. 1.3c has completely

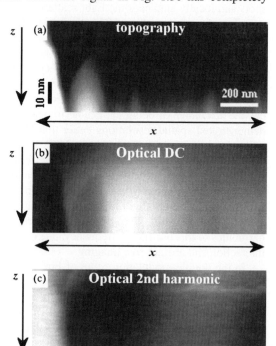

Fig. 1.3. Line scan maps parallel to the sample surface (x-direction) while the tip is approaching the sample (z-direction, *top* to *bottom*) for the topography (**a**), the DC optical signal (**b**) and the AC optical signal demodulated at the the second harmonic (**c**). (Redrawn from [20])

different features from the topography. The surface structure shows a lateral shift of 20 nm with respect to the topography, and starts to be detected about 25 nm away from the feedback-engaged point.

The IR apertureless SNOM design [21] is based on a tapping-mode AFM, which uses a cantilevered, commercial Pt-coated tip as the scattering source. The tip oscillates at its mechanical resonance frequency ($\omega \sim 30$ kHz) with amplitudes of approximately 20 nm. It is illuminated by a sharply focused IR laser with the incident field polarized in the tip axis plane. The backscattered light is analyzed with a homodyne-mode Michelson interferometer, allowing one to record both the amplitude s and the phase θ of the complex scattered field $E_{sca} = s \exp(i\theta) = \alpha_{eff} E_i$, where α_{eff} is the effective polarizability of the system consisting of the tip dipole and its mirror dipole in the sample, and E_i is the incident field. The analytical expression for α_{eff} [41] is given as

$$\alpha_{eff} = \frac{\alpha(1 + \beta)}{1 - \alpha\beta/16\pi(a + z)^3} , \tag{1.2}$$

where z is the tip–sample distance, a is the radius of curvature of the tip and α its its polarizability, $\beta = (\varepsilon - 1)/(\varepsilon + 1)$ in terms of the sample's dielectric constant ε. The nonlinear distance dependence of the near-field interaction in (1.2) is utilized to suppress the large, unavoidable background scattering caused by parts of the tip shaft and sample. The task is accomplished by modulating the tip's vertical position, z, at a frequency Ω, which induces strong harmonics in the near-field scattering $E_{sca}^{(n)}$ and much less in the background $E_{bkg}^{(n)}$ (Sect. 1.3). Therefore, by demodulating the detector signal at the frequencies $n\Omega$ it is possible to extract a background-free nth Fourier coefficient of the near-field interaction $E_{sca}^{(n)}$. As the complex near-field interaction is $s_n \exp(i\theta_n)$ a nonzero phase signal $\theta = \arg(\alpha_{eff})$ arises from a nonvanishing imaginary part of ε, which provides information on the phase of the electric near-field. Figure 1.4 shows an approach curve for the second-harmonic amplitude and phase signals on SiC and Au at $\lambda = 10.6$ µm [21]. When the sample approaches the tip, the amplitude signal s_2 increases strongly for both sample materials on the length scale of the tip radius (approximately 20 nm). Owing to phonon-enhanced near-field interactions, the signal of SiC exceeds that of Au by a factor of 3. At large tip–sample distances the amplitude decreases below the noise level, confirming that background signals are efficiently suppressed. The phase ϕ_2

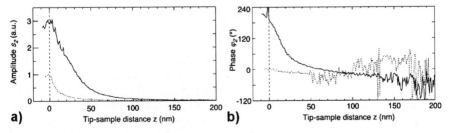

Fig. 1.4. Approach curves for the amplitude (**a**) and the phase (**b**) of the second harmonic demodulated optical signal carried out on SiC (*solid lines*) and Au (*dotted lines*). (Redrawn from [21])

remains nearly constant for Au, while it changes strongly for SiC, as a result of a distance-dependent (phonon) resonance-shift behavior.

1.2.2
The Heterodyne and Pseudo-Heterodyne Apertureless SNOM Concepts

Background suppression in the visible range can be accomplished using heterodyne interferometric detection [18, 33, 42, 43]. The schematic of a typical setup is shown in Fig. 1.2c. The laser beam is split into two parts using an acousto-optic modulator (AOM). One part goes straight through the beam splitter and the microscope objective. The other part is given a frequency shift Ω, usually ranging between 10 and 100 MHz, and is used as the reference beam. The use of two crossed AOMs [43] can be envisaged to reduce the frequency shift Ω to the 100-kHz range, accessible to more commonly available low-frequency lock-in amplifiers. The electric field is polarized in the incidence plane. The backscattered light is collected with the same lens and directed to a fast, high-dynamic-range detector (photomultiplier tube or fast photodiode). The tip of the apertureless scanning near-field optical microscope is dithered vertically, using a tapping-mode scheme, with amplitudes in the few tens of nanometer range. The Ω-shifted component of the detector signal is filtered by a dual-channel lock-in amplifier which determines amplitude and phase simultaneously. As in the homodyne scheme, the near-field tip–sample interaction varies in a nonlinear fashion (1.2) in the narrow near-field range $0 < z < a \ll \lambda$ and the dithering of the tip at frequency Ω and signal demodulation at frequency $n\Omega$ can be used to enhance the near-field over the background contribution. Heterodyne detection directly measures both the harmonic amplitude s_n and the phase θ_n of the complex near-field interaction $E_{sca} = s \exp(i\theta)$. The background reduction power of this technique is clearly evidenced in the approach curves shown in Fig. 1.5. The typical interference oscillations observed in the first harmonic optical signal (Fig. 1.5a) are strongly reduced, they are still visible in the second-harmonic signal (Fig. 1.5b) and are below the detector noise level when demodulating at the third harmonic (Fig. 1.5c). The onset of the pure near-field scattering is clearly visible in all the curves, on length scales of the order of the tip radius.

The main limitation of the heterodyne scheme is that commercially available AOMs operate in a rather limited spectral region and the alignment of the frequency-shifted beam changes with the wavelength, which is a major disadvantage for spectroscopic applications. With the aim of overcoming such limitation, a pseudo-heterodyne technique based on the sinusoidal phase modulation of the reference beam has been developed [44]. The system has been demonstrated in the mid-IR region (10.6 µm) and is based on a homodyne detection setup on which a piezo slab is used to modulate the reference mirror's position (Fig. 1.2b) at frequencies of several hundred hertz. Image acquisition is accomplished with a data acquisition board capable of a real-time fast Fourier transform for the simultaneous recovery of the desired signal amplitude s_n and phase θ_n. By choosing a sufficiently high harmonic n, one can isolate the pure near-field scattering from the overall optical signal: near-field plus background.

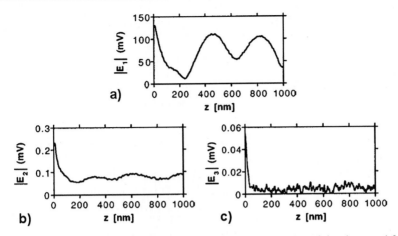

Fig. 1.5. Approach curves for the amplitude of the optical signal E_n demodulated at $n = 1,2$ and 3 (**a, b, c**, respectively) between a PtIr tip and a gold surface. (Redrawn from [19])

1.3
Interpretation of the Measured Near-Field Signal in the Presence of a Background

In an ideal elastic scattering apertureless SNOM experiment, the radiation reaching the detector should originate only from the nanometric region of the probe–sample optical junction. In practice, however, background optical fields that are not related to the field scattered by the tip end are always collected, and dominate the measured optical signals. The probe shaft, the optical elements or portions of the sample "far" from the probe are typical examples of undesirable sources of stray light. A deep comprehension of the mechanisms leading to the formation of the measured signals is therefore fundamental to understanding and improving the background suppression techniques illustrated in the previous section [45, 46].

1.3.1
Noninterferometric Detection

Noninterferometric apertureless SNOM is based on the schematic in Fig. 1.2b in which the reference beam of the interferometer is suppressed. Usually light detection is accomplished in a different direction with respect to the illumination. This approach, combined with tip vibration and demodulation at higher harmonics, works well in the mid-IR region [15, 47, 48], but fails to sufficiently suppress the background in the visible range. Modeling this configuration allows one to gain insight into the nature of this background signal and possible rejection mechanisms at different wavelengths. For purposes of discussion, we assume that the probe is made up of a strongly scattering base (depicted by the ellipsoid in Fig. 1.6), located at a distance $z(t)$ from the average sample surface, ending with a cone having height h_{tip} and apical radius a. Since h_{tip} is on the micrometer scale, the scattering base

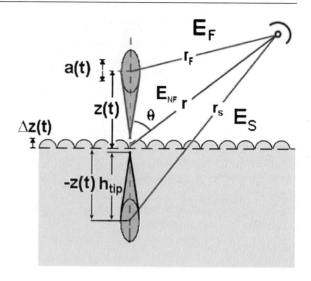

will always be far from the surface. Therefore, any near-field enhancement due to
the dipole image–dipole interaction at close distances is indeed negligible. For the
sake of simplicity, we assume an experimental configuration in which the sample is
scanned in the horizontal plane and the tip moves in the vertical direction z. Upon
external illumination, the field scattered from the tip base $E_F = \xi E_0 \exp(ikr_F)$ will
interfere at the detector with the field owing to its mirror image from the average
sample surface $E_s \propto \varsigma E_0 \exp(ikr_s)$, located at distance $-z(t)$. Here E_0 is the incident
field, ξ and ς are constants accounting for the tip and sample reflectivity, λ is the
laser wavelength, $k = 2\pi/\lambda$, r is the distance between the tip contact point when
the tip first approaches the sample surface and the detector, and θ is the collection
direction. Both r and θ are constants during the scan.

With these assumptions r_F and r_s are given by $r_F = \sqrt{r^2 + z^2 - 2rz\cos\theta}$ and
$r_s = \sqrt{r^2 + z^2 + 2rz\cos\theta}$. The near-field interaction is described by a scattered field
$E_{NF} = \alpha_{eff}(z)E_0$, where $\alpha_{eff}(z) = \alpha_0(z)\exp[i\alpha_1(z)]$ is the complex-valued effective
polarizability of the tip–sample assembly given by (1.2) for the case of p-polarized
excitation. We call $\Delta z = z(t) - h_{tip}$ the separation between the tip apex and the
sample average plane. In constant-gap mode Δz varies during the scan, since the tip
follows the local surface asperities. The total signal measured by the detector is

$$I = |E_{NF} + E_F + E_s|^2 \tag{1.3}$$

and will have the typical form for an interference process

$$I = |E_{NF}|^2 + |E_F|^2 + |E_s|^2 + 2\mathrm{Re}(E_{NF}^* E_F) + 2\mathrm{Re}(E_{NF}^* E_s) + 2\mathrm{Re}(E_F^* E_s) . \tag{1.4}$$

It can be seen that the near-field scattering information is convoluted with the background fields E_F and E_s. The measured signal will contain a near-field contribution
$I_{NF} \sim |E_{NF}|^2$ summed to a much stronger background, $I_{bkg} \sim |E_F|^2 \approx |E_s|^2$, owing
to the fact that $E_F \approx E_s \gg E_{NF}$. The extraction of I_{NF}, however, would be less

problematic if the background were constant with z. Unfortunately this is not the case since, as shown below, I_{bkg} strongly depends on the tip–sample distance. As a consequence, the optical signal measured during a scan of the surface will be just $I(x, y) \approx I_{bkg}[z(x, y)]$, which is an optical readout of the topography, more or less complicated by the functional form of I_{bkg}. To study I_{bkg} we neglect E_{NF} in (1.3), which therefore becomes

$$I_{bkg} \approx |E_F|^2 + |E_s|^2 + 2\mathrm{Re}\left(E_F^* E_s\right) = C + B \cos\left[\Delta\varphi(z)\right] , \tag{1.5}$$

where C and B are constants, and $\Delta\varphi(z) = k\,|\mathbf{r}_s - \mathbf{r}_f|$ is the phase difference between two fields, which for $z \gg r$ becomes $\Delta\varphi(z) \approx 2kz \cos\theta$. Here k is the wavevector $k = 2\pi/\lambda$. What is most important is that demodulation at higher harmonics is not capable to a priori reject such a background, but can only attenuate it. To understand why, let us assume a sinusoidal modulation of the tip. The position of the high-scattering shaft will be given by $z_{tip}(t) = h_{tip} - a_0 \cos(\omega t)$. Such a position will be further modulated by the tip vertical displacement $\Delta z(t)$ during the scan, namely, the sample topography. Inserting the phase $\Delta\varphi\left[z_{tip}(t) + \Delta z(t)\right]$ in (1.5), we finally get

$$I_{bkg}(t) = C + B \cos\left(\varphi_0 + \frac{4\pi \cos\theta}{\lambda}\Delta z(t) - \frac{4\pi \cos\theta}{\lambda}a_0 \cos(\omega t)\right) , \tag{1.6}$$

where $\varphi_0 = 4\pi \cos\theta(h_{tip}/\lambda)$ is a constant phase factor. Equation (1.6) is an expression of the form $\cos[(\alpha + \beta \cos(\omega t)]$ and can therefore be decomposed into a sum of harmonics $I = \sum_n I_n(a_0, \Delta z) \cos(n\omega t)$. The amplitudes of the harmonics are proportional to the nth-order Bessel functions of the first kind J_n as follows [49]:

$$I_{bkg}^{(0)} = C + BJ_0(2ka_0 \cos\theta) \cos(\varphi_0 + 2k \cos\theta \Delta z) ,$$
$$I_{bkg}^{(2n)} = 2B(-1)^n J_{2n}(2ka_0 \cos\theta) \cos(\varphi_0 + 2k \cos\theta \Delta z) , \tag{1.7}$$
$$I_{bkg}^{(2n+1)} = 2B(-1)^n J_{2n+1}(2ka_0 \cos\theta) \sin(\varphi_0 + 2k \cos\theta \Delta z) .$$

Equation (1.7) indicates that a nonvanishing background exists at every harmonic. Its amplitude is modulated by the sample's topography Δz with a sinusoidal law. The odd harmonics are shifted 90° with respect to the even ones. Moreover $I_{bkg}^{(n)}$ depends on the tip vibration amplitude a_0, the wavelength (through k) and the collection angle θ. From (1.7) we see that the amplitude of the background harmonics scales as $J_n(2ka_0 \cos\theta)$. In the small-oscillation approximation, i.e., for $2ka_0 \cos\theta \ll 1$, the first-order Taylor expansion gives

$$I_{bkg}^{(n)}(a_0) \approx \frac{(2\pi \cos\theta)^n}{n!}\left(\frac{a_0}{\lambda}\right)^n . \tag{1.8}$$

The background at the nth harmonic depends on the nth power of the ratio a_0/λ, multiplied by a factor $1/n!$. In particular (1.8) shows that the idea of increasing the oscillation amplitude to better "enter and exit" the sample's near-field is misleading. The only effect of increasing a_0 is an exponential magnification of the far-field background. From the physical point of view this can be understood considering

Fig. 1.7. Plot of the background (*BKG*) modulation amplitudes at the various harmonics as a function of the ratio a_0/λ. (Redrawn from [45])

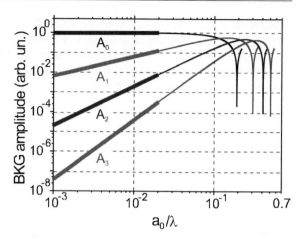

that, since the focal depth of the objectives used is several microns, the tip shaft will always be illuminated, no matter what the value of a_0 is. A good way to reduce the background is, instead, to increase θ owing to the term $\cos^n \theta$. In Fig. 1.7 we plot the background amplitude at the various harmonics as a function of a_0/λ for $\theta = 0$. In the visible range (e. g., for HeNe laser illumination), and for a tip oscillation of 20 nm peak-to-peak ($a_0/\lambda \sim 1.6 \times 10^{-2}$), increasing the harmonic order $n = 1, 2, 3$ leads to an increased rejection of the background, whose modulation amplitude will be reduced to values of $10^{-1}, 5 \times 10^{-3}$ and 2×10^{-4} with respect to the DC background amplitude. Decreasing the tip oscillation by 1 order of magnitude, from 10 to 1 nm (thicker lines in Fig. 1.7) has different consequences depending on the harmonic. While the DC signal is almost unaffected, the amplitude of the signal demodulated at the first, second and third harmonics is reduced, respectively by further 1, 2 and 3 orders of magnitude. In the mid-IR region ($\lambda \sim 10\,\mu$m) the technique is even more powerful, since for $a_0 = 10$ nm [namely, $(a_0/\lambda) \approx 10^{-3}$], first-, second- and third-harmonic demodulation will reduce the background amplitude by factors of 6×10^{-3}, 2×10^{-5} and 4×10^{-8} respectively. The much higher background rejection power of the higher-harmonics demodulation in the mid-IR region allows the near-field scattering intensity ($|E_{\mathrm{NF}}|^2$) in (1.3) to be measured.

1.3.2
Interferometric Detection

As we have seen in Sect. 1.2, interferometric detection has the advantage of permitting a measurement of the amplitude and phase of the electric field. In this section we will focus our attention on the higher background suppression capabilities of such a configuration, which allows apertureless SNOM to be fruitfully employed also in the visible range. The expression of the measured signals is carried out similarly to the noninterferometric case, provided we account for the reference field $E_{\mathrm{ref}} = E_R \exp(i\delta)$. The phase δ can be a constant term $\delta = \delta_0$, a term that varies linearly with time $\delta = \Omega t$ or a sinusoidal function of time $\delta = \delta_0 \cos(\Omega t)$. The first case describes the homodyne detection, in which the reference and the probe beams have the same frequency. The second case describes the heterodyne detection, with the reference field frequency-shifted with respect to the probe beam. The third case is the pseudo-heterodyne detection in which reference and probe beams have the same frequency, but the phase modulation introduces a comb of side-band frequencies which can be used for demodulation purposes. Whatever the detection scheme, the measured signal will be

$$
\begin{aligned}
I &= |E_{ref} + E_{NF} + E_s + E_F|^2 \\
&= |E_{ref}|^2 + |E_{NF} + E_s + E_F|^2 + 2\text{Re}\left[E_{ref}^*(E_{NF} + E_s + E_F)\right] .
\end{aligned}
\tag{1.9}
$$

Apart from the constant $|E_{ref}|^2$, the signal measured in interferometric apertureless SNOM is characterized by a term $I_{lin} = 2Re\left[E_{ref}^*(E_{NF} + E_s + E_F)\right]$, linear in the optical fields, which adds up to the term $|E_{NF} + E_s + E_F|^2$ quadratic in the optical fields which is typical of the noninterferometric detection (1.3). The presence of the linear term allows us to retrieve both the amplitude and the phase of the fields. Only interferometric techniques permit such a term to be isolated from the quadratic one. In particular, in homodyne detection, such a task can be approximately accomplished, provided that we choose a reference field $E_{ref} \gg E_{NF} + E_s + E_F$. In heterodyne and pseudo-heterodyne detection, the reference field is modulated at the frequency Ω ($E_{ref} = E_r \exp(i\Omega t)$ for heterodyne) and multiples of Ω ($E_{ref} = E_r \exp[i\delta_0 \cos(\Omega t)]$ for pseudo-heterodyne); therefore, the linear term can be exactly singled out by lock-in demodulation techniques. Note that in (1.9) the linear term $I_{lin} = 2\text{Re}\left[E_{ref}^*(E_{NF} + E_s + E_F)\right]$ always contains a far-field background contribution. Such a field is amplified by the same factor E_{ref} as the near-field scattering. The advantage brought by the interferometric detection, in fact, does not consist of a different amplification of the near-field with respect to the background, but rather the fact that the near-field signal $S_{NF} \sim |E_{NF}|$ has now to be compared with a background $S_{bkg} \sim |E_F|$, linear in E_F, a much more favorable situation with respect to the noninterferometric configuration where $S_{NF} \sim |E_{NF}|^2$ and $S_{bkg} \sim |E_F|^2$. With the same notations introduced in Sect. 1.3.1, the interferometric term takes the form

$$
I_{int} = 2\text{Re}\left(\begin{array}{l} E_R E_0 \exp\left[-i\delta(\Omega t)\right] \\ \times \{\alpha_{eff}\left[z(\omega t)\right] + \varsigma \exp\left[ikr_s(\omega t)\right] + \xi \exp\left[ikr_F(\omega t)\right]\}\end{array}\right) ,
\tag{1.10}
$$

where we have made explicit the time dependence of the fields on the tip vibration frequency ω. We clearly see that all the three fields do depend on ωt and, therefore, we expect a background component at every harmonic of the tip modulation frequency $n\omega$, exactly as for the noninterferometric configuration. Conversely, any background contribution due to the scattering from the sample surface [43,48] will be totally suppressed by demodulation at $n\omega$ since the sample position is not modulated at ω. In order to avoid phase shifts between the reference field and the field scattered by the tip shaft, the best experimental configuration to be chosen is the one in which the tip only vibrates in the z-direction, while the sample is scanned vertically. We call $\Delta z(t)$ the distance between the tip apex and the local surface, and $z(t)$ the distance between the scattering shaft and the average sample surface. Assuming vertical displacements of both the tip and the sample much smaller than

r_F, and r_s, we can expand (1.10) around the equilibrium positions r_{F0} and r_{s0}. For the heterodyne configuration we get

$$I_{int} \propto \alpha_0 \left[\Delta z + a_0 (1 + \cos \omega t) \right] \cos \{ \alpha_1 \left[\Delta z + a_0 (1 + \cos \omega t) \right] - \Omega t \}$$
$$+ \xi \cos \left(\phi_F + \frac{2\pi \cos \theta}{\lambda} a_0 \cos \omega t - \Omega t \right) \tag{1.11}$$
$$+ \varsigma \cos \left(\phi_s + \frac{2\pi \cos \theta}{\lambda} z + \frac{2\pi \cos \theta}{\lambda} a_0 \cos \omega t - \Omega t \right) ,$$

where we have indicated $\phi_j = r_j (2\pi \cos \theta) / \lambda$. We have assumed ξ and ς to be constants, but any eventual phase shift can be accounted for by redefining ϕ_j. The first term is the near-field interaction, the second and the third ones represent the far-field background. All three terms depend simultaneously on Ωt and on ωt. The dependence on ωt of the near-field scattering is rather complex; therefore, a small-amplitude approximation must be used to well understand the physical meaning of the measured signals. In contrast, the functional dependence of the far-field background is again of the form $\cos[\alpha + \beta \cos(\omega t)]$. Therefore, no matter what demodulation frequency $\Omega + m\omega$ we choose, the measured signal will always contain a part of near-field scattering and a part of the background. Note that the first background term (the second term in 1.11) is independent of the relative position of the the tip and the sample), and therefore its contribution to the signal demodulated at frequency $\Omega + m\omega$ is just a constant offset. Its only drawback is to add electrical noise to the measured signal, or slow changes to the overall signal as a consequence of mechanical or thermal drifts, changing ϕ_F. The third term, conversely, depends on z, and is the responsible for the far-field oscillations visible in the approach curves of Fig. 1.5a and b, and is a potential source of artifacts. Finally, we note that only in the absence of far-field background can the signals measured by lock-in demodulation at frequency Ω be unambiguously assigned to the amplitude α_0 and the phase α_1 of the scattered fields. Only in this case, in fact, (1.11) reduces to the form $\alpha_0 \cos(\Omega t - \alpha_1)$. In the presence of a background, conversely, (1.11) gets the more complicated form

$$I_{int} \propto \alpha_0 \cos(\Omega t - \alpha_1) + \xi \cos(\Omega t - \phi) + \varsigma \cos(\Omega t - \phi') . \tag{1.12}$$

The amplitude and phases provided by lock-in demodulation at frequency Ω are nontrivial functions of α_{eff}, ξ, ς, ϕ and ϕ', and cannot be directly considered as the amplitude and the phase of the scattered field. In pseudo-heterodyne detection the situation is similar. Every signal component modulated at the frequencies $n\Omega + m\omega$, multiples of the mirror vibration and of the tip oscillation frequency, will contain both a part of near-field scattering and a part of far-field background.

1.3.3
Artifacts in Apertureless SNOM and Identification Criteria

In the previous sections we have seen how the high performances of apertureless SNOM are critically subject to the capability of suppressing the far-field background. This signal is due to spurious reflections from the tip shaft and other sources and leads to fictitious optical images called artifacts. Artifacts are specifically due to the cross-talk between the sample topography $\Delta z(x, y)$ to the measured optical signal $I(x, y) \approx I_{bkg} [\Delta z(x, y)]$, namely, the far-field background, through the transfer functions defined by (1.7). Note that we must take into account the rectification effect introduced by the lock-in amplifier on the harmonics, since in real experiments the signal coming from the amplitude channel is monitored. For the noninterferometric detection scheme we therefore have

$$I_{bkg}^{(0)}(x, y) = C + B J_0(2ka_0 \cos \theta) \cos [\phi_0 + 2k \cos \theta \Delta z(x, y)] \ ,$$
$$I_{bkg}^{(2n)}(x, y) = 2B \, |J_{2n}(2ka_0 \cos \theta)| \, |\cos [\phi_0 + 2k \cos \theta \Delta z(x, y)]| \ , \qquad (1.13)$$
$$I_{bkg}^{(2n+1)} = 2B \, |J_{2n+1}(2ka_0 \cos \theta)| \, |\sin [\phi_0 + 2k \cos \theta \Delta z(x, y)]| \ .$$

We see that the functions $I_{bkg}^{(n)}(\Delta z)$ are strongly nonlinear even on scales as small as $\Delta z \sim 50 \, \text{nm}$; therefore, fictitious optical maps are expected to be remarkably different from the topography. This phenomenon has been observed [46] on Al-coated diffraction gratings (Fig. 1.8a, pattern height $\Delta z \sim 55 \, \text{nm}$) for the intensity maps acquired at $n = 1, 2$ (Fig. 1.8b,c). Moreover, since the functional dependence of the odd (and of the even) harmonics is the same, artifacts will produce optical maps that are qualitatively identical (Fig. 1.8b,d and c,e). The latter can be assumed as a criterion for unambiguous artifact identification. We finally note that, for a specific

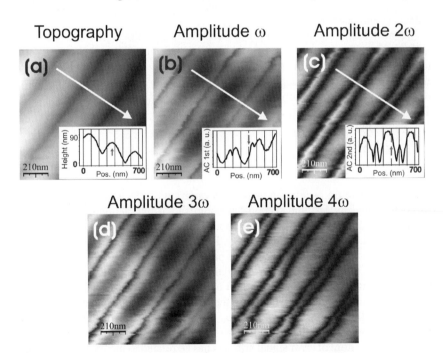

Fig. 1.8. Experimental results with line profiles (*insets*) of the measurements carried out on an Al-coated grating. **a** Topography, **b** first harmonic optical signal, **c** second harmonic optical signal, **d** third harmonic signal, **e** fourth harmonic signal. (Redrawn from [46])

choice of ϕ_0, the derivative of the signals at the contact point can be either positive or negative; therefore, contrast inversion can easily occur in the optical maps at the different harmonics.

Up to now we have focused our attention on the amplitude signal. It is possible, however, to draw interesting conclusions also for the phase signal induced by far-field artifacts. From the expansion of (1.13), we know that, in the small oscillation amplitude approximation, the analytical expression for the nth odd harmonic signal will be of the form

$$I_{\text{bkg}}^{(n)} = \left[2B \frac{(2\pi \cos \theta)^n}{n!} \left(\frac{a_0}{\lambda} \right)^n \right] \left[(-1)^n \sin(\phi_0 + 2k \cos \theta \Delta z) \right] \cos(n\omega t) ,$$

(1.14)

where the sine has to be replaced by a cosine if the even harmonics are considered. The phase signal measured in a real experiment by the lock-in amplifier corresponds to the phase shift Φ_n between $I_{\text{bkg}}^{(n)}$ and the reference $\cos(n\omega t)$. The factor between the large square brackets in (1.14) is a real positive quantity; therefore, all the information on Φ_n is encoded in the term in small square brackets $\Gamma = (-1)^n \sin(\phi_0 + 2k \cos \theta \Delta z)$. Γ, in particular, is always a real quantity which, depending on n and Δz, can be either positive and negative. Therefore, Φ_n is expected to depend on Δz in a steplike fashion assuming only two values: 0 or π, depending on weather Γ is positive or negative. This is a further fingerprint of artifacts occurence.

We now focus our attention on the dependence of the signal strength on the oscillation amplitude a_0 and on possible far-field artifacts induced by a not perfect stabilization of a_0 during the scan. From (1.8) we know that the far-field signals are characterized by a power dependence on the ratio a_0/λ. Equation (1.8) shows as well that any change Δa_0 around a fixed value a_0 is expected to produce a change of the optical signal

$$\frac{\Delta I_{\text{bkg}}^{(n)}}{I_{\text{bkg}}^{(n)}} \approx n \frac{\Delta a_0}{a_0} .$$

(1.15)

That is, a relative tip oscillation amplitude variation of 1% will influence the first-harmonic signal by 1%, the second-harmonic signal by 2%, and so on. Artifacts related to variations of the tip oscillation amplitude have been recently put forward by Billot et al. [50] Such artifacts, named "error signal artifacts", occur when the feedback does not react promptly to the presence of a topographic relief, inducing a change of a_0 during the scan. In particular, the oscillation amplitude changes when the tip encounters a relief, increasing on one edge and decreasing on the other, for a short delay of time before recovering the set point. As a consequence the optical signal will follow the actual value of a_0, resembling the error map. Billot et al. interpreted this effect by means of 2D finite-element numerical methods, but we can assess that the nature of error signal artifacts has be related to the far-field component of the optical signals not properly rejected.

1.3.4
New Techniques for Background Removal

The power-law dependence of the background on the oscillation amplitude a_0 evidenced by (1.8) and displayed in Fig. 1.7 suggests that decreasing a_0 together with increasing n is a good way to reject the far-field background. Decreasing a_0 is, however, not the only way to reduce the background. In fact, when we plot the correct expression $|J_n(2ka_0)|$ for the amplitude of the background harmonics, as in Fig. 1.7, it emerges that there are some particular values of the tip oscillation amplitude $\tilde{a}_0$ that strongly reduce the background. Such values represent the first zeros of the Bessel functions for which $J_n(2k\tilde{a}_0) = 0$. The corresponding values of the ratios $\tilde{a}_0/\lambda$ for $n = 0, \ldots 4$ fall in the interval $a_0/\lambda \sim 0.2 - 0.7$. In order to reject the background at the nth harmonic, we can thus tune the oscillation amplitude a_0 to well-defined values $\tilde{a}_0$ which depend on n and λ. The numerical values of $\tilde{a}_0$ for some typical laser wavelengths in the visible range are shown in Fig. 1.9. In particular, in the near-UV–vis region, the spectrum of values of $\tilde{a}_0$ that null the background at the first three harmonics falls in the 100–200nm range. These values are certainly reasonable for experimental configurations employing free-standing tips or atomic force microscope cantilevers [15, 51], but are quite unusual for systems based on quartz tuning forks, suited for low-oscillation amplitudes. In order to establish how fine the tuning of a_0 must be, we note that the Bessel functions are almost linear around their first zeros. This implies that the maximum discrepancy interval $\Delta a_0/\tilde{a}_0$ allowed to have a background rejection of 3 orders of magnitude is of the order of 0.1%.

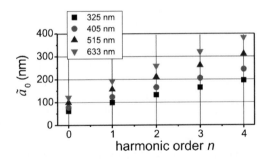

Fig. 1.9. Values of the tip oscillation amplitude $\tilde{a}_0$ corresponding to the first zeros of the nth Bessel functions for various wavelengths λ

1.4
Applications of Elastic-Scattering Apertureless SNOM

The performance of apertureless SNOM and its potential contribution to the understanding of the physics in the nanoscale can be analyzed from the expression of the effective polarization (1.2) of the coupled tip–sample system. Depending on the experimental conditions, one can exploit the tip resonances (polarization α) to enhance the incoming light, sample resonances (parameter β depending on both the real and the imaginary part of the refractive index) to image the light eigenmodes in microstructures and nanostructures and finally new resonances of the coupled system (arising from the zeroing of the effective polarization denominator) in polaritonic material. The first part of this section will be devoted to a review of some applications to material-specific imaging, phase mapping in metal nanostructures and waveguides, and to near-field-induced polariton resonances. In all cases, the tip being a part of the probed system, the scattered signal needs to be analyzed before extracting quantitative information on the sample studied itself. In particular, the tip–sample distance z plays a key role as seen in (1.2) for both the signal amplitude and the spatial resolution of the processed image. This effect and the way to counterbalance it with new meta materials acting as superlenses will be discussed at the end of this section.

1.4.1
Material-Specific Imaging

The light scattered by the coupled tip–sample system depends both on the tip polarizability α and the sample's dielectric constant ε. One consequence is that the scattered signal strongly depends on the material coating the tip but also, for a given tip, on the spectral range covered during the experiment. This latter effect can be even stronger if one takes into account the real geometry of the tip leading to additional resonance effects, depending on both the wavelength and the polarization of the incident electric field [52,53]. If this enhancement (up to several thousands) can advantageously be used for a single-wavelength investigation of nano-sized objects with low scattering efficiency, a tip material with low dispersion in the spectral range studied would be preferred for spectroscopic applications. In that case, apertureless SNOM can be viewed as a nanospectroscopic tool to investigate the optical properties of the sample, providing the tip height z is maintained constant during the imaging process (1.2). Indeed, for a given tip coating, the effective polarization depends on the local dielectric constant of the sample only (through the parameter β). The optical contrast is thus a direct image of the local dielectric function as far as the far-field contribution can be removed from the detected signal. This was achieved by Hillenbrand and Keilmann [18,54,55] (Fig. 1.10) using interferometric detection complemented by demodulation at high harmonics of the tip oscillation frequency, thus rejecting the far-field background below the noise level of the detection. Hence, the apertureless SNOM technique allows one to distinguish material classes such as very low refractive index dielectrics, leading to a weak scattered signal, semiconductors with an intermediate contrast and metal nanostructures having a large optical response. The values obtained experimentally were compared with the values form simulations based on (1.2), and good agreement was obtained despite the rather simple description of the tip shape, reduced to a spherical particle having the same radius of curvature [54]. In addition, the spatial resolution was found to be almost insensitive to the excitation wavelength, but was mainly governed by the tip radius: the bright-to-dark transition occurs within 10 nm ($\lambda/60$) in the visible region and 20 nm ($\lambda/500$) in the IR region for tip radius of about 20 nm [56]. This clearly open the way for a quantitative analyzing tool on a nanometric scale for applications in semiconductor processing and nanotechnologies.

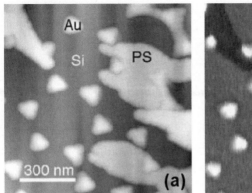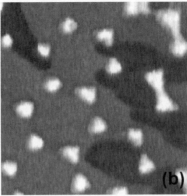

Fig. 1.10. Topography (**a**) and near-field amplitude image (**b**) demodulated at the fourth harmonic of a test pattern of gold triangular islands on silicon, with polystyrene (*PS*) defects. (Redrawn from [55])

1.4.2
Phase Mapping in Metallic Nanostructures and Optical Waveguides

One of the greatest successes of micronologies and nanotechnologies is their ability to manipulate light using either dielectric confinement in microcavities, optical waveguides and photonic crystals or, recently, at the origin of a renewed interest in plasmonics, localization on the nanometer scale by metallic nanostructures e. g., surface-enhanced Raman scattering with single-molecule sensitivity [57,58] and light guiding through nanometric chains of noble metal particles [59–61]. Owing to the potential opportunities in fundamental physics and for applications in opto-electronics the need for methods to image the near-field distribution has become of primary importance. The first examples of near-field images of isolated metal particles were obtained by Krenn et al. [62] in 1995 using a photon STM apparatus, followed by Klar et al. [63] with an aperture scanning near-field optical microscopy and Hamann et al. [64] with an apertureless configuration. A decisive step was then accomplished by Hillenbrand and Keilmann [18] with the application of a hetero-dyne scheme to the SNOM technique. Hence, both the amplitude and the phase of the scattered field can be recorded [65]. In this experiment, the optical maps were processed with a $4\,\mu$m long carbon nanotube (CNT) bundle mounted on the Si tip of a conventional atomic force microscope cantilever in order to reduce the light scattered by the tip itself (increasing the signal-to-background ratio) but also to avoid significant perturbation of the eigenfield pattern by the presence of the tip [66]. In this configuration (Fig. 1.11, panel a), the electric field surrounding the nanoparticle can be decomposed into the sum of the incident field E_i and the electric field E_p generated by the particle. But, owing to the enhancement effects in the vicinity of a metal surface, the local field $E_{loc} = E_i + E_p$ experienced by the tip is almost equal to E_p. The scattered field is thus a direct image of the eigenfield pattern $E_{sca} = s\exp(i\theta) = \alpha_{eff}E_p$, where α_{eff} is again the effective polarization derived in (1.2). As seen in Fig. 1.11, the exact electrodynamics calculations of

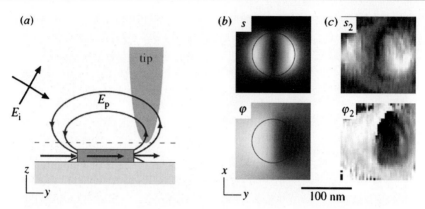

Fig. 1.11. a Dipolar surface plasmon–polariton field pattern of a gold disk on glass at 633 nm. The experiment and the particle eigenfield. Near-field amplitude (*top row*) and phase (*lower row*) obtained by exact electrodynamics calculation (**b**) and measured (**c**) using a carbon nanotube as an optical probe. (Redrawn from [66])

both field amplitude and phase (Fig. 1.11, panel b) are in good agreement with the measured image (Fig. 1.11, panel c), revealing the typical dipolar pattern (two lobes with opposite phases). This demonstrates the robustness of the demodulation technique at high harmonics but also the fact that the presence of the CNT-based tip does not perturb significantly the eigenfield pattern of the nanostructure studied.

The heterodyne demodulation technique has been successfully applied to the mode analysis in optical integrated waveguides [43]. In this case, the incident light is injected in an optical waveguide and the evanescent field resulting from the total internal reflection of light inside the guide is transferred to the far field by the probe. Using the homodyne demodulation at the tip oscillation frequency (Fig. 1.12b) leads to a noisy signal in the overall map resulting from the interference between the background and the field scattered by the tip. In contrast, the heterodyne signal cancels almost totally the unmodulated background contribution, revealing an amplitude image (Fig. 1.12c) in close agreement with the AFM scan (Fig. 1.12a). As expected, the amplitude of the heterodyne signal as well as the signal-to-noise ratio have be considerably enhanced with respect to the homodyne one. The phase information extracted thanks to the heterodyne procedure (Fig. 1.12d) clearly evidences the front waves of the guided mode but also its lateral expansion, which exceeds that of the waveguide (Fig. 1.12a) owing to the presence of evanescent fields decaying exponentially outside the guide. Note that a close inspection of the amplitude profile along the waveguide (Fig. 1.12c) also shows the presence of a partial standing wave arising from the reflection of the guided mode at the output facet of the wave guide.

The phase information can be of primary importance in understanding and characterization of the optical properties at microscale or nanoscale as evidenced in Fig. 1.13 with the example of a very compact wave splitter. This latter is composed of a 1.55 μm input monomode waveguide connected to a multimode interferometer cell [67]. Three kinds of propagating light modes can be clearly discriminated from

AFM   **SNOM f**

SNOM f-Δω/2π   **SNOM f-Δω/2π**
(amplitude)   **(phase)**

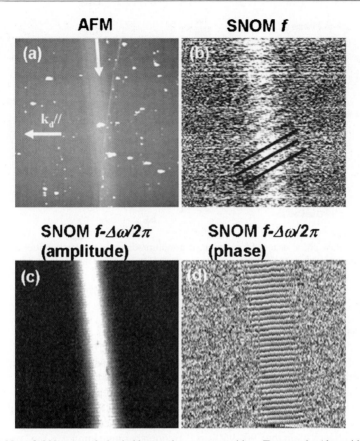

Fig. 1.12. Near-field images of a buried ion-exchange waveguide. **a** Topography (the guide height is 20 nm). **b** Apertureless SNOM image at 1.55 μm demodulated at the first harmonic of the tip vibration. Heterodyne amplitude **c** and phase **d** apertureless SNOM images demodulated at $f - \Delta\omega/2\pi$. (Redrawn from [43])

the shape and spatial extension of the wavefronts: (1) quasi plane wave caused by imperfect light injection in the input guide, (2) circular wave resulting from the losses (scattering) in the multimode interferometer cell and (3) guided modes following the L-shaped monomode output waveguides. Filtering the image using a 2D fast Fourier transform on the complex fields allows one to quantify the losses and coupling efficiency from the amplitudes of the three different waves (cf. Fig. 1.13b), but also to extract geometrical characteristics of the guide and the corresponding dispersion relation from the phase oscillations.

Comparable studies were performed recently to characterize the propagation of phonon-polaritons close to a Au–SiC interface [68] or plasmon-polaritons in metal strips using a photon STM configuration [69], demonstrating the versatility and power of this technique for spatially and frequency resolved optical studies.

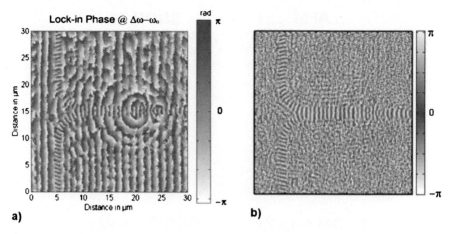

Fig. 1.13. a Raw phase near-field mapping of a multimode interferometer splitter acquired by heterodyne ccanning near-field optical microscopy. **b** Filtered phase map of the guided field after radiation modes have been subtracted. (Redrawn from [67])

1.4.3
Tip-Induced Resonances in Polaritonic Samples

Apart from the tip and sample resonances which are characteristic of the material composition and geometry, new resonances appear from the tip–sample coupling as evidenced by the effective polarization expression (1.2) containing a resonant denominator for $\mathrm{Re}\left[1 - \alpha\beta/16\pi(a + z)^3\right] = 0$. This effect was evidenced by Hillenbrand et al. using a SiC sample [23] which is characterized by a large phonon-induced reflectivity in the $790{-}950\,\mathrm{cm}^{-1}$ frequency range known as the "restrahlen" reflectivity band. However, when excitation is through the near-field interaction with a nonresonant metallic sphere, the scattered signal has a very sharp resonance at $930\,\mathrm{cm}^{-1}$ which is nearly 2 orders of magnitude higher in intensity than the value obtained with a gold reference substrate (Fig. 1.14). Taking into account the demodulation procedure (solid curve), the simulations based on (1.2) reproduce with good agreement the general shape, dynamics (3 orders of magnitude) and frequency of the resonance although a better fitting could be obtained with a refined model. Nevertheless, this spectral response is very different both in amplitude and in line width from what is observed in the far-field experiment, clearly demonstrating the extreme sensitivity of the SNOM technique.

Taking further this concept of tip-induced resonances, the authors predicted a high material discrimination (chemical composition, crystallinity) thanks to the extremely narrow resonance (1% half width at half maximum) and the strong frequency dependence of the longitudinal phonons to a lattice modification. For this purpose Huber et al. [70] studied SiC polytypes 4H and 6H varying only by the stacking sequence in a hexagonal lattice. As evidenced in Fig. 1.15 both amplitude and phase were found to significantly vary between the two polytypes while the excitation energy was changed close to the expected phonon–polariton resonance. A detailed analysis of the prototype spectral signatures clearly evidences two dif-

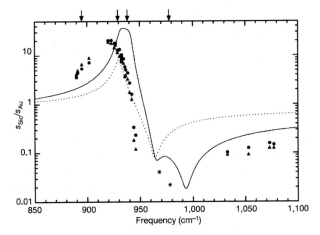

Fig. 1.14. Phonon-enhanced near-field response of SiC normalized to gold. The *symbols* represent the experimental values of the field amplitude demodulated at the second harmonic. The *curves* represent model predictions for a Pt particle in contact with SiC (*dashed line*), and the same (*solid line*) modified to simulate the experimental procedure of sinusoidal tapping motion. (Redrawn from [23])

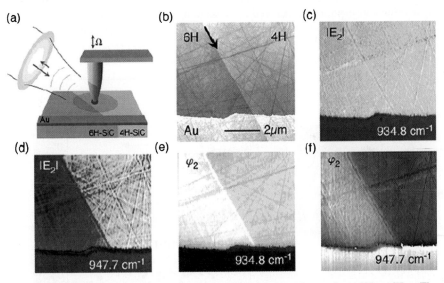

Fig. 1.15. Apertureless SNOM imaging of sharp SiC polytype change from 4H to 6H. **a** The setup. **b** Topography of polytype boundary. **c, d** IR amplitude and **e, f** phase maps demodulated at the second harmonic at two different frequencies. (Redrawn from [70])

ferent amplitudes and phases shifted by about $4\,\text{cm}^{-1}$ in qualitative agreement with corresponding simulations. Hence, apertureless SNOM used in conjunction with heterodyne demodulation can be viewed as a local near-field optical fingerprint of the structural and chemical properties at a nanometer scale.

1.4.4
Applications to Identification of Biosamples

The ability to perform nanoscale spectroscopy in the IR region has possible applications in biology, chemistry and material science thanks to the characteristic vibrational signatures of chemical bounds. The main question to address is how a typical vibrational resonance with a relatively weak oscillator strength translates into a apertureless SNOM optical contrast. As seen if Fig. 1.16b obtained for a single tobacco mosaic virus [71], clear IR amplitude and phase contrast is obtainable with a well-expressed spectral resonance (absorbance and dispersive reflection line shape for the phase and amplitude, respectively). Both line shapes and peak position were well reproduced using the dielectric function extracted from Kramers–Kronig transformation of the conventional IR spectrum of tobacco mosaic virus (although the amplitude was underestimated by a factor 3.3). Nevertheless, the principal outcome of this experiment is that amplitude and phase modifications as large as $\pm 10\%$ and $10°$, respectively, allowed the observed signal to be assigned to the amide I vibrational band of the peptide bonds in the virus cell. This decisive step for quantitative IR nanospectroscopy was further confirmed by comparing the signal of both tobacco mosaic virus and poly(methyl methacrylate) (PMMA) spheres (Fig. 1.16b). For the polymer structures, the resonance band is shifted to $1730\,cm^{-1}$, in full agreement with known absorption spectra, and is found to by almost insensitive to the particle size (Fig. 1.16b) and to the structurally and chemically heterogeneous environment on a nanometric scale.

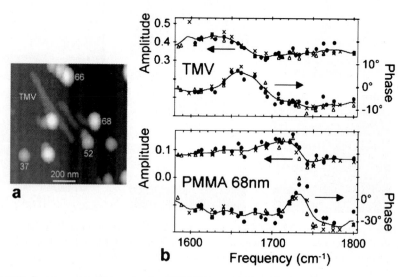

Fig. 1.16. Spectroscopic IR apertureless SNOM investigation of tobacco mosaic virus (*TMV*). **a** Topography. **b** IR spectra of the interior of the TMV marked (*top*) and of the 68nm-high poly(methyl methacrylate) (*PMMA*) particle (*bottom*) extracted from images at different frequencies. (Redrawn from [71])

1.4.5
Subsurface Imaging and Superlensing

The unprecedented optical resolution of a few nanometers obtained with apertureless SNOM justifies the success of scanning near-field approaches, but, in some sense, restricts its application to nanostructures supported on a substrate being or not placed in liquids. The expression for the effective polarizability (1.2) clearly evidences the rapid decay of the near-field signal while the tip is retracted from the sample. To quantify how much the optical contrast and spatial resolution can be affected by this distance, Taubner et al. [72] recorded topographic and optical maps on gold islands deposited on a silicon substrate and partially coated by a PMMA layer. Clearly, the gold islands are still distinguished from the silicon substrate with about 50 nm of polymer spacing, even if the optical signal has considerably reduced its amplitude as expected from a typical approach curved performed at $\lambda = 10.5\,\mu m$ (Fig. 1.17). Since the topography transferred to the PMMA surface does not exactly reproduce the morphology of the gold islands, a direct comparison between the optical image of the buried structure and the topography is difficult. Nevertheless, the spatial resolution can be roughly estimated from the variation of the optical contrast to be smaller 120 nm [56].

Such an "in-depth" probing could be of great importance for technological applications with protection coatings, multilayer structures and submembrane investigation in biology. However, a completely new and revolutionary approach was recently proposed by Taubner et al. [73] by combining the scattering scanning near-field optical microscope in reflection mode with a "superlensing" material. A superlensing

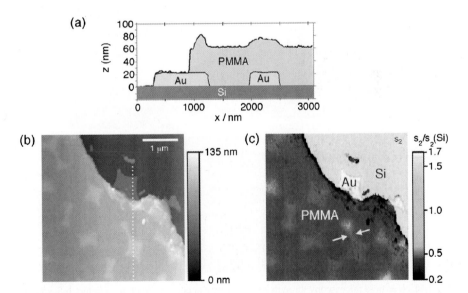

Fig. 1.17. Example of subsurface mid-IR imaging of a sample partly covered with spin-coating PMMA. **a** Structure of the sample. **b** IR amplitude map demodulated at the second harmonic and **c** topography line plot. (Redrawn from [72])

material is one with a negative refractive index [74]. In this case, the propagative part of the light experiences a negative refraction at both air interfaces, leading in some sense to a "conventional" image, while the evanescent part is amplified during its travel through the superlensing material and is able to conserve an appreciable amplitude in the image plane. It is precisely the evanescent component which carries the high special frequencies needed to break through the diffraction limit and would make possible a point image from a point source. Additionally, Pendry [74] showed in his famous paper that such behavior can be approximated using a thin slab of negative-permittivity material, easier to obtain in the optical domain, which leads to an image formation still limited by diffraction but on a subwavelength scale (typically a few nanometers). In the experiment of Taubner et al., the sample to be imaged is separated from the tip by a 440nm SiC superlens coated on both faces by 220nm SiO_2 layers (Fig. 1.18a). When imaging is at a given wavelength where susperlensing occurs, both the amplitude (Fig. 1.18c) and the phase (Fig. 1.18d) images reveal holes formed in a gold layer (Fig. 1.18b) 880 nm underneath/away with an amazing spatial resolution of $\lambda/20$. The test image recorded at lower wavelength does not exhibit noticeable contrast, thus demonstrating the key role of the superlensing layer.

1.5
Conclusions

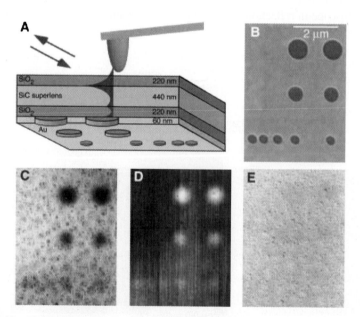

Fig. 1.18. Near-field microscopy through a 880nm-thick superlens structure. **a** Setup. **b** Scanning electron microscopy of the object plane. **c** IR amplitude image at 10.85 μm showing the superlensing effect. **d** IR phase contrast at 11.03 μm. **e** IR amplitude at 9.25 μm where superlensing is not expected to occur. (Redrawn from [73])

Apertureless SNOM represents a promising breakthrough in optical microscopy, with potential applications in a spectral range going from the visible to the mid-IR. In this chapter we have shown applications spanning from materials sciences to living matter. Overall, research in the area of apertureless SNOM has produced promising results and holds promise for high-resolution chemical and materials characterization at finer and finer length scales. As with all near-field methods, sample systems with extended complexity in the z-dimension complicate the analysis of images and limit present applications of these techniques to thin-film samples. Through the continued development of detection techniques, improved probe structures and the refinement of our physical understanding of the factors that govern near-field interactions, we are closer than ever to the realization of a measurement platform that provides artifact-free characterization of complex, heterogeneous nanometer-scale systems.

References

1. Klar TA, Jakobs S, Dyba M, Egner A, Hell SW (2000) Proc Natl Acad Sci USA 97:8206
2. Betzig E, Patterson GH, Sougrat R, Lindwasser OW, Olenych S, Bonifacino JS, Davidson MW, Lippincott-Schwartz J, Hess HF (2006) Science 313:1642
3. Novotny L, Hecht B (2006) Principles of nano-optics. Cambridge University Press, Cambridge
4. Pohl DW, Denk W, Lanz M (1984) Appl Phys Lett 44:651
5. Lewis A, Isaacson M, Harootunian A, Murray A (1984) Ultramicroscopy 13:227
6. Synge EH (1928) Philos Mag 6:356
7. Betzig E, Trautman JK, Harris TD, Weiner JS, Kostelak RL (1991) Science 251:1468
8. Hecht B, Sick B, Wild UP, Deckert V, Zenobi R, Martin OJF, Pohl DW (2000) J Chem Phys 112:7761
9. Gucciardi PG, Trusso S, Vasi C, Patanè S, Allegrini M (2006) In: Bhushan B, Fuchs H, Kawata S (eds) Applied scanning probe methods vol V. Springer, Berlin, p 287
10. Labardi M, Gucciardi PG, Allegrini M (2001) Riv Nuovo Cimento 23:1
11. Hecht B, Bielefeldt H, Inouye Y, Pohl DW, Novotny L (1997) J Appl Phys 81:2492
12. Gucciardi PG, Colocci M (2001) Appl Phys Lett 79:1543
13. Chicanne C, David T, Quidant R, Weeber JC, Lacroute Y, Bourillot E, Dereux A, Colas des Francs G, Girard C (2002) Phys Rev Lett 88:097402
14. Zenhausern F, O'Boyle MP, Wickramasinghe HK (1994) Appl Phys Lett 65:1623
15. Bachelot R, Gleyzes P, Boccara AC (1995) Opt Lett 20:1924
16. Inouye Y, Kawata S (1994) Opt Lett 19:159
17. Novotny L, Stranick SJ (2006) Annu Rev Phys Chem 57:303
18. Hillenbrand R, Keilmann F (2000) Phys Rev Lett 85:3029
19. Hillenbrand R, Keilmann F (2002) Appl Phys Lett 80:25
20. Labardi M, Patanè S, Allegrini M (2000) Appl Phys Lett 77:621
21. Taubner T, Keilmann F, Hillenbrand R (2004) Nanolett 4:1669
22. Stebunova L, Abramitchev BB, Walker GC (2003) Rev Sci Instrum 74:3670
23. Hillenbrand R, Taubner T, Keilmann F (2002) Nature 418:159
24. Ocelic N, Hillenbrand R (2004) Nat Mat 3:606
25. Binnig G, Rohrer H, Gerber C, Weibel E (1982) Phys Rev Lett 49:57
26. Binnig G, Quate CF, Gerber C (1986) Phys Rev Lett 56:930
27. Betzig E, Finn PL, Weiner JS (1992) Appl Phys Lett 60:2484
28. Toledo-Crow R, Yang PC, Chen Y, Vaez-Iravani M (1992) Appl Phys Lett 60:2957
29. Karrai K, Grober RD (1995) Appl Phys Lett 66:1842

30. Shalaev V, Sarychev A (1998) Phys Rev B 57:13265
31. Xu H, Aizpurua J, Kall M, Apell P (2000) Phys Rev E 62:4318
32. Wessel J (1985) JOSA B-Optical Physics 2:1538–41
33. Bek A, Vogelgesang R, Kern K (2006) Rev Sci Instrum 77:043703
34. Patanè S, Gucciardi PG, Labardi M, Allegrini M (2004) Riv Nuovo Cimento 27:1
35. Zenhausern F, Martin Y, Wickramasinghe HK (1995) Science 269:1083
36. Martin Y, Zenhausern F, Wickramasinghe HK (1996) Appl Phys Lett 68:2475
37. Garcia N, Nieto-Vesperinas M (1995) Appl Phys Lett 66:3399
38. Azoulay J, Debarre A, Richard A, Tchenio P (2000) Appl Opt 39:129
39. Maghelli N, Labardi M, Patanè S, Irrera F, Allegrini M (2001) J Microsc Oxf 202:84
40. Formanek F, De Wilde Y, Aigouy L (2003) J Appl Phys 93:9548
41. Knoll B, Keilmann F (2000) Opt Commun 182:321
42. Roy D, Leong SH, Elland M (2005) J Kor Phys Soc 47:S140
43. Gomez L, Bachelot R, Bouhelier A, Wiederrecht GP, Chang S, Gray SK, Lerondel G, Hua F, Jeon S, Rogers JA, Castro ME, Blaize S, Stefanon I, Royer P (2006) J Opt Soc Am B 23:823
44. Ocelic N, Huber A, Hillenbrand R (2006) Appl Phys Lett 89:101124
45. Gucciardi PG, Bachelier G, Allegrini M (2006) J Appl Phys 99:124309
46. Gucciardi PG, Bachelier G, Allegrini M, Ahn J, Hong M, Chang S, Jhe W, Hong SC, Baek SH (2007) J Appl Phys (in press)
47. Formanek F, De Wilde Y, Aigouy L (2003) J Appl Phys 93:9548
48. Formanek F, De Wilde Y, Aigouy L (2005) Ultramicroscopy 103:133
49. Spiegel MR, Liu J (1999) Mathematical handbook of formulas and tables. McGraw-Hill New York
50. Billot L, Lamy de la Chapelle M, Barchiesi D, Chang SH, Gray SK, Rogers J, Bouhelier A, Adam PM, Bijeon JL, Wiederrecht GP, Bachelot R, Royer P (2006) Appl Phys Lett 89:0123051
51. Adam PM, Royer P, Laddada R, Bijeon JL (1998) Ultramicroscopy 71:327
52. Novotny L, Bian RX, Xie XS (1997) Phys Rev Lett 79:645
53. Bachelot R, H'Dhili F, Barchiesi D, Lerondel G, Fikri R, Royer P, Landraud N, Peretti J, Chaput F, Larnpel G, Boilot JP, Lahlil K (2003) J Appl Phys 94:2060
54. Hillenbrand R, Keilmann F (2002) Appl Phys Lett 80:25
55. Hillenbrand R (2004) Ultramicroscopy 100:421
56. Taubner T, Hillenbrand R, Keilmann F (2003) J Microsc Oxf 210:311
57. Kneipp K, Wang Y, Kneipp H, Perelman LT, Itzkan I, Dasari R, Feld MS (1997) Phys Rev Lett 78:1667
58. Kneipp J, Kneipp H, McLaughlin M, Brown D, Kneipp K (2006) Nano Lett 6:2225
59. Salerno M, Krenn JR, Lamprecht B, Schnider G, Ditlbacher H, Felidj N, Leitner A, Aussenegg FR (2002) Optoelectron Rev 10:217
60. Wurtz GA, Im JS, Gray SK, Wiederrecht GP (2003) J Phys Chem B 107:14191
61. Bouhelier A, Bachelot R, Im JS, Wiederrecht GP, Lerondel G, Kostcheev S, Royer P (2005) J Phys Chem B 109:3195
62. Krenn JR, Gotschy W, Somitsch D, Leitner A, Aussenegg FR (1995) Appl Phys Mater Sci Proc 61:541
63. Klar T, Perner M, Grosse S, von Plessen G, Spirkl W, Feldmann J (1998) Phys Rev Lett 80:4249
64. Hamann HF, Gallagher A, Nesbitt DJ (1998) Appl Phys Lett 73:1469
65. Hillenbrand R, Keilmann F, Hanarp P, Sutherland DS, Aizpurua J (2003) Appl Phys Lett 83:368
66. Keilmann F, Hillenbrand R (2004) Philos Trans R Soc Lond Ser A 362:787
67. Stefanon I, Blaize S, Bruyant A, Aubert S, Lerondel G, Bachelot R, Royer P (2005) Opt Express 13:5553

68. Huber A, Ocelic N, Kazantsev D, Hillenbrand R (2005) Appl Phys Lett 87:81103
69. Weeber JC, Lacroute Y, Dereux A, Devaux E, Ebbesen T, Girard C, Gonzalez MU, Baudrion AL (2004) Phys Rev B 70:235406
70. Huber A, Ocelic N, Taubner T, Hillenbrand R (2006) Nano Lett 6:774
71. Brehm M, Taubner T, Hillenbrand R, Keilmann F (2006) Nano Lett 6:1307
72. Taubner T, Keilmann F, Hillenbrand R (2005) Opt Express 12:8893
73. Taubner T, Korobkin D, Urzhumov Y, Shvets G, Hillenbrand R (2006) Science 313:1595
74. Pendry JB (2000) Phys Rev Lett 85:3966

2 Critical Dimension Atomic Force Microscopy for Sub-50-nm Microelectronics Technology Nodes

Hao-Chih Liu · Gregory A. Dahlen · Jason R. Osborne

Abstract. The chapter discusses recent developments within critical dimension atomic force microscopy (CD AFM) and its application to nanostructures having lateral dimensions of 50 nm and less. Challenges in this measurement range call for research and development in probe design, tip–sample interaction control, tip shape characterization, and image reconstruction algorithms. Design considerations and performance improvements are reviewed throughout the chapter and metrological measurement results are presented with emphasis on industrial applications. CD AFM uses novel probes and advanced scan control algorithms to acquire metrological measurements that are (1) NIST-traceable with subnanometer precision and nanometer-level uncertainty, (2) capable of imaging vertical sidewalls and undercut features, (3) direct, nondestructive and fast relative to existing reference metrology systems (RMS), and (4) multiple cross-sectional, with resolution comparable to that of transmission electron microscopy (TEM). Recently, these attributes have enhanced the role of CD AFM as the RMS for other metrology systems.

Key words: Atomic force microscope, Atomic force microscopy, Critical dimension, Image reconstruction, Metrology, Probe tip, Scan algorithm, Tip shape extraction

Abbreviations

AFM	Atomic force microscopy
CD	Critical dimension
CMOS	Complementary metal oxide semiconductor
CNT	Carbon nanotube
DT	Depth trench
ESD	Electrostatic discharge
FIB	Focused ion bean
ITRS	International Technology Roadmap for Semiconductors
IVPS	Improved vertical parallel structure
LER	Line edge roughness
LWV	Line width variation
MEMS	Microelectromechanical system
NIST	National Institute of Standards and Technology
OCD	Optical critical dimension
RMS	Reference metrology system
SAM	Self-assembled monolayer
SEM	Scanning electron microscope
SPM	Scanning probe microscopy
SWR	Sidewall roughness
TEM	Transmission electron microscopy

TUT   Tool-under-test
VEH   Vertical edge height

2.1
Introduction

Since the introduction of the atomic force microscope in 1986 [1], the family of instruments that measure forces by detecting the interactions between a sample and a probe tip at the free end of a microscale cantilever has forged an expanding role in research and development in the microscale and nanoscale world. Atomic force microscopy (AFM) covers a broad spectrum of related techniques, such as topographical imaging, phase imaging, electrostatic force microscopy, magnetic force microscopy, lateral force microscopy, force modulation microscopy, nanoindentation, and nanomanipulation. Among them, the capability to image the sample surface is the most commonly used application.

Critical dimension (CD) AFM is a subgroup within AFM that was developed to measure features of interest by acquiring and analyzing the topography of morpholoically challenging samples. In order to maintain high precision and accuracy, CD AFM is regularly calibrated with traceable dimensional standards and employs advanced scanning capabilities.

2.1.1
AFM for Semiconductor and Data Storage Industries

The miniaturization of device components is one of the most important technological trends since the advent of the transistor. Smaller components usually result in lower manufacturing costs, lower energy consumption, higher performance, and portability. Several industries have benefited from the shrinking dimensions of devices. For instance, the International Technology Roadmap for Semiconductors predicts technological trends of microelectronics and has clearly indicated the continuance of miniaturization for integrated circuits [2, 3]. Changes in fabrication technology and new device design (such as FinFET[1] structures and 3D interconnects) call for new metrology requirements. These in turn provide opportunities for CD AFM.

In recent years, the major use of CD AFM is within research and development departments of semiconductor and data storage industries. Referred to as a complementary or reference metrology system (RMS), CD AFM compensates for limitations of other metrology tools by providing traceable [4], nondestructive calibration.

2.1.2
Scanning Modes: Tapping Versus Deep Trench and CD Mode

AFM provides a fundamental method to image surface topography and, in some cases, to detect localized material properties on a nanometer scale. By monitoring the

---

[1] A multigate field-effect transistor that incorporates more than one gate into a single device.

interaction forces between the specimen and a sharp tip during a scanning operation, the atomic force microscope typically acquires height and phase information of the sample. The primary atomic force microscope operation modes are categorized into contact mode, noncontact mode, and tapping mode on the basis of the tip–sample proximity during scanning. However, as new applications of AFM have been found across industries, more advanced scanning algorithms have been developed to measure 2D and 3D profiles of nanofabricated structures. For instance, two of the widely used scan modes in microelectronics are deep trench (DT) mode and CD mode. CD mode is capable of scanning vertical sidewalls, including undercut features. This enables metrology of feature dimensions, sidewall angle/roughness,

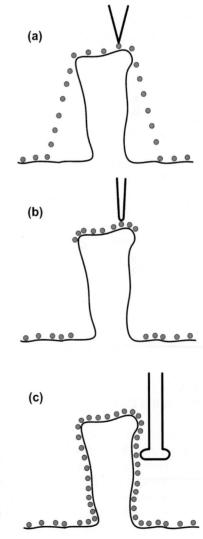

Fig. 2.1. Illustrations show the differences among scan modes: **a** tapping mode; **b** deep trench (DT) mode; **c** critical dimension (CD) mode. In **b** and **c**, the *dots* represent tip–sample contacts during scanning, while in **a**, *dots* indicate the locations of the tip apex when data points are recorded. Note DT and CD modes require special probe tip shapes for proper imaging

and line edge variation [3]; whereas DT mode is tailored for depth measurements on high aspect ratio structures.

A conceptual illustration of the difference among tapping, DT, and CD modes is given in Fig. 2.1. Tapping mode captures data points at a pre-determined time interval while the probe moves at a constant speed over the sample surface; therefore, some measurements (e. g., the sloped sidewall in Fig. 2.1a) are artifacts caused by the data acquisition scheme and the influence of the tip shape. DT (Fig. 2.1b) and CD scanning (Fig. 2.1c) are adaptive modes that record valid data points when the probe tip is on the specimen surface. This approach is superior to conventional scanning modes and entails additional control system complexity. In DT and CD modes, the probe moves in small increments (referred to as "microsteps" but which are actually "nanosteps") where certain criteria must be met prior to each measurement to ensure the probe tip is making contact with the sample surface. The major difference between DT and CD modes is the probe servo direction. In DT mode, the probe servos along the Z-direction (vertically) and moves in the X-direction in uniform intervals after taking each measurement. For CD mode, the probe moves at an angle ranging from less than 45° to 90° to the sample surface. The CD mode approach enables an accurate measurement on vertical or reentrant sample surfaces when using an appropriate tip.

2.1.3
Specialty Probes

As illustrated in Fig. 2.1, CD and DT modes require special probe designs to complement the scanning algorithms. For height and depth measurement applications, DT mode requires straight, long probes to measure high aspect ratio trenches and holes. The typical aspect ratio ranges from 1:4 to 1:10, and in some instances is more than 1:10. Since the space between sidewalls is currently 50–100 nm or smaller, DT-type probes require small apex radii, a small tip half angle, and tilt compensation to reach the bottom of narrow trenches. Two common probes used in DT applications are shown in Fig. 2.2a and b.

The probes used for CD measurements usually have a horizontally flared tip apex region, as indicated in Fig. 2.2c and d. The flared tip apex makes imaging of sidewall and reentrant features possible; however, the accurate measurement of CDs calls for better understanding of the tip shape. Consequently, the topics of tip shape reconstruction and sample image reconstruction will be presented in Sect. 2.3.

2.2
Reference Metrology System and Semiconductor Production

As semiconductor processing advances to the nanometer scale, the need for metrology systems that are scalable to the dimensions of interest is increasingly important, yet ever more challenging. In the nanometer range, metrology tools that were widely used in the micrometer range are no longer suitable or are starting to encounter their physical limitations.

For example, the capability to distinguish two adjacent points depends on the wavelength of the light source. Optical microscopy, utilizing visible light

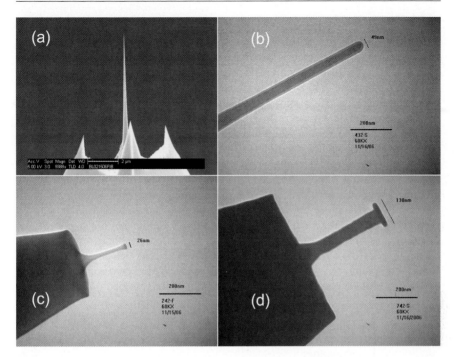

Fig. 2.2. Scanning electron microscopy (SEM) and transmission electron microscopy (TEM) micrographs of specialty probes used in DT and CD modes: **a** FIB3D probe for DT mode; **b** CDP55A probe for DT mode; **c** CD32 probe for CD mode; **d** CDR130SiN (SiN-capped) probe for CD mode

sources, reaches its resolution limit around 0.2 μm. Other beam-based metrology systems, such as scanning electron microscopy (SEM) and scatterometry (or optical CD, OCD, system), have to overcome similar fundamental physical limits as the diffraction, scattering, and interference of light (or electron beam) become significant once the feature size is comparable to the wavelength of the source. Furthermore, there are measurement uncertainties in the optical properties of the materials in the sample structure. Nondestructive beam-based metrology measurements are implemented by directing a beam onto a sample surface in a top-down manner. There is no direct information about the sidewall morphology of the feature. Moreover, this aspect of the feature has become increasingly important in nanoelectronics. On the feature sidewalls, the aforementioned beam interactions are complicated and contribute to the uncertainty of CD metrology measurements.

Many recent improvements have been made to SEM and OCD systems [5–8]. One of the most utilized approaches is the use of a reference library in conjunction with modeling techniques [9, 10]. To generate sidewall profiles of a sample feature more accurately, SEM and scatterometer manufacturers use a separate reference metrology system (RMS) to measure the cross-sectional profiles of a variety of structures (standard artifacts). The metrology data are then fed back to the system to correlate its signal patterns with modeling results. Once a reference library has been built and the tool has been calibrated, the metrology performance of this beam-based (or model-based) tool becomes reliable. Currently, model-based metrology tools remain the "workhorses" in semiconductor production. The high throughput (thus low cost per measurement) and moderate to excellent metrology repeatability and accuracy allow SEM and OCD systems to fulfill metrology needs for current manufacturing.

The choice of an RMS directly impacts the metrology performance of the tool undergoing calibration (i. e., tool-under-test, or TUT). Ideally, the RMS should be an independent and complementary technique that addresses the shortcomings of the TUT. In addition, the RMS needs to be traceable to a metrology dimensional standard. Figure 2.3 illustrates the relationship of dimensional standards, RMS, and in-line metrology tools. Dimensional standards are metrology artifacts whose dimensions (line width, step height, etc.) were calibrated and can be traced back to a fundamental constant (e. g., the lattice structure of silicon atoms). In-line metrology systems are metrology instruments that are part of factorywide automation that provides process control for day-to-day manufacturing of integrated circuits. Historically, transmission electron microscopy (TEM) has been used as the RMS because of its capability to image a cross-sectional profile of the structure of interest, and the measurement

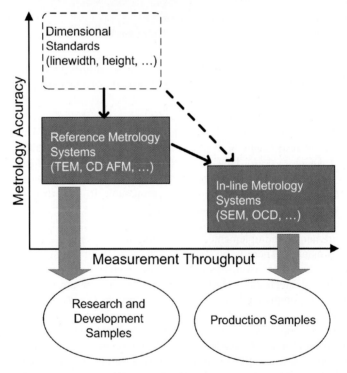

Fig. 2.3. The roles of reference metrology systems and in-line metrology systems in semiconductor and data storage industries. The *arrows* indicate the directions of metrology reference transfer. *CD AFM* critical dimension atomic force microscope, *OCD* optical critical dimension, *SEM* scanning electron microscope, *TEM* transmission electron microscope

accuracy provided by calibration to the lattice structure of atoms [11,12]. The major disadvantages of TEM are the uncertainty of the exact location from which the final "thinned" sample is derived, throughput, and the locally destructive sample removal process. In recent years, CD AFM has been adopted as a complementary RMS to TEM by both metrology tool manufacturers and end users [5,8]. Further discussion on new applications of CD AFM and their complementary role to TEM is presented in Sect. 2.4.2.

2.2.1
Requirements for Metrology Tools

Metrology techniques for integrated circuits are facing challenges as the industry explores the limits of Moore's law, and concurrently introduces new materials and processing techniques. The ITRS reports (2005 edition [2] and 2006 update [3]) clarified requirements for metrology tools needed in the semiconductor industry for both near-term (2005–2013) and long-term (2014–2020) years. The precision of metrology tools is required to be 0.58 nm (3σ value in nanometers) for wafer CD measurements of isolated structures in 2006, and is expected to be 0.27 nm 3σ in 2013. Of particular note is that the precision values include metrology tool-to-tool matching to ensure process control throughout the factory.

The ITRS also lists clear challenges for metrology tools. It is noteworthy that:

1. Three dimensional (CD and depth) measurements are required for technology nodes both before and beyond 32 nm, as process technologies move toward higher aspect ratio structures.
2. Sidewall shape information is increasingly important. Sidewall roughness, sidewall angle, reentrancy, and notching and footing at the bottom of features (bottom CD) are areas that raise challenges for most of today's metrology tools.
3. Desired characteristics of a metrology system are that it be nondestructive, production-worthy for in-die measurement, material-independent, and that it can help determine manufacturing metrology for new device design and materials in development.

Clearly, these challenges provide opportunities for CD AFM systems. Specifically, owing to the fundamental differences in operating principles, CD AFM's "bottom-up" (as compared with the "top-down" measurements from scanning electron microscopy/scatterometry) characterization approach may yield more information on feature sidewalls and the 3D nature of structures. Metrology applications for CD AFM will be discussed in detail in Sect. 2.4.

2.2.2
AFM as a Reference Metrology System

CD AFM demonstrates attributes that are complementary to both model-based and destructive (cross-sectioning) metrology techniques. Unlike electron-beam-based instruments such as the SEM and the transmission electron microscope, CD AFM requires virtually no sample preparation before measurements. In addition, CD AFM measurements are taken at ambient conditions rather than in a vacuum; this advantage eliminates the required pumping time and environmental changes that may induce dimensional distortion of materials.

AFM measurements do not require specimen-slicing or ion-milling of the area of interest. Consequently, no sacrificial wafers are needed in the manufacturing process. It has been reported that features at the same measurement location can be measured repeatedly by AFM through out different processing steps [13]. In addition, CD AFM exhibits a significant advantage over other techniques in the measurement of polymeric and insulating materials. In the front end of line processes of integrated circuits, the characterization of photoresist translates into the dimension control of the final devices. This results in performance control of the circuit designs. For electron-beam metrology tools, it is known that photoresist has both charging and shrinkage issues owing to electron-beam exposure; an accurate measurement of the resist dimensions is difficult for scanning electron microscopy and has to be acquired by CD AFM [6, 14]. In general, dimensional distortions of photoresist, oxides, and nitrides due to a vacuum environment, electron-beam damage, or charging can be eliminated with CD AFM.

Similar to scanning electron microscopy, CD AFM uses raster scanning to collect aerial information of the sample surface. The fundamental difference is each CD AFM scan line in an image provides a cross-sectional profile of the structure. As illustrated in Fig. 2.4, a polycrystalline silicon line structure from a process development department exhibited a wide range of line width variation along a 1-μm distance (22-nm amplitude range with 4.4-nm standard deviation). For metrology measurements of an unknown structure in the tens of nanometer range, reasonable doubt arises when one considers some of the variation as noise artifacts. Using CD AFM to independently measure exactly the same location (a precise sample navi-

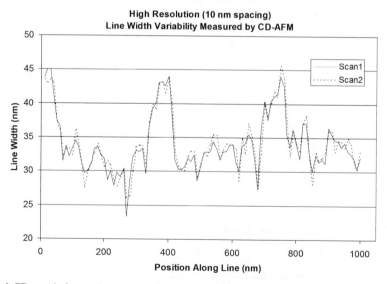

Fig. 2.4. CD atomic force microscopy (*AFM*) line width variation measurement data of a nominal 45-nm line structure in its development stage. (Adopted from [26] with permission)

gation technique called "2D SPM Zoom" [15]) multiple times helps to distinguish "real" variation from noise. In Fig. 2.4, two individual scan results are shown as "Scan1" and "Scan2," which confirm the existence of line width variation of this sample.

Since the nominal line width is 45 nm, the dimensional variation will be difficult to detect if one uses other metrology tools. For instance, if the same structure was characterized by TEM, whose location precision is typically ±0.5 μm (more advanced TEM metrology allows precision as high as ±0.1 μm in sample sectioning [16]), the result would be quite dependent on which axial segment of the line was measured.

2.2.3
AFM as an In-Line Metrology System

Despite the favorable attributes of CD AFM in reference metrology, there are several shortcomings that caused the community to accept CD AFM metrology for in-line manufacturing with reservations. Three of the concerns for using CD AFM in integrated circuits production are:

1. Low measurement throughput
2. Low "ease of use"
3. High cost of ownership (or cost per measurement)

Since CD AFM relies on the detection of tip–sample interaction forces to acquire metrology information, a probe tip has to be in the proximity of a sample surface to invoke physical interactions that can be captured for data analysis. The servoing and travel of a probe on top of sample features takes a longer time than exposing the features to an electron beam. For samples with a high level of 3D information, the adaptive scan modes of CD AFM will take as much time as needed to profile data-rich features (e. g., reentrant sidewall with significant roughness) but will transit quickly when the morphology is less challenging. Consequently, the overall throughput of CD AFM is lower than comparable nondestructive beam-based metrology tools in many production applications.

"Ease of use" is a desirable feature for production equipment in all industries. Because of the breadth of applications, CD AFM tools are designed with great flexibility in scan parameters and various settings. Such flexibility seems overkill for production tools that routinely use only a limited number of applications. Although a simple "go" button for CD AFM measurements is not currently available, tool suppliers are working on limiting the tool complexity presented to the operator.

The semiconductor and data storage industries view the operational cost of capital equipment in terms of its "cost of ownership." The cost of ownership models express a cost per wafer pass as [17–19]

$$ CW = \frac{CF + CV + CY}{TPT \times Y \times U}, \tag{2.1} $$

where CW is the cost per wafer pass, CF the fixed cost, CV the variable cost, CY the cost due to yield loss, TPT the wafer throughput, Y the composite yield, and U the utilization of the tool.

For metrology tools, the cost per measurement becomes a quantifiable specification. Typically, less than ten cents per measurement is a desirable figure in factories. The special probes used for CD AFM measurements are expensive and their lifetime is sometimes less than satisfactory when scanning challenging materials (e. g., bare silicon tips on freshly etched polysilicon). This directly contributes to a higher CV value in (2.1).

The factors that shorten a probe's lifetime can be categorized into three groups:

1. Direct probe damage
2. Probe tip wear
3. Probe contamination

Direct probe damage is defined here as a catastrophic event that can be caused by many issues, such as improper handling of the probe, overly aggressive scan settings, and incompatible probe–sample combinations.

Section 2.5 describes recent advancements in atomic force microscope probe designs that have provided wear-resistant solutions to reduce cost of ownership and increase repeatability in metrology measurements.

The next section will discuss software algorithms that characterize the tip shape. With a thorough knowledge of the atomic force microscope tip shape during measurements, one can eliminate scanning artifacts (in all regions contacted by the tip) and reconstruct the sample morphology with high fidelity.

2.3
Image Analysis for Accurate Metrology

2.3.1
Background

The basic function of an atomic force microscope is to image the morphology of a specimen. Of structural necessity, the tip has a finite size, which results in "dilation" of the sample surface in the acquired image [9, 20, 21]. In order to render the sample surface for metrology measurements with minimal uncertainty, it is desirable that the tip shape distortion be removed from the AFM image (this process is referred to as "erosion" in the terminology of mathematical morphology). With conventional AFM, a convenient alternative approach is to use a tip with as small an apex radius as possible, thereby reducing the scale of the tip-to-sample feature distortion and also permitting higher-resolution imaging. In such cases, erosion of the tip shape from the raw image may even be unnecessary. In contrast and by necessity, CD AFM uses relatively broad tips with horizontal protuberances that allow imaging vertical and even reentrant sidewall features [22]. For this situation, removal of the tip shape distortion becomes essential.

Until recently, attempts to remove CD AFM image distortion were based on subtracting the tip width from dilated feature lines, and conversely, adding the tip width (actually, half the tip width when describing sidewall positions) for trenches [23] in a process referred to as "tip width subtraction." This also corresponded to a period when algorithms were not available to reconstruct reentrant surfaces [10]. In practice, a reconstruction algorithm is required to both reconstruct the tip shape from a characterizer scan, and to reconstruct the sample surface from a raw image once the tip shape is known. Consequently, for CD AFM, neither element was available prior to the introduction of extensions to mathematical morphology that enabled reentrant scanning probe microscopy (SPM) image reconstruction [20, 24, 25].

With the advent of reentrant-capable image reconstruction, two important advances were made. First, the full tip profile could be acquired without the dilation of the reentrant shape characterizer (Sect. 2.3.3). Second, the current tip shape could now be used in reconstructing the sample image using the reentrant erosion algorithm (Sect. 2.3.4).

Subsequently, *in situ* tip reconstruction was provided in a CD atomic force microscope that enabled SEM/TEM resolution images of the current state of the tip as shown in Fig. 2.9 [20, 26]. This advancement allowed monitoring the evolving tip shape during the course of probe use. Importantly, this served to provide a means to replace a tip if a particular shape parameter exceeded a specified threshold (e. g., vertical edge height, VEH, as shown in Fig. 2.7). Prior to this time, a laborious process was used whereby tip edge positions were inferred from scans of a non-reentrant improved vertical parallel structure (IVPS) [27]. Clearly, this was an entirely unsatisfactory process in an automated tool.

The characterization of conventional probe tips and their subsequent removal from AFM images is a mature field that is only briefly surveyed in Sect. 2.3.2. However, the section provides a foundation for understanding the more recent work in reentrant CD atomic force microscope tip characterization and image reconstruction, discussed in the remaining sections.

2.3.2
Conventional Tip Characterization and Image Reconstruction

Villarrubia [9, 10] has presented two thorough reviews of conventional tip characterization and image reconstruction. The methods are presented here only to the extent to provide a foundation for understanding reentrant tip characterization and image reconstruction used with CD AFM. In addition, automated tip characterization is discussed for the first time with conventional AFM. Finally, characterization samples used in convention tip shape reconstruction are discussed.

Most AFM tools use pyramidal-shape probe tips similar to the tapping mode probe depicted in Fig. 2.5. Traditionally, AFM users characterize a probe tip for its "sharpness" [9,28]. A sharp atomic force microscope tip produces images with higher resolution that consequently reveal additional details of a sample surface. Conversely, a blunt tip acts as a "geometrical filter" that blocks fine-surface details and generates scan artifacts. The scanning mode used for convention AFM is generally contact or Tapping^{TM} mode [29]. Neither mode allows vertical or reentrant tip–surface contact and is therefore well suited to conventional parabolic or triangular profile tips.

During the 1990s, SPM tip and image reconstruction methods appeared in the literature that dealt with non-reentrant tip and sample morphologies. The methods can be categorized into two distinct groups:

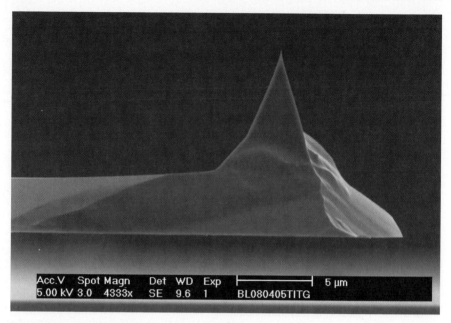

Fig. 2.5. Conventional AFM probe with pyramidal tip shape. The tip apex (radius of curvature) can be estimated by blind tip reconstruction methods

1. Model-shape-based characterizer methods;
2. "Blind" tip reconstruction methods.

The model-shape-based methods rely on assuming the size and geometry of the tip characterizer. The resulting AFM tip characterizer scan is then an image of the characterizer that includes the image dilation due to the tip shape [9, 21]. Uncertainties in the assumed size of the tip characterizer are directly transferred into the reconstructed tip profile.

Two general methods of image reconstruction can be used to "erode" the characterizer shape from the tip–characterizer scan, thus, obtaining the reconstructed tip shape. The first and most general method relies on mathematical morphology [21,28]. From a morphological description, the reflected tip shape (P) is recovered from the characterizer image (I_c) and the reflection of the known characterizer shape (S_c) as

$$P = I_c \ominus S_c \, ,$$

(2.2)

where $\ominus$ is the erosion operator. An alternative approach to morphological erosion is based on the appreciation that at any point in the raw image, the local slope of the AFM image is the same slope as the tip–sample contact point [24,28]. Consequently, if the shape of the characterizer is known (e. g., a sphere or circular cylinder), the distance from a reference point to the contact point can be subtracted from the image, rendering the tip shape. However, the method relies on calculations of derivatives and/or slopes that present challenges when dealing with noise in the AFM image. This in turn generates artifacts in the reconstructed image that are difficult to remove without excessive prefiltering.

The second method for tip shape estimation is "blind" tip reconstruction. The method proceeds on the principle that the outer boundary of the reflect tip shape must fit within the envelope of the sample image at each pixel location. Unlike the model-shape-based method, the "blind" tip characterizer does not necessarily have a known shape but it is desirable to have sharp, needlelike structures to fully image the tip. The method is particularly useful in reconstructing the apex location of tips and imaging tip apex regions below the size uncertainties associated with model-shape-based characterizers. However, noise in the scan image can produce artificially sharp tip reconstructions. This issue has been addressed by "thesholding" noise levels but remains a challenge to the present day [9,30,31]. Moreover, practical use of the method can entail high tip wear rates due to the sharp characterizer samples, and debris generation is an issue due to the high aspect ratio of the characterizer features.

A blind tip reconstruction method was implemented in Nanoscope^{TM} software in 1995 by J. Mosley for automated tip qualification. As shown in Fig. 2.6, the reconstructed tip shape can be evaluated at two elevations from the tip apex. The cross-sectional area of an equivalent circle is calculated. Tip shapes having an effective diameter exceeding a preset limit will be rejected. In Fig. 2.6, the tip under test is estimated to have a diameter of 4.9 nm at 2 nm from the tip apex, and 20.3 nm at 20-nm height. Samples for this method of tip characterization include the

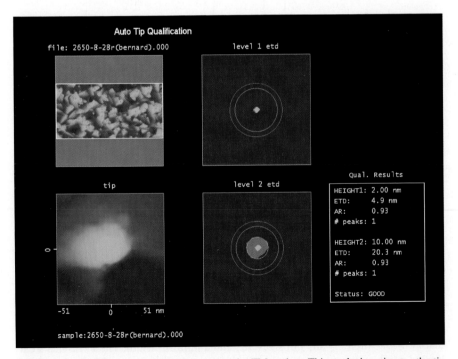

Fig. 2.6. Auto Tip Qual program for conventional AFM probes. This analysis estimates the tip diameter at a fixed height using blind tip reconstruction principles. The sample used for scanning was TipCheck^{TM} from Aurora NanoDevices

Nioprobe™ specimen (its surface consists mainly of niobium with radii less than 5 nm) and the TipCheck™ sample (sharp titanium/titanium oxide debris film on a Si substrate) from Aurora NanoDevices.

2.3.3
CD Tip Shape Parameters

Figure 2.7 depicts the shape parameters that define a typical boot-shape CD probe tip. The tip shape parameters presented here encompassed those used in early work by IBM and TI [32]. Veeco identified additional shape parameters when *in situ* automated reentrant tip reconstruction was first implemented in Nanoscope software in 2003. Of most interest to users are tip width, VEH, and tip overhang.

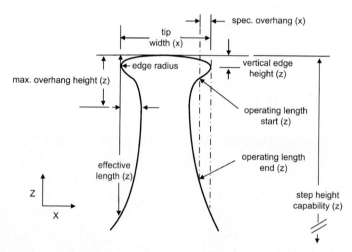

Fig. 2.7. Tip shape parameters of a CD probe (boot-shaped). Note only major parameters are shown here. (Reused with permission from [24]. Copyright 2005, AVS The Science & Technology Society)

2.3.4
CD Tip Shape Characterization Techniques

For CD AFM, the complexity of the tip shape requires a more elaborate characterization technique. The tip shape must first be determined (2.2) in order to recover the sample surface from the imaged data. For current CD AFM, tip characterization proceeds using a "distributed" characterizer system consisting of two structures. The first structure is an "improved vertical parallel structure" (IVPS)—essentially a line feature with extremely smooth sidewalls and a uniform width. The IVPS has a calibrated width and serves to provide a width measurement for the CD tip (Fig. 2.7).

The second structure, shown in cross section in Fig. 2.8a, was specifically designed for high reentrancy and allows imaging of the CD tip sidewalls (it is essential

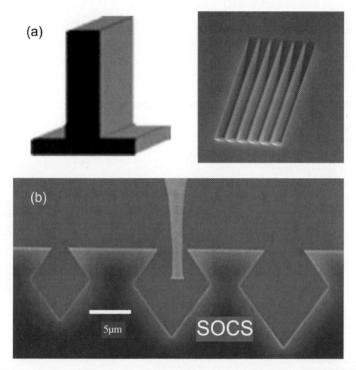

Fig. 2.8. Dual scan tip characterizers: **a** improved vertical parallel structure (*IVPS*); **b** silicon overhang characterizing structure (*SOCS*). (Reprinted from [24, 34] with permission)

that all imaging regions of the tip contact the characterizer for an accurate reconstruction). Known as the silicon overhang characterizing structure (SOCS) characterizer, the device is fabricated by a combination of vertical dry etching and anisotropic wet etching, thus producing a diamond-shaped trench. The critical contact edges of the trench are upwards sloped and sharpened to a radius below 10 nm in subsequent oxidation and etch steps. The large trench undercut, as shown, allows characterization of CD tips with large flare. Left and right sides of the SOCS have an included angle of 54.7°.

Reconstruction of the tip shape proceeds by "eroding" the characterizer edge radius from the tip–characterizer scan data. Next, the profiles are mapped into a line and the left and right tip apex "edges" are captured. Finally, the left and right reconstructed profiles are set to the correct width using the previously measured tip width. An example of the reconstructed tip is shown in Fig. 2.9 along with a TEM image of the same probe. An image of the Nanoscope™ CD tip qualification interface in Fig. 2.10 reveals the *in situ* tip profile and tip shape parameters. This software has been a standard feature of CD AFM since 2003. The reconstructed tip image points are used to determine shape parameters shown in Figs. 2.7 and 2.10. The shape parameters, in combination with the tip width, serve to monitor the state of the tip and flag replacement once any parameter is outside preset bounds. Updating the tip shape occurs periodically as the state of the tip evolves during scanning.

Fig. 2.9. Tip shape profile comparison of a defective carbon nanotube (CNT) probe characterized by CD AFM and TEM. Although most of the tip shape agrees well, the AFM profile shows the limitation of IVPS–SOCS dual scan characterizers (reentrant region below contamination). (Reused with permission from [24]. Copyright 2005, AVS The Science & Technology Society)

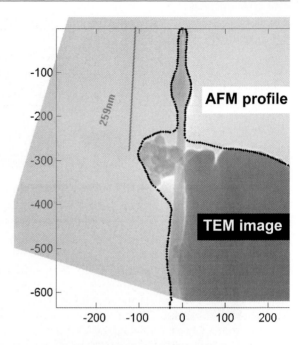

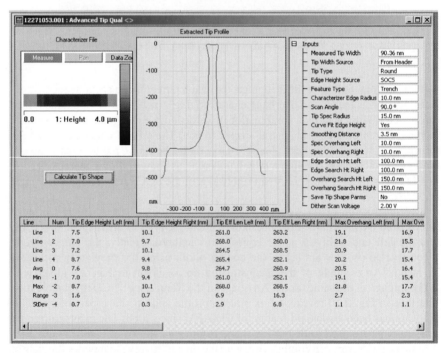

Fig. 2.10. Screen shot of Advanced TipQual analysis of Veeco X3D CD AFM. The analysis generates tip shape profiles based on scan images of tip shape characterizers

Consequently, the current CD tip shape is used for specimen surface reconstruction, described in Sect. 2.3.5.

Figure 2.8 provides images of IVPS and SOCS characterizers. When the IVPS and SOCS structures are scanned, the probe is directed to a series of fiducial marks to acquire a localized region for measurement. With a 2D SPM Zoom procedure [15], a probe can measure the same region on the IVPS or SOCS within a 10-nm range. Since the dimensional variation of the IVPS and SOCS within this range is considered negligible, a tip geometry with low uncertainty can be reliably generated. For tip shape characterization, measurement of VEH typically has 1.2 nm 3σ repeatability; the baseline CD AFM performance for metrology width measurements is 0.7 nm 3σ.

The desire to improve CD atomic force microscope tip qualification throughput and characterizer shape stability motivated further characterizer development. Recently, a carbon nanotube (CNT) across trench concept for CD AFM was developed that allows full 2D tip shape acquisition from a single engagement and scan of a very localized area [33, 34]. As shown in Fig. 2.11a, a single-wall CNT is imaged with CD AFM. In Fig. 2.11b, both the dilated profile of a CDR130S tip is shown and the eroded profile following scanning a multiwall CNT-across-trench. The durability and shape stability of CNT tips are well known [26], further validating the desirability of CNT use as a tip characterizer. Moreover, once the diameter has been calibrated by, for instance, first scanning a NanoCD^{TM} standard, similar levels of uncertainty may be assumed about the circumference of the CNT. Finally, the CNT's conductive properties (and variable conductivity with strain and deflection) allow *in situ* monitoring of the state of the characterizer. This is a desirable feature that has not been feasible with current characterizers.

2.3.5
CD Image (Reentrant) Reconstruction Algorithms

Recently, equations were presented that express CD image bias as a function of tip and feature geometry [20]. The aim of the authors was to conveniently express the relationships between tip–sample geometry and systematic CD bias (known as type-B measurement errors). However, the authors' primary point is that image reconstruction is a more comprehensive method to remove all sources of bias that result from tip dilation of the AFM image (subject to the usual requirement of tip–sample contact). Consequently, methods to eliminate bias for a single measurand are unnecessarily limiting.

Three approaches are presented for reentrant image reconstruction. Each of the methods proceeds with surface reconstruction given inputs of:

1. The dilated CD AFM scan
2. A reconstruction of the CD tip (Sect. 2.3.4)

The essential first step in applying a method is the appreciation that sequenced (x, z) pairs of positions within an AFM scan line allow a departure from single Z-valued (pixel-type) images and enable reentrant specimen region depiction [24]. Consequently, three existing methods of gray-scale image reconstruction [9,28] were

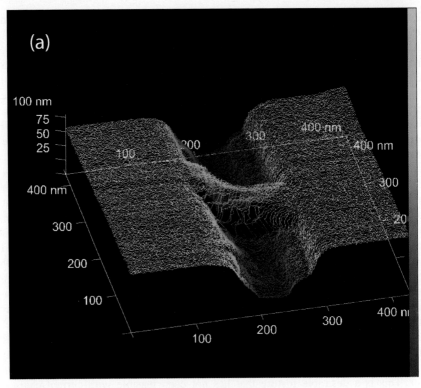

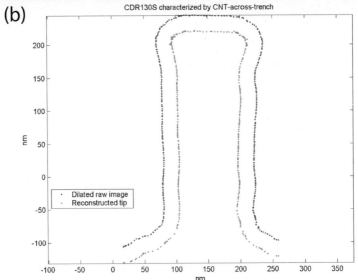

Fig. 2.11. CNT-across-trench characterizers: **a** CD AFM 3D rendering of a single-wall CNT across trench characterizer; **b** tip shape profile of a CDR130S probe characterized by a multiwall CNT across trench characterizer. (Reprinted from [34] with permission)

extended and applied to independently sequenced positional topologies. The focus in this section is the extension of the methods to non-gray-scale morphologies and algorithmic improvements that render them useable. Although the first two methods are presented in terms of 2D scan lines, they are fully applicable in three dimensions incorporating the (Y) slow-scan axis. More recently, Qian and Villarrubia [25] have introduced a Dexel approach to 3D reentrant image reconstruction which is briefly described.

2.3.5.1
Tangent Slope Algorithm

The fundamental concept is that at the point of contact, both the tip and sample surface have the same slope (tangent plane), and simultaneously, the apparent image surface has the identical slope [24, 28]. The offset between the contact point and an image reference point (typically taken to be the minimum Z position of the tip) is a vector defined by the tip boundary position. This is shown in Fig. 2.12 where the local slope associated with image point $I(x_n, z_n)$ leads directly to point $S(x_n, z_n)$ on the reconstructed surface, where contact occurs on a tangent plane. Since the tip shape is known, the correction vector is subtracted from the image point position.

The method, as previously applied, was limited to non-reentrant samples and tips, as surface descriptions were based on single-valued data and scalar "slope" definitions that do not allow reentrant reconstruction. In the more recent development, the method was extended to use:

1. A reentrant ("multivalued") surface description afforded by the CD AFM measurement of sequenced (x, z) data pairs, rather than pixel-type/gray-scale-type data representation
2. A unit normal vector surface description rather than a scalar "slope" description of the surface angle

These two enhancements provide the capability to match undercut image topologies to the reentrant tip contact point and apply the vector correction to generate the reconstructed sample surface. A detailed description of algorithmic improvements to the method, such as removing tip shape restrictions and stability enhancements, will be found in [24].

2.3.5.2
"Erosion" Algorithm

The method presented in this section is equivalent to morphological "erosion" [21], but is not a direct application of erosion in a strict sense as we are concerned with the outer boundary of a surface rather than a complete set of points describing the region.

With a scanning probe microscope, the acquired image can be modeled by the morphological process of "dilation" ($\oplus$),

$$I = S \oplus P , \tag{2.3}$$

Fig. 2.12. Slope image reconstruction method applied to reentrant topology with idealized tip shape (2D example). (Reprinted from [24] with permission)

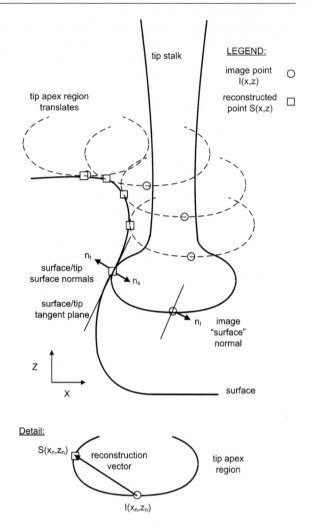

where I represents the image, S is the specimen surface, and P is the reflection of the tip through its origin. This operation is based on Minkowski addition where

$$S \oplus P = \bigcup_{p \in P} S + p \tag{2.4}$$

and p represents each point within the tip region P. The image region is constructed by translating the region S by each element of the set P and then taking the union of all resulting translates. (P is described within this context as a structuring element.)

Recovery of the original surface (that was contacted by the tip) may be implemented with "erosion", or Minkowski subtraction ($\ominus$) once the tip shape is known (Sect. 2.3):

$$S = I \ominus P = \bigcap_{p \in P} I + p . \tag{2.5}$$

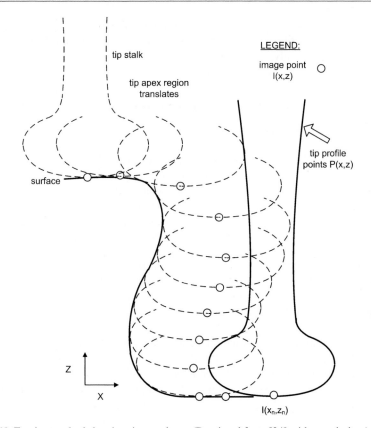

Fig. 2.13. Erosion method showing tip translates. (Reprinted from [24] with permission.)

For this operation, I is translated by every element of P and then the intersection of the sets is retained.

In the method, shown in Fig. 2.13, profiles of the tip are located at each image point. Each profile or translate of the tip is populated with an (x, z) array of points. The reconstructed surface is recovered by sequentially excluding all profile points that fall within the current tip translate, as depicted in Fig. 2.6. Like the tangent slope method, the original image data are prefiltered with a median filter with windowing based on the tip shape parameters. Failing to prefilter the data could result in a method that would bias reconstructed features. (For instance, large x-component noise would reconstruct a line feature with an artificially small width.) Maintaining scan line data density is also important with the method as artifacts can be generated if the original points are sparse with respect to the curvature of the tip.

2.3.5.3
Dexel algorithm

Recently, a dexel approach has been proposed to address reentrant image reconstruction [25]. The dexel representation of an object is based on the concept of pixels, but is more involved. Unlike the pixel representation of an object, which is restricted to a single vertical height (i.e., an "umbra" or upper surface), the dexel can represent multiple heights at a given XY position. Each height represents a transition from inside to outside an object, or vice versa.

2.4
Metrology Applications

The metrology applications of CD AFM can be categorized into depth measurements (DT mode) and CD measurements (CD mode). Typical DT applications include trench capacitor (RC2/RC3), shallow trench isolation (STI), via, divot, and contact hole measurements. CD applications cover line/trench sidewall profiling analysis (sidewall roughness, sidewall slope angle), line width analysis (line edge variation, line width roughness) in complementary metal oxide semiconductor (CMOS) gate, flash logic, perpendicular read/write head, and novel structures such as Fin-FET and strained gate. In general, if the sample feature has sloped sidewalls with no reentrancy, DT mode is sufficient in gathering metrology information at high throughput. However, if the feature surface morphology is uncertain, or is known to have reentrant sidewalls, CD scan mode is needed to image the sample with minimal distortion.

Although other metrology methods may perform measurements at higher speed or lower cost, CD AFM sometimes stands out as the selected metrology tool owing to its nondestructive, lower-bias measurements, and capability to image reentrant profiles. Several examples are given in the following subsection to demonstrate current CD AFM technology.

2.4.1
Examples within Process Control

Figures 2.14 and 2.15 both illustrate typical depth applications in semiconductor processing steps. Figure 2.14 shows a 70-nm trench capacitor reference cell RC2 in 3D rendering. A needlelike focused ion beam (FIB) probe was used for scanning where its sharp tip apex allowed 30-nm travel distance at the bottom of the capacitor. Figure 2.15 compares via measurements from CD SEM and CD AFM. To measure vias, very high aspect ratio tips are required. In this case a 300-nm-long, 30-nm-wide silicon post-type probe tip was used for scanning [35].

The devices for the data storage industry may not shrink as rapidly as those of the nanoelectronics industry; however, their designs evolved toward 3D morphologies relatively early. A perpendicular writer structure, shown in Fig. 2.16, is routinely measured nondestructively by CD AFM.

Several CD applications of AFM are presented in Figs. 2.17–2.19. In Fig. 2.17, CD AFM profiles are depicted of line structures made of various materials (photoresist, hard mask, and silicon oxide) [36]. Each CD AFM scan line within an AFM image provides a cross-sectional, nondestructive profile of the sample. Therefore, in addition to the cross-sectional profile at a specific location, the scan image renders along-the-structure variations which may also be processed with statistical analyses.

Fig. 2.14. AFM 3D rendering of a 70-nm trench capacitor structure (RC2). An FIB3D probe was used for the depth measurements

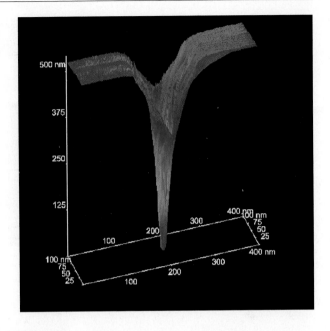

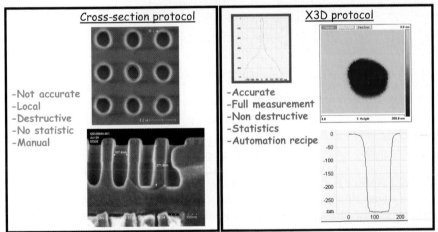

Fig. 2.15. Comparison of via CD measurements from SEM and AFM metrology tools. For CD AFM, the probes used had a 10:1 aspect ratio tip with 30-nm apex diameter. (Data courtesy of Johann Foucher [35], used with permission)

Figure 2.18 gives another example of sidewall roughness, line width variation (LWV), and line edge roughness (LER) measurements. In this case, a predefined (by electron-beam lithography) roughness pattern was made, and CD AFM was adopted as the tool to characterize the roughness structures [37].

For each cross-sectional scan line, detailed analysis on the 2D profile is illustrated in Fig. 2.19. The metrology information on sidewall roughness, reentrancy, and

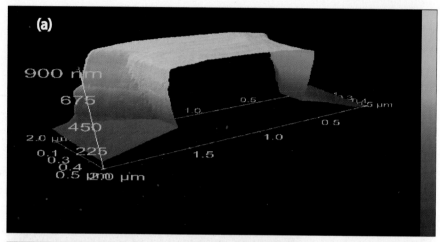

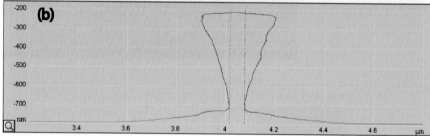

Fig. 2.16. For the data storage industry, the perpendicular writer structures which have significant overhang can only be nondestructively measured by CD AFM tools. **a** 3D rendering of an AFM raw image, **b** cross-sectional analysis of a high-overhang writer structure using a reconstructed AFM image

sidewall angle is essential in process development. The work shown in Fig. 2.19 was part of a process artifact characterization task where CD AFM was the RMS for CD scanning electron microscopy model library development [38].

An advanced application of CD AFM is to utilize its 2D SPM Zoom technique for process monitoring. This concept is illustrated in Fig. 2.20. Cross-sectional measurements such as FIB-SEM, cross-sectional SEM, and TEM require sacrificing samples after each processing step. As a result, one cannot track the profile changes of the same feature across multiple manufacturing stages. CD AFM is capable of measuring the same location of a feature nondestructively; therefore, a study of line width variation and line edge roughness after each processing step became feasible [13].

2.4.2
"Fingerprinting" of Sample Features

The "fingerprinting" technique is a novel application that was recently developed to validate CD AFM [16]. The method employs CD AFM to prescan a region of

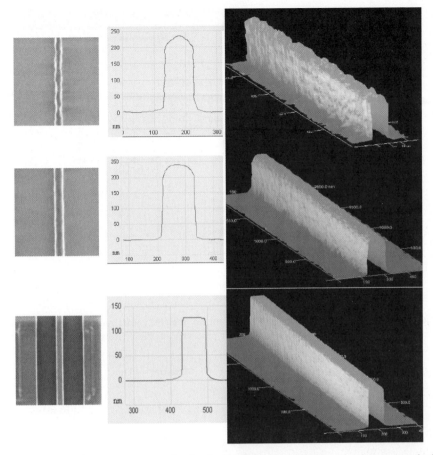

Fig. 2.17. CD AFM measurements of resist (*top*), hard mask (*middle*), and reference standard (*bottom*). SEM top-down views, AFM cross-sectional profile, and profiles along the lines are shown. (Reprinted from [36] with permission)

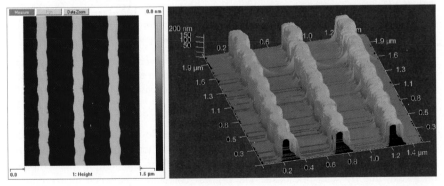

Fig. 2.18. CD AFM image (top-down view and 3D rendering) of a predefined sidewall roughness sample. Line edge roughness (LER) and line width variation (LWV) analyses can be performed from the scan image data after image reconstruction. (Reprinted from [37] with permission)

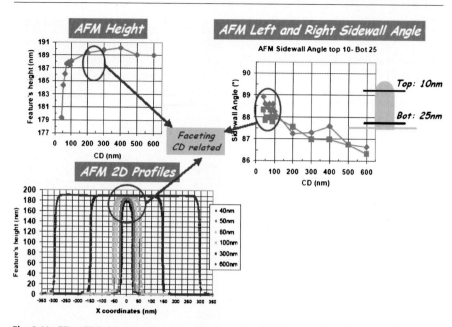

Fig. 2.19. CD AFM data showing the measurements (height, corner radii, 2D profiles, sidewall angles) of an electrom-beam resist under process evaluation. (Reprinted from [38] with permission)

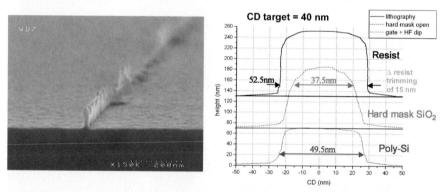

Fig. 2.20. CD AFM was used to track shape evolution during several processing steps. The exact location on the wafer was identified with the 2D SPM Zoom technique, and nondestructive measurements made monitoring of gate formation possible. (Reused with permission from [13]. Copyright 2005, AVS The Science & Technology Society)

a sample, thus obtaining multiple profiles of the sample features. The independent metrology sample (in this case, a TEM specimen) is then extracted from the sample region and the multiple feature measurements from the sample are compared on a scan line by scan line basis. The highest correlation between measurements identifies the "matching" profile between the two metrology systems.

Previously, metrologists were limited by the uncertainty of TEM sampling location when a high-accuracy comparison of cross-sectional measurements was required. As mentioned in Sect. 2.2.2, advanced TEM sample preparation uses a FIB system to make fiducial marks for cross sectioning. The fiducial location accuracy is ±100 nm and typical sample thickness is 80 nm. However, even within the 200-nm range of a feature extraction site, CD AFM measurements, having a spatial resolution of a few nanometers, revealed significant variation (Figs. 2.4, 2.21).

Figure 2.21 depicts interesting characteristics of the CD AFM measurement—the line width variation of each line structure is uniquely different along the slow scan axis (X direction). For each CD AFM scan, several feature structures (in this case, polysilicon lines [7]) are measured simultaneously. Consequently, the "combination" of line width variation values from each individual feature line at a specific X coordinate becomes unique, and it can be used as the "fingerprint" of a feature location.

For example, Fig. 2.22 demonstrates "fingerprinting" applied to a hard mask gate artifact (HMA) structure to identify the TEM sampling site. The middle line width (MCD) and the height of each feature were measured independently by TEM and CD AFM (both tools were calibrated with NIST-traceable standards). The CD AFM scan image had 26 scan lines with 20-nm spacing/resolution across 520 nm in the X direction, each scan line revealed cross-sectional profiles of the sample features (for this HMA sample, 22 features), and it also represented a location along the X direction. Because of the localized variation of feature dimensions, the correlation of CD measurements between AFM and TEM is maximized when the AFM scan line matches the TEM sampling site for the 22 features [16, 33].

Thus with the "fingerprinting" technique, measurements from CD AFM and TEM can be compared at the same location with positional uncertainties below those

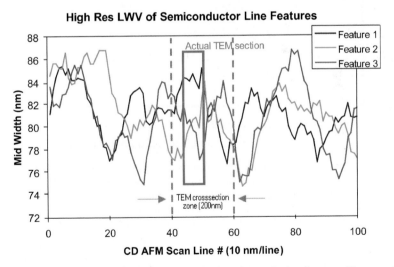

Fig. 2.21. CD AFM data showing the LWV of semiconductor device features. The *gray lines* enclose a 200-nm zone where the TEM cross section is positioned. The *boxed region* indicates the actual cross section where the TEM measurement is acquired (width typically 70–120 nm). For this data set, there are 20 CD AFM scan lines in the 200-nm zone. (Adapted from [16] with permission)

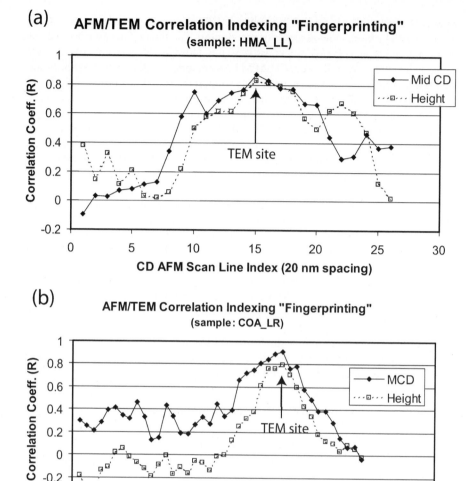

Fig. 2.22. Correlation coefficients between TEM measurements and individual CD AFM scan lines when comparing middle line width (*MCD*) and height measurements: **a** IBM HMA sample; **b** IBM COA sample. Note that each data point represents the correlation coefficient for more than 20 features. From the charts, it is clear that the TEM cross-sectional site matched line no. 15 of the CD AFM image for the HMA sample, and line no. 29 for the COA sample [16, 33] (Copyright IEEE 2007)

obtained with the state-of-the-art fiducial marking (Fig. 2.23). For example, Fig. 2.24 illustrates a comparison of IBM HMA and COA sample cross-sectional profiles from TEM and CD AFM. The agreement of both measurements can be shown qualitatively by superposition of one profile onto the other. Table 2.1 summarizes the measurement

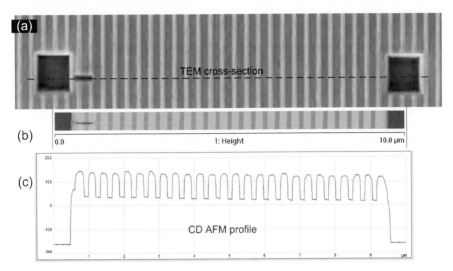

Fig. 2.23. SEM and CD AFM images showing the use of the CD AFM "fingerprinting" technique to pinpoint the location of the TEM sampling site with focused ion beam fiducial marks. (Adapted from [16] with permission)

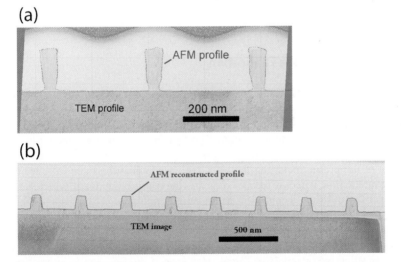

Fig. 2.24. TEM image superimposed with CD AFM "fingerprinting matched site" reconstructed profile (*dotted line*) for AFM/TEM measurement validation: **a** IBM HMA sample; **b** IBM COA sample. This comparison shows excellent agreement between CD AFM and TEM data [16] (Copyright IEEE 2007)

bias between TEM and CD AFM data. Three different probe types were tested on an IBM underetched polysilicon gate sample (COA). Table 2.1 reveals that TEM and CD AFM measurements differ by only a few nanometers in metrology measurements. Work is on going to isolate the uncertainty sources in the study [16, 33]:

Table 2.1. Metrology uncertainty of critical dimension (*CD*) atomic force microscopy

Probe	CDP55A		CDR130SiN		Trident	
	Mid-CD	Height	Mid-CD	Height	Mid-CD	Height
Average bias	−3.169	4.831	1.429	6.666	0.973	3.386
Standard deviation	1.635	2.122	1.708	2.029	1.584	2.437

All units are in nanometers. Transmission electron microscopy is used as the reference metrology tool. Bias = RMS_{TEM} − CD AFM, where RMS is reference metrology system and AFM is atomic force microscopy. CD measurements of IBM COA samples with multiple probe types. Data adapted from [33] with permission

1. Uncertainty of TEM metrology measurements
2. Misalignment of TEM/AFM sample cross section
3. Suboptimal scanning parameters of CD AFM

In summary, the "fingerprinting" technique enables verification of CD AFM metrology with an independent RMS. The comparison of two metrology systems can be implemented with minimal positional uncertainty. As a result, CD AFM was demonstrated as a complementary RMS tool for TEM and a potential replacement when a timely, nondestructive measurement is preferred.

2.5
Developments in Probe Fabrication

The probe tip of an AFM system is a critical sensing component; many functionalities of the tool rely on its design and properties. For example, magnetic force microscopy uses probes that have magnetic properties (coercivity, magnetic moment), while electrostatic force microscopy requires probes that are highly conductive and/or sensitive to electric fields. In this chapter, "tip" refers to the raised and extended portion of the probe, including the apex region which is the intended site for tip–sample interaction, whereas "probe" indicates the whole component, which generally consists of a substrate, a cantilever beam, and a tip. Since AFM uses a probe as the sensor that provides measurements and signals for feedback control, the cantilever of a probe serves as a sensing detector when the tip interacts with the sample. Depending on the scan modes and the detection schemes, the tip–sample interactions are caused by either attractive or repulsive forces. However, if the tip–sample interactions are not detected and fed back for scan control, these forces may cause deformation of the tip shape (tip wear, particle pick-up, breakage), erroneous scanning behaviors (stickiness, floating), or image artifacts.

AFM probes are consumables that have to be tailored for each scanning application. Several materials have been used in the past [39–42], but the majority of probes are made of silicon, which is a reasonable manufacturing path given the rapid development of silicon processing technology. Utilizing the sophistication of the silicon manufacturing technologies, engineers fabricate AFM probes with micromachining processes that involve: photolithography, electron-beam lithography,

Fig. 2.25. Wafer processing of AFM probes. This 4-in. Si wafer consists of close to 400 AFM probes. The missing probe chips were used for processing control in several fabrication steps

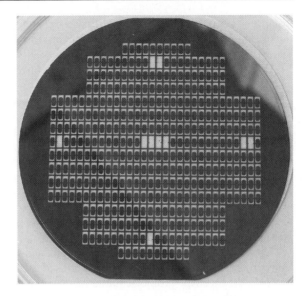

sputtering, chemical vapor deposition, reactive-ion etching, FIB milling, oxidation, and chemical wet etching.

In addition, silicon wafer processing reduces the cost of probe manufacturing by batch production. As illustrated in Fig. 2.25, a 4-in. diameter wafer yields close to 400 probes at the end of fabrication. Like other fields such as microelectromechanical systems (MEMS), the design and development of AFM probes leverages advancements in processing techniques. New materials, higher patterning resolution, and better shaping methods all contribute to the betterment of probe technology that evolves with new applications of AFM.

2.5.1
Tip–Sample Interactions: Tip Shape, Stiffness, and Tip Wear

Traditional silicon AFM probe tips are usually shaped by chemical solutions that anisotropically etch away material. For example, KOH–H_2O–isopropyl alcohol solution is crystallographically selective, and the etchants attack silicon (100) planes faster than (111) planes [43,44]. Since the rate of chemical reaction varies with crystalline orientation, the typical tip has a pyramidal shape. In the unused state, the apex is as sharp as 1–2-nm radius of curvature (Fig. 2.5). However, the moment the tip apex contacts the sample surface, the sharp tip "breaks off" and the sharpness of the tip is typically reduced to around 5 nm (radius of curvature). As the tip is used for an extended time, the tip apex radius can increase to 10–15 nm at a relatively fast rate, and then the rate of wear slows as the tip apex stabilizes. The exact mechanism of tip wear at the nanometer scale is still not well understood, but it is generally believed that tip wear is a combination of physical and chemical reactions [45,46]. One of the contributors of tip wear is the attractive force caused by a thin layer of water on the sample surface. This meniscus force draws the tip to the surface (referred to as "jump to contact" or "snap-in") when it approaches the surface (the actual "snap-in" distance is a function of cantilever stiffness, scanning environment, and tip–sample interaction forces, which are also dependent on materials and tip–sample geometric factors).

Dynamic scan modes such as tapping mode externally excite the cantilever at a fixed frequency near its resonance, providing the tip with sufficient energy to break free from the adsorbed fluid layer.

CD mode is also a dynamic scan mode; the cantilever not only oscillates in a vertical direction (Z-axis), but also dithers horizontally (Y-axis). The dither helps the tip to sense vertical or near-vertical sidewalls as well as removing the tip from a "snap-in" state. The needle-shape or boot-shape probes used in DT and CD modes have shanks that are more likely to be in the proximity of the sample surface during metrology measurements. Moreover, the aspect ratio of these probes also affects the lateral stiffness of a tip. The lateral stiffness and proximity of the surface area of the probe to the surface is a key factor in determining whether a CD atomic force microscope can scan a sample properly in certain applications.

The lateral stiffness of a CD probe is calculated within the Advanced Tip Qual analysis program on a CD atomic force microscope tool. The finite-element calculation uses the tip shape profile acquired from the tip characterization/reconstruction process, and removes tip shape dilation resulting from the dither voltage level (typically 2–4 V). Assuming a circular cylinder model for the tip, one can express the lateral stiffness as [47]:

$$k = F/\delta , \tag{2.6}$$

$$\delta = \frac{FL^3}{3EI} , \tag{2.7}$$

$$I = \frac{\pi}{4}r^4 , \tag{2.8}$$

where k is the stiffness (newtons per meter), F is the force (newtons), δ is the deflection (meters), L is the length (meters), E is Young's modulus (newtons per square meter), I is the cross-sectional moment of inertia, and r is the cylinder cross-sectional radius.

From the above equations, it can clearly be seen that the tip lateral stiffness scales as the fourth power of the tip width. This attribute becomes a technical challenge with the continued miniaturization of sample features (Sect. 2.6). Conversely, tip lateral stiffness scales as the inverse third power of the tip effective length. Thus, the effective length should be limited for a specific application.

The shape of specialty probe tips also calls for additional considerations. For instance, electrostatic discharge (ESD) can damage AFM probes that have shapes more susceptible to electrostatic fields. ESD contributes to 60–90% of the failures of sensitive electronic components; about 70% of these failures result from incorrectly grounded personnel [48]. To mitigate ESD, most probe manufacturers use heavily doped silicon wafers (e. g., antimony as the n-type dopant) that have low resistivity (0.01–0.02 Ωcm).

Generally, ESD damage is not a major contributor to probe damage for conventional AFM probes. However, in a recent study, it was found that ESD during probe handling could cause more than 50% breakage of needlelike, sharp AFM tips (such

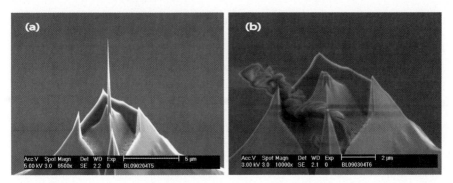

Fig. 2.26. SEM micrographs showing the electrostatic discharge (ESD) damage of a high-aspect ratio AFM probe: **a** a needle-shaped focused ion beam (FIB) probe before ESD; **b** after ESD, the primary spike was melted. (Adapted from [49] with permission)

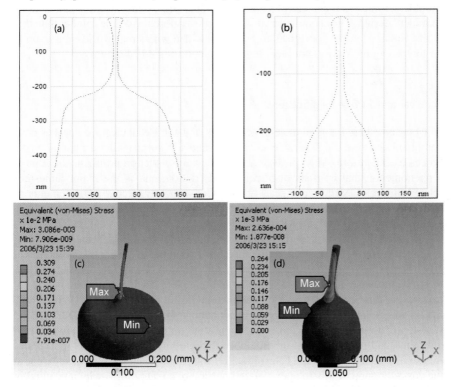

Fig. 2.27. Stress concentration effect of different probe shapes. **a** "large-shoulder" and **b** "small-shoulder" are two profiles of CD32 probes characterized by CD AFM. Their finite-element analysis results (**c, d**, respectively) indicated close to 10 times higher stress concentration on the "large-shoulder" compared to the "small shoulder" probes. Note the red arrows in **c** and **d** indicate locations of maximum stress. The calculated stresses are relative values since scaling rules were followed in the building of the probe models

as FIB probes used for depth measurements) prior to their usage [49,50]. Figure 2.26 provides evidence of ESD damage. The primary spike of the FIB probe was damaged after the user handled it without using a wrist strap and ESD safety gear. With ESD safety procedures in place, the damage rate of the probes was reduced to virtually zero (excluding improper handling). The results recommend ESD safe practice for handling and transportation of all high aspect ratio probe tips. It is noteworthy that the same study also showed an average 15% of probe breakage was caused by mechanical damage, most likely a result of improper handling by the AFM operators. Consequently, AFM tool and probe suppliers have implemented protocols such as automatic probe loading and preloaded probe cassettes to minimize manual handling of AFM probes.

Stress concentration is another important aspect in AFM probe design, especially for sub-50-nm technology. Figure 2.27 gives an example of 32-nm CD probes. Two designs were fabricated: (1) probe with a "large shoulder;" (2) probe with a "small shoulder." Finite-element analysis indicated higher stress concentration on the "large shoulder" design. The estimated maximum stress is close to 10 times higher than that of the "small shoulder" design. This result coincided with a real-world evaluation—prototypes with a "large shoulder" tend to break during AFM scanning.

2.5.2
Tip Wear and Surface Modification

AFM probes share many of the characteristics of a MEMS device; i. e., probes have compliant structures, high surface-to-volume ratio, and similar silicon manufacturing processes. Therefore, AFM probes face similar in-use stiction problems. Stiction is a phenomenon that consists of sticking and friction, and is the major adhesion-related failure of MEMS devices [42–44]. Consequently, recent advancements of surface modification techniques used in MEMS anti-stiction applications can provide a solution to the tip wear issues in AFM.

The flared portion of a CD tip apex wears while scanning on the sidewall of specimen features; wear of the tip flare results in decreased tip width and increased VEH. Periodic *in situ* characterization of CD AFM probes is an optimal method to study the wear behaviors of probes and samples. As illustrated in Fig. 2.28, monitoring the change of tip shape parameters during use helps to understand how the tip wears, and when it occurs during scanning. For this specific CDR50SO tip (CD-type 50-nm probe with small overhang), the tip did not exhibit any wear until after the 22nd measurement site on a polysilicon after etch inspection (AEI) wafer in a semiconductor factory. The profile after the 31st site shows significant wear at the contact region of the tip apex—flared portion of a CD probe—which resulted in reduced tip width and increased VEH values. Further inspection of this probe's use history revealed a freshly etched polysilicon wafer was scanned between the 22nd and 31st measurement sites that resulted in the high-wear event.

The wear mechanisms of CD AFM probes are not well understood. It is evident that a probe wears at a different rate when it scans on different materials for various metrology tasks. Typically, silicon CD probes are used to scan silicon-based structures. However, when the measured features consist of rough silicon or abrasive materials, significant wear of silicon tips can occur.

Silicon CD probes with a silicon nitride capped apex (Fig. 2.2d) are a new wear-resistant design that benefits from high-hardness material, low surface energy, and a unique high-overhang apex region. The thin disk (20–40 nm, various thickness designs are available) of silicon nitride at the apex not only has a low wear rate (0.04 nm per measurement site on polysilicon line/trench; typical bare silicon probes wear 0.12 nm per site), but also has consistent VEH values. The increase of VEH

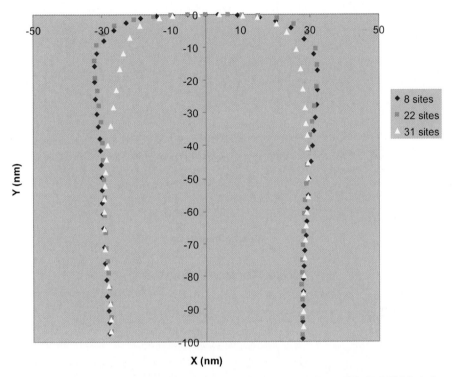

Fig. 2.28. Tip shape profile change on high-wear sample (apex region). This CDR50SO probe exhibited significant wear after scanning 31 sites on a polysilicon after etch inspection (*AEI*) wafer. Note the tip profiles after eight sites and 22 sites are nearly identical; a high-wear event occured between 22 and 31 sites. Each measurement site typically has 32 scan lines; each line is about 1 μm in scan distance

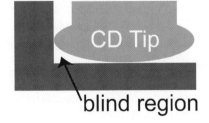

Fig. 2.29. The "blind region" of a CD measurement. Metrological information within the blind region will be lost during a CD scan

(as indicated in Fig. 2.28) is one of the major causes of probe rejection in metrology applications. During CD scanning, the region between the tip apex and both flared edges may not image the bottom corner of the sidewall, as illustrated in Fig. 2.29. The blind region is directly related to VEH; consequently, worn tips, having high VEH and an increased blind region, are rejected for metrology uses.

Surface modification is a promising approach to improve tip–sample wear characteristics for silicon probes. For CD probes, the tip shape parameters are crucial for

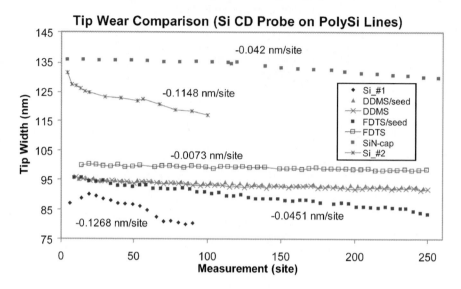

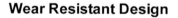

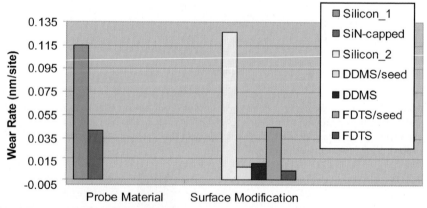

Fig. 2.30. Tip wear rate comparison of various self-assembled monolayer (SAM) coatings and a SiN-capped probe. The *top graph* shows tip width reduction during scanning (Each measurement site typically has 32 scan lines; each line is about 1 μm in scan distance); the *bottom chart* compares wear rate improvement for the SiN-capped tip and SAM coatings. *DDMS* dimethyldichlorosilane, *FDTS* perfluorodecyltrichlorosilane. (Reused with permission from [45]. Copyright 2005, AVS The Science & Technology Society)

Probe-Sample Material Compatibility

sample probe	Resist	Silicon	Poly-Si	Dielectric
Silicon	excellent	good	fair	poor
SiN-cap	excellent	excellent	excellent	good
SiN coating	excellent	excellent	good	fair
Carbon coating or CNT	poor	excellent	excellent	excellent

Excellent: > 1000 sites*
Good: 200 – 1000 sites
Fair: 50 – 200 sites
Poor: < 50 sites or un-scanable
* Tip lifetime is application-dependant, numbers here are typical values

Fig. 2.31. Material compatibility chart shows tip lifetime performances of different probe materials on various semiconductor materials. Each measurement site typically has 32 scan lines; each line is about 1 μm apart

many applications; demands for small tips and low VEH require conformal coatings with minimal addition of materials. Self-assembled monolayers (SAMs) have been used to modify the surface of miniaturized MEMS devices [46, 51]. A recent study has shown that vapor-phase deposition of SAM materials can successfully coat AFM probes without altering critical tip shapes [45].

Figure 2.30 compares tip wear rates of silicon nitride capped probes and CD probes with various SAM coatings. Hydrophobic thin layers (0.5–5 nm) of molecules such as perfluorodecyltrichlorosilane and dimethyldichlorosilane have been successfully applied to the surface of CD probes and tested on polysilicon line structures. When compared with uncoated probes (bare silicon), probes with surface modification exhibit extended tip lifetimes by factors up to 17 [45].

Figure 2.31 presents an empirical tip–sample compatibility matrix for CD AFM metrology measurements. Overall, similar materials (e. g., silicon tip on silicon structures, and carbon-coated tip on 193-nm photoresist) do not work well. Harder materials do not necessarily yield superior wear characteristics. Surface reactivity, hydrophobicity, and process feasibility all take part in the design of wear-resistant probes. Other wear-resistant materials such as diamond-like carbon or tungsten carbide are also, among many other materials, of great interest to probe developers. The development and evaluation of surface-modified probes are currently under way and will soon be offered as options for various metrology applications.

2.5.3
Application-Oriented Probe Designs

The needs of metrology monitoring are highly specific within the microelectronics and data storage industries. The processing technologies advance while introducing new materials and novel structures; new processing steps call for new metrology applications. Naturally, application-oriented probe designs become important.

For example, Fig. 2.32 shows a silicon probe with three extended sharp "fingers" on its apex. This "trident" probe has excellent VEH (as low as 2 nm) and overhang (more than 20 nm each side for 130-nm tip width) that are ideal for special applications such as measuring the "notching" at the bottom of CMOS gate structures, and high-resolution sidewall roughness measurements. As illustrated in Fig. 2.33, trident probes have successfully measured bottom-notched CD of a polysilicon gate

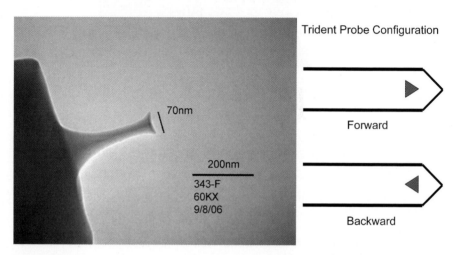

Fig. 2.32. TEM micrograph of a trident probe with 70-nm tip width. Three extended sharp corners have very low vertical edge height to help profile bottom-notching of complementary metal oxide semiconductor (CMOS) gate structures. Trident probes have two configurations: forward design allows for scanning in the X direction (along the probe cantilever); backward design is suitable for Y-axis scan (perpendicular to the cantilever)

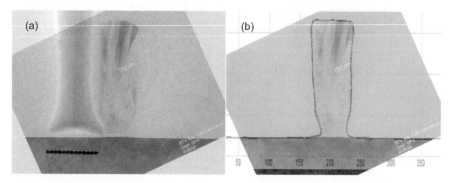

Fig. 2.33. Special application of trident probes. **a** Superposition of TEM micrographs of trident probe and CMOS gate with notching to illustrate that trident probe geometry was designed to measure bottom CD of reentrant features **b** Superposition of CD AFM scan data (after image reconstruction) and TEM micrograph of a similar gate feature to show agreement between TEM and AFM measurements. (Reprinted from [33] with permission)

that was not detectable by SEM, OCD, and even CD AFM with regular CD probes. Note the trident probe was coated with a carbon-based thin film to improve scan performance on abrasive samples such as polysilicon.

Another example of probe design is the offset type of CNT probes in Fig. 2.34. Tilting a vertically aligned CNT at an angle makes CNT probes suitable for CD scan modes. A predefined sidewall roughness sample was scanned with offset CNT probes to resolve detailed sidewall profiles that are usually filtered out when a large CD probe was used. Although the CD scanning can only acquire sidewall information for one side of features, the resulting high-resolution profiles are very valuable since most of the features have symmetric CD profiles.

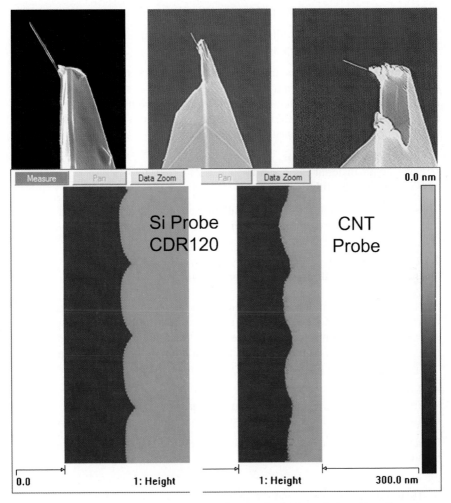

Fig. 2.34. Special offset CNT probes were used to acquire high-resolution sidewall roughness measurements. The CD AFM images compared the same sidewall feature measured by a 120-nm silicon probe and a 20-nm CNT probe. (Data from [26] with permission)

Application-oriented probe designs pose a long-standing question for all probe manufacturers: can a new application market pay for the development of a customized probe? New applications often originated from leading processing groups with a limited initial market volume; and CD AFM seems the only metrology tool for many new processes. Therefore, AFM users are generally willing to share the cost of probe development if the new probes can fulfill the requirements of the task.

2.6
Outlook: CD AFM Technologies for 45-/32-/22-nm Nodes

The major challenges for CD AFM to remain viable at sub-50-nm node sizes are:

1. Understanding and controlling tip–sample interactions
2. Probe technology development

For example, a 32-nm trench structure (assume the trench size is actually $32\,nm^2$) cannot be measured by a 32-nm CD probe. The sticking or snap-in of the probe tip requires that the scan algorithm reverse direction away from the feature. Therefore, a finite distance is needed for the tool to "unstick" the tip from sidewalls. The distance to "pull away" a probe tip depends on several factors:

– Probe scanning algorithms
– Tip–sample material combination
– Probe tip lateral stiffness

These factors again reflect the major challenges mentioned above. Current development status and future outlook will be discussed in the following subsections.

2.6.1
Measuring Sub-50-nm Devices: System Requirements

A metrology instrument has strict system requirements to ensure the measurement uncertainty from the tool itself is minimized to a reasonable level. In sub-50-nm scale, such requirements are even stricter because a 5-nm uncertainty in metrology data reflects 5.6% in a 90-nm node, but 11% variation in a 45-nm structure.

To measure 45 nm precisely and accurately, a CD AFM system needs to have:

– Low stage drift
– Insensitivity to temperature and humidity variation
– Low system noise level

For a system that is used in a production environment, one should consider additional requirements:

– Short feedback response
– Factorywide automation compatability

---

[2] Many so-called 32-nm technology node structures have larger feature sizes. ITRS has stopped using "technology nodes" to address the trend of semiconductor technology.

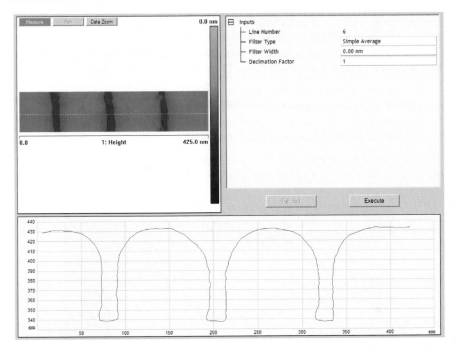

Fig. 2.35. CD AFM scanning image of a 45-nm trench structure. A CD32 probe was used on a Veeco X3D Auto AFM system

– Fast scan control
– Reliable image reconstruction and analysis functions

Many of these requirements are already in place in current CD AFM systems; this makes CD AFM capable of 45-nm measurements (Fig. 2.35). Recent system improvements have focused on new optical and piezo actuator designs, as well as the utilization of fast dither tube actuation to drive a small tip out of a sticky situation [52]. Overall, a complete redesign of a new system is needed for 32-nm structures and beyond, and it is currently under way.

2.6.2
Probe Technology for 45-/32-nm Structures

Making smaller probes with the current silicon probe processing technology has been challenging the limits of physics. With the state-of-the-art approach, 32-nm CD probes and 20-nm DT probes are manufactured in wafer level with success. However, the ability to successfully scan is not decided by tip size but by how the tip interacts with a sample. As the tip lateral stiffness scales as the fourth power of tip size (2.8), a tip loses its stiffness for successful scanning before it is reduced to the proper size. This problem is compounded by the fact that tip lateral overhang requirements are generally expected to remain the same as the node size is reduced.

Consequently, a CD32 tip with 8 nm of overhang would be expected to have 0.8% of the stiffness of a CD70 tip with the same overhang.

To maintain a reasonable stiffness for CD scanning, the empirical guideline is to have a tip lateral stiffness above 1 N/m. There are several approaches for this goal:

- Reduce the overhang of CD probes
- Reduce the probe tip aspect ratio (typically 1:5 and more)
- Redesign probe tip shapes to mitigate stiffness loss and stress concentration
- Use materials with higher mechanical strength

There are also processing issues for small probes. For instance, the capillary forces occurring during wet processing may break tips if their mechanical strength cannot

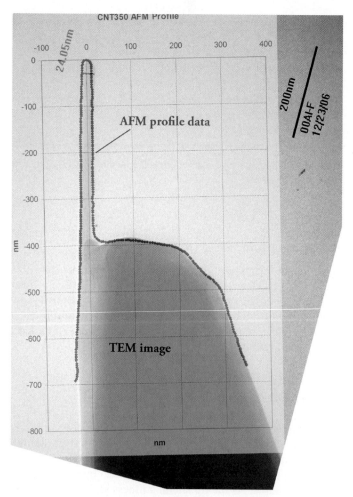

Fig. 2.36. TEM and CD AFM image profiles of a CNT probe. Both profiles were scaled independently with NIST-traceable standards and overlapped for comparison of tip shape. (Reprinted from [34] with permission)

withstand the processing steps. Although silicon-based probes have been dominant for more than a decade, the shrinking technology node calls for new types of probes.

The proposal of using a CNT as an AFM probe has been investigated widely during the past decade [53, 54]. CNTs, having a diameter as small as 1 nm and a unique lattice structure that results in extraordinary mechanical properties (e. g., Young's modulus of 1 TPa), are ideally suited for AFM probes. However, it was not until recently that improvements in CNT probe fabrication and the quantitative statistic data in metrology applications successfully led to factorywide acceptance of CNT probes [26]. The most promising aspect of CNT probes is their superior wear characteristic—an evaluation in a production environment reported more than 5000 measurements in the lifetime of a single tip.

Figure 2.36 shows both TEM and CD AFM images of one of the CNT probes that are commercially available. The CD AFM analysis yielded a tip shape profile comparable to that of TEM, which is essential for accurate metrology. The challenges in CNT probe fabrication are the production scalability and shape modification to enable CD scanning. Nonetheless, the use of CNT-based AFM probes remains a promising solution for sub-50-nm metrology.

Acknowledgements. The authors gratefully acknowledge the contributions of Marc Osborn, Lars Minini and Rohit Jain of Veeco Instruments. Their contributions have helped to advance the CD AFM technologies discussed in this chapter. Many experimental results presented in this chapter come from past and current joint projects with both academic and industrial partners. We would also like to express our appreciations to Johann Greschner and his colleagues at Team Nanotec GmbH for CD probe development work, Bryan Tracy and Amalia del Rosario of Spansion for TEM cross sectional measurements and verification of the CD AFM image reconstruction, Bill Banke of IBM for supplying process-stressed artifacts and innumerable discussions on CD AFM topics, Bharat Bhushan and members of his laboratory at Ohio State University for wear-resistant work, Daiken Chemicals and Sumitomo Corporation for CNT probe developments, Johann Foucher of CEA-Léti for sharing his CD AFM application data, Bernard Mesa of Micro Star Technologies for nondestructive TEM measurements of probes, and, finally, Jeff Chinn and Fred Helmrich at Applied MicroStructures for assistance on SAM probe coatings.

References

1. Binnig G, Quate CF, Gerber C (1986) Phys Rev Lett 56:930
2. International Technology Roadmap For Semiconductors (2005) Edition
   http://www.itrs.net/Links/2005ITRS/Home2005.htm
3. International Technology Roadmap For Semiconductors (2006) Update
   http://www.itrs.net/Links/2006Update/2006/UpdateFinal.htm
4. Dixson R, Fu J, Orgi N, Guthrie W, Allen R, Cresswell M (2005) Proc SPIE 5752:324
5. Banke B, Archie CN, Sendelbach M, Robert J, Slinkman JA, Kaszuba P, Kontra R, De-Vries M, Solecky EP (2004) Proc SPIE 5375:133
6. Gorelikov DV, Remillard J, Sullivan NT (2004) Proc SPIE 5375:605
7. Sendelbach M, Archie CN (2003) Proc SPIE 5038:224
8. Ukraintsev VA (2006) Proc SPIE 6152:61521G
9. Villarrubia JS (1997) J Res NIST 102:425

10. Villarrubia JS (2004) In: Bhushan B, Fuchs H, Hosaka S (eds) Applied scanning probe methods, vol 1. Springer, Berlin, p 147
11. Stenkamp D, Jaeger W (1993) Ultramicroscopy 50:321
12. Tillmann K, Jäger W (2000) J Electron Microsc 49:245
13. Thiault J, Foucher J, Tortai JH, Cunge G, Joubert O, Landis S, Pauliac S (2005) J Vac Sci Technol B 23:3075
14. Wu CJ, Huang W-S, Chen KR, Archie CN, Lagus ME (2001) Proc SPIE 4345:190
15. Osborn M (2005) CD-AFM for advanced process metrology. Technical oral presentation (unpublished) in Semicon West 2005
16. Liu H-C, Dahlen GA, Osborn M, Osborne JR, Mininni L, Tracy B, delRosario A (2007) Measurement uncertainty in nanometrology: leveraging attributes of TEM and CD AFM. Instrumentation and Measurement Technology Conference Proceedings, 2007 IEEE, Warsaw, Poland, pp 1–6
17. SEMI (2001) Equipment Automation Hardware Volume. SEMI E35-0304: Cost of Ownership for Semiconductor Manufacturing Equipment Metrics. SEMI E35-0304
18. SEMI (2005) Equipment Automation Hardware Volume. SEMI E35-0305: Guide to Calculate Cost of Ownership (COO) Metrics for Semiconductor Manufacturing Equipment. SEMI E35-0305
19. Veeco (2006) AN91: 3D atomic force microscopy as an alternative to X-SEM and TEM for advanced process metrology. Veeco Metrology Group, AN91
20. Dahlen GA, Osborn M, Liu H-C, Jain R, Foreman W, Osborne JR (2006) Proc SPIE 6152:61522R
21. Giardina CR, Dougherty ER (1988) Morphological methods in image and signal processing. Prentice Hall, New York
22. Martin Y, Wickramasinghe HK (1994) Appl Phys Lett 64:2498
23. Martin Y, Wickramasinghe HK (1995) J Vac Sci Technol B 13:2335
24. Dahlen G, Osborn M, Okulan N, Foreman W, Chand A, Foucher J (2005) J Vac Sci Technol B 23:2297
25. Qian X, Villarrubia JS (2007) General Three-Dimensional Image Simulation and Surface Reconstruction in Scanning Probe Microscopy using a Dexel Representation. Ultramicroscopy DOI:10.1016/j.ultramic.2007.02.031
26. Liu H-C, Fong D, Dahlen GA, Osborn M, Hand S, Osborne JR (2006) Proc SPIE 6152:61522Y
27. Ukraintsev V (2003) personal communication
28. Keller D (1991) Surf Sci 253:353
29. Elings VB, Gurley JA (1992) In: USPTO (ed) Digital Instruments, Inc., USA
30. Williams PM, Davies MC, Roberts CJ, Tendler S (1998) Apply Phys A Mat Sci Process 66: S911
31. Todd BA, Eppell S (2001) J Surf Sci 491:473
32. Ukraintsev VA (2003) Proc SPIE 5038:644
33. Dahlen GA, Mininni L, Osborn M, Liu H-C, Osborne JR, Tracy B, delRosario A (2007) SPIE Advanced Lithography, Vol 6518. SPIE, San Jose, p 651818
34. Liu H-C, Osborne JR, Osborn M, Dahlen GA (2007) SPIE Advanced Lithography, Vol 6518. SPIE, San Jose, p 65183K
35. Foucher J (2007) Problems with CDP20S tips for via measurements. CEA-Letie, unpublished work
36. Foucher J, Fabre AL, Gautier P (2006) Proc SPIE 6152:61520V
37. Foucher J (2005) Proc SPIE 5752:966
38. Foucher J, Gorelikov DV, Poulingue M, Fabre P, G.Sundaram (2006) Proc SPIE 6152:61521A

39. Bayer T, Greschner J, Martin Y, Weiss H, Wickramasinghe HK, Wolter O (1995) In: USPTO (ed) Ultrafine silicon tips for AFM/STM profilometry. IBM, USA Patent Number 5,382,795
40. Bothra S, Weling MG (1994) In: USPTO (ed) Method of making microscope probe tips. VLSI Technology, USA Patent Number 5,540,958
41. Hantschel T, Vandervorst W (2004) In: USPTO (ed) Probe and method of manufacturing mounted AFM probes. IMEC, USA Patent Number 6,690,008
42. Toda A (1995) In: USPTO (ed) Method of making cantilever chip for scanning probe microscope. Olympus Optical Co. Ltd., USA Patent Number 5,386,110
43. Madou M (1997) Fundamentals of microfabrication. CRC, Boca Raton
44. Plummer JD, Deal MD, Griffin PB (2000) Silicon VLSI technology: fundamentals, practice, and modeling. Prentice Hall, New York
45. Liu H, Klonowski M, Kneeburg D, Dahlen G, Osborn M, Bao T (2005) J Vac Sci Technol B 23:3090
46. Tao Z, Bhushan B (2006) Tribol Lett 21:1
47. Ashby MF (2000) Materials selection in mechanical design. Butterworth-Heinemann, Woburn
48. Namaguchi T, Uchida H (1998) EOS/ESD Symp 20:245
49. Liu H-C (2004) ESD safe probe handling procedure. Veeco Instruments, Inc., Santa Barbara, Document Number 2ENS85002
50. Liu H-C (2004) Probe cassette loading procedure. Veeco Instruments, Inc., Santa Barbara, Document Number 2ENS85003, Report Number 2ENS85003
51. Ashurst WR, Carraro C, Maboudian R (2003) IEEE Trans Devices Mater Reliab 3:173
52. Mininni L, Foucher J (2007) SPIE Advanced Lithography, Vol 6518. SPIE, San Jose, p 651830
53. Dai H, Hafner JH, Rinzler AG, Colbert DT, Smalley RE (1996) Nature 384:147
54. Nguyen CV, Stevens RMD, Barber J, Han J, Meyyappan M, Sanchez MI, Larson C, Hinsberg WD (2002) Appl Phys Lett 81:901

3 Near Field Probes:
From Optical Fibers to Optical Nanoantennas

Eugenio Cefalì · Salvatore Patanè · Salvatore Spadaro · Renato Gardelli ·
Matteo Albani · Maria Allegrini

Abstract. This chapter reports a broad overview of near-field optical probes. They represent the key components for the performance of the scanning near field optical microscope (SNOM). In this frame, we consider the two main classes of sensors: aperture and apertureless probes. In particular, attention is focused on optical fiber probes and on nanoantenna probes. Recent developments in the improvement of optical throughput and in the control of the near field polarization state are reported. The electromagnetic field distributions of the nanometric optical source, as well as the fabrication methods are dealt with. In order to provide a clear, complete and comprehensive description of the technique, brief explanations of the working principles of the SNOM and conventional microscopy are given. Finally, this chapter is tale about recent SNOM applications that have been widened thanks to the improved features of the probes.

Key words: Nanoantenna, Near field, Aperture probe, Apertureless,
Scanning near field optical microscope, Polarization

Abbreviations

AFM	Atomic force microscopy
FDTD	Finite-difference time domain
FIB	Focused ion beam
HPC	Hollow-pyramid cantilevered
MMP	Multiple multipole
PSF	Point-spread function
PSTM	Photon scanning tunneling microscope
SEM	Scanning electron microscopy
SNOM	Scanning near-field optical microscope
SPM	Scanning probe microscope
TEM	Transverse electromagnetic

3.1
Introduction

The ability of light to probe the properties of matter was experienced immediately after the invention of the optical microscope. In fact, important progress in science was possible thanks to the direct observation of what was before invisible to the eyes. Such a trend continued until it was possible to improve the main features of the instruments; i.e., the resolving power, the field of view and the contrast mechanisms. However, the intrinsic limit of the microscope was soon reached. This is due to the diffraction of light that prevents the resolution from being lower than about half a wavelength (Abbé's limit); i.e., $0.2-0.5\,\mu m$ for visible light. Such a limitation, due to the phenomenon of the electromagnetic field propagation, seemed to be unsurpassable up to 1984 when the scanning near-field optical microscope (SNOM) was invented.

The idea of the SNOM was proposed in 1928 by Synge [1]. It basically consists of illuminating a sample by evanescent components of the electromagnetic field arising through a small (subwavelength) aperture in an opaque screen. The screen must be kept a few nanometers away from the surface because of the exponential decrease of the field intensity with distance. The image can be reconstructed by scanning the aperture over the sample surface. The optical resolution can be smaller than that imposed by the diffraction limit owing to the high spectral content (in the wavenumber k-space) of the near fields generated by the aperture. Consequently, the resolution is essentially given by the diameter of the aperture. The SNOM represents the top level for the forefront of science and technology today because it combines the potentials of scanned probe technology with the power of optical microscopy. Two main kinds of SNOMs exist: the aperture SNOM and the apertureless SNOM. The former derives from the technological realization of Synge's idea. The probe is traditionally a tapered glass fiber with a nanometric aperture at the end. The latter is based on Babinet's principle for which the diffraction at a sharp tip or a nanoparticle is the same as that through small hole (the complementary structure), and therefore exhibits the same spectral constituents. As for the others scanning probe microscopes (SPMs), the probe plays a key role in determining the performance of the SNOM. Indeed, the optical spatial resolution depends on the confinement of the optical energy at the apex of the probe. Moreover the efficient use of any optical contrast mechanisms (fluorescence, Raman, etc.), and consequently the applicability range of the technique, is related to a high optical throughput. In this sense, both types of microscopes present specific advantages and disadvantages. For the aperture SNOM, the throughput is a limiting factor; a too small aperture results in a too small, i.e., undetectable, output power density. Such a drawback and the finite skin depth of the metal limits the resolution to about 20 nm [2, 3]. Conversely, with the apertureless SNOM a resolution less than 10 nm has been reached [4]. In this case the limiting factor is to extract the weak tip diffracted field from the total received field dominated by the illumination background [5, 6]. In general, a high-efficiency probe remains one of the major challenges for near-field optics. In this chapter an overview is given of the variety of SNOM probes ranging from the aperture to the apertureless. Special attention is dedicated to report recent developments in terms of performance and capability, and improvements that open the SNOM technique to novel applications. In order to provide a complete and comprehensive description, we start in the next section by illustrating in detail conventional microscopes and SNOMs, remarking on their working principle and the fundamental differences.

3.2
Conventional Microscopy and Near-Field Optical Techniques

Historically, the *conventional microscope* is referred to as the compound optical microscope, mainly composed of an objective lens, an eyepiece lens, and a tube lens. Most of the advances in microscopy focused on improving the resolution. Nevertheless, Abbé's theory of image formation demonstrates that the maximum resolution that can be obtained with conventional optical microscopy is limited by diffraction [7]. Abbé realized that, owing to the finite sizes of lenses, only a part of the propagating light waves can be collected. A higher numerical aperture of the objective lens allows the collection of a larger part of the propagating waves. He introduced the point-spread function (PSF), which gives the intensity profile in the image plane due to a point source in the object plane. Rayleigh pointed out that objects are resolved when the maximum of one pattern coincides with the (first) minimum of the other, thus defining the Rayleigh diffraction limit; namely, the resolution limit is given by

$$d_{x,y} = 0.61 \times \lambda/\text{NA} , \tag{3.1}$$

where d is the distance between distinguishable objects, λ is the wavelength of the light, and NA is the numerical aperture. The most straightforward method to increase the resolving power in conventional microscopy consists of increasing the numerical aperture. Resolution of the order of about half the wavelength, i. e., to 0.2–0.5 μm for visible light, is possible in connection with a numerical aperture of 0.3–0.8.

During the last century important progress in the performance was reached by improving the quality of the optical elements. A smart design of the objective corrects optical and geometrical aberrations and raises the numerical apertures up to 1.25 for homogeneous immersion objectives. The discovery of new fluorescent labels accelerated the expansion of fluorescence microscopy in laboratory applications and research [8]. Also, recent advances in digital imaging and signal processing have allowed microscopists to rapidly acquire quantitative measurements by processing two- and three-dimensional images [9]. Confocal microscopy [10] was one of the first methods developed to increase the resolution and it is the most well-established and widespread method in use today. In this case the imaging is achieved by illuminating a single point by means of a laser and detecting the light coming out from the sample through a pinhole. The entire image is obtained by a raster scan of the surface (Fig. 3.1a). Owing to the pinhole screens, the contribution of out-of focus light above and below the focal plane is reduced $1/z^2$ times, with z denoting the distance to the focal plane, so background and stray light are almost absent in the image [10]. The lateral resolution is improved by a factor of about $\sqrt{2}$ with respect to that of the conventional microscope. As a matter of fact, in the confocal arrangement, the PSF is given by the product of the PSF of the illuminating beam and the PSF of the detection pinhole. Because of this mechanism, a depth resolution is introduced, allowing the imaging of thin slices of the object (sectioning effect) and it is possible to acquire a three-dimensional image of the sample by scanning slice by slice the object under study [10]. Despite the fact that the images have a good signal-to-noise ratio, such a kind of microscopy suffers the weakness of a long measurement time owing to the raster scanning. Such a drawback can be alleviated, as suggested in [11], by using a spinning disk with many holes, however, to the detriment of the resolution.

Several kinds of optical microscopes have been developed by using the interference phenomenon. In particular, the interference of two or more coherent light beams can result in a periodic pattern of light on the sample (object) plane. When

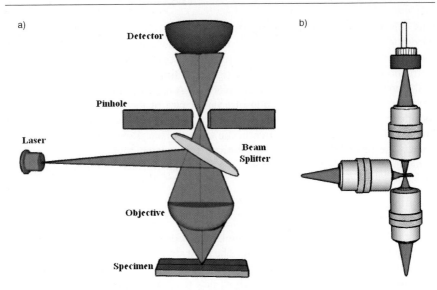

Fig. 3.1. a A confocal microscope setup. **b** A 4Pi microscope

these patterns are used to induce fluorescence, they interact with the sample structure and the recorded emission carries information with higher resolution than that of conventional microscopy. Usually, two objective lenses are used at the front and rear of the sample [12, 13]: the numerical aperture of the setup is increased, resulting in an improved resolution. A similar interference effect can be achieved when the fluorescence is collected from both objectives and combined to interfere on the detector plane. Interference methods are usually limited to thin samples, because the quality of the interference pattern is reduced when the light passes through the sample. A special kind of interference microscope is the 4Pi (Fig. 3.1b). It uses laser light for illumination in a confocal mode. When image restoration is added to the three-dimensional scanned data, the resolution can be improved down to the 100-nm range [14]. The methods so far described try to achieve high resolution by improving the PSF either directly (e. g., by increasing the numerical aperture) or indirectly, by using interference phenomena followed by image processing.

A complete revolution occurred when the lens techniques, like optical microscopy, were substituted by a lensless one: scanning near-field optical microscopy [15–17]. The scanning near-field optical microscopy technique is a member of the scanning probe microscopy family and can be basically sketched as a nanometric optical probe which scans on the sample surface, thus allowing the detection of the optical properties. A common SNOM setup is reported in Fig. 3.2. A light beam (either coherent or noncoherent, polarized or not) of a suitable wavelength is coupled to the SNOM probe. The probe is mounted on a mechanical holder of a system for the coarse approach of the tip to the sample. Generally the sample is mounted on an x–y–z piezo actuator stage which scans the sample under the tip during the experiment. As regards the measurement and the control of the tip–sample distance, a strongly gapwidth dependent signal is required. In the past, the early SNOMs employed an electron tunneling based feedback [15, 18]. Today, most of

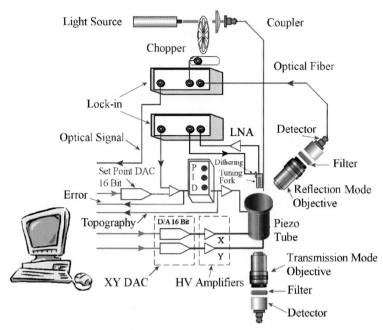

Fig. 3.2. A typical scanning near-field optical microscope (SNOM) setup

the SNOMs utilize a shear-force feedback. It consists of oscillating the probe, at its mechanical resonance frequency, parallel to the sample surface [19, 20]. During the approach, the action of the shear forces detunes the resonance frequency with respect to the driving oscillator leading to a decrease of the amplitude and to a phase shift. Initially, the detection of such quantities was accomplished by an optical system resembling that used for atomic force microscopy (AFM) [19], whereas today it is a piezoelectric detection system based on the tuning fork crystal [21]. Here the probe is glued to one arm of a tuning fork that oscillates at its own resonance frequency. The amplitude of the oscillation is kept constant during the scan by means of a feedback loop acting on the vertical axis of the piezoelectric actuator. The method allows the simultaneous acquisition of both the map of the surface topography and the optical signal. The origin of the shear-force effect has not been completely assessed. A series of mechanisms are involved, comprising viscous damping due to a thin water film, electrostatic forces, and intermittent contact [22–27].

In the SNOM arrangement, the near-field optical source is generally a laser. The choice of coherent light is dictated by the necessity of getting a monochromatic source with a high brightness, able to be coupled to the probe or focused on the sample with high precision. The laser employed can be a gas, solid, or semiconductor one. A critical parameter is the power of the laser since thermal damage to the tip (i. e., fiber probe) and to the sample may appear even when applying laser intensities of a few milliwatts. Using more than 100-mW continuous-wave lasers is tricky and even hazardous, although they allow nonlinear effects. A few experiments have been carried out with white light and with line spectrum sources generated by Raman scattering of a pulsed laser beam in several meters of optical fiber [28, 29].

The near-field light interacts with the sample and can be absorbed, reflected, or locally induce fluorescence or the Raman transitions, depending on the sample properties and the contrast mechanism of interest. In any case, light emerging from the interaction zone has to be collected with the highest possible efficiency. For this purpose, high numerical aperture microscope objectives or mirror systems are frequently used. Moreover filters can be inserted to remove unwanted spectral components. The light so collected is finally detected. Because of the very low level of the signal, the detection is accomplished by very sensitive detectors, such as photomultipliers or avalanche photodiodes. To improve the signal-to-noise ratio, modulation techniques are usually required.

Different possible SNOM configurations are allowed depending on how the near field is generated and how it is detected (Fig. 3.3). The choice of most suitable configuration is often dictated by the sample properties [2]. In the so-called *illumination mode*, probe is coupled to light and produces the near field (Fig. 3.3a). The near field is scattered by the sample and collected in the far field by means of a microscope objective in reflection, in transmission, or in the forbidden angle. The forbidden light collection consists of detecting the light transmitted by the sample at angles greater than the critical one. The reflection geometry is suitable for almost any sample, while the transmission geometry has higher collection efficiency but it is limited to optically transparent samples. A SNOM works in the *illumination collection mode* when the aperture probe is used to generate and to collect the near-field light (Fig. 3.3b). Such a configuration results in the highest lateral resolution. A special collection SNOM called photon scanning tunneling microscope (PSTM), consists in illuminate the sample surface with an evanescent field generated by total internal reflection. The electromagnetic field is collected by the aperture of the probe with a mechanism resembling the electronic tunneling [3,30,31] (Fig. 3.3c). In the apertureless SNOM, the near field is generated by the scattering of a laser beam focused on the apex of a tip with a subwavelength radius [4–6]. The interaction of such a field with the sample sur-

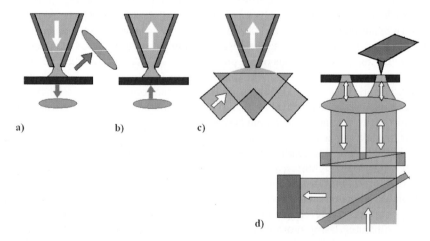

Fig. 3.3. Different types of SNOM: **a** aperture SNOM in illumination mode, the light is collected in transmission or reflection, **b** aperture SNOM in collection mode, **c** photon scanning tunneling microscope, and **d** scanning interference apertureless microscope

face will produce propagating waves which contain the optical properties of the surface on the nanometric scale. The first interferometric apertureless SNOM proposed was the scanning interferometric apertureless microscope realized by the Wickramasinghe group [4]. In this setup the sample was illuminated in transmission mode (Fig. 3.3d) by focusing two laser spots. The interference between the back reflected part of the incident beam and that containing the contribution of the near field scattered by the tip leads to the detection of the amplitude of the electric field (Fig 3.3d). Such a microscope has the advantage of improving the near field to far field ratio but the method is able to detect only modulations of the refractive index on the sample surface. By using a noninterferometric technique, one removes such a limitation. It consists of detecting the modulation of the near field due to the scattering produced by an atomic force microscope or a metallic tip. The intensity of the scattered fields carries also contributes of the inelastic-scattering processes, such as photoluminescence, Raman scattering, and second-harmonic generation. One of the first microscopes based on this principle was realized by the Boccara [5] group and was denominated the "reflection apertureless SNOM". In this scheme the laser beam is reflected by a dichroic beam splitter and focused at the diffraction limit. A tungsten tip approaches from the top in the illuminated area and the backscattered light undergoes spectral filtering and detection. A lock-in amplifier extracts the signal component modulated at the tapping frequency to remove any far-field contribution. An apertureless SNOM resembling a PSTM was introduced by Inouye and Kawata [6]. They scattered the evanescent wave with a metallic tip. The sample is placed on a glass prism with refractive-index-matching gel. The light, locally scattered by the metallic tip, is collected by a normal lens or an objective placed above the prism and sent to the detector.

A special case of an apertureless SNOM is based on the use of a single nanoparticle attached at the apex of a tapered optical fiber or at a tip apex. The nanoparticle works as a nanosource and it is excited either through the fiber aperture [33] or by focusing a laser beam on the tip apex [34, 35].The choice of the right excitation wavelength is a crucial issue in this kind of microscopy and strongly depends on the specific kind of nanoparticle probe [36]. The level of maturity and of performances achieved by the SPMs has triggered the planning and construction of microscopes derived from the combination of different techniques simultaneously: scanning tunneling microscopy, AFM, and scanning near-field optical microscopy. The combined SPMs attract particular interest in materials and surface science since they are able to provide correlation on the local scale among electrical, mechanical, or optical properties that are measured simultaneously. Recently, a tapping-mode atomic force microscope was realized, with simultaneous measurement of the tunnelling current and of the scattered optical signal (AFM/scanning tunneling microscopy/scanning near-field optical microscopy) [37, 38]. Aperture and apertureless SNOMs working at low temperature were also realized [39–41]. They are a powerful tool in the experimental investigation of semiconductors materials and allow the study of quantum wires and quantum dots of composite semiconductors [42–44]. Near-field spectroscopy can therefore "zoom in" on a single semiconductor nanostructure, investigating the interaction between a few photons and a few excitons. The ability of this intriguing technique strongly depends on the probe characteristics; therefore great effort of the scientific community is devoted to improving the probe performances such as throughput, resolution, and polarization properties.

3.3
The Probe

A probe is recognized to be a near-field optical sensor when it is able to generate high spatial optical frequencies (frequencies coded in the nonradiating part of the near field) and/or to collect those contained on a sample surface. Such properties can be exploited, in principle, by nanometric collectors/emitters consisting of a small hole, a small scattering element (such as a small sphere), and the tip of a metallic or dielectric cone. Actually, the main SNOM probes are aperture probes (i.e., metal-coated tapered fiber probes), pointed metal and semiconductor probes, and nanoemitters, such as single molecules or nanocrystals [45]. As will be seen later, continuous efforts, requiring more and more advanced technology, are being made in order to improve the probe performances. In fact, in any scanning probe technique, the quality of the images obtained and the versatility of the technique are largely affected by the probe used. In general the desirable features of a SNOM probe are (1) high optical efficiency, (2) small tip aperture or tip apex dimension (at least a few nanometres in diameter), and (3) controllable polarization properties. In particular, for an aperture probe point 1 is fulfilled by high optical throughput, whereas for an apertureless one it is fulfilled by the electromagnetic field enhancement for improving the signal-to-background ratio. From a historical point of view the first SNOM experiments were attempted with pinholelike probes by Baez [46] with sonic waves, then by Ash and Nicholls [47] in the microwave domain, and finally by Massey [48] in the IR domain and by Fischer et al. [49] in the visible domain. The first examples of employing a bare tapered rod were due to Courjon et al. [30] and Reddick et al. [31]. Whereas the use of a metal-coated tapered rod is due to Pohl et al. [15], who developed a thin conical tip from a quartz rod, and Lewis et al. [16] using a micropipette. Some years later new objects and materials used as nanocollectors/emitters appeared. Moved by AFM technology, they were suggested by Moers et al. [50]. Silicon nitride is highly transparent ($n \sim 2$) and it is quite hard. It has been used either in contact with the sample or in noncontact mode. A short time later the same kind of tip was coated with a metal [51]. Here only the extremity of the apex is metal-free. Several works have been done to create an effective hole, by using semiconductor micromachining techniques by means of the focused ion beam (FIB) technique [52, 53]. As regards nanoscatter probes, Bachelot et al. [54] proposed metallic tips as did Inouye and Kawata et al. [55]. The tip is generally made of tungsten, but also gold and silver have recently been used; it can be either straight or bent in order to be used like an atomic force microscope cantilever [56–58]. Zenhausern et al. [4] have employed a very sharp silicon cantilever tip. Particular tips overlapping aperture and apertureless have been developed too. Danzebrink and Fischer [59] built up a monocrystal GaAs tip that took advantage of the ability of the GaAs single crystal either to emit or to detect light. Inspired by a similar concept, porous silicon tips and fluorescent tips have been fabricated. The former, proposed by Gottlich and Heckl [60], consists of exiting a porous silicon tip by means of light or by means of an electric field. The porous silicon reemits light and behaves as a nanosource. The latter, firstly implemented by Kopelman et al. [61], is based either on filling a micropipette with fluorescent material or sticking fluorescent material on the tip apex. Finally, a tetrahedral glass tip obtained from broken a piece of glass has been realized. Such a tip is special because it can generate or detect plasmonlike signals [62]. This was made possible by a suitable coating of the cleaved glass corner. What we have reported above will be treated in detail by considering the probes as aperture and apertureless ones.

3.3.1
Aperture SNOM Probes

In practice, a near-field aperture probe consists of a dielectric waveguide with a nano-metric aperture at the end. Every probe is composed of three distinct elements, each of them having a well-defined function: a guiding part, a subwavelength segment, and an apex. The guiding part can be a homogeneous dielectric rod, the role of which is to guide the coupled light to the apex or to the remote detector with a minimum of losses. The subwavelength segment is the critical *mesoscopic* transition between the macroscopic guiding part and the apex itself. Up to today, the shapes of the segment produced are, respectively, conical shapes with a sharp or truncated tip, pyramidal, and tetrahedral shapes. The optical efficiency depends strongly on this part; the larger is the tapering angle, the higher is the power throughput. The apex is the diffracting element; a metal coating is needed, basically, to prevent the leakage of light through the sides of the guide and to create the aperture. The spatial resolution of the probe and its ability to maintain a well-defined light polarization depends on the size and the shape of the aperture. Aperture probes have the benefit, with respect to the apertureless counterpart, of low-level background light, which permits a suffi-cient signal-to-noise ratio to be achieved for several applications and the probability of destruction of photounstable dye molecules to be reduced, in fluorescence exper-iments. In order to gain further insight into all these aspects we analyze the light propagation of the most common aperture probes: a metal-coated tapered optical fiber and a hollow cantilever.

As mentioned above, a dielectric probe is generally has a guiding bar followed by a tapered pointed apex. To better understand how such probes work, an analysis of the light propagation into the tapered part and at the aperture must be given. For uncoated probes the waveguide modes HE_{11} of the dielectric rod part of the probe propagate in the tapered zone with no cutoff [63]. Significant simulation has performed on a discrete model that considered the cone as a succession of disks with decreasing diameters and infinitesimal thicknesses (Fig. 3.4) [2]. It turned out that, at each intersection, the HE_{11} field distribution adapts to the distribution appropriate

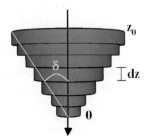

Fig. 3.4. Model of the taper zone of a dielectric aperture probe for transmission power calculation

for the next slimmer section. For each step, however, part of the radiation is reflected, and the transmitted HE_{11} mode becomes less confined as the field extends more and more into the surrounding medium (air).

The consequence is a high throughput but a poor confinement for this type of probe. In particular, it has been found that the superposition of incident and reflected light leads to an intensity maximum at a diameter of about half the wavelength. Furthermore, closer to the cone apex, the light penetrates the sides of the probe, so at the tip apex there is an intensity minimum (Fig. 3.5). Thus, the fiber probe is not useful as a local illumination source. The situation is significantly different when the tapered cone is coated with a metal. An immediate consequence is that the cover confines the fields to the cone, preventing escape through the sides (Fig. 3.6). Such a theme will be discussed later. From the electromagnetic point of view, this probe resembles a hollow metal waveguide filled with a dielectric. A detailed investigation of the propagation of the optical modes in a such structure was carried out by Novotny and Hafner [64]. The mode structure in a tapered hollow waveguide changes as a function of the dimension of the dielectric core [64]. The guided modes present in the straight dielectric core run into cutoff one after the other as the diameter decreases on approaching the apex.

Finally, at a well-defined diameter even the HE_{11} mode runs into cut-off. For smaller diameters of the dielectric core the energy in the core decays exponentially towards the apex since the propagation constants of all modes become purely imaginary. This situation is visualized in Fig 3.7. The mode cut-off is responsible for the low power throughput of aperture probes. Moreover, approximately 30% of the guided light is reflected and the metal coating absorbs 70%. An expression for the power transmitted by the aperture has been obtained [2]. Modeling again the taper

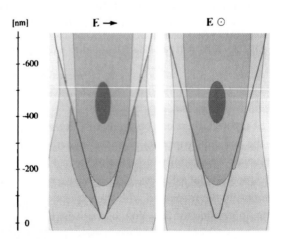

Fig. 3.5. The result of a multiple multipole (MMP) computation of the power density in a conical dielectric probe. The *contours* indicate the constant power on two perpendicular planes through the center of the probe (factor of 3 between successive contours). The darkest contour accounts for the standing wave. The exit of the light at the sides is clearly visible. (Reprinted from [65] copyright 1995, with permission from Elsevier)

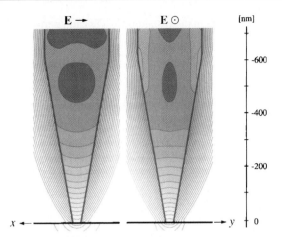

Fig. 3.6. The result of a MMP computation of the power density in a metal-coated aperture probe. The *contours* indicate the constant power on two perpendicular planes through the center of the probe (factor of 3 between successive contours). It can be seen that the light is confined to the taper zone: the light is in part back-reflected and absorbed by the metallization. (Reprinted from [65] copyright 1995, with permission from Elsevier)

zone as a succession of discrete number of small cylinder waveguide (Fig. 3.4), one obtains the following expression for the power loss in the waveguide:

$$P_{\text{loss}}(z) = P(z_0) \exp\left(-2 \int_{z_0}^{z} \alpha_{11}(z)\, dz\right) . \tag{3.2}$$

$P(z_0)$ is the incident power and $\alpha_{11}(z)$ is the attenuation constant of the HE_{11} mode in the z position of the waveguide section. The coefficient α_{11} depends on the diameter of the waveguide section, on the wavelength, and on the properties of the material. Expressing the power as function of the core diameter d, and of the half-cone angle δ, one can obtain useful information for the probe optimization. Considering an aperture with a coating of infinite thickness and a z_0 chosen in the evanescent zone of the HE_{11} mode, the coefficient $\alpha_{11}(z)$ is an exponential function:

$$\alpha_{11}(d) = \text{Im}(n_{\text{metal}}) k_0 e^{-Ad} , \tag{3.3}$$

where n_{metal} is the refractive index of the metal coating, $k_0 = 2\pi/\lambda$, the free-space propagation constant, while A is a constant depending on the characteristics of the

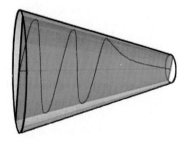

Fig. 3.7. The cutoff of the guided modes into the taper zone of a metal-coated probe. The creation of an evanescent wave also inside the probe should be noted

apex [2]. By inserting $\alpha_{11}(d)$ in the equation for $P(z)$ and performing the integration, we obtain

$$P_{\text{loss}}(z) = P(z_0) \exp\left(a - b\,e^{2Az\tan\delta}\right) , \tag{3.4}$$

in which a and b are two constants depending on n_{metal}, k_0, δ, and d_0 (core diameter). From such an expression it turns out that the transmission power increases with the cone angle δ. Finally it was found that the ratio between the input and the output power is $P_{\text{out}}/P_{\text{in}} \propto e^{-B\cot\delta}$, where B is a constant. Hence, the first important result is found: to improve the optical throughput for a dielectric probe the taper angle has to be as large as possible.

The second point is to understand the role of the aperture. For this aim the distribution of the electromagnetic field in front of the probe aperture has to be calculated. Owing to the complicated boundary conditions imposed by the probe geometry (e. g., truncated conical tip with a metal coating), analytical solutions are not available and the problem has to be attacked by using fully numerical (finite-element method, finite-difference time domain, FDTD, method of moments) or semianalytical (multiple multipole, MMP) methods, requiring extended computational work [66].

Anyway, as a first approach important information can be achieved by approximating the SNOM aperture probe as a circular hole in an infinite perfectly conducting screen. This is the approximation that Bethe [67] and later Bouwkamp [68] used in their theories. Such an approach is, however, most extensively used in near-field calculations because it leads to analytical formulae. It has been accomplished by solving an electrostatic and magnetostatic eigenvalue problem for the static near field existing in the vicinity of the aperture. This solution can be easily recovered by using oblate spheroidal coordinates and harmonic wave functions. The Helmoltz equation $[\nabla^2 + k^2] \cdot \varphi = 0$, which reduces to the Laplace equation for small hole sizes $\nabla^2 \cdot \varphi = 0$, has the solution φ as a product of parametric Legendre wave functions in spheroidal coordinates. In Fig. 3.8 the results of such a calculation for the squared components of the electric field are reported [69, 70].

The calculus was carried out in a plane a z-distance 10 nm (left) or 1 nm (right) along the z-axis below the aperture. The aperture used was 200 nm and the wavelength was 514 nm. The light incident on the aperture is plane wave polarized in the x-direction [69, 70]. Close to the aperture, we observe that all polarization components of the light are present (Figs. 3.8, 3.9). The z-polarized light exists under the metal forming the aperture, and x–y-polarized light exists under the aperture itself. Far from the aperture, the diffracted light will also be predominantly x-polarized. The x-component of the electric field lies primarily under the aperture, spreading with distance; the z-component lies close to the aperture but under the metal, also spreading with distance. The field is most enhanced near the sharp edge of the aperture. There is a small component in the y-direction whose magnitude is much smaller than that of the others (Fig. 3.9).

From Bethe–Bouwkamp theory it also turns out that in the far field, the emission of a small aperture is equal to the radiation of a magnetic and an electric dipole located, respectively, either parallel or perpendicular to the aperture plane [67, 71]. Betzig et al. [33] obtained a fruitful experimental proof of this model, by comparison against measurements of fluorescence for a single molecule. Only the plot of $|E|^2$ is here reported because, at optical frequencies, only the electric field is responsible

Fig. 3.8. The square of the electric field components diffracted by a circular aperture in an infinite perfectly conducting screen. The *left column* was calculated for 10 nm away from the aperture, and the *right column* for 1 nm. **a, b** The $|E_z|^2$ term with the gray ranges 0–0.7 and 0–7.0, respectively. **c, d** The $|E_x|^2$ term with gray ranges 0–1.1 and 0–9.0, respectively. **e, f** The $|E_y|^2$ term with gray ranges 0–0.14 and 0–1.4, respectively. (Reprinted from [69] copyright 2003, with permission from Wiley Interscience)

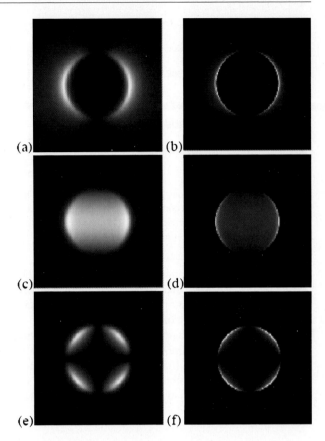

for the interaction with matter (fluorescence, polarization, etc.). Anyway, analogous results apply to the magnetic field components.

As before mentioned, the Bethe–Bouwkamp theory gives correct information on the near-field distribution behind an ideal aperture. A set of corrections have to be introduced to realistically describe an aperture probe (e. g., an optical fiber).

The field of the ideal aperture presents a singularity at the edges in the plane of polarization and zero along the other axis outside the aperture. This is not the case for an aperture probe with a metal coating of finite conductivity. At optical frequencies, the shortest skin depths for a metal are about 6–10 nm, which will widen the effective aperture size and smooth out the singular fields at the edges. Moreover, any realistic metal screen will have a thickness of at least 100 nm. The Bethe–Bouwkamp approximation shows higher confinement of the fields and much higher field gradients than those actually occurring, for example, greater optical forces being exerted on particles next to the aperture. Furthermore, in the approximate model a plane wave was assumed as the exciting field, whereas the actual probe is excited by a waveguide mode in the tapered section. Therefore, a correction has to be introduced to the model for coherent superposition of the magnetic and electric dipole as a radiating system [67, 68]. Obermuller and Karrai [72] proposed an alternative description in which the electric and magnetic dipoles both lie in the

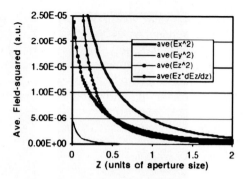

Fig. 3.9. The integral over planes such as in Fig. 3.8 is shown. The average electric field is shown for the different electric field components as a function of distance from the aperture in units of aperture size. In all cases, the electric field is enhanced near the aperture. (Reprinted from [69] copyright 2003, with permission from Wiley Interscience)

plane of the aperture and are perpendicular to each other. On the basis of on these considerations, we show the results of a numerical computation relevant to the field distribution at the circular aperture of a metal coated tapered optical fiber [65] and at the rectangular aperture of hollow cantilever [73].

Figure 3.10 shows the results of a computation performed by the MMP method in order to solve the Helmholtz vector wave equation in three dimensions for the metal-coated fiber probe [65]. The latter has an aperture diameter of 50 nm. Figure 3.10a shows contour plots of the square modulus of the electric field ($|E|^2$) along the taper zone. The HE_{11} waveguide mode ($\lambda = 488$ nm) [65], incident from the

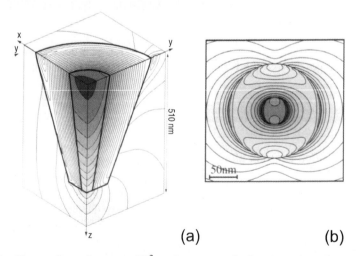

Fig. 3.10. a Contour lines of constant $|E|^2$ on three perpendicular planes through the center of the probes (factor of 2 between successive lines). The exciting HE_{11} mode is polarized along the x-direction.: $\lambda = 488$ nm, $\varepsilon_{diel} = 2.16$, $\varepsilon_{Al} = -34.5 + i8.5$. **b** Electric field evaluated on a plane 1 nm behind the aperture. Contours of constant $|E|^2$ (factor of $\sqrt{2}$ between successive lines). (Reprinted from [65] copyright 1995, with permission from Elsevier)

infinite cylindrical structure and polarized in the x-direction, excites the fields in the tips. We can observe that the contours are discontinuous in the plane of polarization ($y = 0$), as the electric field has a net component perpendicular to the boundaries. In the perpendicular plane ($x = 0$), on the other hand, the electric field is always parallel to the boundaries, leading to continuous contour lines. In Fig. 3.10b the field distribution evaluated in the plane of polarization 1 nm behind the aperture is depicted. The fields are strongly enhanced at the edges of the aperture owing to the high curvature of the metal surface at the rim (tip effect, lightning-rod effect). The resulting near field of the aperture probe therefore consists of two spots, and its strength is given by the aperture diameter [65].

In order to attenuate the fields in the lateral direction, the metal cladding needs to have a minimum thickness. For too thin cladding, the fields transmitted from the core through the cladding can couple with external surface modes [74], producing a spatially extended light source. It is also interesting to visualize the electric field distribution behind the aperture into the spectral domain of spatial frequencies. In Fig. 3.11 the results of a FDTD simulation performed on a probe having a cone angle of 20°, with an aperture diameter of 100 nm, and coated with a 50-nm aluminum layer are shown. The spectrum of the electric field (amplitude) is computed in near field on a plane at $z = 25$ nm from the probe aperture. The scales on the x-axis and and the y-axis are the product of the spatial frequency [defined as $fx = (2\Delta x)^{-1}$ and $fy = (2\Delta y)^{-1}$] times the radiation wavelength. In this *reduced* system of coordinates, $|1| = x$ and $|1| = y$ are equal to $\lambda/2$; therefore, the sub wavelength components lie outside the interval ± 1. The probe is fed with a radiation polarized

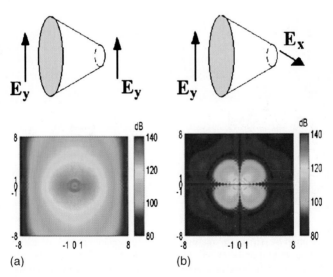

Fig. 3.11. Spectrum of electric field on a the xy-plane at $z = 25$ nm from the probe aperture. **a** The angular spectrum obtained by feeding the probe with the E_y component of the field and calculating the E_y distribution. **b** The angular spectrum obtained by feeding the probe with the E_y component of the field and calculating the E_x distribution The propagative components of the electric field are between -1 and 1, and the remaining part accounts for the evanescent region of the spectrum [75]

first along y and then along x. The output fields are calculated in both polarization directions y and x. Figure 3.11 displays the angular spectra feeding the probe with the E_y component of the electric field. The spectra show an isotropic distribution of the E_y components (Fig. 3.11a) and a narrow spectrum of the E_x component (Fig 3.11b), respectively. In particular the E_x evanescent components have low amplitude. Because of the structure symmetry, the spectrum obtained by feeding the probe with the E_x component would show the same result in the opposite direction. By this calculus we find (1) the clear presence of the evanescent components responsible for the subwavelength resolution and also the presence of the far-field component and (2) a circular aperture is able to maintain the same polarization state as that of guided light, but it is not able to select one.

Concerning the hollow cantilever, there is a different structure on the taper zone. It is usually pyramidal or tetrahedral, whereas the aperture shape can be circular, square, rectangular, or triangular. Moreover, in most cases, the internal part of the tapered structure is unfilled: the dielectric is the air.

A numerical calculation (Fig 3.12, panels a and b) of the electric field behind the square aperture of a hollow cantilever was performed by the Oesterschulze

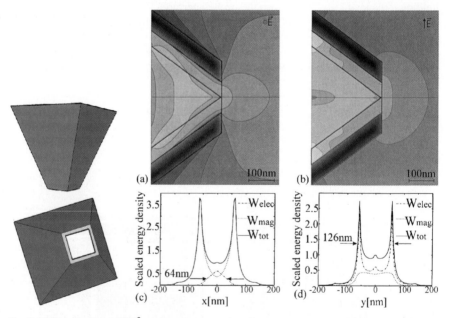

Fig. 3.12. Distribution of $|E|^2$ of an aperture probe of a square aperture of 120-nm size in the two planes of symmetry: *a* in the (x, z) and *b* in the the (y, z) plane with the z-axis orientated horizontally. Values were encoded on a gray scale with a factor of 10 between successive lines. The polarization of the illuminating field is orientated in the y-direction. Profiles of the energy densities in the aperture plane in the x-direction and in the y-direction are shown in *c* and *d*, respectively. The electric energy density is marked by *dash-dotted lines*, the magnetic energy density by *dashed lines* and the total energy density by *continuous lines*. Energy densities in the profiles were scaled with respect to the center value of the total field energy density. (Reprinted from [73] copyright 2001, with permission from Blackwell Publishing)

group [73]. It was found that the incoming mode is nearly totally reflected inside the tip and interferes with the reflected fields, resulting in a standing wave pattern (left-hand side of Fig. 3.12a). Furthermore, $|E|^2$ is damped by about 3 orders of magnitude along the propagation direction while passing the aperture. This accounts for a substantial drop in transmission with decreasing aperture size [73]. Moreover, a confinement of the evanescent electrical field in the aperture was found. In particular, the profiles of the electric energy density in the aperture plane (Fig. 3.12, panels c and d) reveal a saddle-shaped distribution. It peaks on the two opposite rims of the aperture perpendicular to the polarization direction owing to the lightning-rod effect and gives rise to a field enhancement.

A single maximum emerges in the center of the perpendicular direction. Furthermore, the total energy density is dominated by the magnetic field. This corresponds to Bethe's theory of a magnetic dipole dominating the field in a planar subwavelength aperture illuminated by a plane wave in normal incidence [67]. Summarizing, the outcome reported above establishes some of the design goals and limitations of aperture probes:

1. The larger the opening angle of the tapered structure, and the higher the refractive index of the dielectric core, the better the light transmission of the probe. This is because the final cut-off diameter approaches the probe apex [74, 76].
2. In the region of cutoff, the energy is partly dissipated in the metal layer. This can result in significant heating of the metal coating in this region, which consequently might be destroyed. The maximum power that can be sent to such a probe is therefore drastically limited. Improving the heat dissipation in the relevant region or increasing the thermal stability of the coating can increase this destruction threshold [77].
3. The ability of the probe to maintain the polarization state depends on the shape and the symmetry of the aperture.

3.3.1.1
Opical Fiber Probes

The tapered optical fiber was one of the first SNOM probes, and still remains the most commonly used. Thanks to its good performances, the relative ease of fabrication, and its versatility, it has been employed in many SNOM geometries. Apart from the good optical guiding properties (either with continuous or with pulsed waves [78, 79]), the fiber probe has good mechanical properties in terms of stiffness and a high quality factor as an oscillator; therefore it is also an excellent candidate as a sensor for surface topography detection. The fiber probe was fruitfully employed in a pure dielectric probe arrangement, i.e., without a metal coating, above of all as a near-field collector in a PSTM [80, 81]. Beside the traditional coated straight version, also a bended fiber probe was realized. Such a solution was efficiently used for scanning the specimen in the tapping mode regime. However, the probe bent, resulting in a degradation of near-field light polarization properties (in the far field, extinction ratios of approximately 70:1, compared with 100:1 for the straight probe [82, 83]). A further evolution of the probe was introduced by fabricating an optical fiber with a trihedral shape for the tapered zone, instead of the traditional conical one [84]. By FDTD computation it was found that such a configuration has an improved performance in term of the power throughput ascribed to the strong influence of the facet angle on the intensity of power reflected by the probe [84]. Another innovative fiber, to be used like a probe, it is called a photoplastic SNOM probe [85, 86]. It consists of a sharp silicon pyramidal tip glued on a photopoly- mer fiber (Fig. 3.13). The fabrication process starts with the production of the pyramidal mold in silicon by using the anisotropic etchant potassium hydroxide. This results in an inverted pyramid limited by ⟨111⟩ silicon crystal planes having an angle of about 54°. The surface including the mold is covered by a 1.5-nm- thick organic monolayer of dodecyltrichlorosilane and a 100-nm-thick evaporated aluminum film. Two layers of photoplastic material are then spin-coated (thereby conformal filling the mold) and structured by lithography to form a cup for the optical fiber microassembly. Finally, FIB milling was used to form the aperture, which has a diameter on the order of 80 nm with an optical throughput efficiency of 10^{-4}.

To overcome the problem of cutoff in the taper zone, with the aim of increas- ing the throughput power, novel probes were proposed with a coaxial structure (Fig. 3.14) [87, 88].

From electromagnetic theory it is known that coaxial structures have no cutoff frequency and have a power transmission close to unity. Such a concept will be

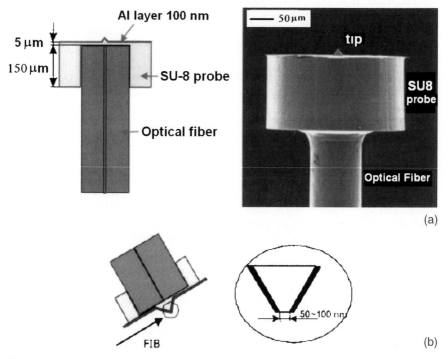

Fig. 3.13. a Schematic (*left*) and scanning electron micrograph (*right*) of a microfabricated and assembled photoplastic SNOM probe. **b** Aperture opening by a focused ion beam (FIB). (Reprinted from [86] copyright 2001, with permission from Blackwell Publishing)

Fig. 3.14. Proposed structure for coaxial SNOM tips. (Reprinted from [87] copyright 2001, with permission from Blackwell Publishing)

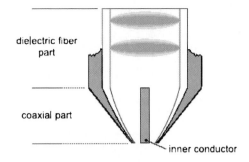

better expounded on later. Another probe developed for the coaxial arrangement was realized by van Hulst group [89] (Fig. 3.15).

Relying on the planar strip-line configuration, 20 nm air lines have been fabricated in the tapered fiber metal coating. The throughput is improved for fields perpendicular to the slit and occasionally they observed profiles with 30–40 nm full width at half maximum for single molecules with the proper dipole orientation directly under the slit.

Recently a study on a similar type of probe with no cutoff, designed with a so-called *double C* cross section, showed that in this kind of structure (no cut-off sections) the throughput improvement is mainly due to the far-field components of the diffracted light (Fig. 3.16). So, the power improvement is obtained with detriment to the near field to far field ratio.

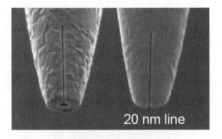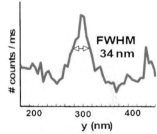

Fig. 3.15. Strip-line probe: a 20-nm-wide air gap, connecting the end facet and cutoff region, was written (FIB milling) in the metal coating of a near-field aperture probe. The fluorescence profile of a single molecule excited by the near-field slit at the facet shows a response as narrow as 34 nm. *FWHM* full width at half maximum. (Reprinted from [89] with permission from the authors)

Fabrication Methods

The tip fabrication typically includes two steps: fabrication of a transparent tapered probe with a sharp apex, and metal coating of the probe to form a transmissive aperture at the apex. As regards the first step, two major methods exist: laser heating–pulling [33] and chemical etching [90–92]. For the coating process, metal evaporation is the main method. Here we report on such techniques and the related innovation referred mostly to the optical fiber probe.

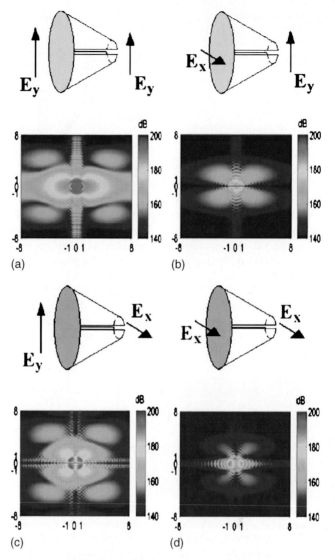

Fig. 3.16. Fourier spectra of the electric field amplitude (E_x and E_y) of the double C probe on an xy-plane at $z = 25$ nm from the aperture. For the *left column* the probe was fed with the E_y component of the field, whereas for the *right column* it was fed with the E_x component. The propagative components of the electric field are between -1 and 1, and the remaining part accounts for the evanescent region of the spectrum. Comparing this result with those reported in Fig. 3.8 clearly shows that such a probe has a power improvement of about 60 dB. However, it has to be stressed that it lies in the far-field part of the spectra [75]

Heating and Pulling.   The heating and pulling method consists of applying axial tension on the fiber and simultaneously heating the area where the fiber is desired to break [93]. The heating, usually performed with a CO_2 laser, makes the fiber softer (the temperature reached is about $1600\,°C$), which then begins to neck under the tension, and finally, when the preset pulling velocity of the free end of the fiber is reached, a strong force engages and the fiber is pulled apart (Fig. 3.17a). In detail, the formation of the tip by laser heating and pulling has two stages in terms of the dynamic behavior: (a) plastic deformation and softening and (2) thinning and breaking. At the beginning of laser heating, the fiber elastic force balances the drawing tension force after a very short acceleration time; then the plastic deformation weakens the fiber and the deformed portion undergoes necking slowly. As the heating and deformation continues, the fiber melts and the viscous force decreases; thus, the acceleration of the rail carriers increases and the heated part continues to thin. The fiber will be ruptured if the external tension force exceeds the tensile strength.

After the break at the point of heating, two tapered fibers are created, having very fine tips at the rupture location. The characteristic extent of this process is less than 1 s. However, the control of the fiber speed enables slower timing of application of the strong force. The typical setup employed for such a method consists of a commercial micropipette puller (as illustrated in Fig. 3.17b). This device is commonly used for creating tapered pipettes for biological uses, such as patch clamp experiments and nucleus transferral. A fiber pulled with such equipment usually has three identifiable taper regions as shown in Fig. 3.18: region 1, where the fiber diameter is reduced from the original to about $10-20\,\mu m$; region 2, needlelike elongated region where the fiber diameter decreases gradually; and region 3, where the fiber diameter is reduced to the final apex diameter. In general, the resulting tip shape depends greatly on the laser power, the timing of the heating and the pulling, as well as the dimensions of the heated area [94].

With such a technique it has been possible to obtain optical fiber tips with apex diameters of 50 nm and less, but with longer tapers (approximately 1 mm) and less

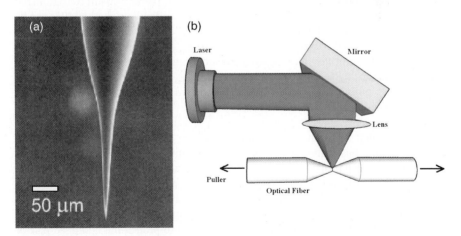

Fig. 3.17. a Tapered optical fiber probe created by pulling. **b** Laser heating–pulling set up. (**a** Reprinted from [3] copyright 2000, with permission from the American Institute of Physics)

Fig. 3.18. The fiber tapering.
(Reprinted from [93] copyright
2003, with permission from the
American Institute of Physics)

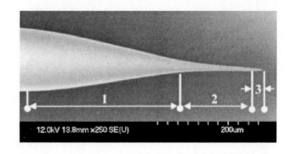

reproducible results than with chemical etching. However, its advantage lies in that it produces excellent surface quality with a very smooth profile in the tapered region, which is favorable for light guiding and the subsequent metallization of the tip. So, to get a competitive heating and pulling technique, the most challenging task is to obtain tips with small apexes and minimized overall tip length. In this sense, several groups systematically studied the fiber-pulling process [92–95]. Especially Valaskovic et al. [94] performed an extensive study on the optimization of the pulling parameters of the fiber probes. They measured transmission efficiencies between 10^{-7} and 5×10^{-5} for tip sizes ranging from 60 to 100 nm. Moreover, generalized optimized parameters was found, consisting in a narrow working range of laser power (1.85–1.95 W) and of the pulling force (0.5–0.9 lb). Finally, combining both ranges, they determined the optimum conditions of 1.90 W and 0.74 lb. Tips with short taper (approximately 300 μm), small apex (approximately 50 nm), and large aperture cone angles (approximately 45°) were achieved [93]. Instead of a laser heating–pulling method the Meixner group [96] used a different approach to reduce the taper length. With use of a foil heater the taper length can be shortened and cone angles on the order of 30° can be reproducibly obtained. Indeed, the length of the molten zone determines the length of pulled optical fiber tips. When is varied the drawing force, an optimum temperature range is found where the taper shape is monotonic throughout the entire tip. The tip end diameter is well below 100 nm for optimized pulling conditions [96]. Moreover, the laser heating and pulling method [93] can be used to produce tips bent to a certain angle to fit particular applications, alternatively to a bending method based on electric arc heating [97]. In this case, at high laser power, the surface of the fiber facing the beam melts before the back surface; thus, melting leads to high surface tension forces in the melted region and the surface tension force pulls the free end of the fiber towards the laser beam. The bending angle can be controlled through the energy dose.

Chemical Etching.   The chemical etching technique is the most intensively studied. Various etching-based methods to produce different configurations of fine optical fiber tips have been proposed in the literature. The basic one is the so-called *Turner method* [90, 98, 99]. It produces wider cone angles and the tip length is reduced to a few hundred microns. This procedure consists of immersing the fiber core, without any jacket, in an hydrofluoridric bath covered by an organic solvent. The HF solution serves as an etchant, and an organic solvent is used to protect the undipped part of the fiber. A meniscus of the HF solution is formed owing to the surface tension difference between the HF solution and the solvent (Fig. 3.19).

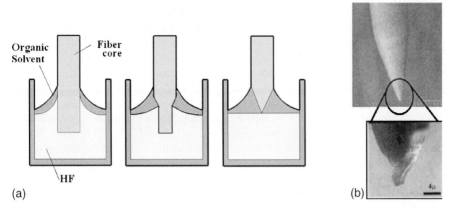

Fig. 3.19. a The Turner etching process. **b** Tapered probe by created etching. (Reprinted from [102] copyright 2003, with permission from the American Institute of Physics)

As the fiber etching proceeds, the meniscus height decreases and its profile changes so that a conical shape is gradually formed. This effect occurs at the local interface where the weight of solution (HF plus the product of the reaction) exceeds the surface tension force (Fig. 3.19). The meniscus then drops down to the next stable point where the amount of the surface tension force, at the new contacting angle, can balance out the weight of the liquid of the new meniscus (Fig. 3.19). During the etching, the fiber is consumed as long as it remains in the acid, i. e., until the meniscus disappears with the consequent complete formation of the tapering (Fig. 3.19). The end of the etched out fiber surface, which was in contact with the HF solution, is now in contact with the solvent. A number of parameters, such as the etchant concentration, the type of solvent, the etching time, and the type of fiber or dopant, influence the etching process [98, 100, 101]. As different dopants and/or dopant concentrations in the fiber are used to create the refractive index profile between the cladding and the core of the fiber, the etching rates of cladding and core are different, which allows one to etch fibers with very high apex angles up to 40° and tip sizes around 30 nm.

For a given fiber, using different organic solvents, which modify the surface tension of the acid/solvent interface and the contacting angle between such an interface and the fiber, changes the final tip shape. As reported in [98], an etching system with an iso-octane layer produces cone angles of 40°, while that with toluene results in angles of 29°. This method is able to produce angles wider than those obtained by pulling together with a larger throughput of about 10^{-3} [103–105].

The main drawback of the Turner etching technique is that the tip shape and its surface quality are quite sensitive to the etching environment. Air movement, temperature fluctuations, vibrations, and etchant concentration can have a strong influence on the shape and the symmetry of the meniscus-etched probe, and on its surface roughness (note the surface of the tip illustrated in Fig. 3.19b). If the surface of the tip is too rough or is marred by other imperfections, then the subsequently applied metal coating can have pinholes that interfere with the definition of a near-field aperture. Irregularities can also become starting points for thermal degradation of the metal coating, thereby lowering the fiber damage threshold and reducing its useful lifetime [106]. Several diversified etching methods based on the Turner etching technique have been developed aiming to achieve a better control of the tip shape and improving the probe performances (surface quality and the cone angle [107–109]). One utilizes different etching rates of different layers in the fiber [110]. Another utilizes a multistep etching procedure including coating of the tip with a polymer to partially protect it from the acid at an intermediate step [111]. In contrast to the static etching, reported up to now, wherein the fiber does not move during the etching procedure, a dynamic etching method was also introduced [112]. It is still a Turner etching based technique, but it introduces a new feature: a vertical movement of the fiber while it is immersed in the etchant. The idea is to alter the recession speed of the meniscus (as happens in the static etching process) by moving the fiber either up or down at a certain speed, constantly or not, whence the name "dynamic". The speed determines the amount of time during which the cross section of the fiber, at each height, is exposed to the acid, and therefore how far the fiber will be etched laterally at that height [102].

Tube Etching and Reverse Tube Etching

Special mention has to given to an alternative interesting etching method called *tube etching* [113]. Such a technique improves the reproducibility and the cone angles, while reducing the roughness of the tip surface, overcoming the drawbacks of the conventional etching techniques. It is quite similar to the Turner etching technique but the fiber is etched without stripping the jacket. The tip grows inside the jacket, which acts as a capillary. As the etching proceeds in the up direction, the tip surface is preserved from external perturbations and becomes smoother, hence ensuring the good performance of the method. The procedure still results in large cone angles. Furthermore (i) tips with reproducible shapes are formed in a high yield, (2) the surface roughness of the taper is drastically reduced, and (3) the tip quality is insensitive to vibrations and temperature fluctuations during the etching process. To characterize the process, a systematic study was carried out by using different fiber types, and by varying the acid concentration percentage HF ranging from 21 to 40), the organic overlayer solvents (iso-octane, *p*-xylene), and the process temperature (from 10 to 50 °C). It turned out that the process is self-limiting, but it is not self-terminating; whereas the etching time does not influence the tip quality. This fact makes the handling of the process straightforward and easy [113]. Furthermore, the cone angle and the roughness of the surface depend mainly on the fiber material, slightly on the acid concentration, and on the process temperature [113, 114]. The smoothness of the glass surface allows us to reduce the aluminum thickness and to refine the coating technique.

A multistep etching method, based on this technique, has created by the Marti group [103]. It consists of two steps. In the first one the fiber undergoes a tube

---

Fig. 3.20. Tube etching (**a**) and *r* reverse tube etching (**b**). The vectorial diagrams describe the fluxes in the two experimental geometries. (Reprinted from [114] copyright 2006, with permission from Elsevier)

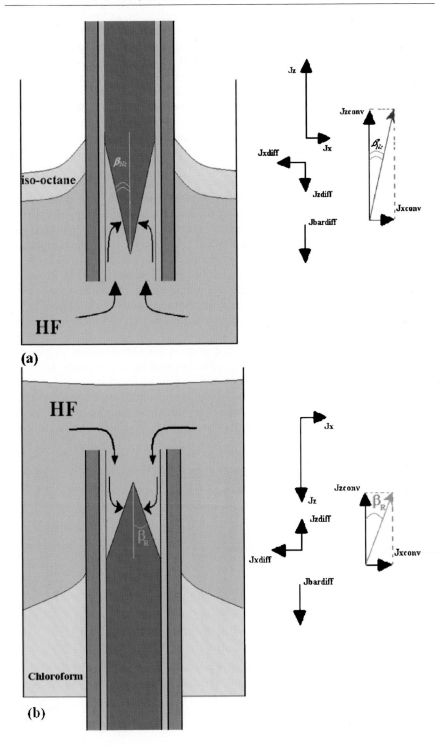

etching. In the second one, the tapered part of the fiber is bevelled using a modified micropipette beveller to obtain a tip diameter in the nanometer range as well as a smooth surface to allow a good metallization by evaporation. By varying the bevelling angle, they could obtain tapered shapes with different cone angles with high reproducibility [103]. SNOM transmission experiments performed with these probes have achieved an optical resolution below 80 nm. In comparison to fiber tips obtained by a standard heating and pulling method, the transmission efficiency of these tips is up to 3 orders of magnitude higher owing to the optimized tapered shape [103]. Recently, Allegrini's group [114] developed a novel improved tube etching technique that consists of inserting vertically the fiber in the etching bath in an upside-down geometry. This single-step method, denominated "*reverse tube etching*", not only preserves the simplicity, the low cost, and the smoothness of the etched surface typically obtained by means of the standard tube etching, but above all it guarantees an increasing of the tip cone angle of about a factor of 2, compared with the tube etching results [114]. A unique model able to explain both the tube etching and the reverse tube etching processes was proposed; it is based on a combination of convection and diffusion processes occurring inside the jacket [113, 114]. Beside the diffusive and the convective components, a component due to a barodiffusion was also added to take into account the gravity effects; as a matter of fact, the presence of a stationary gravitational field in a liquid mixture induces a concentration gradient leading to a mass flux [115].

Such a effect acts in a different way in the tube etching arrangement, with respect to the reverse tube etching (Fig 3.20). In the tube etching it contributes by taking away the reaction product from the fiber core surface, so the effective vertical convective and diffusive fluxes increase with respect to the horizontal ones: in practice it increases the vertical rate attack versus the horizontal one. In contrast for the reverse tube etching, the barodiffusive flux accumulates the reaction products on the

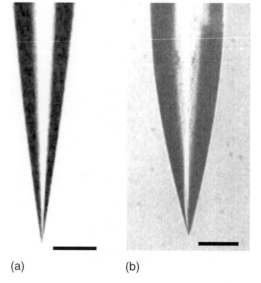

Fig. 3.21. Optical images of SNOM probes produced by tube etching (**a**) and by reverse tube etching (**b**) in the same experimental conditions. *Scale bar* 100 μm. (Reprinted from [114] copyright 2006, with permission from Elsevier)

(a)   (b)

core face. In this case it produces a decrease of the effective vertical convective and diffusive fluxes with respect to the horizontal ones and consequently the respective rates of attack. Recombining the vectors in the two distinct processes the result is that, in the reverse tube etching, the cone angle will be larger than in the tube etching (Fig. 3.20). Figure 3.21 shows two fiber probes obtained with the two methods. It is clearly visible that the fiber fabricated with the reverse tube etching has a cone angle much wider than (almost double) that obtained by tube etching. A detailed description of the process can be found in [114].

Coating Process: Aperture Formation

To form an aperture, the tapered zone of the probe must be coated with an opaque metal to confine the light [33, 116–118]. Indeed, as the diameter of an uncoated fiber in the taper region reduces beyond the waveguide mode-field cutoff, an unacceptable amount of light escapes from the sides of the probe. However, the choice of metal is critical in order to fulfil the creation of as small as possible subwavelength aperture and efficient confinement of the light. In this sense a metal with a small penetration depth should be used. For visible radiation, aluminum has the smallest skin depth (13 nm at 500-nm wavelength) and, therefore, it requires the thinnest coating. The maximal obtainable SNOM resolution is limited by this penetration depth to 20 nm [117, 119]. The metal-coating process is accomplished by vacuum evaporation or alternatively by sputtering. For deposition of aluminum on probe tips, the evaporation process is generally preferable. The directionality of deposition during thermal evaporation (in contrast to the scattering at multiple angles that is characteristic of sputtering) is advantageous because line-of-sight deposition allows the aperture region of the probe to be easily shielded from the metal vapor (Fig. 3.22). The probe rotated at a certain angle in metal vapor (Fig. 3.22). In such a way, the apex is not coated. The size of the aperture so formed depends upon the angle under which the fiber is held (10–20°). To prevent leakage of light, approximately 50–100 nm of

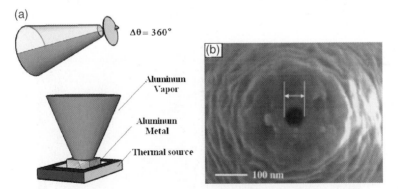

Fig. 3.22. a Thermal evaporation setup. **b** Scanning electron microscopy (SEM) image of the face of the tapered optical fiber apex after aluminum-coating. Note the clear formation of the aperture and the roughness of the surface. (Reprinted from [120] copyright 1998, with permission from the Optical Society of America)

aluminum is coated around the sides of the taper region. As regards the sputtering, it is a multidirectional process, having the characteristic of producing a uniform deposition on irregular surfaces of good-quality metal films. As a consequence of this behavior, plasma sputtering of SNOM probes results in the entire tip being coated, and an aperture must be created in a subsequent process. A possibility is to put the tip and a surface in lateral contact by scanning the tip over a hard surface with a large applied force (low feedback set point voltage). In general, such a procedure is not reliable since it does not produce probes with a reproducible aperture. The limitation of sputtering due to the plugging effect leads the evaporation technique being preferred.

An important problem encountered in the metal-coating process, especially when using thermal evaporation, is grain formation (Fig. 3.22b). In fact, a smooth metal coating is essential because light can leak out of the tapered fiber between metal islands, i. e., pinholes. When this happens, there is no well-defined aperture or there may be multiple apertures, which produce background light that degrades the already low signal-to-noise ratio in SNOM experiments. Furthermore, grains at the end of the tip can block the aperture and/or produce an irregular shape, reducing the symmetry of the aperture and degrading the polarization properties of the light delivered by the tip. At the same time the grains prevent the aperture from being close enough to the sample. Such an effect was found to be more marred with chromium and gold than with aluminum and silver [121–124]. Deeper studies on the evaporation of aluminum process showed that the formation of grains and aluminum oxides can be minimized by reducing the total chamber pressure, increasing the coating rate, and reducing the partial pressure of oxygen and water in the vacuum chamber [94, 125–127].

Moreover, reducing the temperature of the substrate also results in a smoother aluminum coating, though this technique was found to be tricky to execute. Anyway, also when realizing the coating under the above mentioned conditions, the grain imperfections on the face of the probe aperture are not completely eliminated. Therefore the FIB technique was recently employed to "mill" the probe apex with high precision, thus producing a smooth and very flat-end face with a well-defined circularly symmetric aperture down to 40 nm (Fig. 3.23). This technique also improves the polarization behavior of the tip and results in polarization ratios exceeding

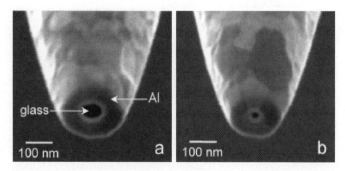

Fig. 3.23. SEM images of two SNOM probes in which the ends were milled flat using a FIB. The aperture diameters are **a** 120 nm and **b** 35 nm. (Reprinted from [52] copyright 1998, with permission from the American Institute of Physics)

1:100 in all directions [52]. Elliptical core fibers, coated with aluminum and with apertures milled by a FIB, yield high throughput and are able to maintain the polarization [128]. These results are satisfactory; however, the FIB technique is rather expensive and complex.

As previously reported, one of the task for optimizing SNOM probes is the improvement of the optical throughput. In order to do that, apart from improving the cone angle, another possibility is to increase the laser power coupled into the fiber. Doing that, one has to take care since a lot of optical power is reflected back and absorbed by the metal coating in the taper zone. Therefore, the power can be increased only up to a certain threshold [94], after which the strong heating destroys the probe aperture. Temperature measurements along a taper of aluminum-coated fiber probes as a function of the optical input power have been performed [129]. It was shown how the heating effect can be reduced by increasing the cone angle. In particular, the temperature coefficients vary from 20 °C/mW for tips with a large cone angle to 60 °C/mW for narrow, long ones. Moreover, the strongest heating occurs as far as about 70 µm from the tip, in the upper part of the taper zone. Here temperatures of 470 °C can be reached for input powers up to 10 mW; after that, thermal damage of the coating occurs. For higher input powers, the aluminum coating usually breaks down, leading to a strong increase of light emission from the structure. Such an effect usually occurs either by straightforward melting of the aluminum layer or by fracture and subsequent rolling up of the metal sheets due to internal stress. Studies on the surface roughness showed that it also plays an influential role in the optical destruction thresholds. For a smoother surface of the taper zone, achieved by tube etching, a destruction limit as large as 120 mJ is obtained. This value is almost twice as large as that of the rougher, conventionally etched fiber probes. Furthermore, the use of additional adhesion layers (Ti, Cr, Co, and Ni) between the glass surface and the aluminum coating produces a significant increase in the optical destruction threshold by up to a factor of 2 [77]. The deposition of alternating thin titanium and thick aluminum layers additionally reduces grain formation and crystallization within the metal coating, leading to firmer optical probes. In this way, an optical destruction threshold of 277 mJ was achieved, corresponding to an improvement in stability of more than 400% compared with that of conventionally fabricated near-field optical tips. Finally, adhesion layers were shown to improve also the mechanical hardness of the metal coating and so the lifetime of the probe [77].

3.3.1.2
The Hollow Pyramid Cantilevered Probes

Recently, a novel class of SNOM probes was introduced, based on a silicon or a silicon nitride cantilever, similar to the ones used in AFM, with a hollow pyramidal tip [130, 131] also called hollow-pyramid cantilevered (HPC) [130, 131] probes. In such probes the pyramid is coated with a metal, usually aluminum or gold, leaving a nanohole at its apex which acts like a scattering centre. The HPC probe is a very promising candidate for the SNOM compared with an optical fiber based probe, in terms of both fabrication and optical performance. In fact these cantilevered tips offer several advantages [132] over tapered fibers, namely:

- Small apertures (size of 40–150 nm depending on the process conditions during fabrication) together with a large taper angle (about 70.5°), ensuring high optical throughput [133, 134].
- The lower absorption allows one to couple higher average power before reaching the onset of thermal damage of the metal coating.
- Silicon dioxide as the tip material guarantees excellent mechanical stability for long-life performance,
- Tip–sample distance stabilization methods used in AFM, such as the tapping mode, can be employed, thus ensuring longer probe lifetimes, as can the shear-force method [135], typically used in fiber-based SNOM setups, resulting in increased control capability and, hence, preserving the tip lifetime.
- Their integration in a conventional SNOM setup is straightforward.

A light beam can be coupled to the subwavelength aperture through the back of the pyramid by means of a microscope objective, as sketched in Fig. 3.24. As a consequence hollow cantilevers are free from the birefringence phenomena usually occurring in the propagation through an optical fiber. Recently, it was demonstrated that the pyramidal geometry does not affect the duration of short laser pulses [132] and preserves a well-defined polarization state [131,136] because of the short optical path inside the tip (the tip length is typically about 12 μm).

Since the production of individual probes is tedious and not always easily re-producible in different laboratories, it would be much more desirable to fabricate standardized probes in large batches, e. g., using established silicon micromachining techniques. This would yield large numbers of probes with the same properties, like aperture size and shape and thus also transmission, obtained by means of standard atomic force microscope cantilever technology. A schematic view of the SNOM probe and the fabrication process are shown in Fig. 3.25.

The probe is based on a Si cantilever with an integrated SiO_2 tip, which is formed by thermal oxidation in wet oxygen at 950 °C for 10 h. A small aperture at the apex of the tip is created by selectively etching the SiO_2 using a thin Cr film as a protective mask. More details of the fabrication process can be found in [137]. Briefly, Si (100) orientation, n-type, resistivity of 10 Ω · cm, 200-mm thick was used as a starting material (step a). After several steps of oxidation, photolithography, and SiO_2 patterning, the structure of the cantilever and a pyramidal etch pit of $15 \times 15 \ \mu m^2$ are formed by anisotropic etching of the Si in a tetramethylammonium hydroxide solution (step b). Next, the wafer is thermally oxidized in wet oxygen to about 1-μm thickness. Owing to the geometric structure of the etch pit and the

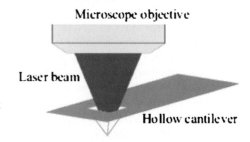

Fig. 3.24. Coupling of laser light to the back aperture of a hollow cantilever by means of a microscope objective

Fig. 3.25. The micromachined hollow cantilevered probe fabrication process. (Reprinted from [138] copyright 2001, with permission from Blackwell Publishing)

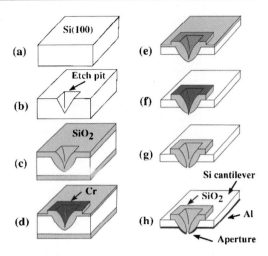

compressive stress generated during oxidizing, the thickness of the SiO_2 at the apex is very much less (about $30 \pm 40\%$) than that at the sidewall of the tip as shown in step c. After lithography, the SiO_2 of 0.6-mm thickness is partly etched and a Cr film approximately 80 ± 120-nm thick is consequently deposited on the upper side of the tip area for protection of the SiO_2 tip during etching of the wafer from the lower side (step d). Next, the wafer is etched from the lower side in the tetramethylammonium hydroxide until the Si cantilever with the SiO_2 tip at the end is released (step e). The wafer is subsequently dipped into a buffered HF for selective etching of the SiO_2 from the lower side until a Cr protrusion appears (step f). After etching-out of the Cr film, a very small aperture at the apex of the tip is created (step g). Finally, an aluminum film about 60 ± 120-nm thick is entirely deposited onto the lower side to form an opaque layer (step h). After the opaque layer has been coated, the size of the aperture is somewhat decreased. Since the SiO_2 tip is formed on the V-groove etch pit on the (100) Si surface, the geometry of the tip is always very stable and the shape of the aperture is always very round and well defined (Fig. 3.26).

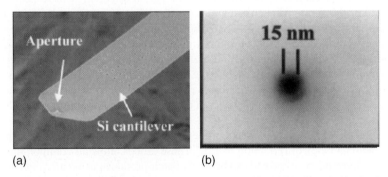

Fig. 3.26. SEM images of the hollow-pyramid cantilevered probe. **a** Cantilever body. On the apex the tip that was grown is visible. **b** The tip aperture. (Reprinted from [138] copyright 2001, with permission from Blackwell Publishing)

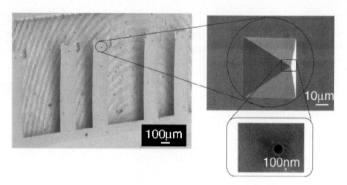

Fig. 3.27. SEM image of an apertured cantilever drilled by a FIB. (Reprinted from [139] copyright 2004, with permission from the American Institute of Physics)

Another way to produce the small apertures, which determine the spatial resolution of the SNOM pyramidal tips, consists of drilling the tip end by a FIB able to produce a grain-free nanoaperture with a perfect circular shape that avoids asymmetric absorption responsible of the dichroic behavior of SNOM optical fibers [138]. This makes hollow cantilevers excellent candidates for polarization-modulation imaging. The aperture size is chosen by changing the ion beam current during the milling process. An example of such an apertured cantilever is shown in Fig. 3.27. Both processes, selective chemical etching and FIB, yield almost perfectly circular apertures.

3.3.1.3
Others Microfabricated Probes

A new type of probe is derived from the previously seen HPC sensor and consists of integrating the microfabricated probe in an optical fiber [140]. In contrast to the previous process, the procedure for mounting microfabricated tips on fiber probes demands an advanced micromanipulator setup. For this purpose a cleaved fiber (125-μm diameter) is attached to an xy stage via a mechanical holder. Thus, the end of the cantilever can easily be adjusted in the lateral direction with respect to the backside of the aperture tip. An aperture cantilever probe is mounted on a z stage to additionally control the longitudinal distance with respect to the cleaved fiber. To facilitate the adjustment process, light is guided through the fiber and illuminates the backside of the tip. When the light transmitted through the aperture tip reaches a maximum value, the tip is fixed onto the fiber. To achieve this, UV light was used to harden a very thin UV-sensitive adhesive layer which was previously deposited on the fiber. One of these probes is shown in the scanning electron microscopy (SEM) image in Fig. 3.28.

Another kind of microfabricated probe consists of tips made of solid amorphous quartz with an index of refraction of 1.51. These probes allow all common imaging modes of force microscopy and can be easily batch-produced using micromachining technology [141]. The fabrication process presented by Shürmann et al. [141] has been reconsidered and optimized in view of reproducibility, efficiency, and cost. The probe tips are still made of SiO_2, 12 μm in height, and are integrated at the

Fig. 3.28. SEM image of a novel
SNOM fiber probe consisting
of a micromachined scanning
force microscopy–SNOM aperture
cantilever mounted on a cleaved
single-mode fiber probe. (Reprinted
from [140] copyright 1999, with
permission from Blackwell Publishing)

end of a silicon cantilever. A 60-nm-thick layer of aluminum completely covers the
tip and the cantilever which appears to be closed at the apex when investigated by
transmission electron microscopy.

The resulting reduction of the cut-off radius for nondissipative light propagation
leads to higher transmission. In Fig. 3.29 an example of such a rectangular-shaped
silicon cantilever with a solid quartz tip coated with aluminum is shown.

A number of cantilever designs was realized for dynamic and contact mode
imaging, with calculated spring constants ranging from 0.5 to 7 N/m for rectangular-
shaped cantilevers of various lengths (Fig. 3.30).

As the cantilevers exhibit a very high stiffness (the spring constants deduced from
the geometry of the levers were in the range from 1 to 7 N/m) no snap-in events
could be observed upon approaching the sample and constant height mode scans
could be performed. Rectangular windows etched into the levers at the position of
the tips were used for light coupling. The mechanism of light coupling through the
metallized tip and the light confinement at the tip apex is not yet clearly understood.
The finite conductivity of aluminum may be responsible for the far-field transmission

Fig. 3.29. a SEM image of a microfabricated optical near-field probe with
a rectangular-shaped silicon cantilever.
The tip is made of amorphous quartz.
b Transmission electron microscopy
image of an unused aluminum-coated
quartz tip. A 60-nm-thick aluminum
film can be clearly distinguished from
the quartz core. The film is also covering the apex of the tip. (Reprinted
from [142] copyright 2001, with permission from Blackwell Publishing)

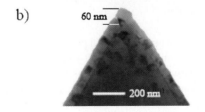

Fig. 3.30. SNOM probes with two different silicon beam designs having various force constants. (Reprinted from [143] copyright 2003, with permission from Blackwell Publishing)

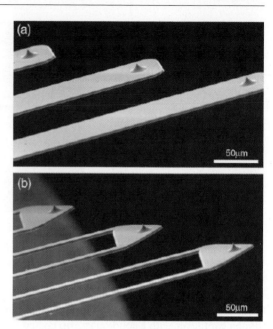

through the thin metal layer. However, a different effect must be responsible for the light confinement necessary for the subdiffraction limited resolution. A potential explanation relies on modes propagating along the air–metal interface in combination with field localization at the tip apex.

The dependence of the light transmission of the cone opening angle of the tips (33°, 55°, and 97°, respectively) has been investigated by experiment [143]. The result is that a smaller tip angle leads to a higher transmittance. This is in disagreement with some earlier calculations [144]. There, the authors studied the influence of cone angle on throughput and spot size of a probe based on a spherical, transparent calotte which was completely coated with metal. However, in that case, the film thickness at the apex was less than the optical skin depth. Pohl et al. [144] found that the transmitted power grows with increasing taper angle.

One possible explanation for this discrepancy may be the evaporation process, which possibly causes less metal deposition on sharper tips. However, in all cases, the minimal thickness of the final coating was much more than the skin depth of the evaporated metal.

3.3.1.4
Polarized Probes

A well-defined and controlled polarization in near field is important for applications such as polarimetry SNOM [145], investigation of magneto-optical effects [146], optical nanowriting on photosensitive polymers [147], and polarization-modulation SNOM, where optical anisotropy associated with material structures and/or strain is measured [148–150]. Polarization employed as a contrast mechanism for SNOM imaging is also useful to distinguish between genuine optical effects and those arising from topographic artifacts [151]. Specific works were devoted to get either linear or circular polarized near-field light coming out from SNOM aperture probes [157]. It should be noted that, since knowledge of the polarization is very important for the interpretation of SNOM images [152], great efforts have been made to elucidate this point by developing two-dimensional [153] and three-dimensional [154] models. Anyway, the polarization behavior and control in SNOM aperture probes is strictly dependent on the kind of the tip and on the aperture shape. In fact, the hollow cantilever is more feasible for obtaining a defined polarization. [131]. Oesterschulze et al. [131] investigated the far-field polarization properties of the hollow cantilever with a rectangular aperture, by measuring the optical transmissivity as a function of the polarization state. The aperture tips were illuminated from the backside by a plane-polarized laser beam and the polarization state of the transmitted light was characterized by an analyzer. As regards the rectangular aperture, it turned out that the polarization state of the incoming light probe is preserved by the probe for those directions parallel to one of the aperture axes (i.e., polarizer angles of 0° and 90°) (Fig. 3.31). Furthermore, an experiment was carried out by illuminating several aperture probes of different size with circular polarized light and measuring the transmitted power as a function of the analyzer angle. It turned out that most of the probes have a polarization contrast better than 0.75 independent of the aperture size. Similar results were found for cantilevered probes, consisting of solid quartz pyramidal and conical tips grown on silicon levers [142].

Recently polarization measurements were performed in near field [155] on hollow cantilevers with circular and elliptical apertures. To probe the in-plane component of the electric near field, a specially developed technique, employing samples in which a fluorescent molecule layer with a thickness comparable to that of the proximity zone is produced inside a host polymer film [149]. The molecular dipoles are oriented along the sample surface by mechanical drawing, producing a strongly anisotropic absorption [156]. By recording the fluorescence induced by the local electric near field for different orientations of the film dichroic axis, one can de-

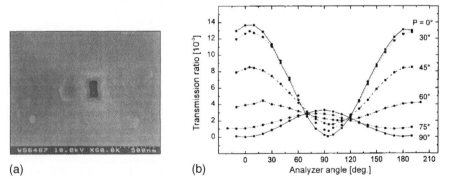

(a)   (b)

Fig. 3.31. a SEM image of an aperture probe with a rectangular shape (160 nm × 80 nm). **b** Transmission ratio between the aperture as a function of the analyzer angle for five different incoming polarization states. (Reprinted from [131] copyright 1998, with permission from Elsevier)

termine the in-plane component of the linear polarization at the aperture. The thin dichroic fluorescent film acts therefore as an effective near field polarization analyzer. Unexpectedly, it was found that the input polarization is always maintained in the near field, independent of the aperture geometry, in spite of the behavior of the transmitted far field, which may be either isotropic or strongly dichroic depending on the ellipticity of the aperture (Figs. 3.32, 3.33).

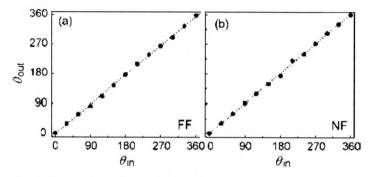

Fig. 3.32. Far-field (*a*) and near-field (*b*) polarization direction θ_{out} as a function of incident polarization direction θ_{in} for an isotropic tip. $\theta = 0°$ refers to one of the pyramid base sides. (Reprinted from [155] copyright 2005, with permission from the American Institute of Physics)

Difficulties to achieve a defined near-field polarization are mainly encountered with optical fiber probes. They are due to fiber birefringence and to grain imperfections on the metal coating of the taper zone. At the fiber apex the polarization is *not linear* because small tensions or torsions caused by even a slight fiber bending provoke strain-induced birefringence. Apart from these effects, external perturbations such as changes in temperature or position will change the polarization of the output beam as a function of time. Uncoated tapered fibers can overcome birefringence problems [157]; however, this is to the detriment of the spatial resolution. The metal-coating process usually is not able to produce a grain-free nanoaper-

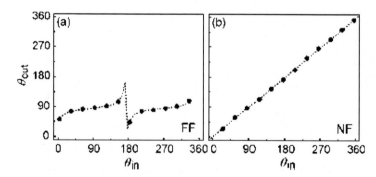

Fig. 3.33. Far-field (*a*) and near-field (*b*) polarization direction θ_{out} as a function of incident polarization direction θ_{in} for an isotropic tip. $\theta = 0°$ refers to the long axis of the ellipse. (Reprinted from [155] copyright 2005, with permission from the American Institute of Physics)

ture with a perfect circular shape. These defects induce highly asymmetric polarization behavior of the near-field light coming out from the probe [158, 159]. A possible solution to such a drawback consists of using a FIB technique. In practice it is applied to "clean" off the grainy coating metal at the end of the tip [158, 160]. This technique produces a smooth and very flat-end face with a well-defined circularly symmetric aperture down to 20 nm that improves the polarization behavior of the tip, resulting in polarization ratios exceeding 1:100 in all directions [158]. These results are satisfactory but the FIB technique is rather expensive and complex. An alternative approach to control the polarization is to use a so-called polarization-preserving optical fiber probes. For instance, Mitsui and Sekiguchi [161, 162] realized and employed a polarization maintaining and absorption reduction fiber.

In Fig. 3.34, a cross section is shown of the single-mode fiber, including its core, cladding, and stress applicators. These parts are made from a Ge-doped silica, pure silica, and B_2O_3-doped silica, respectively. In this type of polarization-preserving fiber, "bow-tie" stress applicators apply tensile stress to the core, thus causing a difference in the refractive indices between the linearly polarized components that are parallel to the fast and slow axes. The linear polarization of those two modes is maintained while they propagate in the fiber, even when the fiber is bent or twisted (Fig. 3.35). The extinction ratio of polarization in this fiber is -32 dB. On this SNOM probe, a detailed study was performed in order to evaluate the influence of probe fabrication processes, such as melting–pulling and bending [162], on polarization properties (Fig. 3.35). It turned out that the extinction ratio always decreases in the linearly polarized mode parallel to the slow axis, and that the decrease is independent of the direction in which the probe is bent. Moreover, it was established that, probably, the cause of the extinction ratio degradation is the isotropic compression stress occurring when the fiber is squeezed for the tapering process; such a stress cancels out the tensile stress applied to the core by stress applicators.

Drew's group [128] instead developed a near-field aperture probe by using an elliptical core fiber. For the tapering process they adopted a chemical etching tech-

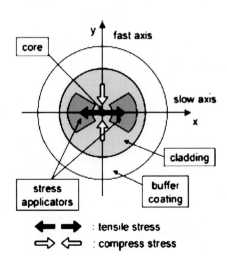

Fig. 3.34. Cross section of a polarization-preserving single-mode fiber. Bow-tie stress applicators apply tensile stress to the core. (Reprinted from [162] copyright 2005, with permission from the American Institute of Physics)

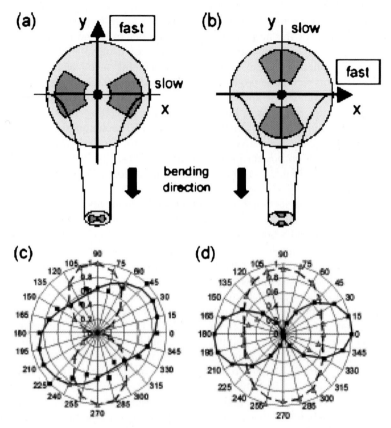

Fig. 3.35. Transmission coefficient of the polarization-preserving fiber probe detected by turning the polarizer: **a, b** the fibers are bent along the fast and slow axes; **c, d** transmission coefficients in the cases of **a** and **b**, respectively. The *squares* and the *solid lines* indicate the result when the linearly polarized light enters parallel to the *x*-direction; the *triangles* and the *dotted lines* show the result for the *y*-direction. Both lines are fitted with the sum of the quadratic cosine function of the polarization angle and a constant. (Reprinted from [162] copyright 2005, with permission from the American Institute of Physics)

nique. For aperture formation the tip was coated with aluminum and refined with a FIB. The probe so processed has an elliptical aperture at its end, thus yielding high throughput and excellent polarization maintenance, with an extinction ratio of 100:1 [128] (Fig. 3.36). In particular, the throughput depends on the incident polarization (Fig 3.36). For polarization parallel to the minor axis, the tip has an insertion loss of only 20 dB for aperture widths of 55 nm (Fig. 3.36). In essence, the probe acts like a polarizer with the polarization axis along the minor axis for very small apertures.

Such results are basically explained recalling that the input light splits into two components: one along the major axis of the elliptical core and the other along the minor axis. These components travel at different velocities, so the output of a fiber with a cleaved end is generally elliptically polarized. However, the aluminum-coated

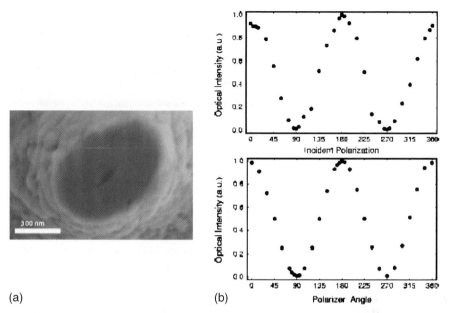

(a)   (b)

Fig. 3.36. a An SEM image of the tip after FIB milling. The aperture has dimensions of about 55 ± 5 nm by 225 ± 5 nm. **b** Output intensity of the probe shown in **a** as a function of input polarization (*top*). The polarization extinction of the same probe for a fixed input polarization (*bottom*). (Reprinted from [128] copyright 2006, with permission from the Optical Society of America)

tapered ends of the etched fiber probes, having subwavelength elliptical apertures, attenuate the component with polarization parallel to the major axis much more than the component parallel to the minor axis because of the boundary conditions imposed by the metal cladding [163]. Recently, a new optical fiber probe design has been proposed whose taper zone is able to select a well-defined polarized near field that does not depend from the polarization state of the light coupled into the fiber. The optical polarized nanosource basically consists of two separated metallic strips of a double C shape followed by a full metal coating at the very end of the dielectric taper zone (Fig. 3.37) [75]. The special geometry of the taper zone is able to maintain the quasi transverse electromagnetic (TEM) mode oscillating perpendicularly to the two metal slabs (the polarization is dictated by the geometry of the taper zone). This is possible because the taper zone of the probe, from an electromagnetic point of view, is a two-conductor structure, able to support the quasi-TEM mode with no cutoff frequencies. Rigorously, the mode would be TEM in presence of perfect conductors. In practice, when conductor conductivity is finite, a dominant hybrid mode exists whose behavior is similar to that of a TEM mode, specifically it does not exhibit a cutoff. Secondary, the geometry of the double connected structure (i.e., the two metallic strips filled with a dielectric material) resembles a bow-tie geometry able to maintain the modes oscillating normally to the strips. Furthermore, pretty close to the apex the two strips are short-cut so that the polarization state is maintained and the near field to far field ratio is increased (Fig. 3.37).

Fig. 3.37. The short-cut double C probe

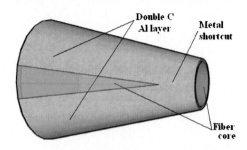

As reported in Sect. 3.3.1.1 (Fig. 3.16) by simulating a double C structure without a short cut at the end, it turned out that the light coming out from the nanoaperture is polarized, but the near field to far field ratio is very low: most of the amplitude of the electric field spectrum lies in the far-field components. So, the structure overcomes the limitation of the common optical fiber probe. In this last case the tapered metal-coated dielectric waveguide corresponds to a one-conductor circular waveguide structure and associated modes experience the cutoff phenomenon [3].

The properties and the performance of such a probe have been simulated by commercial FDTD-based software, showing that the near field coming out from the probe is linearly polarized. In particular, in Fig. 3.38 the angular spectrum of the electric field (amplitude) computed in near field on a plane at $z = 25$ nm (one step in the FDTD mesh grid) from the probe aperture is reported. The probe is fed by a impinging mode polarized first along y and then along x. The output fields were calculated in both polarization directions, y and x (Fig. 3.38). It was observed that on feeding the structure along the y-axis and along the x-axis of the structure, the spectra are pretty similar but one is attenuated with respect to the other. In this case, independent of the polarization direction of the feeding mode, the output spectra have the same distribution of frequencies along y and x, respectively. In particular, the spectra in Fig. 3.38a and b have an isotropic distribution of components, similarly to the standard probe [75]. The spectra calculated in a direction orthogonal to the double C structure (i.e., x) have much lower amplitude and a narrow distribution concentrated around the propagating zone. The ability of the probe to produce polarized near-field light has been confirmed by polarization test measurements carried out on a prototype (Fig. 3.39). The prototype was fabricated by tapering the fiber apex by the tube etching method [113]. Successively, the double C design was realized by thermally evaporating two aluminum layers of 50 nm on the two opposite faces of the apex. A 25-nm layer of LiF was deposited uniformly all around the taper region in order to insulate the double C layer from the successive chromium and aluminum coating (50 nm). For the testing unpolarized light emitted by an LED was coupled to the fiber (Fig. 3.39a). A focusing objective was placed in front of the SNOM probe aperture followed by a rotating holder carrying a linear polarizer. The LED current modulated by the reference signal coming from a digital lock-in amplifier and the light emitted by the probe were detected in phase by a photomultiplier. The results of the optical intensity versus the polarizer angle are reported in the polar diagram shown in Fig. 3.39b. The distribution of the light with the polarizer angle indicates that the light out coming from the SNOM is almost

Fig. 3.38. Spectrum of electric field, on the xy-plane at 25 nm from the probe aperture, for the double C structure of Fig. 3.37. In the *top row* the calculated E_y component is shown and in the *bottom row* the calculated E_x component is shown. In the *left column* and in *right column*, the double C probe is fed with E_y and E_x, respectively. The radiative region of the electric field is between -1 and 1 and the remaining part accounts for the evanescent region of the spectrum [75]

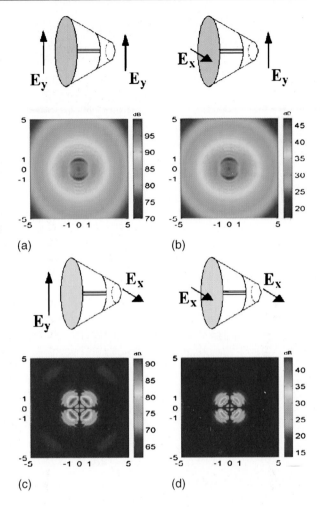

(a)   (b)

(c)   (d)

linearly polarized with an extinction ratio of 6 dB. Such a result shows that the prototype probe is able to produce near-field light with a well-defined polarization state as predicted by the FDTD model. Obviously, better near-field polarization ratios are expected when a polarized source is coupled to the fiber. This is the common situation in a SNOM experiment where usually a polarized laser beam is coupled to the probe. A similar probe design, but more resembling a bow-tie antenna, has just been theorized [73].

The proposed probe consists of a batch-fabricated hollow pyramidal silicon dioxide tip that is partly coated with aluminum to accomplish the tapered dipole antenna. Theoretical calculations of the field distribution were performed to investigate its optical properties. The results, compared with those of conventional aperture tips based on the same silicon dioxide tip configuration, revealed unique properties with respect to a high transmission efficiency [73].

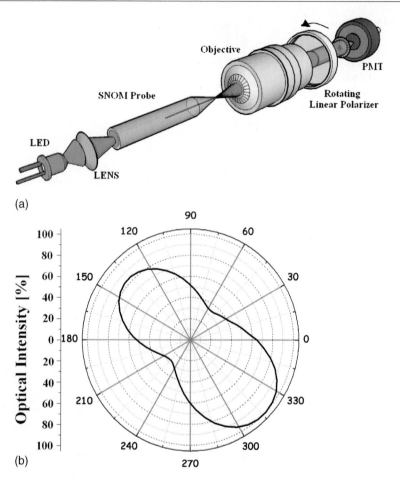

(a)

(b)

Fig. 3.39. a Setup used for far-field polarization measurement. **b** Polar diagram of the SNOM probe prototype. The shape of the diagram indicates that the light coming out from the probe is linearly polarized with an extinction ratio of 6 dB [75]. *PMT* photomultiplier tube

3.3.2
The Apertureless Probe: Optical Nanoantennas

As briefly reported in the Sect. 3.1, invoking Babinet's principle, the near field generated by illuminating a scattering particle of dimension smaller than the radiation wavelength exhibits the same property as that radiated by a subwavelength aperture. This is the basic mechanism of the apertureless probes. They are mainly represented by metallic scanning tunneling microscope, dielectric, or semiconductor atomic force microscope tips (Fig. 3.40). Moreover specially designed probes with a single nanostructure at their apex acting as a nanometer-sized light source or antenna were produced [34, 35]. With respect to the aperture ones, such probes offer higher resolution (below 10 nm) and the ability to confine an unlimited quantity of electromagnetic radiation to the nanometer apex region. However, they have a main drawback related to strong background light derived from the diffraction-limited illumination. Indeed the background power represents a noise for near-field detection, and can cause sample damage (i.e., photodegradation of biomolecules, photobleaching).

The dielectric tips are usually atomic force microscope cantilevers that are commercially available today. They are fabricated by processing silicon nitride (Si_3N_4) by means of wet chemical etching and micromachining procedures. Such probes can have either a conical or a pyramidal shape with a large vertex angle (Fig. 3.40), and can be of standard or sharpened type. In the latter case, the pyramid ends with a small 200-nm height tip having a radius of curvature of about tens of nanometers. This feature, combined with a low aspect ratio, may degrade the lateral resolution and the ability to image abrupt features. Silicon semiconducting tips, instead, can be manufactured to have a conical shape with radii of curvature of less than 10 nm and aspect ratios as high as 1:5 [164]. The force constants of the respective cantilevers (0.2–20 N/m) are suitable for AFM tapping mode operation [165], where the tip oscillates up and down above the surface, and both materials have been successfully used for apertureless SNOM imaging [164].

For use in an apertureless SNOM, however, these tips can be modulated somewhat. For example, if a surface-enhanced Raman scattering effect is needed, atomic force microscope tips can be coated with silver or gold by simple plasma or thermal deposition onto the silicon tip (the grain size of the droplets will be about 40 nm) [166]. Other metal-coated tips have been produced using nickel, platinum, chromium, titanium, and cobalt [167]. A precise thickness with a minimal grain size was realized by ion-beam sputtering. The radius of curvature of the probe so obtained ranges between 10 and 20 nm. Regarding the metallic tips, they can be easily fabricated by using the assessed procedure for scanning tunneling microscope tips. It is an electrochemical etching process that starts from thin metal wires with diameters in the 50–300 μm range and can produce tungsten [168,169], silver [170], gold [171], or platinum [172] tips. In electrochemical etching, a metal wire is dipped into the etching solution and a voltage is applied between the wire and a counter-electrode immersed in the solution. The surface tension of the solution results in a meniscus being formed around the wire. Etching proceeds most rapidly at the meniscus. The technique is distinguished as DC or AC depending on whether the

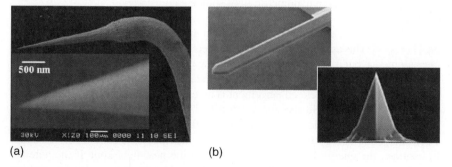

(a)   (b)

Fig. 3.40. SEM images of a bent tungsten tip (**a**) and of an atomic force microscope silicon cantilever (**b**). (Reprinted with permission from Nanosensors TM)

applied voltage is constant or periodic. In the first case, the tapering action will naturally stop with the drop off of the lower part of the wire. It is a slow process but it allows the sharpest tips to be obtained. However, it was observed that for some materials such as gold and silver, DC etching produces relatively rough tip surfaces. In the AC attack the tip formation will occur when the current decrease below approximately two thirds of the initial value. In this case, it is crucial to switch off the etching voltage at the precise current value. Many process factors influence the probe quality. The concentration of the etching solution, the applied voltage, the waveform (for AC mode), the depth and the shape of the counter electrode, and length of wire are much more important factors. These vary from setup to setup and have to be determined empirically. With a good set of parameters, tip diameters of less than 20 nm can be achieved with a yield of 50%. Also for apertureless probes FIB lithography represents a useful additional technique. Its sculpting and milling action improve the sharpness of the etched probes, with a consequent improvement of the performance [173]. This technique has been used more recently to sharpen silver apertureless SNOM tips [174] produced by electrochemical etching, decreasing the radius of curvature to 10–15 nm.

It is important to stress that the apertureless probe has two different roles: local scatterer and local exciter. In the former case, the pointed probe has to locally perturb the electromagnetic fields at a sample surface; the response to this perturbation is detected in the far field at the same frequency of the incident light. This scheme relies on the fact that the field near any laser-irradiated object is made of propagating waves and also of evanescent waves that remain confined to the object's surface. The evanescent waves can be partially converted to propagating radiation by using a pointed probe as a local scatterer. Scattering-type measurements typically have a high signal-to-noise ratio, but the interpretation of the optical information acquired is difficult to achieve owing to the topographic artifacts [175]. The problem is related to the fact that the scattering efficiency depends sensitively on the separation between the probe and the sample, making the differentiation of true optical contrast (related to the local dielectric constant) tricky.

The apertureless probe can act as a local light source as well. This is accomplished under the appropriate polarization and excitation conditions using suitable probe geometries and probe materials [176, 177]. Under these conditions, the field close to the tip apex can become strongly enhanced over the field of the illuminating laser light, thus establishing a nanometric light source.

Instead of using a tip to locally scatter a sample's near field, one can use the tip to provide a local-excitation source for a spectroscopic response of the sample [178,179]. This approach enables simultaneous spectral and subdiffraction spatial measurements, but it depends sensitively on the magnitude of the field-enhancement factor. In principle, the two schemes, local excitation and local scattering, are not independent. A strong field enhancement is an essential condition for efficient scattering and vice versa. So the above-described behavior of an apertureless probe resembles that of a classic antenna. In fact, at radiofrequency, considering a receiver, electromagnetic energy has to be canalized to the near-field zone of the antenna. In contrast, the energy has to be released from the near-field zone if the antenna operates as a sender. The correspondence of the field enhancement for the antenna is immediate. It is intrinsic to the working principle: an antenna concentrates electromagnetic energy into a tight space so creating a zone of high energy density. In the context of near-field optics one expects to make use of this property to generate a highly confined light source. An important point for an antenna is that to get efficient coupling between the near-field and the far-field an impedance matching is needed. In other words, the antenna locally modifies the density of electromagnetic states, thereby increasing the modes into which the source can radiate and vice versa. The most efficient antenna designs implemented at optical frequencies are the lambda-half antenna [180] and the bow-tie antenna [181–183]. However, by making use of electromagnetic resonances associated with surface plasmons, any metal nanostructure can be viewed as an optical antenna. Of course, the efficiency depends on the material composition and the geometry of the nanostructure.

It is useful to go deeper to understand the origin of the tip enhancement and, hence, the ability to produce highly confined electromagnetic fields (i.e., nanoscale light sources). To this aim we can analyze the electromagnetic field distribution of one of the most widely used forms of antenna: a sharp metal tip [6, 184, 185]. In general, the enhanced electromagnetic field originates from a combination of (1) an electrostatic lightning-rod effect, which is a direct result of the geometric singularity of sharply pointed metal structures, and (2) surface plasmon resonances, which depend on both the excitation wavelength and the tip geometry [186, 187].

The confined electric field is readily produced if the electric field vector of the incident light field is polarized along the direction of the tip shaft. This light field periodically drives the free electrons on the surface of the metal along the direction of polarization i.e., along the longitudinal axis of the tip. The resultant enhanced electric field is confined to a volume on the order of $(20\,\mathrm{nm})^3$, with the confinement volume dependent on the end diameter of the metal tip (Fig. 3.41). If the incident electric field vector is not polarized along the direction of the tip shaft, the field enhancement effect is significantly weaker (Fig. 3.41) [184]. In the case of Raman scattering induced by nanoscale metal structures [188, 189] the strongest contribution to surface-enhanced Raman scattering is electromagnetic. The observed enhancement is caused by the enhancement of the incident field E_i at ω_i and the scattered field E_s at $\omega_i - \omega_v$ (ω_v is the vibrational frequency). The enhancement factor M is then expressed as the product with the total electric field E_t:

$$M = \left(\frac{E_t(\omega_i)}{E_i(\omega_i)}\right)^2 \left(\frac{E_t(\omega_i - \omega_v)}{E_i(\omega_i - \omega_v)}\right)^2 \approx \left(\frac{E_t(\omega_i)}{E_i(\omega_i)}\right)^4 , \qquad (3.5)$$

where we have taken $\omega_v \ll \omega_i$.

It has been determined that the local field near a laser-irradiated tip can be approximated by the electric field of an effective dipole $p(\omega)$ located at the center of the tip apex. The magnitude of the effective dipole can be written as [190]:

$$p(\omega) = \begin{bmatrix} \alpha_\perp & 0 & 0 \\ 0 & \alpha_\perp & 0 \\ 0 & 0 & \alpha_\| \end{bmatrix} \cdot E_o(\omega) , \qquad (3.6)$$

where we chose the z-axis to coincide with the tip axis and E_0 is the exciting electric field in the absence of the tip. The transverse polarizability $\alpha_\perp$ is identical to the

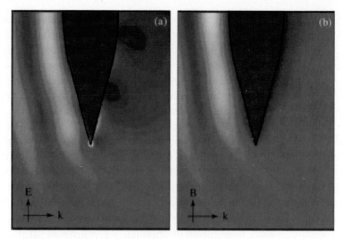

Fig. 3.41. Calculated intensity distribution near a laser-irradiated gold tip. The exciting wave is incident from the left and forms a standing wave pattern as it interferes with the reflected field from the tip shaft, and at the end of the tip, the wave diffracts. Two different excitation polarizations are used. The field-enhancement effect is observed only if the incident wave is polarized along the tip axis (**a**). In the case of an incident wave polarized perpendicular to the tip axis, the field near the tip is attenuated (**b**). (Reprinted from [36] copyright 2006, with permission from the Annual Review of Physical Chemistry)

quasistatic polarizability of a small sphere, whereas the longitudinal polarizability $\alpha_{\parallel}$ is given by $\alpha_{\parallel}(\omega) = 8\pi\varepsilon_0 r_0^3 f_e(\omega)$, where r_0 is the tip radius and f_e the complex field enhancement factor. Thus, the enhancement at the tip generates locally a light source which can be accurately represented by a single on-axis oscillating dipole [190,191]. The dipole strength is a direct measure for the field enhancement factor.

Transposing the antenna concept from radiofrequency to the optical region, the first requirement to remember is that the size of optical antenna is in the range of the detected wavelength and fabrication techniques with nanoscale spatial resolution are required (e. g., electron-beam lithography).

The efficiency of an antenna depends critically on its resonances, as dictated by the operation frequency and by the shape, material, and dimensions of the antenna (to get efficient coupling between the near-field and the far-field by use of impedance matching). Optical antennas have proved and potential advantages in the detection of light, showing polarization dependence, tunability, and rapid time response. They also can be considered as point detectors and directionally sensitive elements. So far, these detectors have been thoroughly tested in the mid-IR region with some positive results in the visible. Different designs have been proposed and tested [192–194].

In Fig. 3.42a, a SEM image of a dipole optical antenna designed to resonate in the IR region is shown [192]. The dipole is represented by the horizontal structure having a total length of 6.7 μm. The transductor is located in the junction appearing in the center of the dipole. This case corresponds to a metal oxide metal diode. The V-shaped metallic structures are the connection lines used to extract the rectified signal from the antenna. The metallic structure was fabricated on a Si/SiO$_2$ wafer.

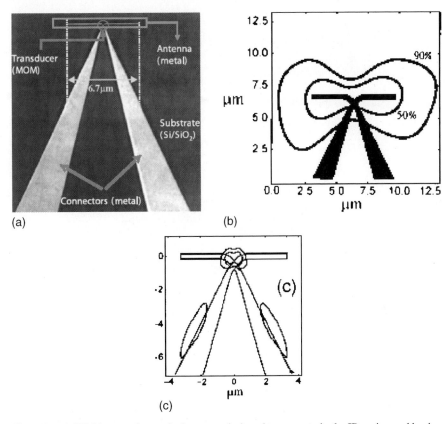

Fig. 3.42. a A SEM image of an optical antenna designed to resonate in the IR region and having a dipole antenna design. Spatial responses of a dipole optical antenna in the IR region (**b**) and in the visible region (**c**) have been measured. *MOM* metal oxide metal. (**a** Reprinted from [195] copyright 2005, with permission from Institute of Physics Publishing; **b** reprinted from [193] copyright 1999 and **c** reprinted from [196] copyright 1999, with permission from the Optical Society of America)

The light is incident perpendicularly to the plane represented in the image. It can come from the air or from the substrate [193]. By calculation, it turned out that such an optical antenna probe has polarization sensitivity, is tunable, and has directionality. It was found that the area containing about 90% of the spatial responsivity of the antenna is about the square of the detected wavelength, both in the IR and in the visible spectrum (Fig. 3.42b,c) [192, 193].

Moreover further important information has been obtained: the optical responsivity map. It requires the accurate measurement of the signal when the antenna scans a given light beam distribution (CnO_2 laser), and a postprocessing to implement the deconvolution (Fig. 3.43).

A special antenna design is the metallic bow-tie one. It is a special antenna since it has almost perfect impedance matching [197]. The probe consists of two opposing tip-to-tip metal triangles allowing the optical field to be further confined to

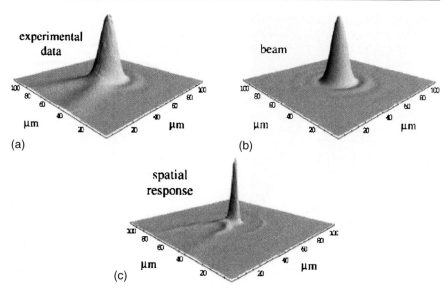

Fig. 3.43. The elements used for the characterization of the responsivity map of the optical antenna $[S(x, y) = R(x, y)I(x, y)]$. **a** Experimental measurement of the responsivity after scanning the detector under a CO_2 laser beam waist $R(x, y)$. **b** The laser beam waist used for deconvolution $I(x, y)$. **c** Spatial response of the antenna after deconvolution $S(x, y)$. (Reprinted from [193] copyright 1999, with permission from the Optical Society of America)

the nanometer scale. It acts as a coupled-dipole system where the gap determines the coupling strength and the field enhancement. For a given metal with certain plasmon resonance, the excitation wavelength and the incident polarization determine the effective resonance conditions [198, 199]. An approximately 10 nm optical spot is expected owing to the resonant surface plasmon excitation localized at the tips of the bow tie. In order to work as a SNOM probe, the bow tie needs to be fabricated on the sides of a sharp dielectric tip, for instance, an atomic force microscope probe. So far, most research has focused on optical antennas supporting geometrical resonances and fabricated on planar substrates, thus restricting their applications. A few examples of such kinds of antennas mounted on a probe apex are reported in the literature [194, 200–202]. Here we report a meaningful example: an aluminum bow-tie antenna [194,203]. The aluminum bow-tie antenna was fabricated at the apex of a pyramidal Si_3N_4 atomic force microscope cantilever tip. The tip was coated by evaporation with a homogeneous aluminum layer (40-nm thickness). A bow-tie-shaped metallic structure was sculptured from the film by means of FIB milling. This includes, in a final step, the cutting of a narrow gap between the two arms of the antenna right at the apex. The fabricated bow tie antenna is shown in Fig. 3.44. The inset in Fig. 3.44b indicates a gap width between the bow-tie arms not quite as small as it appears in the top view (Fig. 3.44a). The antenna is rounded at the top and slightly tilted with respect to the axis of the pyramid. Blaze angle, radius of curvature of the antenna arms at the feed gap, and feed gap width were determined (Fig. 3.44). The overall antenna length is 170 nm, which is in-between the length of the half wave dipole (i. e., about 120 nm owing to the average dielectric properties of

Fig. 3.44. Optical antenna at the apex of an atomic force microscope tip. **a** Top view, **b** side view. The dimensions are as follows: aluminum thickness, 40 nm; antenna overall length, 170 nm; blaze angle, 40°; feed gap width, approximately 50 nm; radius of curvature at the feed gap, 30 nm. (Reprinted from [194] copyright 2005, with permission from the American Physical Society)

the environment) and the first minimum of the antenna responsivity at about twice that length [204]. Such a probe was successfully tested for visible light on a single quantum dot [194].

A similar antenna design consisting of two opposing tip-to-tip gold nanotriangles has been theoretically and experimentally checked (Fig. 3.45). The enhancement of the electromagnetic fields has been verified.

Currents, field distributions, and scattering efficiencies in the antennas at optical wavelengths have been obtained from FDTD simulations using realistic wavelength-dependent dielectric constants. The experimentally measured resonant wavelengths and intensity enhancements from individual bow-tie antennas turned out to be in excellent agreement with the FDTD simulations (Fig. 3.45) [203].

Another type of antenna probe, deriving from the integration of a metallic tip grown on the aperture of a fiber probe, was fabricated. It was denominated "tip on aperture" (Fig. 3.46) [205, 206].

The tip concentrates the light coming through the aperture, combining the advantages of aperture SNOM and apertureless SNOM. For this aim, the tip has to be

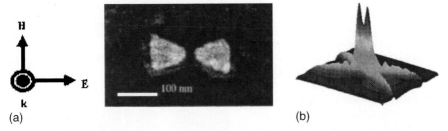

Fig. 3.45. a SEM image of two representative gold bow-ties on the fused silica–indium tin oxide substrate. The polarization of the incident beam ($\lambda = 56$ nm) and the coordinate system are shown. The gap width for the bow tie is 16 nm. **b** Three-dimensional surface plot of the finite-difference time domain calculated $|E|^2$ enhancement at the top surface of the antenna. The illumination is from the substrate side. (Reprinted from [203] copyright 2005, with permission from the American Physical Society)

Fig. 3.46. SEM images of a "tip on aperture". **a** View of the antenna and the fiber aperture (diameter of 70 nm). **b** The grown aluminum tip is 50 nm in width and 100 nm in length. (Reprinted from [206] copyright 2007, with permission from the American Physical Society)

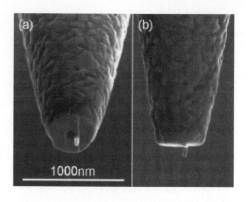

positioned and tailored to optimally concentrate at its apex the light fields provided by the aperture, fully exploiting plasmon and antenna resonances. The probe fabrication consists of two main steps: a conventional typical fiber tapering, and growing of the metal tip. The latter was accomplished by electron-beam deposition or using a dual-beam (FIB and scanning electron microscope) machine and then covered with metal [206].

Recently Taminiau et al. [206] fabricated a tip on aperture tunable to different resonances by controlling the tip length (Fig. 3.46b). The antenna response was simulated, giving insight into the antenna excitation conditions and resonance behavior and showing the equivalence of the antenna to the standard monopole antenna (Fig. 3.47). Experimental optical measurements were performed on fluorescent molecules with a resolution down to 25 nm.

The near field appears strongly localized within 5 nm in the z-direction, thus promising even higher resolution with sharper tips. Finally, the experiment was in agreement with the simulation in terms of resonances and field localization of probe-based optical monopole antennas. Moreover, clear polarization-dependent excitation conditions were observed [206].

Fig. 3.47. Monopole antennas and excitation configurations. **a** Cross section of the probe-based antenna configuration. **b** Conceptual description of a base-fed $\lambda/4$ monopole antenna and the associated characteristic current $|I|$ and field $|E|$ distributions. **c** Coaxially fed monopole. **d** Far-field illuminated monopole. The instantaneous field of the characteristic antenna response is shown and is equivalent for all the excitation schemes. (Reprinted from [206] copyright 2007, with permission from the American Physical Society)

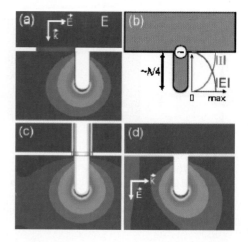

3.4
Applications and Perspectives

The superresolution makes the SNOM a very useful tool in nanotechnologies and especially it finds applications in the study of photonic components as well as in their nanofabrication. Moreover, owing to its properties, the SNOM is a well suitable for use in physics, chemistry, material and life sciences. Unfortunately near field means low-level signal, i. e., it is very hard to have a good signal-to-noise ratio from a standard near-field microscopy experiment and using a standard probe. It is sometimes unrealistic to set up a spectroscopy experiment. In the last few years, the experimental activities of near-filed optics have been considerably extended and became one of the guidelines that have stimulated the evolution of nanophotonics. Developments in near-field optical probe design and sometimes a smart optical setup are widening the range of application of the near-field technique. It has been applied in studies, on the nanometer scale, of the optical properties of nanophotonics devices, and, recently, it has been easy to find a number of examples where the SNOM probe has been used as a pencil in simple and fascinating nanolithography experiments [147,207]. Aperture probes are widely applicable to almost any kind of sample. Recently, the fruitful application of the "hollow cantilever" was carried out. These probes have found application in nanolithography experiments on PMA4 polymers and their ability to preserve a particular polarization state has been demonstrated [155]. Figure 3.48 shows the topography and the optical signal of a Fischer pattern sample obtained

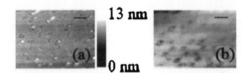

Fig. 3.48. Topographic (**a**) and optical (**b**) images of an aluminum pattern deposited on a glass cover slip collected in transmission mode using a hollow cantilever. *Scale bar* 500 nm. (Reprinted from [208] copyright 2006, with permission from the American Institute of Physics)

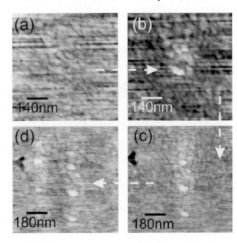

Fig. 3.49. Four frames showing the consecutive writing of five dots (protrusions) on the free surface of a PMA/PMA4 copolymer film. (Reprinted from [208] copyright 2006, with permission from the American Institute of Physics)

by a hollow cantilever. In this experiment the probe-to-sample distances were stabilized by means of a shear-force technique [208]. The method allows stability and
repeatability that are comparable to the qualities of AFM.

The technique was used for nanomachining of the PMA4 surface on which
a number of single dots were realized one after another and imaged step by step with
a topography map (Fig. 3.49).

In some cases the properties of special probes allow novel experiments and results otherwise not obtainable. The standard aperture technique is able to achieve
resolutions down to 50–100 nm. The use of a sharp apertureless tip (nanoantennas) to
locally perturb the fields at the sample has allowed higher spatial resolution in elastic
scattering [32, 209], Raman scattering [210–212], and fluorescence excitation [213].
Improving the resolution has opened a new view on the nanoworld that allows the
imaging of a single nanostructure or molecule [214]. Unfortunately, it has been
a challenge [215, 216] to image fluorescent molecules with an apertureless SNOM
owing to the limited number of photons available before photobleaching. Therefore,

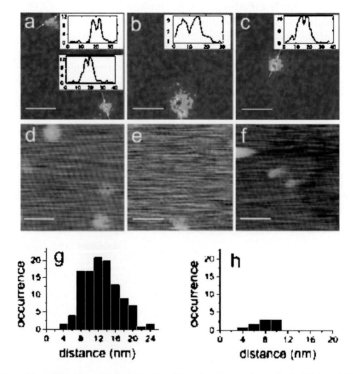

Fig. 3.50. a–c Near-field images of Cy3 pairs. The signal-to-noise ratio is 12.4 and 15.9 for **a**,
16.1 for **b**, and 20.4 for **c**. The *insets* show the profiles with a line cut through the image centers
(indicated by *arrows*), where the horizontal axis is in pixels (1 pixel = 1.95 nm) and the vertical
axis is the pixel signal. **d–f** Atomic force microscopy images corresponding to images **a–c**,
respectively. *Scale bars* 50 nm. **g** Histogram of distances between the resolved Cy_3 molecules.
h Histogram of distances between the two artifactual lobes of single Cy_3 molecules. (Reprinted
from [219] copyright 2006, with permission from the American Physical Society)

the use of the apertureless technique has been limited to inorganic samples that usually have a higher damage threshold. As reported in the previous section, properly designed nanoantennas enhance the power of the optical near field by several orders [217,218] or reduce nonradiative energy transfer, thus holding promise for imaging single molecules. Nevertheless the background signal in a nanoantenna-based experiment has to be reduced to avoid molecular photobleaching. This target was recently achieved by vertically oscillating the nanoantennas and phase-filtering the optical signal [219], and the method allowed the imaging of isolated Cy3 molecules (Fig. 3.50).

A different approach in using nanoantennas to improve the optical resolution consists of second-harmonic apertureless SNOM imaging. The technique allows one to achieve strongly confined sources of second-harmonic light at the probe tip apex and/or surface area under the tip [220]. The problem of the strong background coming out from the laser pump has been solved by filtering out the laser light and collecting the light at the second-harmonic wavelength. The improvement in the optical signal here is paid for in terms of instrumentation, because a pulsed laser focused on the apex of the nanoantennas is required [221]. Moreover, the second-harmonic intensity is determined by the fourth power of the local fundamental field; therefore, the second-harmonic apertureless SNOM is a much more sensitive tool for mapping the electromagnetic field distribution over surfaces than the linear apertureless SNOM techniques. In addition to the high-resolution surface imaging, the technique finds application in studying the localized electromagnetic excitation at surfaces, surface plasmon interaction, and their modifications, which are of importance in the context of surface-enhanced nonlinear phenomena. Finally, plasmon excitation has been used to improve the Raman signal in apertureless SNOM experiments [222] and the second harmonic generation technique may be used to excite plasmons in metal-coated nanoantennas at wavelengths not easy to handle such as UV or DUV 20.

References

1. Synge EH (1928) Philos Mag 6:356
2. Novotny L, Hecht B (2006) Principles of nano-optics. Cambridge University Press, Cambridge
3. Hecht B, Sick B, Wild UP, Deckert V, Zenobi R, OJF Martin, Pohl DW (2000) J Chem Phys 112:7761
4. Zenhausern F, O'Boyle MP, Wickramasinghe HK (1994) Appl Phys Lett 65:1623
5. Bachelot R, Gleyzes P, Boccara AC (1995) Opt Lett 20:1924
6. Inouye Y, Kawata S (1994) Opt Lett 19: 159
7. Abbé M (1873) Arch Mikroscop Anat Entwicklungsmech 9:413
8. Hermann B (1998) Fluorescence microscopy, 2nd edn. BIOS, Oxford
9. Schaefer LH, Schuster D, Herz H (2001) J Microsc 204:99
10. Shotton DM (1989) J Cell Sci 94: 175
11. Stephens DJ, Allan V J (2003) Science 300:82
12. Gustafson MGL, Agard DA, Sedat JW (1999) J Microsc 195:10
13. Wells WA (2004) J Cell Biol 164:337
14. Egner A, Verrier S, Goroshkov A, Soling HD, Hell SW (2004) J Struct Biol 147:70
15. Pohl DW, Denk W, Lanz M (1984) Appl Phys Lett 44:651
16. Lewis A, Isaacson M, Harootunian A, Muray A (1984) Ultramicroscopy 13:227

17. De Serio M, Zenobi R, Deckert V (2003) Trends Anal Chem 22:70
18. Durig U, Pohl DW, Rohner F (1986) J Appl Phys 59:3318
19. Toledo-Crow R, Yang P, Chen Y, Vaez-Iravani M (1992) Appl Phys Lett 60:2957
20. Betzig E, Finn P, Weiner S (1992) Appl Phys Lett 60:2484
21. Karrai K, Grober RD (1995) Ultramicroscopy 61:197
22. Okajima T, Hirotsu S (1997) Appl Phys Lett 71:545
23. Davy S, Spajer M, Coujon D (1998) Appl Phys Lett 73:2594
24. Gregor M, Blome P, Schofer J, Ulbrich R (1996) Appl Phys Lett 68:307
25. Smolyaninov II, Atia WA, Pilevar S, Davis CC (1998) Ultramicroscopy 71:177
26. Wei CC, Wei PK, Fann W (1995) Appl Phys Lett 67:3835
27. Durkan C, Shvets IV (1996) J Appl Phys 79:1219
28. Seidel J, Grafström S, Loppacher C, Trogisch S, Schlaphof F, Eng LM (2001) Appl Phys Lett 79:2291
29. Aigouy L, Andréani FX, Boccara AC, Rivoal JC, Porto JA, Carminati R, JJ Greffet JJ, Mégy R (2000) Appl Phys Lett 76:397
30. Courjon D, Sarayeddine K, Spajer M (1989) Opt Commun 71:23
31. Reddick RC, Warmack RJ, Ferrel TL (1989) Phys Rev B 39:767
32. Zenhausern F, Martin Y, Wickramasinghe H (1995) Science 269:1083
33. Betzig E, Trautman JK, Harris TD, Weiner JS, Kostelak RL (1991) Science 251:1468
34. Sqalli O, Bernal MP, Hoffmann P, Marquis-Weible F (2000) Appl Phys Lett 76:2134
35. Kalkbrenner T, Ramstein M, Mlynek J, Sandoghdar V (2001) J Microsc 202:72
36. Novotny L, Stranick SJ (2006) Annu Rev Phys Chem 57:303
37. Nakajima K, Micheletto R, Mitsui K, Isoshima T, Hara M, Wada T, Sasabe H, Knoll W (1999) Appl Surf Sci 144 520
38. Cefalì E, Patanè S, Gucciardi PG, Labardi M, Allegroni M (2003) J Microsc 210:262
39. Grober RD, Harris TD, Trautman JK, Betzig E (1994) Rev Sci Instrum 65:626
40. Durand Y, Woehl JC, Viellerobe B, Gohde W, Orrit M (1999) Rev Sci Instrum 70:1318
41. Gray MH, Hsu JWP (1999) Rev Sci Instrum 70:3355
42. Hess HF, Betzig E, Harris TD, Pfeiffer LN, West KW (1994) Science 264:1740
43. Emiliani V, Lienau C, Hauert M, Coli G, DeGiorgi M, Rinaldi R, Passaseo A, Cingolani R (1999) Phys Rev B 60:13335
44. Lomascolo M, Vergine A, Johal TK, Rinaldi R, Passaseo A, Cingolani R, Patanè S, Labardi M, Allegrini M, Troiani F, Molinari E (2002) Phys Rev B 66:041302
45. Michaelis J, Hettich C, Zayats A, Eiermann B, Mlynek J, Sandoghdar V (1999) Opt Lett 24:581
46. Baez A (1956) J Opt Soc Am 46:901
47. Ash EA, Nicholls G (1972) Nature 237:510
48. Massey GA (1984) Appl Opt 23:658
49. Fischer UC, Durig U, Pohl DW (1987) Scanning Microsc Suppl 1 47
50. Moers MHP, Tack RG, Noordman OFJ, Segerink FB, van Hulst NF, Bolger B (1993) In: Near field optics. NATO ASI series E, vol 242. NATO/Kluwer, Dordrecht
51. Baida F, Courjon D, Tribillon G (1993) In: Near field optics. NATO ASI series E, vol 242. NATO/Kluwer, Dordrecht
52. Veerman JA Otter AM, Kuipers L, van Hulst NF (1998) Appl Phys Lett 72:3115
53. Muranishi M, Sato K, Hosaka S, Kikukawa A, Shintani T, Ito K (1997) Jpn J Appl Phys 36:942
54. Bachelot R, Gleyzes P, Boccara AC (1997) Appl Opt 36:2160
55. Inouye Y, Kawata S (1997) Opt Commun 134:31
56. Jin RC, Chao YW, Mirkin CA, Kelly KL, Schatz GC, Zheng JG (2001) Science 294:1901
57. Mock JJ, Barbic M, Smith DR, Schultz DA, Schultz S (2002) J Chem Phys 116:6755
58. Gole A, Murphy CJ (2004) Chem Mater 16:3633

59. Danzebrink HU, Fischer UC (1993) Near field optics. NATO ASI series E, vol 242. NATO/Kluwer, Dordrecht
60. Gottlich H, Heckl WM (1995) Ultramicroscopy 61:145
61. Kopelman R, Tan W, Shi ZY, Birnbaun D (1993) Near field optics. NATO ASI series E, vol 242. NATO/Kluwer, Dordrecht
62. Fischer UC (1993) In: Near field optics. NATO ASI series E, vol 242. NATO/Kluwer, Dordrecht
63. Marcuse D (1989) Light transmission optics. Krieger, Malabar
64. Novotny L, Hafner C (1994) Phys Rev E 50:4094
65. Novotny L, Pohl DW, Hecht B (1995) Ultramicroscopy 61:1
66. Girard C (2005) Rep Prog Phys 68:1883
67. Bethe H (1944) Phys Rev 66:163
68. Bouwkamp C (1950) Philips Res Rep 5:321
69. Hallen HD, Jahncke CL (2003) J Raman Spectrosc 34:655
70. Hallen HD, Ayars EJ, Jahncke CL (2003) J Micros 210:252
71. Drezet A, Woehl JC, Huant S (2002) Phys Rev E 65:046611
72. Obermuller C, Karrai K (1995) Appl Phys lett 67:3408
73. Oesterschulze E, Georgiev G, Muller-Wiegand M, Vollkopf A, Rudow O (2001) J Microsc 202:39
74. Novotny L, Pohl DW (1995) In: Marti O, Muller R (eds) Photons and local probes. NATO ASI series E. Kluwer, Dordrecht
75. Patanè S, Cefalì E, Spadaro S, Gardelli R, Albani M, Allegrini M. to be published on J Microsc
76. Novotny L, Pohl DW Hecht B (1995) Opt Lett 20:970
77. Stockle RM, Schaller N, Deckert V, Fokas C Zenobi R (1999) J Microsc 194:378
78. Smith S, Orr B, Koppelman R, Norris T (1994) CLEO Anaheim MD Tech Dig Ser 8:147
79. Smith S, Orr B, Parent G, Van Labeke D Baida F I (2001) J Microsc 202:296
80. Meixner AJ, Bopp MA, Tarrach G (1994) Appl Opt 33:7995
81. Muller R, Lienau C (2001) J Microscopy 202:339
82. Xie XS, Dunn RC (1994) Science 265:361
83. Muramatsu H, Yamamoto N, Umemoto T, Homma K, Chiba N, Fujihira M (1997) Jpn J Appl Phys 36:5753
84. Spajer M, Parent G, Bainier C, Charraut D (2001) J Microsc 202:45
85. Genolet G, Despont M, Vettiger, Staufer PU, Noell W, de Rooij NF, Cueni T, Bernal MP, Marquis-Weible F (2001) Rev Sci Instrum 72:877
86. Kim BJ, Flamma JW, Ten Have ES, Garcia-Parajo MF, van Hulst NF, Brugger J. (2001) J Microsc 202:16
87. Demming F, Jersch J, Klein S, Dickmann K (2001) J Microsc 201:383
88. Lapchuk AS, Kryuchyn AA (2004) Ultramicroscopy 99:143
89. Taminiau T, Overman J, Segerink F, van Hulst N (2005) In: Proceedings of ICONIC 2005 UPC, Barcelona
90. Puygranier BA, Dawson P (2000) Ultramicroscopy 85:235
91. Essaidi N, Chen Y, Kottler V, Cambril E, Mayeux C, Ronarch N, Vieu C (1998) Appl Opt 37:609
92. Williamson R, Miles M (1996) J Appl Phys 80:4804
93. Lazarev A, Fang N, Luo Q, Zhang X (2003) Rev Sci Instrum 74:3684
94. Valaskovic G, Holton M, Morrison G (1995) Appl Opt 34:1215
95. Garcia-Parajo M, Tate T, Chen Y (1995) Utramicroscopy 61:155
96. Gallacchi R, Kolsch S, Kneppe H, Meixner AJ (2001) J Microsc 202:182
97. Muramatsu H, Chiba N, Homma K, Nakajima K, Akata T (1995) Appl Phys Lett 66:3245

98. Sarah A, Philipona C, Lambelet P, Pfeffer M, Marcuis-Weible F (1998) Ultramicroscopy 71:59
99. Hoffmann P, Dutoit B, Salathe RP, (1995) Ultramicroscopy 61:165
100. Matsumoto T, Ohtsu M (1996) J Lightwave Technol 14:2224
101. Mononobe S, Ohtsu M (1996) J Lightwave Technol 14:2231
102. Lazarev A, Fang N, Luo Q, Zhang X (2003) Rev Sci Instrum 74:3679
103. Held T, Emonin S, Marti O, Hollricher O (2000) Rev Sci Instrum 71:3118
104. Novotny L, Pohl DW, Hecht B (1995) Opt Lett 20:970
105. Zeisel D, Nettesheim S, Dutoit B, Zenobi R (1996) Appl Phys Lett 68:2491
106. Wolf JF, Hillner PE, Bilewicz R, Kolsch P, Rabe JP (1999) Rev Sci Instrum 70:2751
107. Drews D, Ehrfeld W, Lacher M, Mayr K, Noell W, Schmitt S, Abraham M (1999) Nanotechnology 10:61
108. Noell W, Abraham M, Mayr K, Ruf A, Barenz J, Hollricher O, Marti O, Gunther P (1997) Appl Phys Lett 70:1236
109. Yatsui T, Kourogi M, Ohtsu M (1998) Appl Phys Lett 73:2090
110. Mononobe S, Saiki T, Suzuki T, Koshihara S, Otsu M (1998) Opt Commun 146:45
111. Mononobe S, Naya M, Saiki T, Ostu M (1997) Appl Phys 36:1496
112. Muramatsu H, Homma K, Chiba N, Yamamoto N, Egawa A (1999) J. Microsc 194:383
113. Stockle R, Fokas C, Deckert V, Zenobi R, Sick B, Hecht B, Wild UP (1999) Appl Phys Lett 75:160
114. Patanè S, Cefalì E, Arena A, Gucciardi PG, Allegrini M (2006) Ultramicroscopy 106: 475
115. Landau LD, Lifshitz EM (1987) Fluid mechanics, 2nd edn. Pergamon, Oxford
116. Betzig E, Trautman JK (1992) Science 257:189
117. Pohl DW (ed) (1991) Scanning near-field optical microscopy (SNOM). Academic, London
118. Pohl DW, Denk W, Lanz M (1984) App Phys Lett 44:651
119. Hsu JWP (2001) Mater Sci Eng 33:1
120. McDaniel EB, SC McClain, Hsu JWP (1998) Appl Opt 37:84
121. Buratto SK (1996) Curr Opin Solid State Mater Sci 1:485
122. Weston KD; Buratto SK (1997) J Phys Chem B 101:5684
123. De Aro JA, Weston KD, Buratto SK, Lemmer U (1997) Chem Phys Lett 277:532
124. Weston KD, De Aro JA, Buratto SK (1996) Rev Sci Instrum 67:2924
125. Hollars CW, Dunn RC (1996) Rev Sci Instrum 69:17
126. Curran JE, Page JS, Pick U (1982) Thin Solid Films 97:259
127. Krueger WH, Pollack SR (1972) Surf Sci 30:263
128. Adiga VP, Kolb PW, Evans GT, Cubillos-Moraga MA, Schmadel DC, Dyott R, Drew HD (2006) Appl Opt 45:2597
129. Stahelin M, Bopp MA, Tarrach G, Meixne AJ, Zschokke-Granacher I, (1996) Appl Phys Lett 68:2603
130. Mihalcea C, Scholz W, Werner S, Münster S, Oesterschulze E, Kassing R (1996) Appl Phys Lett 68:3531
131. Oesterschulze E, Rudow O, Mihalcea C, Scholz W, Werner S (1998) Ultramicroscopy 71:85
132. Labardi M, Zavelani-Rossi M, Polli D, Cerullo G, Allegrini M, De Silvestri S, Svelto O (2005) Appl Phys Lett 86:031105
133. Vollkopf A, Rudow O, Oesterschulze E (2001) J Electrochem Soc 148:G587
134. Georgiev G, Müller-Wiegand M, Georgieva A, Ludolph K, Oesterschulze E (2003) J Vac Sci Technol B 21:1361.
135. Ambrosio A, Cefalì E, Spadaro S, Patanè S, Allegrini M, Albert D, Oesterschulze E (2006) Appl Phys Lett 89:163108
136. Biagioni D, Polli M, Labardi A, Pucci G, Ruggeri G, Cerullo M, Finazzi Duò L (2005) Appl Phys Lett 87:223112; (2006) Appl Phys Lett 88:209901 (erratum)

137. Minh PN, Ono T, Esashi M (1999) Appl Phys Lett 75:4076
138. Minh PN, OnoT, Tanaka S, Esashi M (2001) J Microsc 28:202
139. Masaki T, Inouye Y, Kawata S (2004) Rev Sci Instrum 75:3284
140. Vollkopf A, Rudow O, Leinhos T, Mihalcea C, Oesterschulze E (1999) J Microscopy 194:344
141. Schürmann G, Noell W, Staufer U, de Rooij NFR, Eckert R, Freyland JM, Heinzelmann H (2001) Appl Opt 40:5040
142. Eckert R, Freyland JM, Gersen H, Heinzelmann H, Schurmann G, Noell W, Staufer U, De Rooij NF (2001) J Microsc 202:7
143. Aeschimann L, Akiyama T, Staufer U, De Rooij NF, Thiery L, Eckert R., Heinzelmann H (2003) J Microsc 209:182
144. Pohl DW, Novotny L, Hecht B, Heinzelmann H (1996) Thin Solid Films 273:161
145. Fasolka MJ, Goldner LS, Hwang J, Urbas AM, De Rege P, Swager T, Thomas EL (2003) Phys Rev Lett 90:016107
146. Ono M, Sone H, Hosaka S (2005) Jpn J Appl Phys 44:5434
147. Likodimos V, Labardi M, Pardi L, Allegrini M, Giordano M, Arena A, Patanè S (2003) Appl Phys Lett 82:3313
148. Lacoste T, Huser T, Prioli R, Heinzelmann H (1998) Ultramicroscopy 71:333
149. Ramoino L, Labardi M, Maghelli N, Pardi L, Allegrini M, Patanè S (2002) Rev Sci Instrum 73:2051
150. Ambrosio A, Alderighi M, Labardi M, Pardi L, Fuso F, Allegrini M (2004) Nanotechnology 15:S270
151. Valaskovic GA, Holton M, Morrison GH (1995) J Microsc 179:29
152. Betzig E, Trautman JK, Weiner JS, Harris TD, Wolfe R (1992) Appl Opt 31:4563
153. Novotny L, Pohl DW, Regli P (1994) J Opt Soc Am A 11:1768
154. Huser T, Novotny L, Lacoste T, Eckert R, Heinzelmann H (1999) J Opt Soc Am A 16:141
155. Biagioni P, Polli D, Labardi M, Pucci A, Ruggeri G, Cerullo G, Finazzi M, Duò L (2005) Appl Phys Lett 87:223112
156. Tirelli N, Amabile S, Cellai C, Pucci A, Regoli L, Ruggeri G, Ciardelli F (2001) Macromolecules 34:2129
157. Adelmann C, Hetzler J, Scheiber G, Schimmel T, Wegener M. Weber B, Lohneysen HV (1999) Appl Phys Lett 74:179
158. Veerman JA, Otter AM, Kuipers L, van Hulst NF (1998) Appl Phys Lett 72:3115
159. Veerman JA, Garcia-Parajo MF, Kuipers L, van Hulst NF (1999) J Microsc 194:477
160. Pilevar S, Edinger K, Atia W, Smolyaninov I, Davis C (1998) Appl Phys Lett 72:3133
161. Mitsui T, Sekiguchi T (2004) J Electron Microsc 53:209
162. Mitsui T (2005) Rev Sci Instrum 76:043703
163. Zakharian A, Mansuripur M, Moloney (2004) J Opt Express 12:2631
164. Patanè S, Gucciardi PG, Labardi M, Allegrini M (2004) Riv Nuovo Cimento 27:1
165. Zhong Q, Inniss D, Kjoller K, Elings VB (1993) Surf Sci Lett 290:L688
166. Anderson MS, Pike WT (2002) Rev Sci Instrum 73:1198
167. Melmed AJ (1991) J Vac Sci Technol B 9:601
168. Kulawik M, Nowicki M, Thielsch G, Cramer L, Rust HP, Freund HJ, Pearl TP, Weiss PS (2003) Rev Sci Instrum 74:1027
169. Fotino M (1992) Appl Phys Lett 60:8
170. Dickmann K, Demming F, Jersch J (1996) Rev Sci Instrum 67:845
171. Billot L, Berguiga L, de la Chapellea ML, Gilbert Y, Bachelot R (2005) Eur Phys J Appl Phys 31:39
172. Libioulle L, Houibion Y, Gilles JM, (1995) J Vac Sci Technol B 13:1325
173. Vasile MJ, Biddick C, Huggins H (1994) Appl Phys Lett 72:575
174. Hartschuh A, Sanchez EJ, Xie XS, Novotny L (2003) Phys Rev Lett 90:095503

175. Hecht B, Bielefeldt H, Inouye Y, Pohl DW, Novotny L (1997) J Appl Phys 81:2492
176. Novotny L, Bian RX, Xie XS (1997) Phys Rev Lett 79:645
177. Martin OJF, Girard C (1997) Appl Phys Lett 70:705
178. Novotny L, Sanchez EJ, Xie XS (1998) Ultramicroscopy 71:21
179. Sanchez EJ, Novotny L, Xie XS (1999) Phys Rev Lett 82:4014
180. Muhlschlegel P, Eisler HJ, Martin OJF, Hecht B, Pohl DW (2005) Science 308:1607
181. Schuck PJ,Fromm DP, Sundaramurthy A, Kino GS, Moerner WE (2005) Phys Rev Lett 94:017402
182. Grober RD, Schoelkopf RJ, Prober DE (1997) Appl Phys Lett 70:1354
183. Farahani JN, Pohl DW, Eisler HJ, Hecht B (2005) Phys Rev Lett 95:017402
184. Novotny L, Bain RX, Xie XS (1997) Phys Rev Lett 79: 645
185. Zenhausern F, Martin Y, Wickramasinghe HK (1995) Science 269: 1083
186. Hillenbrand R, Keilmann F (2000) Phys Rev Lett 85:3029
187. Aigouy L, Andreani FX, Boccara AC, Rivoal JC, Porto JA, Carminati R, Greffet JJ, Megry R (2000) Appl Phys Lett 76:397
188. Wessel J (1985) J Opt Soc Am B 2:1538
189. Nie S, Emroy SR (1997) Science 275:1102
190. Bouhelier A, Beversluis MR, Hartschuh A, Novotny L (2003) Phys Rev Lett 90:13903
191. Bouhelier A, Beversluis MR, Novotny L (2003) Appl Phys Lett 82:4596
192. Alda J,. Rico-García JM, López-Alonso JM, Boreman G (2005) Egypt J Solids 28:1
193. Fumeaux C, Alda J, Boreman GD (1999) Opt Lett 24:1629
194. Farahni JN, Pohl DW, Eisler HJ, Hecht B (2005) Phys Rev Lett 95:017402
195. Alda J, Rico-García JM, López-Alonso JM, Boreman G (2005) Nanotechnology 16:S230
196. Alda J, Fumeaux C, Codreanu I, Schaefer JA, Boreman GD (1999) Appl Opt 38:3993
197. Grober RD, Schoellkopf RJ, Prober DE (1997) Appl Phys Lett 70:1354
198. Fromm DP, Sundaramurthy A, Schuck PJ, Kino G, Moerner WE, (2004) Nano Lett 4:957
199. Schuck PJ, Fromm DP, Sundaramurthy A, Kino G, Moerner WE (2005) Phys Rev Lett 94:017402
200. Kalkbrenner T, Hakanson U, Schadle A, Burger S, Henkel C, Sandoghdar V (2005) Phys Rev Lett 95:200801
201. Anger P, Bharadwaj P, Novotny L (2006) Phys Rev Lett 96:113002
202. Kuhn S, Hakanson U, Rogobete L, Sandoghdar V (2006) Phys Rev Lett 97:017402
203. Sundaramurthy A, Crozier KB, Kino GS, Ginzton EL, Fromm DP, Schuck PJ, Moerner WE (2005) Phys Rev B 72:165409
204. Balanis C (1997) Antenna theory: analysis and design, 2nd edn. Wiley, New York
205. Frey HG, Keilmann F, Kriele A, Guckenberger R (2002) Appl Phys Lett 81:5030
206. Taminiau TH, Moerland RJ, Segerink F, Van Hulst N (2007) Nano Lett 7:28
207. Labardi M, Coppedè N, Pardi L, Allegrini M, Giordano M, Patanè S, Arena A, Cefalì E (2003) Mol Cryst Liq Cryst 398:33
208. Ambrosio A, Cefalì E, Spadaro S, Patane S, Allegrini M, Albert D, Oesterschulze E (2006) Appl Phys Lett 89:163108
209. Hillenbrand R, Keilmann F (2002) Appl Phys Lett 80:25
210. Stockle RM, Suh YD, Deckert V, Zenobi R (2000) Chem Phys Lett 318:131
211. Hartschuh A, Sánchez EJ, Xie XS, Novotny L (2003) Phys Rev Lett 90:095503
212. Ichimura T, Hayazawa N, Hashimoto M, Inouye Y, Kawata S (2004) Phys Rev Lett 92:220801
213. Gerton JM, Wade LA, Lessard GA, Ma Z, Quake SR (2004) Phys Rev Lett 93:180801
214. Betzig E, Chichester RJ (1993) Science 262:1422.
215. Azoulay J, Débarre A, Richard A, Tchénio P (2000) Europhys Lett 51:374
216. Trabesinger W, Kramer A, Kreiter M, Hecht B, Wild UP (2002) Appl Phys Lett 81:2118

217. Schuck PJ, Fromm DP, Sundaramurthy A, Kino GS, Moerner WE (2005) Phys Rev Lett 94:017402
218. Muhlschlegel P et al (2005) Science 308:1607
219. Ma Z, Gerton JM, Wade LA, Quake SR (2006) Phys Rev Lett 97:260801
220. Labardi M, Zavelani-Rossi M, Polli D, Cerullo G, Allegrini M, De Silvestri S, Svelto O (2004) Opt Lett 29:62
221. Zayats AV, Sandoghdar V (2001) J Microsc 202:94
222. Anderson N, Bouhelier A, Novotny L (2006) J Opt A Pure Appl Opt 8:S227

4 Carbon Nanotubes as SPM Tips: Mechanical Properties of Nanotube Tips and Imaging

Sophie Marsaudon · Charlotte Bernard · Dirk Dietzel ·
Cattien V. Nguyen · Anne-Marie Bonnot · Jean-Pierre Aimé · Rodolphe Boisgard

Abstract. In this chapter, a thorough investigation of the use of carbon nanotubes as nanoprobes of an atomic force microscope is presented. Because of their mechanical robustness, their controlled geometry with high aspect ratio, their small size and well-defined chemical composition, carbon nanotubes as probes solve many experimental and modeling problems of local probes methods. Over the last decade, many attempts were dedicated to use of carbon nanotubes as nanoprobes. However, in spite of a large number of works, there are still many questions concerning the proper use of carbon nanotubes, such as the type of more appropriate growth methods and an accurate interpretation of the mechanical properties. We present two growth methods based on chemical vapor deposition: (1) anchoring of multiwalled nanotubes to a commercial Si tip and (2) direct growth of a single-walled nanotube on the tip apex. Control parameters such as radius, length, angle with the sample and anchoring are discussed. The mechanical properties of those nanotubes anchored to the tip are modeled and experimentally probed by dynamical atomic force microscopy in frequency modulation mode. Most of the nanotube mechanical behavior can be understood with a flexural elasticity and an adhesive force. We particularly focus on evaluation of the nanotube equivalent stiffness and on its adhesion force and energy. We demonstrate that the balance between adhesion and elastic energy can be altered by changing the oscillation amplitude. Finally, the nanotube adhesion to the surface is used to image a heterogeneous sample, demonstrating the ability of the nanotube to be chemically sensitive.

Key words: Carbon nanotubes, Single-walled carbon nanotubes, Multiwalled carbon nanotubes, Fabrication of carbon nanotube tips, Chemical vapor deposition, Mechanical properties, Elasticity, Adhesion, Evaluation of carbon nanotube stiffness, Evaluation of carbon nanotube adhesion, Dynamical atomic force microscopy, Frequency modulation, Imaging, Chemical contrast

Abbreviations

AFM	Atomic force microscopy
AM	Amplitude modulation
CNT	Carbon nanotube
CVD	Chemical vapor deposition
FM	Frequency modulation
HFCVD	Hot filament assisted chemical vapor deposition
HRTEM	High-resolution transmission electron microscopy
MWCNT	Multiwalled carbon nanotube
SEM	Scanning electron microscope
SPM	Scanning probe microscopy
SWCNT	Single-walled carbon nanotube

4.1
Introduction

Since the very beginning of near-field microscopy, including scanning tunneling microscopy, atomic force microscopy (AFM) and scanning optical microscopy, the probe tips have ben recognized as key elements in the performance of these near-field techniques. Consequently, many efforts were immediately focused on conceiving apexes of probe tips with a high aspect ratio structure and mechanical robustness. In general, the tip governs the strength as well as the type of interaction with the sample and thus determines the experimental resolution. For example, charging the tip transforms the microscope to an electromagnetic sensor [1, 2] with long-range forces,[1] whereas grafting the tip allows the the microscope to act as a chemical [3,4] or a biological [5] sensor.

It was soon recognized that the tip geometry greatly influences imaging resolution, particularly for imaging steep edges. Focused ion beam or acid attack have been employed to sharpen commercial Si tips [6, 7]. In dynamical AFM, for producing high-resolution images, bumping the tip intentionally into the surface to cause the transfer of surface atoms to the tip apex has been used to control the chemistry of the tip.

The discovery of nanotubes [8] offered a new route to obtain tips with small apexes with a high aspect ratio. Carbon nanotubes (CNTs) are rolled-up graphene sheets that form long cylinders. The cylinder can be made of a single sheet—referred to as a single-walled CNT (SWCNT)—or of many sheets forming concentric tubes—referred to as a multiwalled CNT (MWCNT) (Fig. 4.1). The radius of the CNT can vary from less than 1 nm to a few tens of nanometers, and the length van vary from a few hundred nanometers to some millimeters. CNTs offer even more than size and geometry to scanning probe methods.

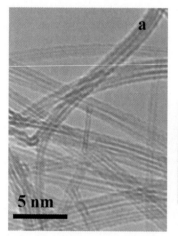
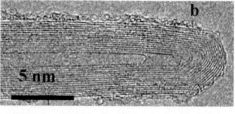

Fig. 4.1. Transmission electron microscopy micrographs of carbon nanotubes (CNTs): **a** single-walled CNTs (SWCNTs); **b** multiwalled CNT (MWCNT)

---

[1] A discussion about long-range and short-range forces can be found in [1].

In AFM, one important problem is the tip instability during the experiment. In comparison with conventional Si tips, CNT tips are less reactive and thus more likely to reduce surface contamination, particularly when scanning soft samples. In addition, tip wear also changes the physical characteristics of a tip; thus, quantitative analysis in AFM is a challenge. CNT AFM probes have great advantages because of their chemical and mechanical robustness, and thus may offer the potential for quantitative AFM methods. CNT tips are stable and thus have prolonged imaging lifetime compared with commercially available and very delicate Si probes with radius from 5 to 15 nm. CNTs have exceptional mechanical properties, with a Young modulus[2] higher than 1 TPa (compared with the value for diamond of 1.2 TPa) [9,10]. CNTs are able to buckle reversibly [11–13], and thus reduce damage to the CNT tip.

CNTs also present interesting conducting properties [14, 15], enabling their use as scanning tunneling microscope tips [16, 17] and as scanning nanolithography tools [18–20].

In addition, CNTs offer a promising route for new developments of chemical or biological tip functionalization. The small radius limits the effect of tip enlargement usually encountered with chemical modification. Oxidized CNT tips with terminal carboxyl groups enable selective functionalization of the CNT, thus providing an interesting development for chemical force spectroscopy [21,22]. Scanning tunneling microscope tips with a carboxyl-terminated CNT tip provided chemical selective imaging [23].

In addition to scanning probe microscopy (SPM) applications, CNTs with all their exceptional properties have attracted scientific attention in many fields, including molecular electronics, mechanical reinforcement, and chemical and biological sensors [24]. Among the various CNT applications, the application of CNTs as atomic force microscope tips has been suggested to be the closest for commercialization [24].

In early 2000, many works on CNT tip demonstrated the possibility of using bundles of SWCNTs or MWCNTs for imaging. Almost a decade later, the time has been reached for more quantitative studies to come to the fore, in particular works on competing adhesion and elastic forces acting on CNTs (an important issue for nanofabrication [25]) are beginning to emerge [26, 27]. Many previous studies on CNT tip fabrication focused on the important requirements of highly controlled properties of CNT tips. The ability to control integrated nanostructures is a common and general challenge for nanotechnology and must be solved in order to fully realize the technological advantages of CNT tips as well as successful commercialization of CNT tips. More specific to the scanning probe microscope tip application, we need absolute control of the mechanical properties of the CNT tip. Basic requirements must be fulfilled: the CNT must be tightly fixed on the tip or cantilever; its geometrical properties, especially its length, must be controlled as well as its orientation relative to the sample surface.

We present first two opposite approaches for CNT tip fabrication: the direct growth on silicon tips for SWCNTs and the manual attachment technique of MWC-

---

[2] The Young modulus is a coefficient of elasticity of a substance. It expresses the ratio between a stress that acts to change the length of a body and the fractional change in length caused by this force.

NTs on silicon tips by fusing the MWCNT to the Si surface through an applied electrical field. A presentation of complementary methods to tailor the properties of the CNT tips and a discussion on the use of CNTs in liquid environments will follow.

The next section of this chapter focuses on direct evaluation of the mechanical properties of CNTs using dynamical SPM. In particular, we address a constant nanoscale question: the competition between bulk and surface forces. The experimental signals recorded while the CNT tip approaches a surface reveal a contribution of both elastic and adhesion forces for a CNT interacting with a surface. The energy losses are due to mechanical hysteresis originating mainly from the adhesion forces between the CNT tip and the surface. We evaluate the elastic properties of the stiffness and adhesion of a CNT by measuring a pull-off distance in AFM approach–retract curves from which energy losses can also be deduced.

In the last section, an example of the effect of mechanical properties on AFM imaging is given, along with a review of imaging studies in SPM dynamical modes.

4.2
CNT Tip Fabrication

Since the discovery of CNTs, many efforts have been focused on synthesizing CNTs. The first CNTs were produced by arc discharge using graphitic electrodes in an inert noble gas environment. This method is simple, relatively low cost, and enables large quantities of CNTs to be produced. However, it produces impurities such as amorphous carbon without accurate control of the radius and the length of the CNTs [28]. CNTs can also be synthesized by a laser ablation technique, producing high-quality CNTs, however at a lower yield rate. A simple CNT synthesis method is by thermal chemical vapor deposition (CVD) at a high temperature, typically greater than 700 °C. This method was first developed by Endo and Kroto [29] in 1992, where hydrocarbon gases are decomposed on catalytic surfaces at an elevated temperature to produce CNTs. The CVD process has the ability to control precisely the location of CNT growth by using patterned catalysts [30]. A lower temperature plasma-enhanced CVD process can also be used to grow carbon nanostructures on patterned catalyst surfaces. Here, the plasmas supply additional energy to produce highly reactive carbon ion species by dissociating the hydrocarbons in the gas phase, which then nucleate and grow into carbon nanostructures on metal catalyst surfaces.

For AFM application, CNT tip fabrication requires precise positioning of the CNT at the apex of a conventional Si tip, with a good mechanical contact. Fabrication techniques for CNT tips can be classified by two approaches: (1) direct growth of the CNT on the Si tip or (2) manual attachment of the CNT on Si tips after the separate growth of the CNT.

The direct growth on the Si tip is usually achieved by selective positioning and size control of the catalyst particle at the tip apex [31].

Different methods have been developed to attach previously grown CNTs on tips. The so-called pick-up technique developed by Hafner et al. [32] uses scanning probe microscope tips scanning a flat surface with some CNTs grown perpendicular to the surface. The tip eventually picks up a CNT, which then binds to the tip through attractive van der Waals forces. Another technique uses manual attachment controlled by optical microscopy by local fusing of a metal layer to fix the CNT on the tip [19]. The current for fusing also induces electrical breakdown of the CNT, detaching it from its previous surface. When dealing with SWCNTs, manual assembly or the pick-up technique may lead to SWCNT bundles rather than unique SWCNTs. To limit assembly of bundles, mechanical attachment has also been performed inside a scanning electron microscope (SEM) [33–35] but it increases the time required for the attachment process. More recently, room-temperature liquid-phase dielectrophoresis has been demonstrated as an alternative technique for production of MWCNT [36] and SWCNT [37] tips.

The two different fabrication strategies, namely, manual CNT attachment and direct growth of CNTs on Si tips will be presented in detail: (1) manual attachment and electrical-induced fusing of MWCNT to Si atomic force microscope tips employing an optical microscope and (2) direct catalytic CVD growth of SWCNTs and double-walled CNTs on the Si tip apex utilizing localized heating with a tungsten filament. These two fabrication techniques illustrate the variety of processes developed for CNT atomic force microscope tips.

4.2.1
MWCNTs and Fusing

MWCNT tips have been successfully fabricated by manually attaching MWCNTs to Si pyramidal tips. The first CNT tip was fabricated by literally gluing the CNT onto one of the sides of the pyramidal-shaped tip of a conventional Si cantilever [11]. A source of MWCNTs was created by transferring as-grown MWCNTs onto a strip of adhesive tape. Wit use of an inverted optical microscope equipped with two X–Y–Z microtranslators/manipulators, the MWCNT on the adhesive tape was attached to the Si tip of a conventional cantilever

Subsequent to the first manual attachment, the technique was further refined through a number of improvements. For example, an electric field has been applied to facilitate the process of attaching the MWCNT to a Si pyramidal tip. This electric field improved the efficiency of the transfer and orientation of the MWCNT for better alignment on the Si tip. By employing an applied DC field between the MWCNT and a Co-coated Si probe, this process greatly enhanced the optical microscopic manual method of attachment [38, 39].

In the work of Nakayama et al. [33, 40, 41], a low-density MWCNT aligned on a knife edge was produced by solution electrophoretic method. The process involved placing two knife edges, with a 500-μm gap, in a solution of isopropyl alcohol containing dispersed MWCNTs derived by an arc discharge synthesis. Upon applying an AC electric field of 5 MHz and 1.8 kV cm^{-1}, the MWNT moved onto the knife edges and aligned almost parallel to the electric field. This produced knife edges with a density of less than one MWCNT per micron, with some of the MWCNTs reported to protrude as much as 1.3 μm from the knife edge. Because of the short protruding length, the MWCNTs were not observable by optical microscopy and therefore the MWCNT had to be manipulated using a SEM. An individual MWCNT was transferred from the knife edge to the Si tip by applying a DC bias voltage between the knife edge and the Si probe in a SEM.

Another improvement was made by using a CNT source composed of low-density, long and individually separated MWCNTs on a wire. This was accomplished by coating a Pt wire with a viscous liquid catalyst solution composed of an Fe metal complex and a methanol solution of block copolymer P123 [19]. CVD growth at 750 °C using 1000 sccm of ethylene produced MWCNTs on the Pt wire, with areas of low density, greater than 10 μm long, with individually separated MWCNTs protruding from the wire. This allowed for easier sorting and optimal alignment (i. e., manipulation) of individual MWCNTs by optical microscopy and hence provided a very efficient means of attaching a single MWCNT to a Si probe. It is important to note that a scanning probe with an individual MWCNT, rather than a bundle as the tip, enables tracking of deep and narrow features at the sub-100-nm length scale.

The manipulation of MWCNTs for the fabrication of scanning probes is made possible owing to a long CNT extending its length from the substrate. In other words, because MWCNTs are less flexible, MWCNT with lengths up to tens of micrometers can grow normal to the surface. This is mainly due to the large radius of the tube, as will be discussed in detail in the following. In contrast, SWCNTs and even MWCNTs with very small diameters and similar lengths are more flexible owing to their smaller diameter. Therefore, these types of CNTs do not individually protrude from a surface and hence manual manipulation as a method for fabricating an individual SWCNT is not possible.

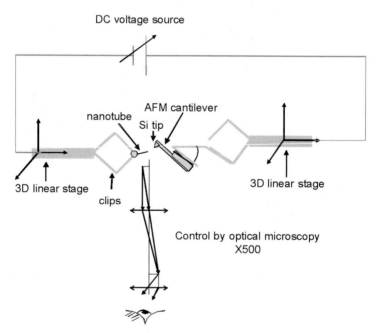

Fig. 4.2. The experimental setup used to fuse a MWCNT on a Si tip: all the experiment is controlled by an optical microscope (magnification ×500). The tip and the wire with the CNT can be moved by a 3D precision linear stage; they are linked to a DC voltage source

Figure 4.2 details the different parts of the experimental setup used to manually attach and electrically induced fuse MWCNT to Si atomic force microscope tips employing an optical microscope.

The key steps involved during the attachment of a MWCNT to a Si tip are shown in Fig. 4.3. These photographs were captured at ×500 magnification with an inverted Zeiss X100 optical microscope. On the left side of these photographs, a MWCNT with a length of about 20 μm protrudes from the surface of a wire. The triangular-shaped tip of a Si probe coated with a 5-nm Ni film is clearly seen on the right side of the photographs.

Figure 4.3a shows the initial alignment of the MWCNT with the apex of the Si tip. Next, the MWCNT is moved to make contact with the surface of the Si tip as seen in Fig. 4.3b. It is worthwhile pointing out that the alignment and contact of the MWCNT with the surface of the Si tip can be clearly seen by positioning both structures in the same focal plane of the optical microscope. Once in contact, further refinement of the alignment of the MWCNT relative to the apex of the Si tip is achieved by adjusting the X, Y, Z-positions of the two structures. Clearly seen

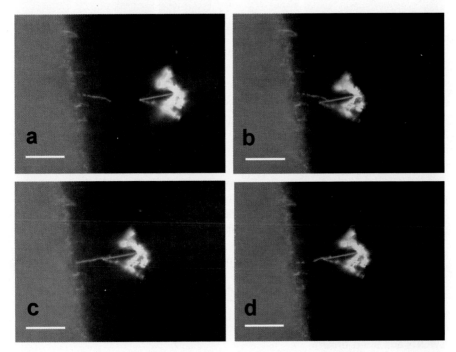

Fig. 4.3. Optical digital micrographs (the wire with many CNTs is at the *left*, the pyramidal Si tip is at the *right*) of the key steps in the manual manipulation process for fabricating a MWCNT tip. **a** Finding an individually separated and long MWCNT on a Pt wire and aligning it with the Ni-coated Si tip. **b** Positioning the MWCNT to make physical contact with the surface of the Si tip in the same optical focal plane. **c** Adjusting the Si tip and the respective MWCNT to align with the apex of the Si tip. **d** Increasing the applied potential to fuse the MWCNT to the Ni-coated Si surface by Joule heating and then subsequent breaking of the MWCNT from the Pt wire CNT source. *Scale bars* 20 μm

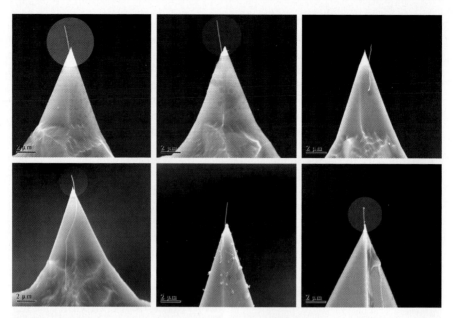

Fig. 4.4. Examples of scanning electron microscopy (SEM) images of MWCNTs fused on pyramidal Si tips

in Fig. 4.3c is the straightened MWCNT aligning perfectly with the apex of the Si probe. Applying a voltage across the well-aligned MWCNT in contact with the Ni-coated Si probe at this stage induces fusion of the MWCNT to the Si probe; presumably Joule heating at the high contact resistance between the MWCNT and the Ni coating strengthens this junction. Increasing the applied voltage further causes the MWCNT to break from the Pt wire source, as seen in Fig. 4.3d. The particular MWCNT tip produced here is about 5 µm in length as seen from the photographs. The limit of the resolution at ×500 magnification is a MWCNT length of about 2 µm.

Examples of MWCNT obtained by this fusing method are displayed in Fig. 4.4. The clear advantage of the fusing relies on its robustness: the fused Ni coating ensures firm attachment of the MWCNT onto Si tips. Some MWCNT probes could be used for hundreds of SPM experiments without a sign of failure. The mechanical properties of some of those MWCNT probes are detailed in Sect. 4.3.5.

4.2.2
SWCNTs and Direct Growth

The possibility to grow CNT by CVD techniques using catalyst particles deposited on a surface has opened up the ability to localize CNTs at determined places [42, 43] and thus to prepare devices by a self-assembly process [44].

First used to grow MWCNTs, hot filament assisted CVD (HFCVD) leads also to the growth of well-crystallized SWCNTs, and has the ability to localize and self-assemble suspended isolated SWCNTs [45].

The HFCVD apparatus has been built for diamond thin film growth [46]. For the synthesis of SWCNTs, substrates are covered with a 2–8-nm thick Co catalytic layer deposited by standard evaporation techniques. The vapor is composed of 5–20 vol % methane in hydrogen. Typical deposition parameters are a substrate temperature of 750–850 °C and a total pressure of 30–100 mbar. The tungsten filament, placed 1 cm above the substrate, is heated to 1990–2100 °C. The specificity of this HFCVD technique is to take advantage of this hot tungsten filament to decompose the vapor into active species. It also helps to control the catalyst particle formation during the initial temperature rise. During the synthesis, in situ measurements of the reflectivity and elastic scattered light of 633-nm laser radiation give insight into real-time growth kinetics as presented in Fig. 4.5.

The thickness of cobalt catalyst appears to be the critical parameter. A thickness of the cobalt layer greater than 9 nm leads to the growth of too many CNTs at the tip apex, while for too thin a layer (less than 4 nm), the yield of SWCNT growth at the apex of the Si tip becomes negligible. With an optimal catalyst thickness of the order of 5–8 nm, the yield of production of a unique SWCNT bundle at the apex of a commercial Si tip is about 30%. Figure 4.6 shows self-assembled CNT tips grown on a Si probe covered with a 7-nm-thick Co layer. The tips appear wrapped with Co particles and a single CNT emerges from the tip apex. Its diameter lies in the nanometer range, while the length is about a few hundred nanometers. The vibration occurring at the free CNT end under electron illumination points out its flexibility.

The peculiarity of this process is to enhance the growth of a single CNT at the tip apex with a preferential orientation given by the edge of the cone (Fig. 4.7a–d).

SWNT growth by HFCVD

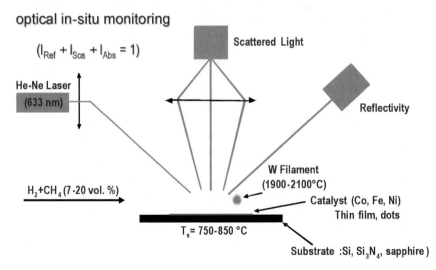

Fig. 4.5. The experimental setup used to grow SWCNTs and double walled CNTs by hot filament assisted chemical vapor deposition (*HFCVD*) with optical in situ monitoring

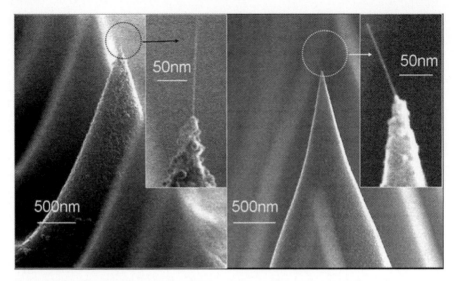

Fig. 4.6. SEM images of Si tips with a CNT emerging at the apex

Figure 4.7e is an illustration of the CNT growth path: a CNT originating from or near the tip base is stuck by adhesion forces and grows along the cone edge, until it reaches the vicinity of the apex and emerges, while keeping the same orientation. Figure 4.7f and g shows other CNT growth arrangements. In Fig 4.7f, the CNT escapes before reaching the tip apex and is perpendicular to the Si tip edge. In Fig. 4.7g, a CNT escapes from the apex, but bends up to form a loop. It must be noted that the formation of a loop has sometimes been observed in real time under electron irradiation during SEM scanning[3]. Therefore, it should be useful to select with care the scan speed to limit the electron illumination time, which also can create defects along the CNT.

The CNT structure is determined by on-site high-resolution transmission electron microscopy (HRTEM) and Raman spectroscopy (Fig. 4.8). HRTEM observations (Fig. 4.8a) demonstrate the formation of predominantly SWCNTs and double-walled CNTs [48]. The CNTs are either isolated or in small bundles of two to three CNTs, thus ensuring the CNT diameter is less than 5 nm.

The quality of the structure is confirmed by Raman spectroscopy using 633-nm He–Ne laser radiation focused on the tip apex on a 1-m^2 area. Figure 4.8b shows an example of a Raman spectrum with a Raman resonance radial breathing mode at $\omega_{RBM} = 128\,cm^{-1}$ and three tangential Raman peaks at $\omega_T = 1597$, 1579 and $1568\,cm^{-1}$ [49]. From the relation $d(nm) = 248/\omega_{RBM}\,(cm^{-1})$ between the tube diameter d and the radial breathing mode frequency, a Raman resonant semiconducting tube [50] with $d = 1.9\,nm$ is identified. Unfortunately, the Raman resonance condition is not fulfilled for every tube diameter, so it cannot allow for distinguishing between SWCNTs and double-walled CNTs. The narrowness of the most intense tangential peak at $\omega_T = 1597\,cm^{-1}$ and a width at half maximum of $\Delta\omega_T = 7\,cm^{-1}$,

---

[3] A detailed study of the electron radiation effect on CNTs has been provided by Kis et al. [47]

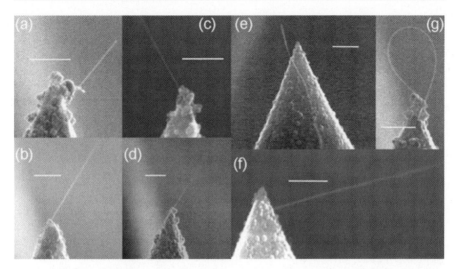

Fig. 4.7. SEM images of CNT conformations: the majority emerge at the apex parallel to the Si tip cone edge (**a–d**), a few emerge before reaching the apex and escape parallel (**e**) or perpendicular to the side of the Si tip (**f**) or closed on themselves in a loop (**g**). *Scale bars* 100 nm

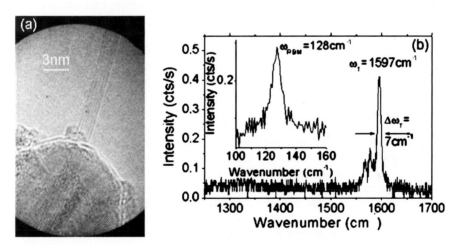

Fig. 4.8. On-site high-resolution transmission electron microscopy image of a double-walled CNT at the Si tip apex (**a**) and a typical Raman spectrum (**b**)

with the nearly indiscernible band D attributed to disordered carbon at $1320\,cm^{-1}$, are also strong indications of the excellent crystallinity and purity of the CNTs grown.

The HFCVD technique appears to be powerful for growing highly crystalline and pure SWCNTs at the apex on Si tips by taking advantage of the catalytic properties of a thin cobalt layer. In particular, adjustment of the substrate preparation and the synthesis conditions allows isolated SWCNTs to self-assemble. The exceptional advantage of this method appears to be the growth of a unique bundle extending from the apex of the Si tip. HRTEM and Raman spectroscopy show that the process leads to mainly SWCNTs and double-walled CNTs with excellent structural properties. Moreover, SEM observations indicate that during the growth, adhesion forces stick CNTs along the tip until one of them reaches the end of the tip. Conical tips guide the growth of the CNT to the tip apex, from where it escapes in a direction parallel to the cone edge. The SWCNTs or double-walled CNTs (hereafter called SWCNTs for simplicity) presented in the following were all fabricated with this HFCVD technique.

4.2.3
Controlling and Tailoring the Properties of CNT Tips

As for all CNT applications, the biggest challenge is to control the location of the CNT, its length, its diameter and its angle. They are essential parameters for tip use, as will be shown in Sect. 4.3 and 4.4. Up to now, many different techniques have been developed that try to circumvent precise drawbacks but none has emerged to solve them all. We detail different methods to tailor the angle of the CNT with the surface, its diameter and its length and finally to sharpen its end. We then discuss an important issue for imaging biological samples: the use of CNTs in a liquid environment. First, we describe the specific situation encountered for use of CNTs as scanning tunneling microscope tips where not only firm mechanical contact is needed but also electrical conductivity.

To achieve both mechanical robustness and stable electrical conductivity, CNT tips were recently synthesized by dielectrophoresis on tungsten tips, then the adhesion between the tip and the CNT was reinforced by electron-beam deposition of amorphous carbon which was annealed, followed by coating with a PtIr layer [17].

CNTs misaligned with respect to the tip can be corrected by a focused ion beam [51,52]. Using a SEM to attach CNTs, while time-consuming, enables control of direction and length when using a specially designed field-emission SEM [34].

The control of the diameter is a difficult task. It is usually possible to achieve by having SWCNTs and double-walled CNTs or MWCNTs, but the diameter distribution generally varies from 1 nm to a few nanometers for SWCNTs and from a few nanometers to a few tens of nanometers for MWCNTs. Usually one adjusts the length of the CNT as a function of its diameter. This is what matters, as the mechanical properties of the CNT depend on a combination of radius and length. We will see in Sects. 4.3 and 4.4 that to avoid too flexible probes, the smaller the radius, the smaller the length.

For the length control, an in situ SPM technique can be used to shorten the CNTs after their fabrication and first characterization: electrical pulse etching is used for SWCNT attached by the pick-up method [53–55] and for MWCNTs fused on tips [39]. The CNTs are oxidatively shortened by applying a DC bias between the tip and the substrate while they are monitored in force versus distance mode. The CNT picked up held only by van der Waals forces can also be pushed by decreasing incrementally the tip–sample distance [55]. Recently a method combining defect formation via electron irradiation and simultaneous resistive heating and electromigration has been used to shrink CNTs inside a transmission electron microscope [56].

In situ sharpening of the MWCNT ends by removing outer layers has also been demonstrated [57], by locally stripping away the outer layers at the CNT end [58]. The process involves applying a lower voltage than that required for shortening.

For biological samples, an important issue is the ability to image in a liquid environment; thus, the question of the behavior of CNTs a liquid arises. The wetting of such a hydrophobic nanoobject is not obvious. For picked up CNTs, van der Waals interactions are not strong enough to sustain immersion in water [32], but it is possible to prevent this by gluing the picked up CNT on the tip by applying UV-curable glue prior to scanning for picking up the CNTe [59]. To strengthen the attachment, sputter-coating of chromium film after attachment of a MWCNT bundle with acrylic adhesive has also been reported [60]. MWCNTs attached under a SEM have also been used in dynamical frequency modulation (FM) mode in a liquid to reduce hydrodynamic squeeze-damping between the tip apex and the surface [34]. This question has also been addressed for MWCNTs fused on tips [61]. Most of the long MWCNT probes failed after immersion in water because of bending back on the tip. Lessening the hydrophobic character of the MWCNTs by chemically coating them with ethylenediamine enabled imaging in a liquid.

4.3
Understanding the Mechanical Properties of CNT Tips: A Dynamical SPM Frequency Modulation Study

CNTs are new probes with specific properties, an it is particularly important to characterize their mechanical properties before their use for complex samples such as soft molecules or nanostructures with steep edges. This section details the mechanical properties of CNT probes with an evaluation of an effective stiffness and adhesion energy.

4.3.1
Mechanical Properties of CNTs

Despite their nanometer-scale size, the mechanical properties of CNTs can be described with a classical continuum mechanical model, using concepts like the spring constant and a simple relationships between the Young modulus and geometrical characteristics of the nanomaterial. One of the reasons comes from the fact that Hooke's law, which defines the Young modulus, is based on a parabolic approximation of any smooth potential for small displacements around the equilibrium position. As a result, for mechanical properties, there is no characteristic length down to atomic scale. Some approaches use quantum mechanics with atomistic molecular dynamics to be able to parameterize the CNT properties and then use those properties in classical mechanics [62].

A first and common approximation assumes the CNT to be a homogeneous bar. The theoretical framework used in the present review is based on this hypothesis as we shall see that this simple assumption enables one to understand most of the behavior of a CNT used as a tip. The main challenge is to be able to discriminate between surface forces and bulk ones (here adhesion and elastic forces, respectively).

Detailed descriptions of the mechanical properties of CNTs, taking into consideration the CNT sheet thickness, can be found elsewhere [10, 63–68].

The representation of a CNT as a homogeneous bar has been used in many SPM studies [9, 12, 69] where the CNTs were the samples probed with regular tips. The earliest experiments of Wong et al. [12] enabled the experimental evaluation of the Young modulus and the lateral stiffness of MWCNTs pinned at one end and moved laterally by the tip. Pinned ropes of SWCNTs were deformed by the tip in an experiment by Walters et al. [70], allowing the measurement of the maximal strain and lower-bound tensile strength.

4.3.2
Mechanical Properties of CNT Tips

The following paragraphs will focus on direct measurements of the mechanical properties of CNT tips using sensitive dynamical SPM modes. Except when mentioned, firm binding of the CNT on the tip is assumed.

Even when using a simple mechanical model to describe the CNT, when a surface approaches the CNT, many different mechanical situations can happen.

Figure 4.9 describes two opposite situations: either the end of the CNT is pinned on the surface (Fig. 4.9a) and the largest deformation is in the middle of the CNT or the end of the CNT can slide on the surface (Fig. 4.9b) and the largest deformation will be undertaken by the free end of the CNT.

The pinned and sliding situations correspond, respectively, to different stiffness [71, 71]:

$$k_{b,p} = 64\pi E \frac{r^4}{L^3} \quad \text{and} \quad k_{b,s} = \pi E \frac{r^4}{L^3} \,. \tag{4.1}$$

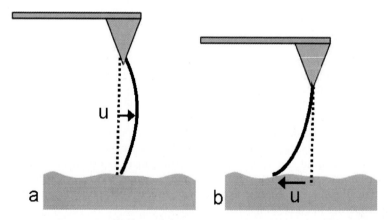

Fig. 4.9. Two situations for a CNT squeezed between a tip and a sample surface. **a** The end of the CNT is pinned on the surface. A simple geometrical consideration leads to a square-root dependence of the elastic displacement of the CNT as a function of the vertical displacement [71]. **b** The end of the CNT slides on the surface. A linear relationship is considered between the vertical displacement and the CNT deformation [71]

As expected, the stiffness depends on the Young modulus E of the CNT and varies rapidly with the radius r and the length L of the CNT. Consider a radius of 2 nm, a length of 500 nm and a Young modulus of 1 TPa, the bending stiffness values are $k_{b,p} \approx 1.9 \times 10^{-2}\,\mathrm{N\,m^{-1}}$ and $k_{b,s} \approx 4.01 \times 10^{-4}\,\mathrm{N\,m^{-1}}$.

Often, the vibration amplitude of the free CNT end is expressed assuming a thermal energy of $1/2 k_B T$:

$$X_{CNT} = \sqrt{\frac{4 k_B\, T L^3}{3 \pi E r^4}} \,. \tag{4.2}$$

With the previous numerical values, it gives a vibration of 3.7 nm^4 at ambient temperature. This value is interesting as it gives an idea of the influence of the geometrical parameters of the CNT, but it does not describe the situation of a CNT interacting with a surface. We will illustrate that this formula describing a free CNT end vibration is not valid when the CNT images a surface.

It is also informative to evaluate the elongation stiffness:

$$k_e = \pi E \frac{r^2}{L} \,. \tag{4.3}$$

With the numerical values of r and L, $k_e \approx 2.5\,\mathrm{N\,m^{-1}}$; thus, nearly 400 times the bending stiffness $k_{b,s}$.

In the experiments, complex mechanical response can quickly occur as many different contacts between the free end of the CNT and the surface can happen with any combination of stiffness.

4.3.3
Mechanical Properties of CNT Tips in Dynamical Experiments: Competition Between Elasticity and Adhesion

To measure the properties of the CNT, dynamical scanning probe experiments use the behavior of an oscillating cantilever–tip described as a spring and a mass. The interaction of the CNT with the surface modifies the oscillator properties of the atomic force microscope—namely, its resonance frequency ν_0 and its energy dissipation.[5] The changes can be derived from numerical modeling and simulations [59,68,75,76]. In the present review, experimental results are analyzed with the help of a variational principle a giving tractable analytical expression [71] in the case of a CNT deformation similar to that sketched in Fig. 4.9b during the sliding on the surface. As this theory helps to understand many different situations, it will be used throughout the chapter. The main analytical expressions are given and the parameters defined here will be used to characterize the CNT.

When taking only the elasticity of the CNT into account, the normalized frequency shift is given by

---

[4] This vibration was used to evaluate the mechanical properties of CNTs in [64, 66].

[5] The resonance curve shape is also highly changed by the interaction with the surface; the (cantilever–tip) oscillator becomes nonlinear [72–75].

$$\frac{\Delta v}{v_0} = \frac{1}{2}\frac{k_{NT}}{\pi k_C}\left(-d\sqrt{1-d^2}+\cos^{-1}d\right), \tag{4.4}$$

where $\Delta v = v - v_0$ is the frequency shift, v_0 is the free frequency value when no interaction takes place, k_{NT} and k_C are the stiffness of the CNT and that of the cantilever respective stiffness and $d = D/A$ is the normalized distance with respect to the oscillation amplitude A. D is the CNT end–surface distance at the equilibrium position, as depicted in Fig. 4.10.

Note that (4.4) predicts a universal curve identical to the predicted curve with a frequency shift scaling close to $A^{3/2}$ [74, 77].

In addition to conservative forces, dissipative interactions must also be considered, resulting in additional energy being lost during the oscillation cycle. Dissipation originates from electrical losses, viscoelastic processes [78–81] or mechanical instabilities due to adhesion [71, 78]. A basic phenomenological development expresses the cantilever–tip oscillator dissipation with a damping coefficient γ giving the linear relationship for the dissipative force $\boldsymbol{F} = -\gamma\boldsymbol{v}$, where $\boldsymbol{v}$ is the oscillator velocity.

The total damping coefficient of the scanning probe oscillator is [82]

$$\gamma_{tot} = \gamma_0 + \gamma_{int}, \tag{4.5}$$

where $\gamma_0 = \frac{m\omega_0}{Q} = \frac{k_c}{2\pi v_0 Q}$ is the damping coefficient without interaction, m is the effective mass of the oscillator, $\omega_0 = 2\pi v_0$ is the oscillator pulsation and Q is the quality factor.[6]

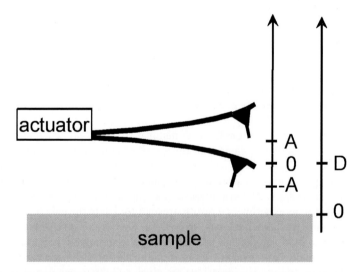

Fig. 4.10. Definition of the experiment parameters: the CNT tip oscillates with an amplitude A, leading to a instantaneous position varying between $-A$ and $+A$ around the equilibrium position of the end of the CNT. The distance D is the equilibrium distance of the end of the CNT to the sample. The first time the CNT touches the sample during an oscillation cycle is when the end of the CNT is at the lowest position and when the distance is equal to the oscillation amplitude

---

[6] The quality factor of an oscillator is the ratio between the energy stored in the oscillating system and the input energy required to maintain stationary motion. It is related to the half

The total damping coefficient is related to the mean energy dissipated per oscillation cycle:

$$E_{\text{diss}} = \pi \gamma_{\text{tot}} \omega A^2 . \tag{4.6}$$

A model based on the variational principle has been proposed, taking into account both the repulsive elastic and attractive adhesive contributions to the frequency shift from which dissipative interactions can be derived.

Figure 4.11 depicts the forces taken into consideration during the interaction of the CNT with sample. When approaching the surface, the CNT experiences no forces until it touches the surface, where the elastic repulsive force starts. The slope of the force curve is the stiffness of the CNT k_{NT}. On retraction, the repulsive force still occurs, followed by an attractive force modeled as an attractive stiffness k_{aNT} acting up to the CNT unsticking from the surface at a location $Z = \Delta$, for a force value $F_{\text{adh}} = k_{\text{aNT}} \Delta$. The hysteresis between approach and retraction for Z values between 0 and Δ induces energy loss, given by the shaded area in Fig. 4.11:

$$E_{\text{adh}} = \frac{1}{2} k_{\text{aNT}} \Delta^2 . \tag{4.7}$$

Analytical expressions for the frequency shift are derived for two different regimes: intermittent contact of the CNT with the surface from $D = A$ to $D = -A + \Delta$ (in the case sketched in Fig. 4.9a, with the positions as defined in Fig. 4.10) and permanent contact when the CNT never leaves the sample from $D = -A + \Delta$ to $D = -A$.

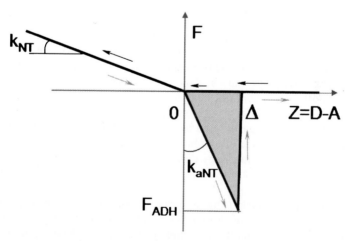

Fig. 4.11. Variation of the force sensed by the CNT during sample approach and retract plotted against the difference of the distance minus the oscillation amplitude

width of the Lorentzian resonance curve: the larger the quality factor, the sharper the curve, the smaller the energy dissipated per oscillation cycle. A typical value for quality factors of a commercial cantilever–tip is 500 in air, when far from the surface. The value drops to 1–10 for oscillation in liquids and increases to a few thousand at millibar pressure (Fig. 4.12a). The quality factor is an essential parameter for dynamical experiments [73, 83–85].

For intermittent contact, with the normalized distances $\delta = \Delta/A$ and still $d = D/A$,

$$\frac{\Delta \nu}{\nu_0} = \frac{1}{2\pi k_c} \left\{ k_{NT} \left[\left(-d\sqrt{1-d^2} + \cos^{-1} d \right) \right] \right.$$
$$+ k_{aNT} \left(\frac{\cos^{-1}(d-\delta) - \cos^{-1}\delta}{2} \right.$$
$$\left. \left. + \frac{d}{2}\sqrt{1-d^2} - \frac{(d+\delta)}{2}\sqrt{1-(d-\delta)^2} \right) \right\} . \tag{4.8}$$

The first repulsive term of (4.8), is identical to (4.4). The added term (second line) gives the contribution of the attractive adhesive interaction depicted in Fig. 4.11 which intervenes only when leaving the surface, thus during the backward motion of the tip.

For permanent contact,

$$\frac{\Delta \nu}{\nu_0} = \frac{1}{2\pi k_c} \left\{ k_{NT} \left[\left(-d\sqrt{1-d^2} + \cos^{-1} d \right) \right] \right.$$
$$\left. + k_{aNT} \left(2\cos^{-1} d - d\sqrt{1-d^2} - \pi \right) \right\} . \tag{4.9}$$

The repulsive contribution has not changed but the adhesive contribution during the oscillation cycle now happens when the repulsive force does not.

No additional damping occurs in the permanent contact according to the model.

The measurement of force gradient through resonance frequency shifts (conservative forces) and dissipative forces can be done using two dynamical modes that catch the oscillator properties of the atomic force microscope.

The most common mode is the amplitude modulation (AM) mode, usually called tapping mode, where the frequency of excitation and the driving are kept constant. Two signals are available: the oscillation amplitude and the phase delay of the oscillator with the excitation. For imaging, most commercial apparatus record the piezoceramic displacement necessary to maintain the amplitude constant. This mode is convenient and can be easily used to image. But this mode is more difficult to analyze because any change of amplitude can be due to a conservative attractive or repulsive interaction or to energy dissipation. Experimental tapping results thus cannot be interpreted straightforwardly either concerning images or approach–retract curves.

The other mode available is the frequency modulation (FM) mode, sometimes called noncontact resonance mode or near-contact resonance mode [86]. In this mode an electronic loop maintains the amplitude and the phase delay at the resonance value of the oscillator. Two signals are available: the resonance frequency shift and the damping, which is the DC output of the proportional/integral gain that maintains the amplitude constant. If the FM mode is correctly set—locked at the resonance phase shift [87]—the conservative forces will change only the frequency shift, whereas dissipative forces will affect the damping signal alone. This damping signal is proportional to the energy dissipated to maintain the oscillation, and is thus proportional to the oscillator damping coefficient γ. Attractive conservative forces induce a decrease of the resonance frequency, whereas repulsive interaction increases the resonance frequency. Because the FM mode has transient times much lower than those of the amplitude detection technique [86], it was first used under ultrahigh vacuum with a high quality factor, thus long transient times, with true atomic (or maybe even subatomic) resolution [88]. To ease the data interpretation, we will present FM results in the following. Detailed AM studies can be found in [67,68,75].

4.3.4
Experimental Signals

This section contains the description of typical experimental results, starting with resonance curves of commercial cantilever–tips with a CNT attached (Fig. 4.12a). When the tip is far from the surface, the quality factor is close to 500. Under a pressure of a few millibars, it becomes close to 2000, with a shift of resonance frequency. This result indicates that the treatment (in this case the catalyst deposit followed by CVD direct growth) done to the tip–cantilever system does not affect the quality of the resonance curve, an essential parameter for the data interpretation.

FM frequency shift and damping signals recorded during approach of a CNT tip to a hard surface are shown in Fig. 4.12b and c. The following paragraphs explain the different domains of the approach curves and emphasize the magnitude of the forces and energies at play.

4.3.4.1
Resonance Frequency Shift: Conservative Part of the CNT–Surface Interaction

The variation of the frequency curve (Fig. 4.12b) contains three different domains highlighted in different colors from black (far from the sample) to light gray.

In the black part, there is no interaction of the CNT with the surface; it oscillates at its free resonance frequency and the frequency shift is zero. When the oscillation amplitude of the CNT (25 nm) is close to the CNT–surface distance, the CNT starts to interact with the surface, the frequency slightly decreases as the overall force balance is attractive, and then increases when the CNT moves further toward the surface to positive values, indicating an increase of the repulsive interaction. This dark gray part corresponds to the intermittent contact area: as the CNT goes toward the surface, the repulsive force acts for a longer time in the oscillation period and thus has a larger contribution to the positive frequency shift. The transition from the intermittent contact domain to the next permanent contact is marked by the frequency jump. The end of the CNT never leaves the surface in this domain colored in light grey.

It is possible to evaluate the effective interaction stiffness k_{int}^{eff} corresponding to the frequency shift by using the general expression describing the resonance frequency shift [77–79]:

$$\Delta \nu = \nu_0 \left(\sqrt{1 + \frac{k_{int}^{eff}}{k_c}} - 1 \right) . \tag{4.10}$$

Fig. 4.12. Typical experimental results obtained with a CNT tip. **a** Variation of the reduced oscillation amplitude with the excitation frequency in air (*diamonds*) and under a mild vacuum (8 mbar; *circles*). The oscillator consisting of the cantilever and the tip with a CNT is defined by its resonance frequency (143,420 and 143,800 Hz in air and in a vacuum, respectively) and their quality factor Q given by the Lorentzian fits (*solid lines*): 490 and 2070. **b**, **c** CNT approach curves for approach toward a hard and flat surface obtained in dynamical frequency mode. D is the distance of the end of the CNT at the equilibrium position to the surface. The free oscillation frequency is 143,800 Hz; the oscillation amplitude is 25 nm. The CNT is 500 nm long with a radius of about 2 nm. The experiments were conducted on freshly cleaved graphite at 8-mbar pressure. **b** Variation of the resonance frequency shift $\Delta\nu$ with CNT to sample distance D. Three domains are delimited with colors: in *black* the noninteracting part; in *dark gray* the intermittent contact area; and in *light gray* the permanent contact area. **c** Variation of the damping signal with CNT to sample distance D. Note that the transitions between the domains always happen always at a sharp damping variation

In Fig. 4.12b, the maximum of frequency shift is around 100 Hz, giving

$$k_{int}^{eff} = k_c \left[1 - \left(\frac{\Delta\nu}{\nu_0} + 1 \right)^2 \right] \approx 4.2 \times 10^{-2} \, \text{N} \, \text{m}^{-1} \; .$$

Evaluating the frequency shift for permanent contact at approximately 30 Hz, using (4.4) with $d = -1$ gives an idea of the stiffness of the CNT: $k_{NT} \approx 4 \times 10^{-3} \, \text{N} \, \text{m}^{-1}$, a value about 10 times the expected sliding bending stiffness (4.1) for this SWCNT of about 500-nm length and radius of 2 nm (example given previously): $k_{b,s} = 4 \times 10^{-4} \, \text{N} \, \text{m}^{-1}$.

4.3.4.2
Damping Signal: Dissipative Part of the CNT–Surface Interaction

The variation of the damping signal recorded simultaneously is plotted in Fig. 4.12c against the CNT end to surface distance. Far from the surface (at large D values, in black), the damping signal is about 0.37 V, a value depending on the electronic feedback loop settings, but proportional to the oscillator damping coefficient γ_0, thus noted Damp$_0$. Energy loss is calculated from the experimental data:

$$\gamma_0 = \frac{k_c}{2\pi\nu_0 Q} = \frac{30}{2\pi \times 143,800 \times 2070} \approx 1.6 \times 10^{-8} \text{kg s}^{-1} \; ,$$

$$E_0 = \pi\gamma_0 2\pi\nu A^2 \approx 2.8 \times 10^{-17} \, \text{J} \approx 178 \, \text{eV} \; .$$

These values are typical for an experiment conducted under a vacuum of 8 mbar. As the percentage of variation is easily measured, this gives a sensitivity threshold of about 2 eV.

The change from noncontact to intermittent contact is marked by the sharp increase in the damping. As soon as the CNT touches the surface, the CNT has to unstick from it: in each oscillation period, the oscillator looses the energy corresponding to the mechanical hysteresis (Fig. 4.11). The damping level is nearly constant

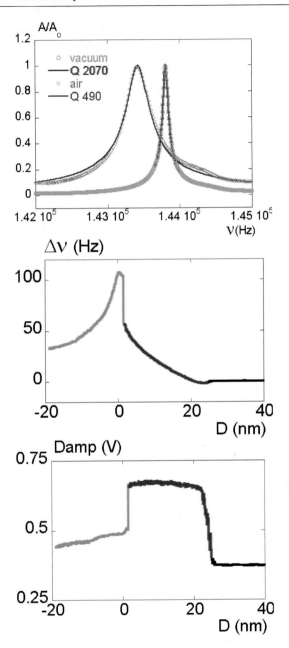

throughout the intermittent area in dark gray, which is consistent with a constant energy needed to unstick. We can evaluate the damping coefficient and the energy associated with the damping value of 0.67 V:

$$\gamma_{\text{int}} = \gamma_{\text{tot}} - \gamma_0 \propto \left(\frac{\text{Damp}_{\text{tot}}}{\text{Damp}_0} - 1 \right) \gamma_0 \approx 1.3 \times 10^{-8}\,\text{kg s}^{-1}\,,$$

$$E_{\text{int}} = E_{\text{tot}} - E_0 \propto \left(\frac{\text{Damp}_{\text{tot}}}{\text{Damp}_0} - 1 \right) E_0 \approx 144\,\text{eV}\,.$$

Assuming the energy losses are solely due to adhesion, we can also evaluate the unsticking force:

$$F_{\text{adh}} = \gamma_{\text{int}} A \omega \approx 250\,\text{pN}\,.$$

This value can be compared with values deduced from molecular mechanics simulations on SWCNTs [68] where the force response was calculated using a Lennard-Jones interaction: assuming the initial negative force to be mostly adhesive gives the same value of about 0.3 nN.

Assuming a van der Waals force acting between a spherical CNT end and a flat surface gives a rough evaluation of this adhesion force:

$$F_{\text{sp-plan}} = \frac{Hr}{6d_{\text{c}}^2} \approx \frac{10^{-19} \times 10^{-9}}{6 \times \left(0.165 \times 10^{-9} \right)^2} \approx 600\,\text{pN}\,,$$

with the CNT end radius $r = 1$ nm, the Hamaker constant of 10^{-19} J and a closest contact distance d_{c} of 0.165×10^{-9} m. It gives the same order of magnitude as the experimental data.

As soon as the CNT does not leave the surface, the damping signal rapidly decreases toward lower values. This permanent contact part of the curve is subject to slight variations, probably due to different sticking and sliding conditions between the CNT and the surface (see the results with a loop reported later).

The permanent contact can never be reached with most of the usual pyramidal Si tips without CNTs, as the resonance frequency shift (due to very high cantilever stiffness) and the energy lost (and thus the damping signal) increase too rapidly and exceed the higher limits of the apparatus.

4.3.5
Mechanical Properties of MWCNTs

4.3.5.1
Helical MWCNTs: Elastic Contribution

After this brief survey on FM signals and evaluating the forces in play, we focus now on the data obtained with a helical MWCNT shown in Fig. 4.13a. From this large-diameter (70-nm) MWCNT, a significant contribution of the elastic force is expected. Variations of the frequency shifts are plotted against the MWCNT–surface distance for different amplitudes in Fig. 4.13b. As in Fig. 4.12b, the three domains—

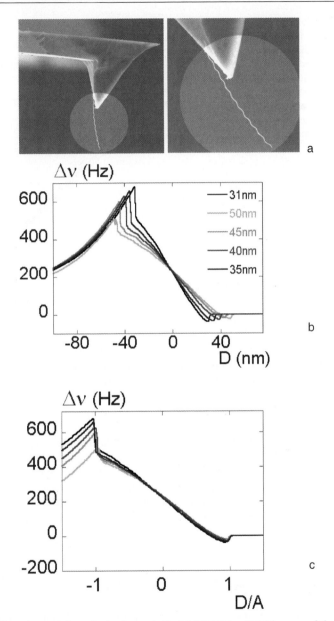

Fig. 4.13. Experimental data obtained on a helical MWCNT. **a** SEM images of the MWCNT. Its length is 12 μm, and it has a radius of 35 nm. The MWCNT is slightly helical with a helix radius of 50 nm and a pitch of 1.2 μm. **b** Frequency shifts recorded with this helical MWCNT on silica at ambient pressure for oscillation amplitudes varying from 31 to 50 nm plotted against the MWCNT to sample distance D. **c** Same as for **b**, but plotted against the normalized distance $d = D/A$. The uniqueness of the variation of the curves indicates that the main contribution to the frequency shift is elastic

Fig. 4.14. Experimental and modeled (4.4) frequency shift of the helical MWCNTe plotted against the normalized distance. Three values of stiffness are used for the fits. The best fit is obtained for a MWCNT stiffness of about $0.07 \, \mathrm{N\,m^{-1}}$

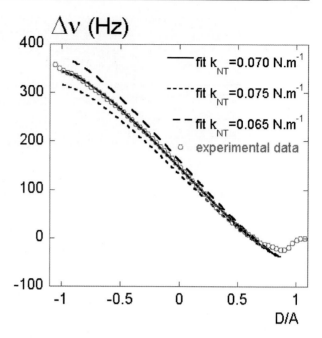

no interaction, intermittent contact and permanent contact—can be delimited with the frequency decrease and jump. With those data, it is very easy to test if the main contribution to the frequency shift comes from elastic behavior of the MWCNT. Remembering (4.4) is independent of the amplitude, we plot the same data against the normalized distance in Fig. 4.13c. All the curves merge into one, indicating that the MWCNT behaves mainly elastically.

This enables us to evaluate the MWCNT stiffness with (4.4)[7]: $k_{NT} \approx 0.07 \, \mathrm{N\,m^{-1}}$ as plotted in Fig. 4.14. Compared with the value obtained with (4.1), $k_{b,s} \approx 0.003 \, \mathrm{N\,m^{-1}}$, it is about 25 times larger, indicating maybe some sticking on the surface or a contribution of the spring constant of the helical structure [71].

In the following examples, the adhesive properties of the CNT tips have to be taken into account to model their mechanical behavior, as it is encountered for most of the CNTs.

4.3.5.2
Mechanical Properties of Straight MWCNTs: Elasticity and Adhesion

While to first order previous results suggest a resonance frequency shift mostly due to an elastic response, in most cases the mechanical response of the CNT tips contains more influences than in this particular example of a helical structure.

Figure 4.15 shows the experimental data (normalized frequency in Fig. 4.15b, interaction energy dissipated in Fig. 4.15c) collected on the straight MWCNT shown in Fig. 4.15a.

---

[7] For all the stiffness values of the CNTs determined in the review, it is the ratio k_{NT}/k_c which is given by the fits. A cantilever stiffness k_c of $30 \, \mathrm{N\,m^{-1}}$ is assumed hereafter.

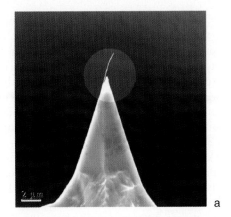

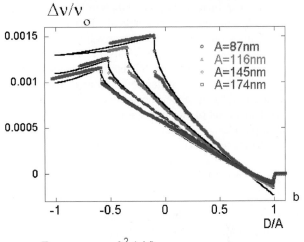

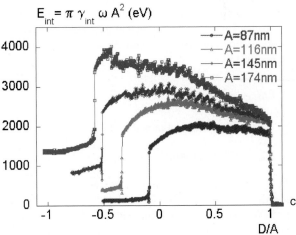

Fig. 4.15. Experimental data obtained on a straight MWCNT. **a** SEM image of the MWCNT: length L about 3.2 μm and radius r about 30 nm. **b** Frequency shifts recorded with this MWCNT on graphite at 8-mbar pressure for oscillation amplitudes varying from 87 to 174 nm plotted against the normalized distance $d = D/A$. **c** Corresponding dissipated energy

In contrast to the helical MWCNT, the representation of the frequency shifts against the normalized distance for different oscillation amplitudes (Fig. 4.15b) does not lead to a unique curve. This result means that (4.4) is unable to give a comprehensive analysis of the variation of the conservative force. Thus, it indicates that the elastic contribution is not the unique origin of the resonance frequency shift. Moreover, the position of the frequency jump is not anymore at a fixed D/A location; it indicates that the size of the intermittent contact area depends now on the amplitude. This point, directly correlated to the strength of the adhesion force, is discussed in Sect. 4.3.7.

We can now use the model that includes adhesion ((4.8), (4.9)). Figure 4.15b shows the comparison of the experimental frequency shifts (colored symbols) with the analytical expressions of the model (black lines) for four amplitudes.

It is possible to interpret the main features of the experimental frequency curves recorded at four different amplitudes using the simple analytical model with the following parameters: $k_{NT} \approx 0.04\,\mathrm{N\,m^{-1}}$, $k_{aNT} \approx 0.1\,\mathrm{N\,m^{-1}}$ and $\Delta \approx 74\,\mathrm{nm}$. Compared with the previous helical MWCNT, k_{NT} for this straight MWCNT is about 2 times smaller, giving a larger weight for the contribution of adhesion forces in the frequency shift signal. Using (4.1) with the MWCNT geometrical values shown in Fig. 4.15a gives a CNT repulsive stiffness of twice k_{NT}: $k_{b,s} \approx 0.08\,\mathrm{N\,m^{-1}}$. For the fitting process, it has to be pointed that the value of the pull-off distance Δ is imposed by the experimental data: it is given by the position of the frequency jump. Thus, a set of two adjustable parameters describes the frequency signals for four amplitudes.

The adhesion force can be derived as

$$F_{adh} = k_{aNT}\Delta \approx 7.5\,\mathrm{nN}\ .$$

To track the origin of the energy dissipation, the interaction energies for four oscillation amplitudes are plotted together in Fig. 4.15c: at the beginning of the curve the energy jump does not depend much on the amplitude, indicating again a mainly adhesive origin for energy dissipation for this MWCNT.

To check the autoconsistency of the model, we can use (4.7) that links dissipation energy and parameters (k_{aNT} and Δ) derived from the frequency shift (conservative interaction) fit

$$E_{adh} = \frac{1}{2}k_{aNT}\Delta^2 \approx 1710\,\mathrm{eV}$$

and compare the value obtained with the experimental interaction energy E_{int}:

$$E_{int} \sim 1790 - 2090\,\mathrm{eV}\ .$$

The dissipated adhesion energy as given by the model is close to the experimental dissipated energy. The model not only describes the main conservative features of this MWCNT when interacting with a surface through the resonance frequency shift, but it also gives coherent values compared with the experimental damping signal (thus dissipative contribution) data.

4.3.6
Mechanical Properties of SWCNTs: Main Adhesive Contribution

As the SWCNT is very much thinner than the MWCNT but is still long, we recall the spring constant scales as r^4/L^3, it will be more flexible, and its expected stiffness (calculated previously, $k_{b,s} \approx 4 \times 10^{-4}\,\mathrm{N\,m^{-1}}$) is much smaller. The elastic contribution to the frequency shift is thus expected to be 2 orders of magnitude lower than for MWCNTs, leading to a higher adhesive contribution.

Despite this difference with MWCNTs, we will see that the same theoretical expressions ((4.8), (4.9)) can capture the main experimental features.

Figure 4.16 shows data recorded with a SWCNT or a double-walled CNT shown in Fig. 4.16a. The resonance frequency shifts and dissipated energy variation are plotted in Fig. 4.16b and c for different amplitudes as a function of the normalized distance. The transition from intermittent to permanent contact marked by the rapid increase of the frequency shift coincident to the energy decrease happens again at a location depending on the amplitude, indicating a distance necessary to unstick the CNT from the sample.

The analytical expressions (black lines) can reproduce the main experimental frequency features for four amplitudes with a set of three parameters: $k_{NT} \approx 0.001\,\mathrm{N\,m^{-1}}$, $k_{aNT} \approx 0.06\,\mathrm{N\,m^{-1}}$ and $\Delta \approx 30\,\mathrm{nm}$. As for the helical CNT, k_{NT} is about 30 times larger than the expected stiffness in the case of the sliding end of the CNT, $k_{b,s} \approx 4 \times 10^{-4}\,\mathrm{N\,m^{-1}}$. The two other parameters give an evaluation of the adhesion force,

$$F_{adh} = k_{aNT}\Delta \approx 1.8\,\mathrm{nN}\ ,$$

a value slightly smaller than for the straight MWCNT.

The value is about 6 times the value of 0.3 nN deduced from the experimental data and 3 times the rough evaluation given by the sphere–plane interaction described in Sect. 4.3.4.

As the tube is quite soft, one might consider the tube lying on the surface on its side (Fig. 4.17). The adhesion force model can be implemented considering a tube interaction [89]:

$$F_{adh} = \frac{HL_c}{12\sqrt{2}d_c^{5/2}}r^{1/2}\ ,$$

where L_c is the length of the contact of the tube with the surface, $H = 10^{-19}\,\mathrm{J}$,[8] $r = 1\,\mathrm{nm}$ and $d_c = 0.165\,\mathrm{nm}$. L_c has to be approximately 3.5 nm for an F_{adh} value of 1.8 nN, a rather reasonable value.

This value can be compared with values obtained by Walkeajärvi et al. [26], who evaluated adhesion forces with gold lines for MWCNTs from 0.2 to 2.3 nN, or adhesion force measurements between a CNT and a silicon dioxide substrate [27] for a SWCNT embedded in oxide: the experimental value is 10 nN, a slightly larger

---

[8] An experimental value has been determined by Akita et al. [94] by transmission electron microscopy work on manipulated CNTs: the authors deduced a Hamaker constant for nanotube sides of $H \approx 60 \times 10^{-20}\,\mathrm{J}$.

Fig. 4.16. Experimental and modeled data obtained on a SWCNT. **a** SEM image of the SWCNT. The free length is evaluated to be 500 nm, with a radius of about 2 nm. **b** Variation of experimental and model frequency shifts against the normalized distance for four different amplitudes (Eqs. (4.8), (4.9) were used to fit the data). The fit parameters value are $k_{NT} = 0.012 - 0.009\,\mathrm{N\,m^{-1}}$, $k_{aNT} = 0.06 - 0.18\,\mathrm{N\,m^{-1}}$ and $\Delta = 30$ nm. The smallest stiffness that can be evaluated with this method is about $0.001\,\mathrm{N\,m^{-1}}$. For sake of clarity only a fraction of the experimental data are shown. **c** Corresponding variation of dissipated energy due to interaction with the sample deduced from the experimental data

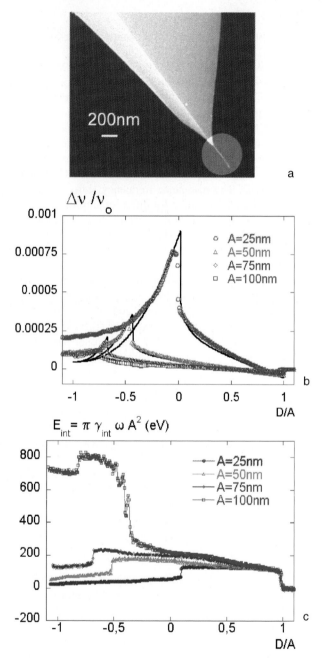

Fig. 4.17. A CNT lying on its side on the surface, with definition of the parameters used to evaluate its adhesion to the surface

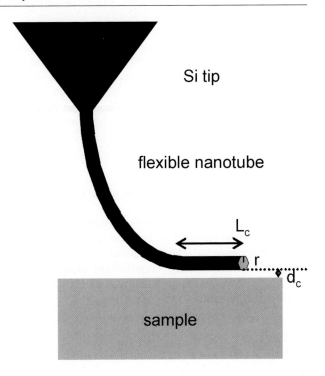

value, as expected for an embedded CNT in opposition to our SWCNT side-to-plane adhesion.

Some SWCNTs are lost during the approach to the sample, indicating a weak attachment point. We expect the adhesion force with the surface to be larger than the strength of the CNT attachment on the Si tip: the CNT detaches from the tip.

The main origin of dissipation can be deduced from the experimental data on the interaction energy: if it is linked to adhesion of the CNT to the surface, the energy needed to unstick the CNT should be independent of the oscillation amplitude. Using (4.6), we display interaction energy in Fig. 4.16c. For the three smallest amplitudes, the energy dissipated varies slightly with the CNT displacement. For the largest amplitude, the energy increases suddenly without any sudden apparent frequency change. This could be due to complex sticking conditions between the CNT and the surface. But in all cases, the first energy jump has nearly the same height (about 125 eV) whatever the oscillation amplitude.

As for the MWCNT, the validity of the model can be tested by using (4.7) that links parameters deduced from conservative interaction obtained from frequency shift fits ((4.8), (4.9)) to energy dissipation,

$$E_{\text{adh}} = \frac{1}{2} k_{\text{aNT}} \Delta^2 \approx 168 \text{ eV} ,$$

a value rather close to the experimental one. Again, the simple analytical model gives a coherent picture of the interaction of the CNT with the surface with a set of two adjustable parameters (the pull-off distance is excluded as its value is determined without any choice) explaining both frequency and damping curves (thus conservative and dissipative interactions) for a set of four oscillation amplitudes.

At the beginning of the CNT contact, the main energy dissipation source is adhesive loss. When the CNT goes closer to the surface, many complicated situations can happen with, for example, energy dissipation due to CNT friction on the surface or a drastic change in the contact between the CNT and the surface as a function of the vertical displacement, thus a CNT force acting on the contact.

4.3.7
Comparison of Mechanical Properties of CNTs

It is informative to compare the mechanical properties of different CNT tips. We first discuss the stiffness and adhesion energies of a SWCNT, a MWCNT and a silicon nanoneedle. Then we focus on a different way to evaluate the weights of the adhesive and elastic contributions of the different CNTs. In particular, we explore the transition from large to small elastic forces corresponding to large to small oscillation amplitudes, respectively.

4.3.7.1
Stiffness and Adhesion Energy

Figure 4.18 compares the experimental frequency shifts (Fig. 4.18d) and dissipated energies (Fig. 4.18e) recorded for a SWCNT (Fig. 4.18a), a MWCNT (Fig. 4.18b) and a silicon nanoneedle carved with a focused ion beam (Fig. 4.18c) [90].

For a given CNT–surface distance, the largest frequency shift is for the nanoneedle, indicating a high force gradient, whereas the frequency shift is the smallest for the SWCNT, corresponding to the smallest stiffness. The fit stiffness values are: 3.3×10^{-3}, 4.5×10^{-2} and $0.9 \, \mathrm{N \, m^{-1}}$ for the SWCNT, MWCNT and nanoneedle, respectively. For the corresponding energy variation, the nanoneedle shows a viscoelastic energy increase while approaching the sample [90], whereas the energy of the CNTs is mostly constant before the CNTs reach the permanent contact domain, indicating adhesive losses due to surface unsticking. The adhesion energy of the SWCNT is about one tenth that of of the MWCNT, in agreement with the expected smaller CNT diameter.

In conclusion, both SWCNTs (except for a few of them) and MWCNTs were shown to sustain multiple approach–retract curves. In some cases, the Si tip could be reached in the approach and it could touch the surface without any apparent sign of mechanical behavior change of the CNT. In most cases, too flexible probes should be avoided. Whereas SWCNTs offer a smaller radius and a batch growing process, the question of their attachment to the Si tip is open. For fused MWCNTs, their advantage is the solidity of the attachment, with a larger radius.

AFM in dynamical FM mode appears to be a powerful tool for measurements, enabling the evaluation of the effective stiffness of CNTs when combined with a simple analytical model and enabling energy measurements with a high sensitivity when working under a vacuum.

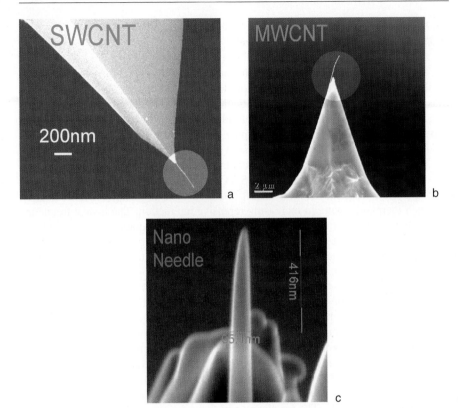

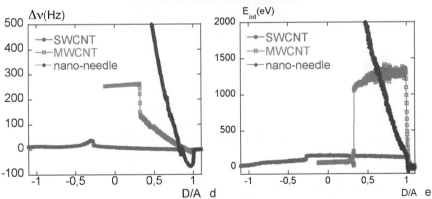

Fig. 4.18. Comparison of data obtained in frequency modulation mode for a 50-nm amplitude with a SWCNT, a MWCNT and a Si carved nanoneedle all shown as SEM images in **a–c**, respectively **d** Variation of the frequency shift. **e** Corresponding energy variation during the approach of the CNT to the surface

4.3.7.2
Oscillation Amplitude and Pull-Off Distance

An easy way to balance the bulk elastic force stored by the CNT is to change the oscillation amplitude: the smaller the amplitude, the lower the elastic energy. It is thus possible to achieve experimental situations where the adhesion energy becomes larger than the elastic one at small amplitudes. We first illustrate this situation with experimental frequency shift and damping curves, then we compare the different CNT plots of the intermittent domain size variation with the oscillation amplitudes.

This transition to mostly adhesive contribution is illustrated with experimental data in Fig. 4.19 for a 250-nm-long SWCNT (Fig. 4.19a). At large amplitudes (dark gray) the frequency (Fig. 4.19b) first decreases owing to attractive interaction and then increases owing to increasing repulsive interaction. The lowest frequency shift value increases for amplitudes decreasing as for a frequency varying with the amplitude power law $A^{-3/2}$ [72, 74, 77]. Also the damping curves (Fig. 4.19c) of the largest amplitude have the usual steplike shape, with a step height increasing for decreasing amplitudes, as expected for constant energy proportional to the product of damping signal and the square of the amplitude , i.e., γA^2.

For the two smallest amplitudes however, the shape of the curve changes: the frequency slope is much larger, the lowest frequency shift nearly vanishes, the height of the damping signal is not constant anymore or its value is not larger than the higher amplitude damping signal one. This change is due to an oscillation amplitude lower than the pull-off distance Δ. As soon as the CNT reaches the surface, its elastic force is not sufficient to overcome the adhesion force with the surface: the CNT sticks to the surface and reaches the permanent contact domain without passing through the intermittent contact one. The frequency and damping curves for 12-nm oscillation amplitude reveal a very small intermittent contact domain on a piezo movement of about 3–4 nm, before the transition to the large permanent domain.

When the oscillation amplitude is less than the pull-off distance Δ, the CNT thus changes its behavior to a complete sticking one. In practice, to avoid the CNT sticking on the surface when the attachment to the tip is not secured, large amplitudes should be favored.

This transition from the usual intermittent contact domain with both elastic and adhesive contributions to purely adhesive behavior can be highlighted with the plot of the variation of the intermittent contact domain size with the oscillation amplitude. Figure 4.20 presents this plot for the helical MWCNT presented in Sect. 4.3.5.1, the MWCNT of Sect. 4.3.5.2 and the SWCNT of Sect. 4.3.6.

For the helical CNT, the MWCNT and the SWCNT, the domain size is proportional to the oscillation amplitude, with an identical slope of 2, but the extrapolation of the curve for the helical CNT crosses the origin, whereas the other two CNTs have a null domain size for an oscillation amplitude smaller than 25 and 78 nm for the SWCNT and the MWCNT, respectively. From this amplitude value and lower ones, the CNT cannot leave the surface as soon as it touches it because the elastic force is not sufficient to overcome the adhesion force. This amplitude oscillation value corresponds to the pull-off distance Δ in the analytic model. This confirms again the larger contribution of the elastic properties in the case of the helical CNT, and the mixing of adhesive properties with elastic ones in most cases. Another example

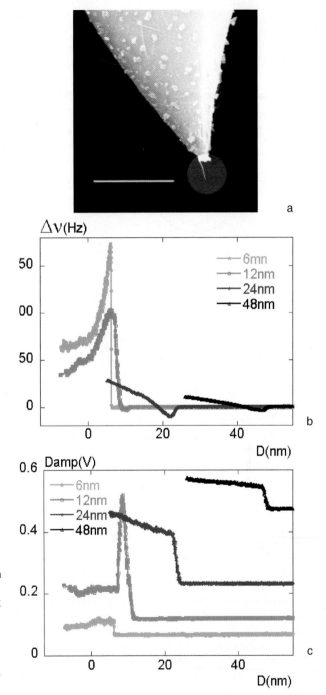

Fig. 4.19. The transition from elastic and adhesive contributions (large amplitudes) to complete adhesive contribution for a SWCNT shown in **a** (*scale bar* 1 μm). **b** Variation of the resonance frequency shift with distance *D* for four amplitudes. **c** Corresponding damping variations. The transition happens for an amplitude between 12 and 6 nm where the intermittent contact domain disappears

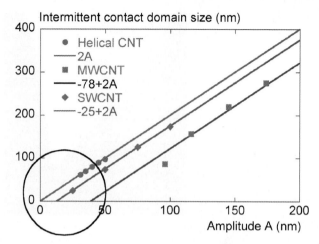

Fig. 4.20. Variation of the intermittent contact domain size as measured from experimental curves with the oscillation amplitude for a helical CNT, a MWCNT and a SWCNT. The *circle* highlights the interpolation curves (*straight lines* with equations shown, values are in nanometers) crossing the origin for the helical CNT, but not for the straight CNTs with significant adhesive contribution. The value at the origin is the pull-off distance

of the main elastic behavior for CNTs was demonstrated by molecular simulation in [68], where the beam model of elasticity gives nearly an identical force to the molecular mechanics force at the beginning of the approach curve. In this case, the CNTs were single-walled but very short ones, 13.6 nm, leading to very large stiffness values of the order of $1\,\mathrm{N\,m^{-1}}$. The elastic contribution will thus lead to ultrashort SWCNT behavior.

We will see that the sensitivity of the CNTs to sticking can be exploited for imaging purposes.

4.3.8
Special cases

4.3.8.1
Periodic Helical MWCNT Signals

The helical structure exhibits specific mechanical properties owing to the periodicity, which can be highlighted in approach curves with very large scan sizes (Fig. 4.21).

Those curves recorded with helical MWCNTs of 1.2-μm pitch length (Fig. 4.21a) reveal new periodic sawtooth features both in frequency shift (Fig. 4.21b) and damping (Fig. 4.21c). Each sawtooth frequency shift or damping signal is coherent with a sticking CNT scenario when the bulk elastic energy of the CNT cannot overcome the surface adhesion energy. The distance between two neighboring frequency or damping peaks is about 1.2 μm and is thus very similar to the helical pitch.

The scenario for those long-distance approach curves is thus contact, then adhesion of the end of the CNT, followed by the usual CNT sliding on the surface, but in

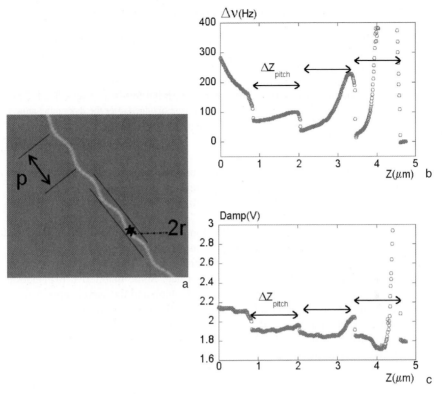

Fig. 4.21. Helical MWCNT data. **a** Close-up of the helical CNT. The pitch is 1.2 μm, its curvature is 50 nm. **b** Variation of the resonance frequency shift for a large scan size showing periodic events. **c** Corresponding variation of the damping

contrast to the straight CNT, after one pitch length, the next spire comes into contact with the surface, sticks and then slides, followed by other spires, giving the periodic signals in Fig. 4.21b and c.

4.3.8.2
SWCNT Loop

In the case of a loop (Fig. 4.22, panel a) formed by the SWCNT at the end of the tip apex, the damping and frequency shift curves reveal complex phenomena. However, a scenario can be built based on previous results qualitative analysis. Frequency and dissipation curves for a loop are shown in Fig. 4.22, panels b and c, respectively.

As for a straight CNT, the resonance frequency first decreases owing to attractive forces then increases owing to an increasing repulsive contribution with a sharp variation of frequency in coincidence with damping variation. But the damping signal increases after the first jump in the "intermittent contact domain" as defined previously and then decreases after the sharp frequency increase together with the

Fig. 4.22. Experimental data obtained with a single-walled loop at the tip apex. **a** SEM image of the loop emerging from the tip. This CNT was directly grown on the tip, using the method described in Sect. 4.2.2. The CNT is a single or a double CNT with the growing end which is stuck on the tip forming a loop. The length of the loop is close to 850 nm. **b** Variation of the resonance frequency shift $\Delta\nu$ with loop to surface distance D. First curve recorded after engagement of the microscope tip. **c** Corresponding variation of the damping signal. **d** Tenth curve recorded: the frequency variation has changed as it did for each curve. **e** Tenth curve recorded: corresponding damping variation. **f** SEM image obtained after the dynamical atomic force microscopy experiments showing that the loop has changed

⸻▶

decrease of the damping. Those coincident variations were used previously to define the beginning of the permanent contact domain. The intermittent and permanent domain lengths are nearly similar: about 160 nm. The damping curve can be schemed by the diagram presented in Fig. 4.22, panel c′, showing six different zones, labeled from 1 to 6, linked to the deformation of the loop during the approach to the surface. The increase of the damping in the intermittent contact domain (zones 2–3) and the decrease in the permanent contact domain (zones 3–4) are compatible with variation of the loop contact area with the surface. Some qualitative explanation can be given. New points of sticking between the loop and the sample appear from zones 2–3, increasing the contact area and thus the dissipation. Note that curve hysteresis between approach and retraction (not shown) is often observed for the loop in contrast to usual SWCNT curves. At point 3, a large part of the loop is in permanent contact with the surface, but not all the loop: some parts are still in intermittent contact and it is only in position 5 that all parts of the loop never leave the surface during the overall period. This speculative differentiation in two parts of the loop is compatible with the double sharp frequency increase (Fig. 4.22, panel b) associated with the damping decrease, and with the similar length of intermittent and permanent domains.

Using the same procedure as in Sect. 4.3.4.1, one can evaluate the interaction energy at the first damping jump transition between zones 1 and 2 (E_{int12}) and just before the damping decrease (E_{int23}) between zones 2 and 3:

$$E_{int12} \approx 1580 \, \text{eV} \quad \text{and} \quad E_{int23} \approx 2252 \, \text{eV} \ .$$

Those energy values are much larger than the usual SWCNT ones; E_{int23} is the largest value ever obtained for a CNT.

When loops are formed during direct growth of SWCNTs by CVD, one side of the loop is linked to the tip by van der Waals interactions only; the CNT attachment is thus weak on one side. This could be assumed from the experiments as the curves changed from one to another, as displayed in Fig. 4.22, panels d and e. The same energy evaluation gives

$$E_{int12} \approx 1270 \, \text{eV} \quad \text{and} \quad E_{int23} \approx 5019 \, \text{eV} \ ;$$

thus, a final value twice the previous one.

The loop slide on the tip was confirmed on the scanning electron microscopy image taken after the experiments: at the end of the experiment, the loop is folded on the tip.

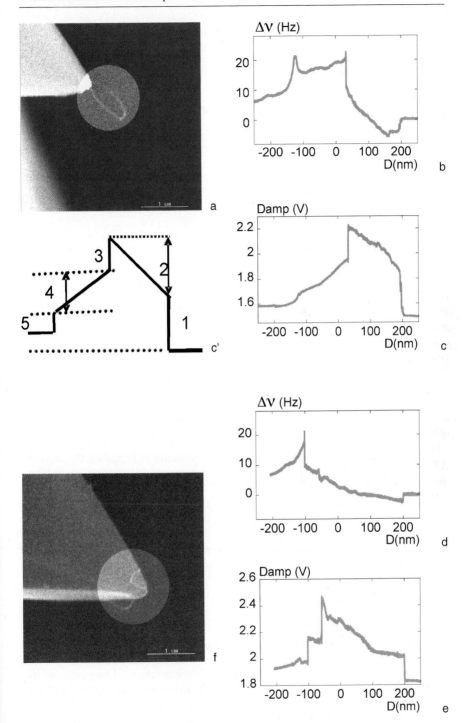

Those results confirm that from experimental data we can differentiate whether the CNT slides on the surface or on the tip: if the attachment on the tip is not firm, the curves evolve from one to the other.

4.4
Imaging

Besides evaluating forces and interactions at play, the approach curves shown previously can also help to understand the image contrast, assuming a steady state for the CNT–cantilever oscillator at each image point and a flat feedback signal. We will see an example where the sensitivity of the CNT to sticking conditions on different surface materials will enable chemical recognition.

4.4.1
Literature Tour

Earlier studies of imaging with CNTs as probes illustrated the benefit compared with conventional imaging with Si tips in AM (tapping) mode. The samples imaged included biological macromolecules—higher resolution [31,91–93], deep features—height contrast closer to topography [39, 94], or heterogeneous grain samples—longer stability during long and continuous scanning [39].

Joint studies of the mechanical properties of CNTs and their ability to image pointed out the influence of probe stiffness. As expected, too thin and too long CNTs lead to image blurring and instability [76,94], even though they might be useful in very specific cases [95]. The influence of the angle of the CNT with respect to the surface has also been analyzed mostly numerically [68, 76, 96] and experimentally by comparing transmission electron microscopy and SPM results on multiple CNT probes [55]. All the study agreed drew the same conclusion regarding the angle of the CNT with the surface: the closer to 90° the better.

As for regular SPM tips, the geometry and the size of the CNT are not the only factors enabling high-resolution imaging. Several other factors may have an important influence on the image contrast: balance between elastic forces and adhesion forces (inducing imaging in attractive or repulsive force balance), the quality factor of the CNT–cantilever oscillator,and the force range [59, 71, 96, 97]: in the end, it is the interaction of the CNT with the sample that governs the imaging. An example of particular image contrast linked to the sensitivity of the CNT to surface sticking is given in next section.

4.4.2
Using the Mechanical Properties of CNTs for Imaging

An illustration of the usefulness of knowledge of the mechanical properties of the CNT probe is given in Fig. 4.23 as SPM images of a heterogeneous surface. InAs dots epitaxially grown on GaAs were imaged in repulsive force balance and dynamical FM mode at a positive constant frequency shift with a regular Si tip (Fig. 4.23a) and with a helical MWCNT (Fig. 4.23b).

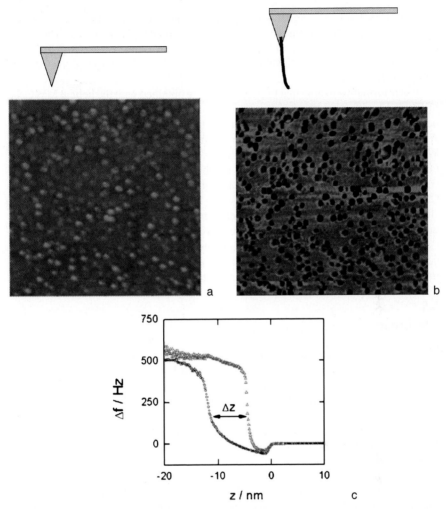

Fig. 4.23. Data recorded on In As dots. **a** Height image (size 1 μm × 1 μm) recorded in repulsive force balance with a regular commercial Si tip. **b** Height image (size 1 μm × 1 μm) recorded with a helical MWCNT probe in the frequency modulation mode, at a positive frequency shift of 150 Hz. **c** Variation of the frequency shift during approach of the MWCNT towards the InAs surface (*triangles*) and GaAs islands (*circles*): the different sticking on In/As explains the height difference of **a** and **b**

The InAs dots of about 3-nm height in Fig. 4.23a appear as holes when imaged with the helical CNT. The origin of this height contrast inversion between regular Si tip and CNT probe images can be found in the variation of the frequency shift recorded on the InAs islands and on the GaAs surface displayed in Fig. 4.23c. The sticking of the MWCNT is greater on the GaAs surface than on the InAs islands, inducing a smaller intermittent contact area (corresponding to a smaller sticking distance Δ) for the GaAs surface. To maintain the frequency shift at a positive value of the set point, the feedback has to move the surface a distance $\Delta Z \approx 8$ nm, resulting in a piezoceramic movement being recorded in the height image. The island dots of 3 nm will then appear as holes of approximately 5 nm. This sensitivity of the CNT to different material stickability illustrates the possibility to use the CNT as a mechanical sensor.

It is informative to note that those results were obtained with the helical MWCNT of Fig. 4.13 with a length of 12 μm and a radius of 35 nm: the ability to distinguish between dots at a distance of a a few nanometers could be surprising at first glance. As an example, using (4.2) to evaluate the free CNT end vibration gives 1.4 nm. We clearly see that this formula is not valid when the CNT interacts with a surface. One element could give a starting clue: this helical CNT has a very small adhesive contribution (Sects. 4.3.5, 4.3.7) coherent with a small apex size. Moreover, when CNTs are fused on the tip, a current is used to break down the CNT usually at defect sites. Because the outermost carbon shells are in contact with the electrodes, they will be the first to break: this process thus favors a smaller CNT apex than the main CNT diameter. It could explain the high image resolution compared with the CNT geometrical considerations together with the negligible adhesive contribution to its mechanical properties compared with its elastic properties.

4.5
Conclusion

CNTs are powerful scanning probe microscope tips enabling not only higher resolution owing to their high aspect ratio, but also longer lifetime because of their exceptional mechanical properties.

Among the numerous ways developed to produce CNT tips, we have developed two different strategies: either attaching MWCNTs one by one under an optical microscope on tips by nanofusing or directly growing SWCNTs with a batch process. The MWCNT attached by fusing and the SWCNTs directly grown on tips are stable enough to record multiple approaches and retractions from the surface. One MWCNT even enabled multiple experiments over a period of 6 months.

The mechanical properties of CNTs can be measured directly in SPM. They can be described with a bulk force coming from linear elasticity and a surface force: adhesion. Those two forces are in competition. The elastic contribution can be adjusted with the choice of amplitude: the smaller the amplitude, the lower the elastic energy stored in the CNT. In practice, a larger adhesion contribution means that the CNT sticks to the surface: it may lead to the CNT being lost when the attachment of the CNT to the tip is weak. We used a simple analytical model able to describe mechanical properties of tips as different as helical MWCNT, straight MWCNT and SWCNT, SWCNT loops and even a carved nanoneedle to quantify the effective stiffness of the CNT and also adhesive parameters such as an effective adhesive stiffness. The data give directly the pull-off distance. The energy losses measured with FM mode indicate that the main origin is adhesive, enabling adhesion energy measurements. The effective stiffness of CNTs typically varies from 0.001 N m^{-1} (SWCNT) to a few tenths of newtons per meter (MWCNT) compared with a value close to 1 N m^{-1} for a silicon nanoneedle. The adhesion energies range from a few tens of electronvolts to a few thousand electronvolts. SPM dynamical FM mode has proven to be a powerful tool for measurements of the of adhesion energy of a CNT and, when combined with an analytical model, for evaluation of the effective stiffness of a CNT when the CNT is squeezed between the tip and the surface. As a result, it is also possible to discriminate between surface and bulk forces (here adhesion and elasticity, respectively) that are important for nanomaterial processing and use.

The sensitivity of CNTs to different sticking surfaces has been used for mechanical imaging on a heterogeneous sample.

Acknowledgements. The authors are grateful to D. Mariolle and F. Bertin from CEA LETI France for carving of the pyramidal silicon tip to create a nanoneedle, for some SEM images and fruitful discussion. S.M., C.B. and J.P.A. benefit from financial support of the Région Aquitaine and C'Nano Grand Sud Ouest.

Symbols

Symbol	Meaning	Unit
A	(Tip–cantilever) oscillation amplitude	nm
δ	Normalized distance necessary to unstick from the sample	–
Δ	Distance necessary to unstick the nanotube from the sample	nm
$\Delta v = v - v_0$	Resonance frequency shift	Hz
d	Reduced distance $d = D/A$	–
D	Nanotube end-surface distance at the equilibrium position	nm
E	Young modulus	Pa
E_0	Energy lost per cycle without any interaction with the sample	J or eV
E_{adh}	Energy loss of the nanotube due to adhesive hysteresis	J or eV
E_{int}	Energy lost per cycle due to the interaction with the surface	J or eV
γ_w	Damping coefficient of the oscillator without interaction	$kg\,s^{-1}$
γ_{int}	Damping coefficient of the oscillator with interaction	$kg\,s^{-1}$
k_{aNT}	Attractive nanotube stiffness deduced from a fit of the experimental data	$N\,m^{-1}$
k_B	Boltzmann constant	$1.38062 - 23\,J\,K^{-1}$
$k_{b,p}$	Nanotube bending stiffness when the end of the carbon nanotube is pinned on the surface	$N\,m^{-1}$
$k_{b,s}$	Nanotube bending stiffness when the end of the carnon nanotube slides on the surface	$N\,m^{-1}$
k_c	Cantilever stiffness	$N\,m^{-1}$

Symbol	Meaning	Unit
k_e	Nanotube elongation stiffness	$N\,m^{-1}$
k_{int}^{eff}	Interaction effective stiffness	$N\,m^{-1}$
k_{NT}	Nanotube stiffness deduced from a fit of the experimental data	$N\,m^{-1}$
L	Nanotube length	nm
m	Effective mass of the oscillator	kg
ν_0	(Tip–cantilever) oscillator resonance frequency without interaction with the surface	Hz
ω_w	Pulsation of the oscillator	$rad\,s^{-1}$
Q	Quality factor of the oscillator	–
r	Nanotube exterior radius	nm
T	Temperature	K
X_{tip}	Vibration amplitude at the free end of the nanotube induced by thermal excitation	nm

References

1. Giessibl FJ (2003) Rev Mod Phys 75:949–989
2. Nonnemacher M, O'Boyle MP, Wrickramasinghe HK (1991) Appl Phys Lett 58:2921–2923
3. Frisbie CD, Rozsnyai LF, Noy A, Whrighton MS, Lieber CM (1998) Science 265:2072–2074
4. Noy A, Sanders CH, Vezenov DV, Wong SS, Lieber CM (1998) Langmuir 14:1508–1511
5. Kueng A, Kranz C, Mizaikoff B (2003) Sens Lett 1:2–15
6. Nanoworld (2007) http://www.nanoworld.com
7. Nanosensors (2007) http://www.nanosensors.com
8. Iijima S (1991) Nature 354:56–58
9. Salvetat JP et al. (1999) Phys Rev Lett 82:944–947
10. Lu JP (1997) Phys Rev Lett 79:1297–1300
11. Dai HJ, Hafner JH, Rinzler AG, Colbert DT, Smalley RE (1996) Nature 384:147–150
12. Wong EW, Sheehan PE, Lieber CM (1997) Science 277:1971–1975
13. Iijima S, Brabec C, Maiti A, Bernholc J (1996) J Chem Phys 104:2089–2092
14. Lemay SG, Janssen JW, van den Hout M, Mooij M, Bronikowsky MJ, Williw PA, Smalley RE, Kouwenhoven LP, Dekker C (2001) Nature 412:617
15. Ouyang M, Huang JL, Cheung CL, Lieber CM (2001) Science 291:97–100
16. Shimuzu T, Tokumoto H, Akita S, Nakayama Y (2001) Surf Sci 486:L455
17. Konishi H, Murata Y, Wongwiriyapan W, Kishida M, Tomika M, Motoyoshi K, Honda S, Katayama M, Yoshimoto S, Kubo K, Hobara R, Mastuda I, Hasegawa S, Yoshimura M, Lee J-G, Mori H (2007) Rev Sci Instrum 78:013703-1–013703-6
18. Dai H, Franlin N, Han J (1998) Appl Phys Lett 73:1508
19. Stevens R, Nguyen C, Cassel A, Delzeit L, Meyyappan M, Han J (2000) Appl Phys Lett 77:3453–3455
20. Kuramochi H, Ando K, Tokizaki T, Yasutake M, Pérez-Murano, Dagata JA, Yokoyama H (2004) Surf Sci 566:343–348
21. Wong SS, Joselevitch E, Wooley AT, Cheung CL, Lieber CM (1998) Nature 394:52–55
22. Miyauchi Y, Chiasi S, Murakami Y, Hayashida Y, Maruyama S (2004) Chem Phys Lett 387:198–203
23. Nishino T, Ito T, Vezenov DV (2002) Anal Chem 74:4275–4278

24. Collins PG, Avouris P (2000) Sci Am 62–70
25. Avouris P, Hertel T, Martel R, Schmidt T, Shea HR, Walkup RE (1999) Appl Surf Sci 141:201–209
26. Walkeajärvi T, Leivonen J, Ahlskog A (2005) J Appl Sci 98:104301
27. Whittaker JD, Minot ED, Tanenbaum DM, McEueun PL, Davis RC (2006) Nano Lett 6:953
28. Ajayan PM, Ebbesen TW (1997) Rep Prog Phys 60:1025–1062
29. Endo M, Kroto HW (1992) J Chem Phys 96:6941–6944
30. Dai H, Rinsler AG, Nikolaev P, Thess A, Colbert DT, Smalley R (1996) Chem Phys Lett 260:471–475
31. Cheung CL, Hafner JH, Lieber CM (2000) Proc Natl Acad Sci USA 97:3809–3813
32. Hafner JH, Cheung CL, Oosterkamp TH, Lieber CM (2001) J Chem Phys 105:743–746
33. Nishijima H, Kamo S, Akita S, Nakayama Y, Hohmura KI, Yoshimura SH, Takkeyasu K (1999) Appl Phys Lett 74:4061–4063
34. Jarvis S, Ishida T, Uchihashi T, Tokumoto H, Tokumoto H (2001) Appl Phys A 72(Suppl):S129–S132
35. Martinez J, Yusvinsky TD, Fennimore AM, Zettl A, Garcia R, Bustamante C (2005) Nanotechnology 16:2493–2496
36. Lee HW, Kim SH, Kwak YK, Han CS (2005) Rev Sci Instrum 76:046108:1–046108:5
37. Tang J, Yang G, Zhang Q, Parhat A, Maynor B, Liu J, Qin LC, Shou O (2005) Nano Lett 5:11–14
38. Stevens RMD, Frederick NA, Smith BL, Morse DE, Stucky GD, Hansma PK (2000) Nanotechnology 11:1–5
39. Nguyen CV, Chao KJ, Stevens RMD, Delzeit D, Cassel A, Harper JD, Han J Meyyapan M (2001) Nanotechnology 12:363
40. Nakayama Y, Nishijima H, Akita S, Hohmura KI, Yoshimura S H, Takeyasu K (2000) J Vac Sci Technol B 18:661–664
41. Yamamoto K, Akita S, Nakayama Y (1996) Jpn J Appl Phys 35:L917
42. Dai H, Kong J, Zhou C, Franklin N, Tombler T, Cassel A, Fan S, Chapline M (1999) J Phys Chem B 103:11255
43. Homma Y, Kobayashi Y, Ogino T, Yamashita T (2002) Appl Phys Lett 81:2261–2263
44. Kong J, Zhou C, Morpurgo A, Soh T, Marcus C, Quate C, Dai H (1999) Appl Phys A 69:305
45. Marty L, Bouchiat V, Naud C, Chaumont M, Fournier T, Bonnot AM (2003) Nano Lett 3:1115–1118
46. Bonnot AM, Mathis B, Moulin S (1993) Appl Phys Lett 63:1754–1756
47. Kis A, Csany G, Sulvetat JP, Lee TN, Couteau E, Kulik AJ, Benôit W, Brugger J, Forro L (2004) Nat Mater 3:153
48. Kociak M, Suenaga K, Hirahara K, Saito Y, Nakahira T, Iijima S (2002) Phys Rev Lett 89:155501:1–155501:4
49. Saito R, Dresselhaus G, Dresselhaus MS (1998) Physical properties of carbon nanotubes. Imperial College Press, London
50. Kataura H, Kumazawa Y, Maniwa I, Umezu I, Suzuki S, Ohstua Y, Achiba Y (1999) Synth Met 103:2555
51. Park BC, Jung KY, Song WY, O BH, Ahn SJ (2006) Adv Matter 18:95–98
52. Stevens R, Nguyen C, Meyyapan M (2006) IEEE Trans Nanotechnol 5:255–257
53. Wong SS, Harper JD, Lansbury PT Jr, Lieber CM (1998) J Am Chem Soc 120:603–604
54. Cooper EB, Manalis SR, Fang H, Dai H, Minne ZC, Hunt T, Quate CF (1999) Appl Phys Lett 75:3566–3568
55. Wade LA, Shapiro IR, Ma Z, Quake SR, Collier CP (2004) Nano Lett 4:725–731

56. Yuzvinsky TD, Mickelson W, Aloni S, Begtrup GE, Kis A, Zettl A (2006) Nano Lett 6:2718–2722
57. Nguyen CV, So C, Stevens RM, Li Y, Delziet L, Sarrazin P, Meyyapan M (2004) J Phys Chem B 108:2816–2821
58. Collins PH, Arnold MS, Avouris P (2001) Science 292:706–709
59. Chen L, Cheung CL, Ashby PD, Lieber CM (2004) Nano Lett 4:1725–1731
60. Li J, Cassel AM, Dai H (1999) Surf Interface Anal 28:8–11
61. Stevens RM, Nguyen CV, Meyyapan M (2004) IEEE Trans Nanobiosci 3:56–60
62. Rotkin SV, Subramoney S (2005) Applied physics of carbon nanotubes. Springer, Berlin
63. Yacobson BI, Brabec C, Maiti A, Bernholc J (1996) Phys Rev Lett 76:2511–2514
64. Treacy MMJ, Ebbesen TW, Gibson JM (1996) Nature 381:678–680
65. Krishnan A, Dujardin E, Ebbesen TW, Yianilos PN, Treacy MMJ (1998) Phys Rev Lett 58:14013–14019
66. Poncharal P, Wang ZL, Ugarte D, Herr VD (1999) Science 283:1513–1516
67. Solares SD, Esplandiu MJ, Goddard WA, Collier CP (2005) J Chem Phys B 109:11493–11500
68. Kutana A, Giapis KP, Chen JY, Collier CP (2006) Nano Lett 6:1669–1673
69. Salvetat JP, Bonard JM, Thomson NH, Kulik AJ, Forró L, Benoit W, Zuppiroli L (1999) Appl Phys A 69:255–260
70. Walters DA, Ericson LM, Casavant MJ, Liu J, Colbert DT, Smith KA, Smalley RE (1999) Appl Phys Lett 74:3803–3805
71. Dietzel D, Marsaudon S, Aimé JP, Nguyen CV, Couturier G (2005) Phys Rev B 72:035445-1–035445-16
72. Aimé JP, Boisgard R, Nony L, Couturier G (1999) Phys Rev Lett 82:3388–3391
73. Nony L, Boisgard R, Aimé JP (1999) J Chem Phys 111:1615–1627
74. Garcia R, Pérez R (2002) Surf Sci Rep 47:197–301
75. Lee SI, Howell SW, Raman A, Reifenberger R, Nguyen CV, Meyyappan M (2004) Nanotechnology 15:416–421
76. Snow ES, Casavant MJ, Novak JP (2002) Appl Phys Lett 80:2002–2005
77. Giessibl FJ (1997) Phys Rev B 56:16011–16015
78. Dürig U (1999) Appl Phys Lett 75:467–473
79. Dürig U (2000) New J Phys 2:5-1–5-12
80. Boisgard R, Aimé JP, Couturier G (2002) Appl Surf Sci 188:363–371
81. Dubourg F, Aimé JP, Marsaudon S, Couturier G, Boisgard R (2003) J Phys Condens Matter 15:6167–6177
82. Cleveland JP, Anccksykowsky B, Schmid AE, Elings VB (1998) Appl Phys Lett 72:2613–2615
83. Rodriguez TR, Garcia R (2003) Appl Phys Lett 82:4821–4823
84. Anczykowski B, Cleveland JP, Krüger D, Elings V, Fuchs H (1998) Appl Phys A 66:S885–S889
85. Moreno-Moreno M, Raman A, Gomez-Herrero J, Reifenberger R (2006) Appl Phys Lett 88:193108-1–193108-3
86. Albrecht TR, Grutter P, Horne HK, Rugar D (1991) J Appl Phys 69:668–673
87. Couturier G, Boisgard R, Dietzel D, Aimé JP (2005) Nanotechnology 16:1346–1353
88. Hembacher S, Giessibl FJ, Mannhart J (2004) Science 305:380–383
89. Israelachvili JN (1992) Intermolecular and surface forces. Academic, New York
90. Dietzel D, Bernard C, Mariolle D, Iaia A, Bonnot AM, Aimé JP, Marsaudon S, Bertin F, Chabli A (2006) J Scan Probe Microsc 1:45–50
91. Hafner JH, Cheung CL, Lieber CM (1999) Nature 398:761–762
92. Woolley AT, Guillemette C, Cheung CL, Housman DE, Lieber CM (2000) Nat Biotechnol 18:760–763

93. Bunch JS, Rhodin TN, McEuen PL (2004) Nanotechnology 15:S76–S78
94. Akita S, Nishijima H, Nakayama Y (2000) J Phys D Appl Phys 33:2673–2677
95. Shapiro IR, Solares SD, Esplandiu MJ, Wade LA, Goddard WA Collier CP (2004) J Phys
    Chem B 108:13613–13618
96. Solares SD, Matsuda Y, Goddard WA III (2005) J Phys Chem B 109:16658–16664
97. Strus MC, Raman A, Han C-S, Nguyen CV (2005) Nanotechnology 16:2482–2492

5 Scanning Probes for the Life Sciences

Andrea M. Ho · Horacio D. Espinosa

Abstract. Scanning probe based patterning techniques have the unique ability to deposit biological material into specific architectures on substrates and read and analyze the patterns using an atomic force microscope. Such devices are able to make much smaller biomolecule patterns, on the order of nanometers, than conventional techniques such as microcontact printing and optical lithography. A reduction in patterned feature size allows for greater sensitivity in biological studies and in life sciences applications such as drug screening and immunoassays. A variety of tools for the fabrication of nanoarrays are discussed. These include open- and closed-channel devices and pipette-based devices. Their potential for the integration of active components or augmentation to large-scale arrays for high-throughput deposition are examined. The mechanisms for deposition and biomolecule transport are also explained.

Key words: Nanoarrays, Patterning, Atomic force microscopy, Dip-pen nanolithography, Fountain probe, Protein array, Water meniscus

5.1
Introduction

In recent years novel techniques for depositing very small amounts of material, on the order of picoliters or less, onto surfaces at precise locations have been developed. While such efforts are clearly relevant in the fabrication of miniature electronics devices, it is now apparent that such techniques have many applications in biology and the life sciences as well. For example, the ability to spatially orient and immobilize biomolecules on a solid substrate is useful in the development of genomic or proteomic profiles of cells, drug screening, as well as biosensing, which requires precise and high-density arrays of biological material. Because cell functions are often mediated by the binding of a ligand to its membrane receptor, the study of cellular functions of living cells at the nanometer scale requires a device capable of delivering proteins to a particular location at a specific time [1]. This must be followed by an observation of the response.

In the design of a biomolecule-patterning device it is necessary to consider a multitude of factors. These include its resolution—the minimum feature size that can be patterned on the substrate; its reproducibility—small features must be repeatable and reliably patterned; and its ability for precise positioning. For use in high-throughput applications, a device should be easily scalable, i.e., its design should enable it to pattern arrays of biological material on the scale of a few centimeters. Devices that could deliver multiple solutions would have the advantage of depositing different molecules within one set of nanostructures spaced nanometers apart. For applications in the life sciences, a device must be able to function in ambient conditions. Finally, the ability to verify the successful deposition of biological material is also useful.

A key aspect in nanopatterning is the careful control of substrate surface chemistry. Nonspecific binding must be avoided otherwise intended signals will be masked in biosensing applications. Protein patterns must also be stable because in biosensors or bioanalytical devices immobilized proteins are often rinsed or washed with water, buffer solutions or surfactants [2]. This can be controlled by an appropriate choice of interactions used for immobilization. Immobilization via electrostatic interactions is normally reversible; proteins may be removed using certain buffers or surfactants. For long-term stability, covalent binding, which involves the formation of disulfide, imine or amide bonds, is often used.

In this chapter we review micro and nano patterning technologies with a particular emphasis on bioapplications. We begin by looking at microscale techniques for DNA and protein studies and then proceed to a review of nanoscale technologies, focusing on probe-based device designs and patterning results. The theoretical aspects behind nanoscale deposition is discussed followed by an overview of advances in the parallelization of the aforementioned devices.

5.2
Microarray Technology

Molecular spot arrays can be used for massively parallel determination and measurements of binding events and open up possibilities for automation. A major benefit of microarrays and nanoarrays is that they require only minute amounts of sample material. In essence, microarrays consist of many microscopic spots, each containing identical molecules, attached to a solid support. In a typical microarray experiment, or immunoassay, these spots contain one binding partner, such as a receptor or ligand. A sample containing targets to be investigated is then added to the array, and binding between the probe spots and the targets may occur. Fluorescent labels are often used to detect such binding events. DNA is likely the most interesting biomolecule because of its role in information storage [3]. Its stability also makes it useful in directing the immobilization and assembly of nanostructures. Patterned in arrays, DNA can be used as probes to analyze genetic defects and single nucleotide polymorphisms using lab-on-a-chip approaches.

There are two main methods for microarray production. The first involves synthesis on the chip to create libraries of short oligonucleotides (short sequences of nucleotides, the structural units of DNA and RNA) or peptides (compounds of two or more amino acids). This may be accomplished using techniques such as optical lithography. The second method involves separate synthesis with subsequent deposition on the chip, which is required for larger biomolecules such as polynucleotides or proteins. This type of deposition has been accomplished with restricted reproducibility via contact printing using pin tools dipped into a sample solution and brought into contact with the support material, thereby dispensing the solution. Optical lithography and microcontact printing will be discussed in greater detail in the following section. The experimental variability using microspotting techniques is expected to decrease with the recent advances being made [4]. Other devices are noncontact equipment such as inkjet printers which use the piezo effect to deposit nanoliter volumes of solution [3]. The distance between deposited drops has been minimized to 200 μm, resulting in arrays that are not very dense. Nevertheless, inkjet methods allow for in situ synthesis and can therefore deposit longer oligonucleotides than methods such as photolithography. Longer oligonucleotides offer sufficient specificity to detect genes using fewer probes [5].

5.2.1
Microcontact Printing

Microcontact printing is one method used to create microarrays. The technique covers an existing patterned surface, the master, with a liquid prepolymer. This is then cured to create an elastomeric stamp [6]. Typically, poly(dimethylsiloxane) (PDMS) is used. The hardened stamp is then peeled off the master. Ink solution is applied, the solvent is allowed to evaporate and the stamp is brought into contact with a substrate, at which point the ink is transferred. The ink forms a self-assembled monolayer (SAM) which is a replica of the master pattern.

Microcontact printing was first introduced by Whitesides and coworkers in 1993 as an alternative to photolithography [7]. The technique can be used to modify the adsorption properties of gold substrates for the attachment of proteins. Because proteins adsorb preferentially on some materials and are repelled by others, specific materials such as alkanethiols can be transferred via microcontact printing onto gold in order to functionalize it for protein patterning. However, proteins can also be directly transferred by the elastomeric stamp. Untreated PDMS provides a hydrophobic surface very much like the polystyrene used for adsorption of proteins in immunoassays. The process for protein patterning is as follows. A PDMS stamp is covered with protein solution for inking. This creates a monolayer of protein on the stamp surface. The stamp is rinsed and dried, and the pattern is then transferred onto a substrate. Contact need only be made for a few seconds and pattern transfer occurs only where the stamp contacts the substrate. Immunoglobulin G (IgG) protein was transferred onto a silicon wafer using a stamp made of the siloxane Sylgard 184. Features with dimensions as small as 500 nm were replicated. Atomic force microscopy (AFM) imaging verified that patterns had high contrast and resolution because the stamp was mechanically stable and proteins did not diffuse significantly across the surface. The features also retained their biological activity after printing.

For proteins that may not survive adsorption processes at surfaces, additional steps may be taken to immobilize them onto a substrate although this lengthens and complicates the process [8]. Aside from proteins, other patterned materials include lipid bilayers and poly(amino acids). In the case of lipid bilayers however, the printed lines were expected to be 9-μm wide, but were actually 19-μm wide. Because PDMS is deformable it would not be surprising if areas surrounding the intended pattern also made contact with the substrate.

Microcontact printing is suitable for printing over large areas at once. However, resolution is limited by the feature size of the master, mechanical deformation of the stamp and ink diffusion around the contact areas. The resolution may be improved by using stamps of higher stiffness. However, high-resolution masters require electron-beam patterning and reactive ion etching of a silicon-on-insulator (SOI) wafer. Michel et al. [9] used a high-stiffness PDMS stamp with 80-nm-diameter posts to transfer single antibody molecules. Although they claim 99% transfer efficiency, the resulting spots were not regular in size and shape. Finally, a major feature of microcontact printing is that the flexible silicone stamps require a casting mold, the master, with a predefined layout. This may be costly and may necessitate a series of photolithography processing steps [10].

5.2.2
Optical Lithography

The photolithographic technique can create DNA patterns with feature sizes as small as 18 μm [5]; however, it is expensive and places a limit on the allowable oligonucleotide length. For reliable detection of each gene, ten to 20 probes are needed, thereby limiting arrays to about 12000 genes per square centimeter. Smaller feature sizes are possible using polymeric photoresists used in the semiconductor industry which exhibit a nonlinear response to illumination intensity. Features as small as 8 μm have been constructed in this manner. UV lithography can be used to directly pattern alkanethiol SAMs with micron-scale resolution [11]. It is then possible to attach biomolecules to these patterns.

Lithographic patterning of DNA operates as follows (Fig. 5.1). Synthetic linkers modified with photochemically removable protecting groups are attached to a glass substrate [12]. The substrate is exposed to light directed through a mask to selectively deprotect and activate certain sites. Protected nucleotides can then attach to these

Table 5.1. Techniques suitable for use in the life sciences

Deposition method	Best resolution	Ink	Substrate	Reference
Microcontact printing	500 nm	IgG protein		[7]
	19 μm	Lipid bilayers		[8]
Optical lithography	8 μm	Alkanethiols		[11]
	500 nm	Streptavidin		[16]
Dip-pen nanolithography	30-nm lines	Collagen	Gold	[43]
	45-nm dots	IgG protein		[44]
Surface patterning tool	2–3-μm dots	Cy3-streptavidin		[51]
	150-nm lines	Quantum dots conjugated to streptavidin		
Microspotters	30-μm dots	IgG protein and oligonucleotides	Glass	[53]
Nanopipettes	440-nm dots	IgG protein	Glass	[61]
	510-nm dots	Biotinylated DNA		
Nanofountain probe	40-nm lines	MHA	Gold	[35]
	200–300-nm dots	DNA		[67]
	200–300-nm dots	IgG protein		Unpublished

IgG immunoglobulin G, *MHA* 16-mercaptohexadecanoic acid

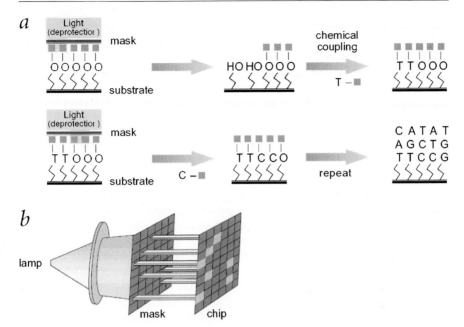

Fig. 5.1. a Photolithographic oligonucleotide synthesis. Light directed through a mask activates certain sites and protected nucleotides then couple to these sites. The process is repeated, with different sites being activated and different bases being coupled. **b** A lamp, mask and array [12]

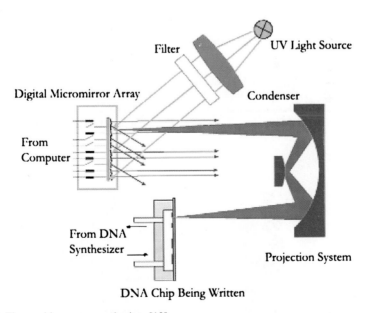

Fig. 5.2. The maskless array synthesizer [13]

sites. The process is repeated; coupling different bases allows arbitrary DNA probes to be constructed at each site. A drawback is that a changing set of monitored genes requires a new design and new masks. To overcome this, a maskless array synthesizer (MAS; Fig. 5.2) was developed [13]. Computer-generated virtual masks were relayed to a digital micromirror array which used 1:1 imaging to address pixels in a 10 mm × 14 mm area. The MAS system was used to print checkerboard patterns of 16-μm features (the size of each micromirror).

5.2.3
Protein Arrays

Protein patterning is useful for miniaturizing biochemical tests for high-throughput screening of new drug candidates and for performing studies in protein expression, protein–protein and protein–enzyme interactions and cell adhesion. Whereas DNA microarrays have been well established for genomics, no such high-throughput technologies yet exist for proteomics [14]. However, advances are being made and there are many protein array suppliers for the wide range of applications for which they are needed. (Listings are found in [14, 15]). For the photolithographic production of submicron protein patterns, pH-responsive films were patterned by exposure to 365-nm light. This allowed for the conjugation of aminooxy-modified biotin and the subsequent immobilization of streptavidin. Protein patterns as small as 500 nm were produced [16].

A conventional assay technique is the enzyme-linked immunosorbent assay (ELISA; Fig. 5.3), used to detect peptides, proteins, antibodies or hormones [17,18].

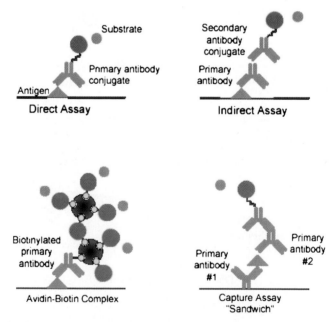

Fig. 5.3. ELISA immunoassay formats [17]

For antibody detection, an antigen is immobilized on a solid support. The sample for inquiry (any body fluid) is applied, and if the antibody in question is present it will bind to the antigen. Detection is accomplished via an enzyme directly linked to this primary antibody or a secondary antibody that recognizes the primary antibody. Or, if the primary antibody has been labeled with biotin, it can then be incubated with streptavidin. Incubation of this complex with an appropriate substrate produces a detectable product; typically a color change is imparted. The amount of color is measured proportional to the amount of antibody present in the sample. The most commonly used enzymes are horseradish peroxidase and alkaline phosphatase. ELISA is typically performed in 96-well or 384-well polystyrene plates coated with a ligand which will passively bind antibodies and protein; thus, unbound materials can be easily separated from the bound material during the assay. False positives are possible owing to the nonspecificity of protein–antibody interactions, and some analytes such as HIV require retesting using western blot, an electrophoretic technique where antibodies are directed against a number of viral proteins.

5.3
Nanoarray Technology

5.3.1
The Push for Nanoscale Detection

The delivery of biological material in increasingly smaller volumes has the potential to advance many applications. Studies of protein function are made possible, for instance, protein clustering in cell focal adhesion occurs at 5–200-nm length scales; delivery of material at this relevant length scale allows proteins of interest to be bound to a substrate [19]. Lateral control in creating specific adhesive and inert sites can create rigid ligand templates for cell binding. The specific interactions between binding pairs can be used for protein immobilization; these include affinity capture ligands (streptavidin–biotin binding) and antigen–antibody recognition for immunoassays [20]. Phospholipid deposition can be used to make model systems that can mimic the structural complexity of biological membranes [21]. It is also useful for studies investigating the binding of a protein and drug with supported lipid bilayers.

In the fabrication of arrays, it is important to reduce the sample volume, particularly in applications where limited sample amounts are available, such as in the analysis of multiple tumor markers from biopsy material [14]. Laser capture microdissection for cancer biomarker screening can obtain a few cells from a cancer tissue section, but it is very difficult to analyze the protein content of these cells using microarrays [22]. On the other hand, nanoarrays require only minute volumes of sample, approaching the volume of a single cell, so solid-phase testing on single cells can be performed.

A reduction in an array spot size suggests that statistically every molecule in an analyte has the opportunity to sample the entire capture surface in a reasonable amount of time [24]. In microarrays, not all of the sample will reach the capture surface unless energy is added to enhance the movement of the analyte, such as by agitation, mixing or electromotive force; thus, part of the analyte is effectively wasted.

Smaller arrays imply that biochemical reactions may not be diffusion-limited. In effect, nanoarrays offer the possibility of greater sensitivity in diagnostic tests, since disease progression is often correlated with protein levels [23]. Protein nanoarrays would be particularly useful, as protein signals cannot be amplified via methods such as polymerase chain reaction (PCR). Since many protein biomarkers are present in concentrations of only 10–100 pg/ml, improving assay sensitivity could lead to the discovery of new biomarkers that currently go undetected [24]. Essentially, the fabrication of nanoarrays will allow the screening of smaller volumes in shorter amounts of time. It will allow for higher-density arrays; one assay could then screen a greater number of targets.

Microarray analysis relies on optical readout methods, either by the observance of a color change as in ELISA, or the reading of fluoresence signals, which requires fluorescently labeled molecules for the detection of protein–protein interactions [25]. However, such labeling may cause deformational changes of the protein molecule, which may in turn affect the protein function. Radioisotopes are also sometimes used for labeling, but this requires the subsequent management of radioactive materials [26]. Labeling efficiency varies, so quantitation may not be reliable, and the labeling process is time-consuming and labor-intensive. For all these reasons, label-free detection is preferable. It also allows for real-time detection and in situ identification.

The atomic force microscope (Fig. 5.4) is an instrument capable of label-free sample analysis. It is a member of the family of scanning probe microscopes (SPM), which makes use of specialized probes to scan a sample surface to produce maps of topography, conductivity, binding force or friction among many others. These data sets can also be obtained all at once. The resolution of the technique is highly dependent on the probe quality and sharpness. Nanoarrays are well suited to atomic force microscope readout. For example, arrays for virus detection contain domains of

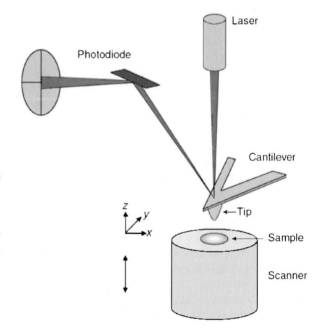

Fig. 5.4. Principle of atomic force microscopy (AFM). The sample is mounted on a piezoelectric scanner for three-dimensional positioning. The force between the tip and the surface is determined by measuring the cantilever deflection optically (with a laser and a photodiode) [28]

antibodies directed against specific viral species [22]. Any of these viruses present in a test solution will attach to certain domains and can be detected via AFM. A variety of solutions can be tested, including serum, sputum, sludge, coffee and urine, some of which inhibit other methods such as PCR. Perhaps most importantly, AFM is applicable in almost any environment, including liquids. This is required for studying cells and membranes in their native environment. Lastly, AFM-based methods are amenable to subsequent manipulation by the probe. For example, DNA can be induced to fold by controlled pushing by an atomic force microscope tip [27].

5.3.2
Probe-Based Patterning

The following section discusses a variety of scanning-probe-based lithography techniques suitable for use in the life sciences. They each have their own advantages and disadvantages. Their resolution and typical examples are given in Table 5.1. Many factors influence the ability to pattern very small features. These include environmental conditions, appropriate ink and substrate chemistry and the material and geometry of the probe tip.

5.3.2.1
Dip-Pen Nanolithography

Dip-pen nanolithography (DPN; Fig. 5.5) was first introduced to the research community in 1999 as a tool for patterning nanostructures, in which an atomic force microscope probe tip (approximately 20 nm radius) is dipped into a solution to coat it with the desired molecules [29, 30]. Patterning is typically accomplished with a commercial atomic force microscope which controls the movement of the probe. When the probe is brought into contact with a substrate, molecules diffuse from the probe to the substrate. In most cases, the ink molecules produce a local reactive functionalization of the surface, such that specific biochemical adhesion experiments can be conducted at these length scales. In general, the deposited material and the substrate must be paired, such that a chemical reaction occurs upon delivery, or a surface SAM is formed so that readout is possible using the atomic force microscope.

DPN is characterized by simplicity and high writing resolution (less than 100 nm). A large amount of literature is available on the use of single-probe DPN. Apertured probes [32, 33] can store and dispense larger numbers of ink molecules, but the writing speed is not increased substantially and the writing resolution is inferior to that of DPN. The resolution is affected by the geometry and aperture size of the probe tip and the wetting properties of the tip and the substrate. The smallest features produced were typically twice the diameter of the aperture; the smallest apertures produced were 35 nm in diameter [34]. Lastly, because such apertured probes are milled by a focused ion beam, they are not easily mass produced. Pulled-glass nanopipettes offer continuous ink delivery, but lower resolution (approximately 1 μm), and cannot be integrated by microfabrication into larger systems. Nanofountain probes (NFP) [35, 36] exhibit continuous ink delivery for a long-range writing capability, resolution close to that of DPN, and can be integrated in arrays and systems, although the probes and AFM systems become more complicated and costly.

Fig. 5.5. Dip-pen nanolithography (DPN) patterning [31]

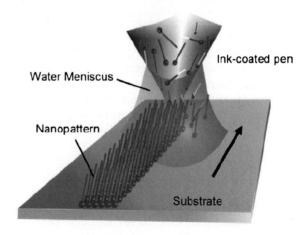

Indirect DPN

The most common demonstrations of DPN patterning have involved the deposition of n-alkanethiols onto gold surfaces because thiols self-assemble and strong thiol–gold bonds are formed [38]. The fabrication of patterns of biological materials began with indirect adsorption of the molecules of interest onto DPN-generated templates. 16-Mercaptohexadecanoic acid (MHA) could be deposited onto gold substrates via DPN; features as small as 100 nm were possible [39]. Nonpatterned areas were then passivated with 11-mercaptoundecyl tri(ethylene glycol) (PEG-SH). The desired proteins, lysozyme or rabbit IgG were attached by submersing the substrate into a 10 μg/ml protein solution for 1 h. No detectable nonspecific binding to the passivated areas was observed. It was also demonstrated that retronectin could adsorb specifically to the MHA-patterned dots 200 nm in diameter, spaced 700 nm apart. In turn, 3T3 Swiss fibroblast cells could be attached to these retronectin patterns, indicating that nanoarrays could be used for cell adhesion studies as well.

Metal ion-affinity templates can also be used to immobilize antibodies [40]. As metal ions are not susceptible to denaturation, they can be used as a linker to immobilize many unmodified polyclonal and monoclonal antibodies. First, DPN was used to pattern MHA spots. The surrounding areas were passivated with PEG-SH and the carboxylic acid groups of the MHA were coordinated to Zn^{II} ions. The substrates were then exposed to solutions of the desired antibody.

Indirect DPN techniques have the potential to be used for the creation of sandwich assays with the purpose of disease detection. For example, DPN can be used to deposit an array of MHA spots which then attract the p-24 antibody for HIV [41,42]. The surrounding areas are passivated with bovine serum albumin. The array could then be used to detect HIV in patient samples, exceeding the detection limit of conventional ELISA-based immunoassays.

Direct DPN

Direct deposition of biological material eliminates the additional step of first patterning a linker molecule. It also eliminates cross-contamination of array features because the desired chemistry is carried out only at specific locations [29]. This is essential for the prevention of nonspecific adsorption, which becomes an increasing problem as arrays get smaller—a few nonspecifically bound molecules may overwhelm the entire intended signal [26]. Silicon atomic force microscope tips were used to deposit thiolated native collagen (1 mg/ml in 1 mM HCl) and collagen-like peptides (40 mg/ml) onto gold substrates via DPN [43]. Line widths down to 30 nm were achieved. Collagen arrays could conceivably be used to induce an assembly network of collagen scaffoldings that would direct cell attachment or as guest–host systems for other biological entities.

In most cases, however, direct deposition by DPN typically requires prior chemical modification of the atomic force microscope tip. For example, commercial atomic force microscope tips functionalized with 0.1 mM thiotic acid could be used to deposit lysozyme and rabbit IgG (Fig. 5.6) [44]. The modified atomic force microscope tip was immersed in protein solution (500 µg/ml) for 1 h and then used to pattern immediately. The patterned IgG dots were 45–200 nm in diameter, and the biomolecules maintained their biorecognition properties after being patterned. The study demonstrated that multicomponent arrays could be patterned without cross-contamination. However, high humidity (80–90%) was required for successful patterning; a relative humidity below 70% resulted in inconsistent ink transport properties. Furthermore, the transport rate depended on the composition of the protein.

2-[Methoxypoly(ethyleneoxy)propyl]trimethoxysilane can also be used to functionalize atomic force microscope tips. The material forms a biocompatible and hydrophilic surface layer on the tip, and prevents protein adsorption and denaturation on the tip surface. It also reduces the activation energy for transporting proteins from the tip to the substrate. Without it, the protein solutions (500 µg/ml) do not wet the silicon nitride (Si_3N_4) cantilevers and inconsistent or low-density protein patterns are produced. Lim et al. [45] immersed such an atomic force microscope tip in protein solution for 1 min and used it to pattern in 60–90% relative humidity. Rabbit IgG features ranging from 55 to 550 nm were fabricated. It was also demonstrated that human, goat and mouse IgG, and anti-human, anti-goat and anti-mouse IgG could be patterned in this manner.

Nickel-coated Si_3N_4 atomic force microscope tips were used to deposit histidine-tagged ubiquitin (300 µg/ml) and thioredoxin (250 µg/ml) proteins onto nickel oxide surfaces [46]. As in the previous cases, bare Si_3N_4 cantilevers could not be coated homogeneously with the proteins, resulting in inconsistent or low-density protein patterns.

The proteins angiogenin and integrin $\alpha_v\beta_3$ were deposited using gold-coated silicon cantilevers functionalized for 30 min in 1 mM mercaptoundecanoic acid [25]. The resulting SAM increased tip hydrophilicity and facilitated protein adsorption on the tip surface. The tip was then immersed in the desired protein solution for 1 h and the molecules were patterned via DPN. The smallest spots created were 120-nm wide and protein transfer was found to be affected not only by the tip–substrate contact time but also by the contact force. In order to produce regular patterns, high humidity was required, as in the work in [44].

Demers et al. [47] used DPN to directly pattern hexanethiol-modified oligonucleotides on polycrystalline gold, and oligonucleotides bearing 59-terminal acrylamide groups on derivatized silica. The atomic force microscope tip had to be

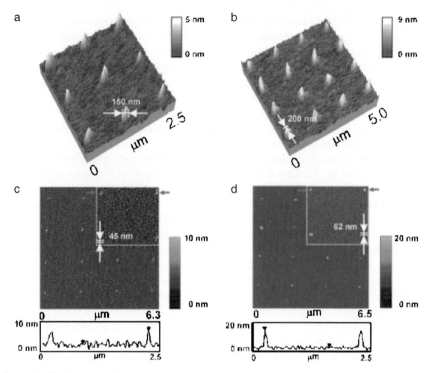

Fig. 5.6. AFM images of protein arrays generated via direct-write DPN. **a** Nanodot array of lysozyme; **b** array of immunoglobulin G (IgG); IgG dots before **c** and after **d** treatment with a solution of anti-IgG-coated nanoparticles [44]

first silanized by immersing it in a solution of 39-aminopropyltrimethoxysilane in toluene, to render it positively charged and hydrophilic.

Although protein delivery was not tested, PDMS-coated atomic force microscope tips were found to deliver MHA, octadecanethiol (ODT) and cystamine [48]. As cystamine is very volatile, its patterning via conventional DPN with Si_3N_4 tips is problematic. However, the PDMS-modified tips successfully patterned 200–300-nm-wide lines. The packing density of cystamine could be influenced by the writing speed. The PDMS acts as a reservoir that absorbs inks and allows for 1–2 h of continuous MHA patterning. The pattern resolution was not quite as good as that of regular DPN—the thinnest lines generated were 55-nm wide; however, these are much smaller features than can be obtained via methods such as microcontact printing.

It was demonstrated that a monolayer of succinic acid succinimidyl ester 5-thioyloxy-2-nitrobenzyl ester (SSTN) could be used as a photocleavable cross-linker to chemically link avidin to a gold-coated atomic force microscope tip [1]. When the tip was brought into contact with a biotinylated mica substrate, UV irradiation was applied to cleave the SSTN, releasing avidin onto the substrate. The protein-free atomic force microscope tip could then be used to image the deposited material. The delivered avidin was confined in a range of $100 \times 90 \, nm^2$ and maintained its affinity to biotin.

Most DPN writing has been accomplished using the contact mode operation of the atomic force microscope. This is the most common method of AFM operation, in which the probe tip and the substrate remain in close contact while the substrate is being scanned. A drawback of this method is the exertion of large lateral forces on the substrate as the probe is dragged across it. In tapping mode AFM, a cantilever probe of lower stiffness is oscillated at its resonance frequency and taps the substrate for a small fraction of its oscillation period, thereby reducing the lateral force exerted on the substrate as well as the tip–substrate adhesion forces. Agarwal et al. [49] deposited a synthetic peptide MH_2 via tapping mode AFM, which allows for gentle imaging of deposited material as well as deposition on soft substrates. The drive amplitude was found to be a critical factor. When coated with peptide, the drive amplitude of the AFM probe decreased significantly, and as the peptide was being deposited, the drive amplitude required for imaging increased. In order to successfully deposit peptides in tapping mode, the drive amplitude had to be increased by a factor of 2–5 from its imaging value, causing the probe to exert a greater contact force on the substrate. Nevertheless this force is still less than that in contact mode DPN.

5.3.2.2
Open-Channel Pens

Reese et al. [5] micromachined stainless steel fountain pens (Fig. 5.7) with open trenches which narrowed to a width of 30 µm at the pen tip. The trench width dictated the patterned spot size; spots 10–30 µm wide and 20–140 µm long were deposited. The highest density arrays were 25,000 spots per square centimeter of fluorescent dye. With a single loading, a pen could print five to 20 spots. Printed oligonucleotides with a mean spot size around 3500 µm² were successfully hybridized. Nevertheless, such fountain pens are not capable of nanoscale patterning.

A device named the surface patterning tool (SPT) was developed by Henderson et al. [50]. An SPT consists of a cantilever with a split gap at the end, a reservoir on the handling chip, and a 1-µm-deep open transportation microchannel connecting the gap and the reservoir. Sample loading is carried out by filling the reservoir with sample solutions as well as by dipping the cantilever end into sample fluid. These designs, dedicated to biomolecular patterning, allowed reliable patterning of large molecular species and reduced reloading requirements. The length of the SPT cantilevers ranges between 200 and 300 µm, and the width ranges between 20 and 40 µm. The split gap is approximately 1 µm wide and approximately 40 µm long. At the fixed end of the cantilever, a 10-µm-deep rectangular reservoir is located on the handling substrate. The depth and the width of the microchannel are 1 µm and 1–10 µm, respectively.

Testing of the fabricated SPTs was performed using a dedicated commercial instrument called a NanoArrayer (BioForce Nanosciences, Ames, IA, USA). This instrument is equipped with a precision motion control system and an environmental chamber. Although this instrument uses the same optical lever deflection scheme employed in AFM, it does not scan or acquire images. SPTs are mounted to form a 12° angle with the deposition substrates such that only the tip end is in contact with the substrate. Patterning was demonstrated at a relative humidity of 35–50% using

Fig. 5.7. Stainless steel fountain pens [5]

Cy3–streptavidin [50]. A 10×10 array of spots with a diameter of $2-3\,\mu m$ was routinely obtained; this value was limited by the width of the SPT's split-gap. A single loading of the tool printed at least 3000 spots in about 1 h. It was also demonstrated that quantum dots conjugated to streptavidin could be deposited in patterns of lines and spots using the SPT. These features had line widths of approximately 150 nm to $7\,\mu m$ and spot diameters of $3-5\,\mu m$ [51].

5.3.2.3
Microspotters

In microspotting technologies, a biochemical sample is loaded into a spotting pin by capillary action, and a small volume is transferred to a solid surface by physical contact between the pin and the solid substrate [52]. Other than microspotting pins, capillaries or tweezers can act as a printhead of biochemical samples. Printheads can be moved by the XYZ motion control system of an atomic force microscope and brought into a contact with a surface to transfer the sample.

Belaubre et al. [53, 54] fabricated microspotters (Fig. 5.8) on SOI substrates using conventional micromachining techniques.

Arrays of 2-mm-long, 210-μm-wide and 5-μm-thick cantilevers, spaced 450 μm apart, were microfabricated. These cantilevers were used to pattern a glass slide with 1-pl volumes of a solution containing cyanine3-labeled oligonucleotides (15-mers) [53]. Anti-goat IgG (rabbit), microarrays were also generated on a glass slide coated with dendrimer molecules as cross-linkers. In both cases, 30-μm-diameter spots were obtained. It was also demonstrated that no cross-contamination was observed for two different biological samples deposited with the same cantilevers if a cleaning procedure was used. Using a similar microspotter, Leichle et al. [55] deposited colloidal solutions containing poly(ethylene glycol) 600 and aminopropy-

Fig. 5.8. Microspotter array [53]

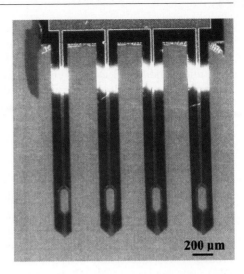

ltriethoxysilane nanoparticles with diameters of approximately 300 and approximately 150 nm, respectively, on surfaces to form spots with diameters ranging from approximately 10 μm to more than 100 μm.

To further reduce droplet size, Saya et al. [56] introduced a technique to fabricate an in-plane sharp nanotip incorporated into the channel of a cantilever (Fig. 5.9). Tips with a curvature radius less than 100 nm were fabricated, and were connected to V-shaped microchannels 5 μm wide. The cantilevers were used to deliver water–glycerol droplets. The surface wettability of the nanotip and the substrate were important factors in determining the droplet size. For a hydrophilic tip contacting a hydrophobic substrate, droplets 3–4 μm in diameter were achieved, and the contact time did not significantly affect the drop size. Smaller droplets 2 μm in diameter could be achieved with high uniformity using a hydrophobic tip and a hydrophilic substrate; however, the contact force and time affected the spot size.

Pens with open microchannels integrated on cantilevers have the advantage of being clog-free and allow for easy cleaning and simple microfabrication. However, such open microfluidic elements such as microchannels and reservoirs are prone to cross-contamination via vapor from different types of samples, especially when loaded in arrays of cantilevers [57]. Evaporation may be critical in

Fig. 5.9. Scanning electron microscopy (SEM) images of a cantilever array (*left*) with a nanotip (*center* and *right*) [56]

some applications although its rate can be reduced with environmental condition-
ing. Enclosed microchannels are beneficial in such cases, although they are more
difficult to microfabricate and are subject to clogging. Pipettes are conventionally
used microfluidic devices with enclosed channels. Microneedles with embedded
microchannels were also demonstrated to deliver liquid materials [58]. In the following subsections, microcantilever devices with enclosed microchannels are described.

5.3.2.4
Pipettes

Bruckbauer et al. [59] developed a delivery system based on scanning ion-
conductance microscopy. In this voltage-controlled nanopipette, an ion current flows
between an electrode in the pipette and one in a bath; the pipette acts as the ink reservoir. Deposition of molecules to a surface occurs in the presence of aqueous buffer.
Protein G was successfully delivered using the pipette. Biotin and single-standed
DNA with spot sizes of 830–860 nm were also deposited.

Following this, goat anti-rabbit IgG antibody was delivered onto charged glass,
with spot sizes of about 2 μm [60]. The flow of molecules began at an applied voltage
of −0.5 V and increased linearly with applied voltage between −1.0 and −1.5 V.
Although voltage control allows for the control of ink delivery at the single-molecule
level, the diffusion of molecules in solution and on the substrate ultimately increases
the patterned feature size. Because the nanopipette deposits in solution, protein
denaturation may be prevented. However, the use of voltage subjects the sample
to a high electric field. Nevertheless, owing to the small current, very little heating
(less than 1 K) occurs. Indeed, the biomolecules were demonstrated to maintain their
functionality after deposition. This was demonstrated via binding experiments with
IgG and hybridization studies with DNA.

In order to reduce the patterned feature size, a double-barreled nanopipette
(Fig. 5.10) operating in air was fabricated from 1.5-mm-diameter pulled-glass capillaries with a septum down the center [61]. The device was used to deliver fluorescently labeled rabbit IgG to a polyethyleneimine-coated glass surface, with the
smallest spot produced being about 440 nm in diameter. Biotinylated DNA was delivered onto streptavidin-coated glass with an average spot size of 510 ± 40 nm for
a 5-s deposition time. In both cases, molecules were delivered out of one barrel at
any one time. Distinguishing itself from the original nanopipette, the double-barreled
pipette could write with two different inks. Two different, noncomplementary sequences of biotinylated DNA were loaded into the barrels. The DNA to be patterned
could be selected by changing the sign of the applied voltage. The two inks could
be delivered to the same location on a substrate. The double-barreled pipette is also
capable of topographical scanning but its resolution is not as good as that of AFM;
it cannot track up steep slopes or tight grooves as effectively. Most recently, the
double-barreled pipette was used to deposit water droplets, working under oil [62].

A major drawback of this device is that it is not easily scalable because the
pipettes are individually fabricated by laser pulling. The smallest pipette that can be
made using this approach has a radius of approximately 20 nm, limiting the pattern
resolution [63].

Fig. 5.10. Principle of double-barreled pipette deposition. The voltage between the two barrels controls the molecular delivery [61]

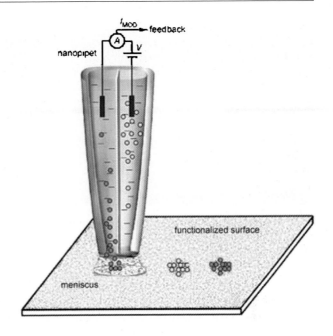

5.3.2.5
Closed-Channel Cantilevers

The concept of the nanofountain probe (NFP) was first introduced by Espinosa and coworkers in [64–66]. The device allows for both high-resolution patterning and continuous sample feeding through closed microchannels. The chip without its cantilevers has an overall size of 1.8 mm × 3.2 mm to fit easily into commercial AFM equipment. In the latest version of this chip, the two opposing sides of the chip each have 12 cantilevers; one set is 520 μm long and the other is 430 μm long. A volcano-like dispensing tip (Fig. 5.11) exists at the end of each cantilever and has a ring-shaped aperture and is able to generate sub-100-nm lines routinely.

For fast-evaporating solutions such as alkanethiols, the writing mode is similar to that of DPN, and pattern resolution is controlled by the radius of the core tip, not the aperture size. MHA (1 mM) in ethanol with line widths as small as 40 nm was patterned onto a gold substrate at a humidity of 60% using the NFP [35]. The NFP's imaging capability in topography was similar to results obtained using commercially available atomic force microscope tips, and its sensitivity in lateral-force imaging was better than with commercial tips, an advantage when imaging SAM patterns. This imaging capability is essential for immediate examination of deposited patterns and also serves as a means for realigning probes when writing multiple materials.

In typical operation, however, the NFP preserves the liquid state of the ink during the patterning process [36]. The liquid present near the writing tip provides a continuous source of solvent vapor, preserving a high local solvent vapor pressure, which may contribute to the formation of a more vigorous capillary condensation meniscus [65]. Solvent vapors may also condense on the substrate, leading to a solvent

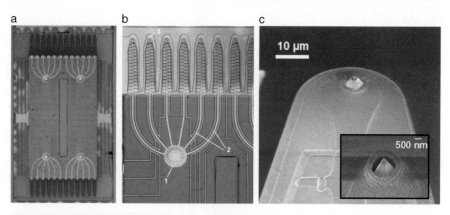

Fig. 5.11. Second-generation nanofountain probe (NFP) chip. **a** Optical image of the front view of an NFP chip **b** Closeup view of the chip showing *1* on-chip reservoir, *2* microchannels and *3* volcano tip **c** SEM image of a cantilever integrated with a volcano-like tip at the end; *inset* closeup of the volcano tip [36]

prewetting layer in the vicinity of the tip. The transport of ink molecules may also occur by coevaporation with the solvent and recondensation. This mechanism may lead to feature-growth dynamics different from that of DPN.

Batch fabrication allows for straightforward scaling to NFP arrays. Indeed, 12-cantilever NFP arrays have been fabricated, and their multiple on-chip reservoirs allow simultaneous patterning of two different solutions. Because the NFP can make use of two different reservoirs, it does not require a second tip to deposit an additional species. Simultaneous patterning with two different thiolate solutions, MHA and $1H,1H,2H,2H$-perfluorododecane-1-thiol in acetonitrile was demonstrated with the NFP [36]. Also, use of the NFP does not require repeated reinking and realignment of the tip during patterning, the latter of which is time-consuming and difficult to accomplish precisely. Finally, deep channels (more than 500 nm) allow the delivery of larger particles, including nanoparticles [37]. High compliance of the core tip relative to the volcano shell structure was observed in the induced deformation during scanning electron microscopy (SEM) observation of dry tips and is helpful in preventing clogging of the probe orifice and mechanically aids the transfer of larger species from the volcano cavity to the substrate. However, this may negatively affect alignment accuracy.

What sets the NFP apart from other devices is that while nanopipettes, apertured pyramidal tips and quill-type SPTs function by the formation of an outer meniscus between the probe and the substrate, the NFP forms a meniscus between the ultra-sharp atomic force microscope-like tip and the substrate, as in DPN. However, the NFP does not require modification of the tip surface in order to deliver biomolecules, because solutions are directly delivered from the on-chip reservoir to the tip. The NFP was capable of *direct* deposition of hexanethiol-modified oligonucleotides on a gold substrate [58, 67].

As shown in Fig. 5.12, the NFP was used to pattern a gold substrate with alkanethiol-modified oligonucleotides. After passivation of the unpatterned areas, the DNA spots were hybridized with complementary DNA-functionalized gold

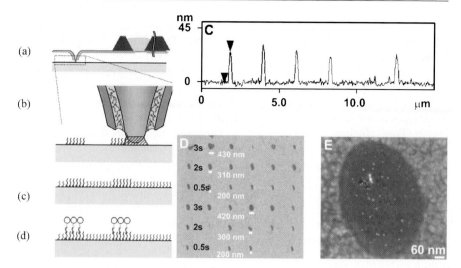

Fig. 5.12. Experimental procedure for DNA patterning: **a** molecular ink feeding, **b** direct patterning of a gold surface with alkanethiol-modified oligonucleotides, **c** passivation of the unpatterned areas with C6 thiol to avoid unspecific binding, **d** hybridization of the linker and probe DNA strands. **C** Height profile of the same array, acquired in tapping mode operation. **D, E** SEM image of a dot array and a single dot, respectively. Multiple gold nanoparticles are visible in **E** [67]

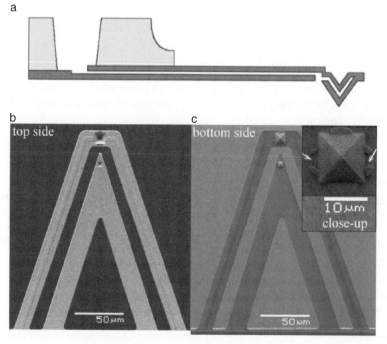

Fig. 5.13. a Cross section of the micromachined fountain pen; **b** *top view* and **c** *bottom view* of the fountain pen; *inset* closeup view of the pen tip, with outlet holes indicated by *arrows* [68]

nanoparticles (approximately 15 nm in diameter), demonstrating that the patterned DNA maintained its biological activity. Features 200–300 nm in diameter were routinely achieved.

A variation of the NFP concept introduced by Espinosa and coworkers [64, 65] was implemented by Deladi et al. [68, 69] in a fountain pen consisting of two V-shaped cantilevers, one that is "multifunctional," and the other an inner cantilever solely for detection. A channel embedded in the first cantilever connected the reservoir to the tip. Two variations of this design were fabricated. In the first design, the outlet hole was located at the apex of the pyramidal probe tip, and had to be fabricated by focused ion beam milling, a process which must be done individually on each pen. The second design made use of batch processing, which dictated an opening at the base of the pyramid (Fig. 5.13). The pen allowed for continuous fluid supply. It was demonstrated that etchant could be delivered to etch chromium; the thinnest line etched was 350 nm wide and 14 nm deep. The thinnest features constructed were 500-nm-thick ODT lines on gold substrates.

5.3.3
Alternative Patterning Methods

5.3.3.1
Nanografting

Nanografting is a technique similar to DPN, but involves material removal as well as deposition. The basic procedure involves scanning a SAM on a substrate with an atomic force microscope tip using a force greater than the threshold force needed to displace SAM molecules [70]. SAM molecules are thus removed, and thiol inks previously adsorbed on the atomic force microscope tip can be adsorbed to these areas of the substrate. Care must be taken not to cause plastic deformation of the substrate beneath the SAM. Xu et al. [70] nanografted thiol solutions with concentrations ranging from $2 \, \mu M$ to $2 \, mM$; the concentration did not appear to be a critical parameter in grafting. On the other hand, the scan rate was important. Slow scans often resulted in distorted patterns due to thermal drifts, but fast scans did not produce patterns with sufficient coverage. Liu et al. [71] demonstrated that alkanethiol features as small as $2 \times 4 \, nm^2$, and 10-nm-wide lines could be produced via nanografting. Wadu-Mesthrige et al. [2] used nanografting to prepattern SAMs with thiols terminated with protein-adhesive groups such as aldehyde and carboxylate, which would then dictate the subsequent adsorption of proteins. Nanopatterns of lysozyme, bovine serum albumin and rabbit IgG were fabricated, with lateral dimensions ranging from 10 nm to 1 μm.

5.3.3.2
Conductive DPN

In AFM charge writing, positive or negative surface charge patterns are created on an insulating substrate such as a polymer or silicon dioxide by applying voltage to an atomic force microscope tip. Mesquida et al. [10] created microarray charge patterns on poly(methyl methacrylate) substrates which were then immersed in water-in-perfluorinated oil emulsions where the water microdroplets were filled with TTR$_{105-115}$ peptide fibrils. The fibrils then attached to the charge patterns. The resolution was mainly limited by the size of the water droplets, which can typically be up to 5 μm in diameter. The use of water droplets as "containers" is advantageous in that the attachment process depends little on the content of the droplets, allowing the patterning of a variety of nano-sized biological materials.

Naujoks and Stemmer [72] used the same method to attach IgG–biotin stabilized emulsion droplets onto positively charged patterns. After drying, the protein molecules and salt from the buffer solution were left behind. The resulting spot sizes varied from 0.5 to 1.5 μm (Fig. 5.14). For these experiments, an atomic force microscope was operated in Kelvin probe force microscopy mode, where the lateral resolution was known to be in the range 50–100 nm, defined by the long range of electric forces and the geometry of the atomic force microscope tip. The pattern resolution might be enhanced by finding optimal emulsifying parameters and process conditions such as immersion time and drying ambiance.

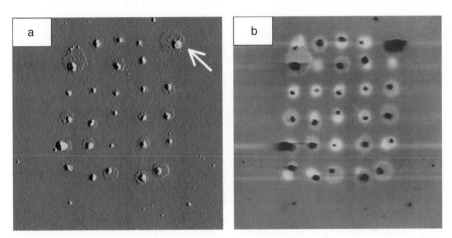

Fig. 5.14. a AFM (topography) image of IgG–biotin deposited on poly(methyl methacrylate). **b** Kelvin probe force microscopy (surface potential) image of positive charge pattern [72]

5.3.3.3
Electrochemical DPN

For the direct patterning of biological material, Agarwal et al. [73] used electrochemical DPN (e-DPN) to deposit a histidine-tagged peptide MH$_2$ and the histidine-tagged protein TlpA in tapping mode. An ionized nickel-coated substrate was held at ground potential and a negative bias was applied to the AFM probe tip. The method takes advantage of nickel–histidine binding and requires that a potential of about −2 V be applied. No functionalization of the silicon AFM probes was necessary. They were simply coated by immersing them in the desired solution for 15–30 s and then air-dried before patterning.

5.4
Nanoscale Deposition Mechanisms

Reproducibility is a concern for any fabrication technique. Nanopatterning results are not always repeatable and the resulting features are not necessarily well formed. An understanding of the ink deposition process is essential; however, in DPN, the mechanism for ink transfer is not completely understood and there is much controversy. It has been generally accepted that DPN patterning involves two processes: first, the transport of the molecular species from the atomic force microscope tip to the substrate through a water meniscus and, second, the adsorption of the ink onto the substrate. Control of these processes then would ultimately control the patterned feature resolution. Most studies on the patterning mechanism thus far have focused on the deposition of alkanethiols.

Theoretical studies on the dynamics of self-assembly have been performed in order to investigate the effects of deposition and atomic force microscope tip scanning rates on the resulting patterns generated via DPN [74]. The DPN process was treated as a two-dimensional Fickian diffusion process with a fixed source. Ink molecules on the atomic force microscope tip were driven by a concentration gradient to be deposited on a substrate. They then diffused over a monolayer of existing ink, and the diffusion was terminated by binding of the molecules to the bare substrate. In the patterning of dots by holding an atomic force microscope tip in contact with a substrate, the constant flux model predicted a $t^{1/2}$ dependence of the dot diameter, where t is the tip–substrate contact time. For the same number of molecules deposited, the final spot radius converged to the same value at long times, independent of the deposition rate. Increasing the diffusivity over the bare substrate resulted in circles that were fuzzier and not as well defined. For the generation of more complex patterns such as lines, a moving tip was needed and the tip scan rate had to be considered. In general, a fast scan or a slow deposition rate enhanced resolution, but as the scan rate increased further, patterned lines were no longer continuous. Such studies underline the necessity of determining optimal scan and deposition rates in order to produce coherent patterns using DPN. However, the effects of temperature, relative humidity and the presence of a water meniscus were not accounted for in these studies.

It is known that in ambient conditions, a water meniscus exists between an atomic force microscope tip and a substrate, and its volume increases with increasing relative humidity [75]. There have been suggestions that ink molecule transport occurs through this meniscus; however, Sheehan and Whitman [75] found that facile ODT deposition occurred even at 0% relative humidity in dry nitrogen. Since ODT is water-insoluble, its transport is not easily explained using a water meniscus model.

Cho et al. [76] proposed a "double-molecular layer" model in which the molecular ink on the atomic force microscope tip consists of a bulk solid region covered by a thin mobile layer on the surface (Fig. 5.15). A melting transition is suggested to explain the observed temperature dependence of the growth rate of patterns. The solid region serves as a reservoir of ink molecules which regulates the density of the surface layer, and the mobility and the number density of the mobile layer determine the molecular transport rate in DPN. It was proposed that three properties affect the mobility of the surface molecules—thermal energy, which increases the diffusion constant, residual solvent, which enhances diffusion of surface molecules, and ad-

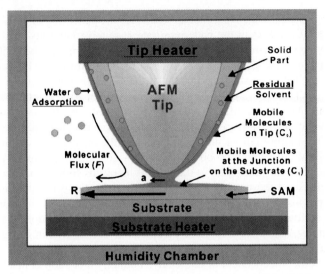

Fig. 5.15. Double molecular layer model for molecular ink transport [76]. *SAM* self-assembled monolayer

sorbed water, which enhances the mobility of hydrophilic molecular inks such as MHA and biomolecules.

In the general case where the diffusion and deposition rates are on the same order,

$$R \propto t^{\nu} \quad \text{and} \quad 1/3 \leq \nu \leq 1/2 \,,$$

where R is the radius of a patterned dot, t is the tip–substrate contact time and ν is the scaling parameter. ν approaches $1/3$ as the deposition rate becomes comparable to the diffusion rate, and $\nu = 1/2$ in a constant deposition case. In the patterning of ODT, a hydrophobic ink, on gold substrates, there are unstable fluctuations in the scaling parameter. These experimental results were explained using the double-molecular layer model. Above the melting temperature of ODT, the solid region melts and a single molecular layer forms. There is no longer a solid "reservoir" that regulates the density of the surface layer, leading to the unstable fluctuations in ν. Essentially there is a phase transition of the molecular species moving through the nanoscale junction. In the patterning of MHA, the scaling parameter increases with increasing relative humidity. It is suggested that adsorbed water enhances the diffusion constant on the substrate more than it enhances the transport rate at the tip–substrate contact point, implying that water molecules are also adsorbed on the SAM regions, not just on the probe tip.

In any case, direct environmental SEM (ESEM) observations verified that the height of the water meniscus between a silicon nitride probe tip and a gold or silicon substrate increases exponentially with increasing relative humidity [77]. In addition, for gold substrates, there was no observable meniscus below a relative humidity of about 70%, although, since the ESEM resolution in the imaging conditions used was approximately 50 nm, it is possible there was still a small meniscus. Still, these results are consistent with experimental results for the patterning of MHA, in which there was little increase in the patterning rate at relative humidity between 0 and 50%, while the rate increased significantly at a relative humidity above 50%.

AFM studies by Rozhok et al. [78] of water meniscus formation between a probe tip and a NaCl substrate suggested that even at 0% relative humidity as defined by Sheehan and Whitman, water on the substrate will collect at the point of contact form to a meniscus. As it is very difficult to remove adlayers of water from surfaces, a small meniscus will typically form, even when a hydrophobic atomic force microscope tip is used. Only in ultra-high-vacuum conditions can the meniscus be eliminated.

Weeks et al. [79] found no observable MHA deposition at a relative humidity below 15%. This was attributed to the lack of meniscus formation. Increasing the humidity increased the patterned dot size for a given tip–substrate contact time. The group proposed that the DPN deposition process transitions from a dissolution-dominated regime to a diffusion-dominated one. For the fabrication of small alka-nethiol features, the most important parameter is the activation energy of thiol detachment from the atomic force microscope tip. This transition occurs at a partic-ular contact time, independent of the relative humidity. For short contact times, the transfer process is dominated by surface kinetics, whereas for long contact times, it is controlled by diffusion. Furthermore, no effect of contact force was observed; the water meniscus causes a capillary force between the tip and the substrate that is much greater, so capillarity dominates the total load force. This is in agreement with Zou et al. [80], who concluded that feature size is independent of contact force over a range from 0 to 10 nN.

It is thus fairly well agreed upon that the molecule transport rate is different in the fabrication of small and large patterns. It is slower for large features (longer tip dwell times) because the cantilever continues to deposit molecules on an area that has already been patterned. Ink depletion on the surface of the atomic force microscope tip also contributes to this phenomenon. Moreover, there is an equilibration time when patterning begins. This may imply that the solvent plays an important role in determining the mobility of the ink on the cantilever [81]. It may also be due to the fact that a water meniscus takes time to reach an equilibrium width when patterning begins, as observed via ESEM [82]. Moreover, the transport of thiol inks is affected by the state of the substrate, not only the ink already present on the substrate, but also adsorbates that may unintentionally be present. Uncontrolled surface chemistry may be one factor leading to differing results found in the DPN literature.

Schwartz [83] contends that the water meniscus is not universally responsible for molecular transport in DPN. In contrast to many other experimental results, both ODT and MHA were readily patterned on a gold substrate at 0.0% relative humid-ity. Two different but compatible models were proposed. For less polar molecules such as ODT, molecular transport occurs via molecular surface diffusion, known as "reactive spreading" in microcontact printing. This process depends on thermal energy. The second model is relevant to the transport of small ions such as DNA, as a solution in adsorbed water, and is compatible with the humidity dependence of the patterning rate of certain molecules. The degree of polarity of a molecule will deter-mine its patterning behavior between the two extreme models. It has been suggested that MHA transport is due to two mechanisms. At low to moderate (50%) relative humidity, transport is caused by thermally activated surface diffusion [84]. As humidity is increased, dissolution and bulk transport enhance the transport rate; above 70% humidity, the water meniscus begins to have an effect as well. Furthermore, surface water is not responsible for molecular transport.

This is consistent with MHA dot patterns generated at high humidity [85]. At relative humidity less than 80%, filled MHA dot patterns were always observed, but increasing the humidity to approximately 84% resulted in the formation of hollow ring structures. A bulk meniscus transport model cannot explain such patterns because it predicts only filled circles. Rather, an annular diffusion model proposed that ink molecules are transported at the air–interface of the meniscus.

Whereas the models mentioned previously could describe the deposition of molecules which bind strongly to their substrates, Manandhar et al. [86] investigated the anomalous patterns formed by the weak binding of 1-dodecylamine (DDA) on mica, which cannot be explained by random walks. Anomalous patterns are sometimes formed when surface binding is weak and interactions between molecules dominates, determining the final pattern. For DDA, small patterns are isotropic, but become more anisotropic as the pattern size increases. Thus, to create large, regular patterns it may be necessary to move the atomic force microscope tip to fill the area, instead of relying on diffusion to produce the final pattern. Finally, a constant deposition rate was observed, implying that stable DPN writing is possible even for weak-binding cases. Lee and Hong [87] developed a theoretical model to explain these irregular patterns. The corresponding computer simulations showed that in the presence of strong intermolecular interactions and preferential substrate–molecule interactions while the diffusive motion of individual molecules is suppressed, irregular star patterns and fractal-like structures can be formed. On the other hand, if the intermolecular interactions between deposited molecules are weak relative to the thermal energy, diffusive motion becomes important and regular circular patterns are formed as predicted by random-walk models.

Beyond the physical capability of patterning dots with specific sizes, there is the question of the appropriate feature size for a particular application. In biodetection, there exists an optimal spot size for the attainment of a robust diagnostic readout. Individual molecules are dynamic in nature; in order to obtain a good signal there must be many sampling events or long sampling times for a single molecule. Array spots smaller than a certain size will not have enough active and correctly oriented capture molecules, leading to inaccurate quantitation and reduced dynamic range [24]. It was estimated that for typical IgG, the number of molecules in a 1-μm-diameter spot is on the order of 10^4 whereas a 250-nm-diameter spot will have on the order of 10^3 molecules. Spots below this 250-nm threshold may lack a sufficient number of molecules for reliable quantitative data to be obtained. Further studies must be conducted to determine the optimal feature size for particular bioassays.

5.5
AFM Parallelization

One requirement for a deposition technique is its capability for high-throughput patterning. A particular device design should be scalable, i.e., its ink-delivery probe should be arrayable and it should be possible to fabricate the entire device in mass quantities.

5.5.1
One-Dimensional Arrays

Xu et al. [88] reported a second version of the quill-type SPT described in Sect. 5.3.2.2, in which five cantilevers were arranged linearly (Fig. 5.16) [88]. The microcantilevers and corresponding microfluidic network were capable of transporting multiple fluid samples from macroscale reservoirs located on the SPT substrate through microscale channels to the distal end of the cantilevers. Five cantilevers and five reservoirs were arranged in a chip so that multiple biological samples could be transferred from the reservoirs to the SPT cantilever array. The overall size of an SPT chip was 3 mm × 6 mm. Each cantilever was 250 μm long, 30 μm wide and 2 μm thick, while consecutive cantilevers were separated by a spacing of 50 μm. The microchannel on each cantilever was 15 μm wide and 1 μm deep. Multiple ink loading and patterning were tested using two different fluorescent protein solutions, Cy2–donkey anti-goat IgG and Texas Red–donkey anti-rabbit IgG in phosphate-buffered saline, and alternatively loaded into the five reservoirs by hand-pipetting. The solutions transferred from the reservoirs to the distal end of each channel by capillary action and the fluids were confined inside the microchannels without observed cross-contamination. A dithiobis(succinimidyl undecanoate)/gold surface was patterned to generate 10 × 10 multiple-ink dot arrays, with the mean spot diameter being about 12 μm. The SPTs generated biological arrays with a routine spot size of 2–3 μm. Several thousand spots could be printed without reloading. The minimum spot size of the SPT was mainly limited by its gap width, which could be further reduced with a higher-resolution lithography technique.

Using single AFM probes in DPN limits a pattern to the 90 μm × 90 μm scan size of the AFM instrument. In order to create patterns that span macroscopic distances on the order of centimeters, it is necessary to use multiple-pen cantilever arrays. A 26-pen array of cantilevers was used to pattern a 10 mM $NHSC_{11}SH$ acetonitrile solution

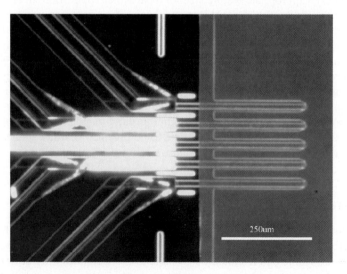

250um

Fig. 5.16. Surface patterning tool cantilever array [88]

(NHSC$_{11}$SH is an amine-reactive alkyl thiol molecule) [89]. The surrounding areas on the substrate were passivated with PEG–SH. The dot array was then used as a template to immobilize protein A/G through covalent coupling. Highly dense arrays of 23,400 dots spaced 1 μm apart could be generated in this manner. The protein A/G templates could then be used for adsorption of human IgG, anti-β-galactosidase and anti-ubiquitin. It was demonstrated that the anti-β-galactosidase and anti-ubiquitin retained their biological activity after being deposited.

The second and third generations of the NFP introduced by Espinosa and coworkers allowed for multi-ink patterning using a linear array of 12 cantilevers [36, 66]. The device successfully deposited alkanethiols and oligonucelotides as described in Sect. 5.3.2.5, as well as bovine serum albumin and IgG proteins and gold nanoparticles [37].

5.5.2
Two-Dimensional Arrays

The IBM Millipede was likely the first example of a massively parallel cantilever array. The chip was an array of 32×32 cantilevers conceived for data storage using polymer thermal indentation [90, 91]. Feedback control in the z direction brought the entire array into contact with a substrate. This simplified the system but imposed a strict requirement on the uniformity of tip height and cantilever bending in order to minimize tip wear due to force variations over the array. Another two-dimensional cantilever array includes a 25×40 array of nickel probes fabricated by Zou et al. [80], with a tip-to-tip distance of 100 μm in the x and y directions.

In an initial attempt at a very large scale array for DPN, a 55,000-pen two-dimensional array occupying 1 cm^2 was fabricated using lithographic techniques [92]. The pens were spaced 90 and 20 μm apart in the x and y directions, respectively. As is the case for all arrays, the tips must all be aligned prior to patterning. Rather than implementing feedback for each individual cantilever, a gravity-driven alignment method was used for this large array. Complex patterns of 80-nm ODT dots on a gold substrate were successfully generated. One challenge that remains with DPN arrays is the difficulty in realigning the array after reinking or switching inks. Another is the variability in the inking process. In most cases, when the DPN tip is inked, excess ink is removed by a high-pressure blast of air. In an effort to increase reproducibility in inking the tip, Rosner et al. [31] developed arrays of microfabricated inkwells to supply ink to specific cantilevers in a probe array. A noninked probe could then be used for pattern inspection. A slower scan speed could be used to obtain better image quality while minimizing contamination onto the pattern. The inkwells employ open-channel meniscus-driven flow in microchannels to bring fluid stored in 1-mm-diameter reservoirs to a series of appropriately spaced inkwells. Different inks can be loaded onto different tips on the same cantilever array, allowing for multicomponent patterning.

The 55,000-pen array was used to deposit 1,2-dioleoyl-sn-glycero-3-phosphocholine (DOPC), a phospholipid, via DPN onto substrates of silicon wafers, glass slides and evaporated metal films [21]. Line features as thin as approximately 93 nm were patterned. The probe tips were coated with ink by immersing them into inkwells for at least 30 min. It was observed that in ambient humidity (30–50%), the DOPC did not flow onto the tips, but at a humidity greater than 70% the phospholipid ink became sufficiently fluid to easily coat the tips.

Although most cantilever probe arrays used in the experiments described are made of inorganic materials such as silicon, silicon nitride or other metals, polymer probes may be of interest because they offer different biochemical properties [80]. Furthermore, they have a smaller Young's modulus; thus, cantilevers can be made shorter while maintaining the same force constant. Processing of polymers does not require sophisticated equipment, so such probe arrays could be manufactured more cheaply. Zou et al. fabricated an array of more than 1,000,000 probes on a 7.5-cm glass slide. Tips and cantilevers made of photodefinable polyimide film were individually supported on SU-8 bonding structures; each of these structures had a footprint of $10\,\mu m \times 10\,\mu m$.

Passive pen arrays command that all pens in the array create the same pattern. Different inks must be amenable to the same environmental conditions in order to be deposited at the same time. Although they are more complicated to fabricate, active pens can create more complex structures, with multiple inks over large areas, and inks that do not necessarily deposit under the same environmental conditions can be deposited in series.

Rosner et al. [31] fabricated thermal bimorph actuators from evaporated Cr/Pt/Au on silicon nitride cantilevers. The gold thin-film resistor essentially acts as a heater and concentrates the heat away from the probe tip. When power is delivered, cantilevers are actuated in the direction normal to the heated surface. Bullen et al. [93] fabricated a similar array of ten thermal bimorph actuators, with cantilevers $300\,\mu m$ long spaced $100\,\mu m$ apart. Resistive heating actuates the cantilevers away from the substrate. Each cantilever was individually addressed by two lead wires. A typical actuation current of 10 mA resulted in an 8-μm deflection and an average probe temperature of 298 K above ambient. A ten-probe individually addressed, thermally actuated array of cantilevers $1400\,\mu m$ long, $20\,\mu m$ wide, $8\,\mu m$ thick and spaced $100\,\mu m$ apart was used to pattern ODT on gold substrates with sub-50-nm resolution [94]. Zou et al. [80] fabricated an array of 45 thermal bimorph probes in which cantilever alignment was achieved via contact sensing based on the detection of electrical continuity between the tips and the substrate. This required that both elements be partially or entirely conductive.

Limitations to thermal bimetallic actuation include the potential negative effects of the heat on inks such as biomolecules [95]. There may also be thermal crosstalk between probes—heat conduction and convection through the air between probes leading to unwanted tip deflection, effects which increase as the distance between probes decreases. Bullen and Liu [95] found that the smallest pitch in thermal bimetallic arrays where crosstalk was manageable was $100\,\mu m$. This led to the fabrication of an array of electrostatically actuated DPN probes with an array pitch of $30\,\mu m$ (Fig. 5.17). The probes act as electrodes separated from the counter electrode, the array holder, by SU-8 photoepoxy acting as an insulator. The probes are grounded via conducting paste. Actuation occurs by applying a voltage (typically around 190 V) to the counter electrode; the probe tips are then pulled off the surface. Grounding the lithography surface ensures there is no electric field or actuator force between the tips and the substrate. While thermal crosstalk is avoided with electrostatically actuated probes, there may be crosstalk due to fringe electric fields.

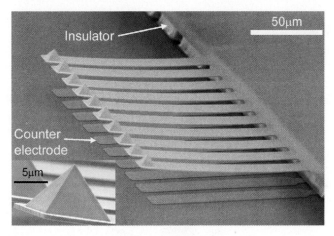

Fig. 5.17. SEM image of an electrostatically actuated DPN array [95]

Crosstalk deflection with these probes was similar to that in thermal bimetallic actuators, but at the substantially reduced array pitch of 30 µm.

A third method for probe actuation uses piezoelectric films. Lead zirconate titanate (PZT) has a high piezoelectric coefficient and can be integrated into cantilever probes. Zhu et al. [96] described a method to fabricate $50 \times 100\ \mu m^2$ microcantilevers with 1-µm-thick PZT membranes with almost 100% yield. Currently, the integration of piezoelectric bending actuators onto the NFP array is under development, and preliminary actuation results confirmed the feasibility of the operation with active probes [97]. With active probes capable of lifting the tips off the substrate, independent patterns can be made on individual writing sites, in one scanning stroke. Since scalability was proven with the one-dimensional array, the NFP is currently being augmented to a two-dimensional array with an integrated microfluidic network. This would bring the NFP one step closer to a production tool to generate, in a massively parallel fashion, templates that consist of broad range of analytes. While the augmentation to two-dimensional arrays is a matter of scalability and wafer-level-yields engineering, the operation of two-dimensional NFP arrays would require specially designed AFM systems rather than the presently available AFM instruments. With the adoption of piezoelectric precision stages or microelectromechanical system stages [98,99], it is feasible that the two-dimensional NFP array could pattern large areas to produce multicomponent templates. In this case, leveling between the array and the substrate can be achieved with a three-probe feedback scheme, as reported for scanning probe data storage [100]. By avoiding the implementation of individual sensors on each probe, this approach can reduce the complexity of the system. With this scheme, the two-dimensional array can be maintained parallel to the substrate in the same way as an air table. Like one-dimensional arrays, passive probes are adequate if duplicate patterns, for example, nanoarrays, are to be fabricated. Alternatively, individual active probes allow the writing of different patterns with each probe.

For the operation of two-dimensional arrays of active probes with independent bending actuation, each cantilever requires one electrical connection for its electrode, assuming one set of electrodes shares the electrical path, while the others are indepen-

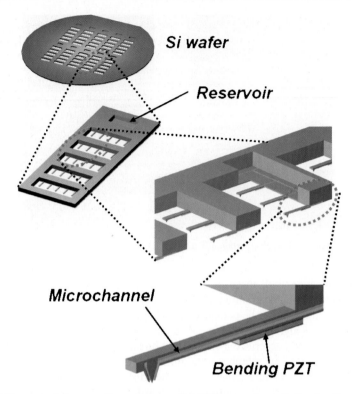

Fig. 5.18. Conceptual diagram of massively parallel NFP arrays for wafer-level patterning. *PZT* lead zirconate titanate

dently driven. Such direct drive is straightforward and simple when the size of the array is small. However, for a large array, such as a massively parallel two-dimensional array with a large number of cantilevers, a multiplexed addressing scheme needs to be employed to reduce the number of connections. This scheme is widely used in computer memories and liquid-crystal displays. Since the probe must be in continuous contact with the substrate between command signals, a bistable (memory) element must be integrated at the root of each cantilever. Furthermore, with multiple reservoirs on the wafer, individual reservoirs may contain one type of molecular ink to feed a cantilever or a group of cantilevers. Through this approach of multiple inks and massively parallel tips, NFP arrays have the potential to manufacture nanoscale features of chemical/biochemical materials at the wafer level (Fig. 5.18).

5.6
Future Prospects for Nanoprobes

Cantilevered nanoprobes offer the possibility of flexible protein deposition for diagnostic applications and drug screening. In addition to being compatible with SPM-readout methods, nanoprobe devices can be used in concert with classic protein analysis techniques such as ELISA. The probes can also deliver DNA for the purposes of genomic studies or nanoconstruction. The most recent advances in nanoprobes for biomolecule deposition involve the integration of microfluidics into cantilevers and the arraying of multiple probes in one and two dimensions. Higher-throughput patterning and a reduction in feature size have been achieved, and a better theoretical understanding of the patterning process has been acquired.

Aside from the direct and indirect patterning of biological materials, there are several other applications offered by cantilevered probes, such as in-depth studies of cell adhesion. Cells were able to adhere to DPN-generated patterns [39], although the pattern resolution was well below that of the length scale at which protein clustering in focal adhesion occurs [19]. Increasing the resolution attainable via probe-based patterning methods will improve the quality of rigid ligand templates for such studies. Controlled delivery can also be used for the study of cell communication: local perturbation of one cardiac myocyte in a small cluster of cells was achieved with the insertion of less than five α-toxin channels into the cell membrane using the nanopipette [101]. It could be observed that perturbation of one cell in a small cluster can affect the neighboring cells.

The atomic force microscope can also be exploited for its force sensitivity; it is able to measure piconewton forces associated with single molecules. In fact, AFM is currently the only force technique capable of mapping and analyzing single molecules with nanoscale lateral resolution [28]. AFM force spectroscopy involves measuring the interaction force between the AFM probe tip and the sample as the tip is pushed towards the sample and then retracted. The AFM probe deflection is measured as a function of the vertical displacement of the piezoelectric scanner. The technique can interrogate the forces and dynamics of interaction between individual ligands and receptors; this is essential to fundamental studies in molecular recognition, protein folding and unfolding, DNA mechanics and cell adhesion. The development of smaller cantilevers such as nanotube tips functionalized with single biomolecules should improve the force resolution, allowing for the measurement of smaller unbinding forces.

Finally, functionalized cantilevers may also operate as biosensors. For example, specific binding events cause changes in cantilever deflection [102]. Grogan et al. [103] demonstrated that antibody-coated cantilevers could be used to detect myoglobin concentrations in the range of normal physiological concentration in human serum. The change in frequency of a resonating cantilever can also be measured; this change is due to adsorption of biological material onto an appropriately functionalized tip. For example, Gupta et al. [104] fabricated arrays of silicon cantilever beams 20–30 nm thick to detect the mass of individual virus particles.

Cantilevered probes have the potential to enable new investigations in the life sciences, and improve diagnostic abilities in the health sciences. With continuing advancements, high-throughput assays performed rapidly and with high sensitivity are becoming a reality.

Acknowledgements. This work was supported primarily by the Nanoscale Science and Engineering Initiative of the National Science Foundation under NSF Award Number EEC–0647560. The authors acknowledge the Microfabrication Applications Laboratory of the University of Illinois at Chicago, the Cornell NanoScale Science & Technology Facility of Cornell University, the

Materials Processing and Crystal Growth Facility of Northwestern University, Northwestern University Atomic- and Nanoscale Characterization Experimental Center, and the Electron Microscopy Center of Argonne National Laboratory for the fabrication and characterization facilities.

References

1. Tang Q, Zhang Y, Chen L, Yan F, Wang R (2005) Nanotechnology 16:1062–1068
2. Wadu-Mesthrige K, Amro NA, Garno JC, Xu S, Liu G-Y (2001) Biophys J 80:1891–1899
3. Bier FF, Andresen D, Walter A (2006) In: Urban GA (ed) BioMEMS. Springer, Dordrecht, pp 167–198
4. Auburn RP, Kreil DP, Meadows LA, Fischer B, Matilla SS, Russell S (2005) Trends Biotechnol 23(7):374–379
5. Reese MO, van Dam RM et al (2003) Genome Res 13:2348–2352
6. Sotomayor Torres CM (ed) (2003) Alternative lithography: unleashing the potentials of nanotechnology. Kluwer/Plenum, Dordrecht
7. Bernard A, Renault JP, Michel B, Bosshard HR, Delamarche E (2000) Adv Mater 12(14):1067–1070
8. Hovis JS, Boxer SG (2000) Langmuir 16(3):894–897
9. Michel B, Bernard A, Bietsch A, Delamarche E, Geissler M, Juncker D, Kind H, Renault JP, Rothuizen H, Schmid H, Schmidt-Winkel P, Stutz R, Wolf H (2001) IBM J Res Dev 45(5):697–719
10. Mesquida P, Blanco EM, McKendry RA (2006) Langmuir 22(22):9089–9091
11. Tarlov MJ, Burgess DRF, Gillen G (1993) J Am Chem Soc 115(12):5305–5306
12. Lipshutz RJ, Fodot SPA, Gingeras TR, Lockhart DJ (1999) Nat Genet 21:20–24
13. Singh-Gasson S, Green RD, Yue Y, Nelson C, Blattner F, Sussman MR, Cerrina F (1999) Nat Biotechnol 17(10):974–978
14. Stoll D, Templin MF, Bachmann J, Joos TO (2006) In: Urban GA (ed) BioMEMS. Springer: Dortrecht, pp 245–268
15. Anonymous (2006) Nature 444(7121):963–964
16. Christman KL, Requa MV, Enriquez-Rios VD, Ward SC, Bradley KA, Turner KL, Maynard HD (2006) Langmuir 22(17):7444–7450
17. Pierce Biotechnology (2006) ELISA. Pierce Biotechnology, Rockford
18. Saliterman SS (2006) BioMEMS and medical microdevices. SPIE, Bellingham
19. Arnold M, Cavalcanti-Adam EA, Glass R, Blümmel J, Eck W, Kantlehner M, Kessler H, Spatz JP (2004) Chem Phys Chem 5(3):383–388
20. Ngunjiri JN, Li J-R, Garno JC (2006) In: Kumar CSSR (ed) Nanodevices for the life sciences, vol 4, 1st edn. Wiley-VCH, Weinheim pp 67–108
21. Lenhert S, Sun P, Wang Y, Fuchs H, Mirkin, Chad A (2007) Small 3(1):71–75
22. Henderson E, Lynch M, Nettikadan S, Johnson J, Huff J, Xu J, Mosher C, Vengasandra S, Radke K, Cao C, Ryckman K, Kristmundsdottir A (2004) Microsc Microanal 10:1432–1433
23. Christman KL, Enriquez-Rios VD et al (2006) Soft Matter 2:928–939
24. Lynch M, Mosher C, Huff J, Nettikadan S, Johnson J, Henderson E (2004) Proteomics 4:1695–1702
25. Lee M, Kang D-K, Yang H-K, Park K-H, Choe SY, Kang C, Chang S-I, Han MH, Kang I-C (2006) Proteomics 6(4):1094–1103
26. Yu X, Cheng DXQ (2006) Proteomics 6(20):5493–5503
27. Hu J, Zhang Y, Gao H, Li M, Hartmann U (2002) Nano Lett 2(1):55–57
28. Hinterdorfer P, Dufrene YF (2006) Nat Methods 3(5):347–355

29. Piner RD, Zhu J, Xu F, Hong S, Mirkin CA (1999) Science 283:661–663
30. Ginger DS, Zhang H, Mirkin CA (2004) Angew Chem Int Ed Engl 43(1):30–45
31. Rosner B, Duenas T, Banerjee D, Shile R, Amro N, Rendlen J (2006) Smart Mater Struct 15:S124–S130
32. Meister A, Jeney S, Liley M, Akiyama T, Staufer U, De Rooij NF, Heinzelmann H (2003) Microelectron Eng 644:67–68
33. Meister A, Liley M, Brugger J, Pugin R, Heinzelmann H (2004) Appl Phys Lett 85(25):6260–6262
34. Fang A, Dujardin E, Ondarcuhu T (2006) Nano Lett 6(10):2368–2374
35. Kim K-H, Moldovan N, Espinosa HD (2005) Small 1(6):632–635
36. Moldovan N, Kim K-H, Espinosa HD (2006) J Micromech Microeng 16:1935–1942
37. Wu B, Ho A, Moldoval N, Espinosa H (2007) Langmuir 23(17):9120–9123
38. Nafday OA, Weeks BL (2006) Langmuir 22:10912–10914
39. Lee K-B, Park S-J, Mirkin CA, Smith JC, Mrksich M (2002) Science 295:1702–1705
40. Vega RA, Maspoch D, Shen CKF, Kakkassery JJ, Chen BJ, Lamb RA, Mirkin CA (2006) Chembiochem 7(11):1653–1657
41. Lee KB, Kim EY, Mirkin CA, Wolinsky SM (2004) Nano Lett 4(10):1869–1872
42. NanoInk (2007) Protein arrays application note. NanoInk, Skokie. http://www.nanoink.net/d/appnote_protein_arrays.pdf
43. Wilson DL, Martin R, Hong S, Cronin-Golomb M, Mirkin CA, Kaplan DL (2001) Proc Natl Acad Sci USA 98(24):13660–13664
44. Lee K-B, Lim J-H, Mirkin CA (2003) J Am Chem Soc 125:5588–5589
45. Lim J-H, Ginger DS, Lee K-B, Heo J, Nam J-M, Mirkin CA (2003) Angew Chem Int Ed Engl 42(20):2309–2312
46. Nam J-M, Han SW, Lee K-B, Liu X, Ratner MA, Mirkin CA (2004) Angew Chem Int Ed Engl 43(10):1246–1249
47. Demers LM, Ginger DS, Park S-J, Li Z, Chung S-W, Mirkin CA (2002) Science 296:1836–1838
48. Zhang H, Elghanian R, Amro NA, Disawal S, Eby R (2004) Nano Lett 4(9):1649–1655
49. Agarwal G, Sowards LA, Naik RR, Stone MO (2002) J Am Chem Soc 125:580–583
50. Xu J, Lynch M, Huff JL, Mosher C, Vengasandra S, Ding G, Henderson E (2004) Biomed Microdevices 6(2):117–123
51. Vengasandra SG, Lynch M, Xu J, Henderson E (2005) Nanotechnology 16:2052–2055
52. Schena M, Heller RA, Theriault TP, Konrad K, Lachenmeier E, Davis RW (1998) Trends Biotechnol 16(7):301–306
53. Belaubre P, Guirardel M et al (2003) Appl Phys Lett 82(18):3122–3124
54. Belaubre P, Guirardel M et al (2004) Sens Actuators A 110:130–135
55. Leichle T, Silvan MM et al (2005) Nanotechnology 16(4):525
56. Saya D, Le T et al (2007) J Micromech Microeng 17(1):N1
57. Bietsch A, Zhang J, Hegner M, Lang HP, Gerber C (2004) Nanotechnology 15(8):873
58. Espinosa HD, Kim K-H, Moldovan N (2006) In: Kumar CSSR (ed) Nanodevices for the life sciences, vol 4, 1st edn. Wiley-VCH, Weinheim, pp 109–149
59. Bruckbauer A, Ying L, Rothery AM, Zhou D, Shevchuk AI, Abell C, Korchev YE, Klenerman D (2002) J Am Chem Soc 124(30):8810–8811
60. Bruckbauer A, Zhou D, Ying L, Korchev YE, Abell C, Klenerman D (2003) J Am Chem Soc 125(32):9834–9839
61. Rodolfa KT, Bruckbauer A, Zhou D, Korchev YE, Klenerman D (2005) Angew Chem Int Ed Engl 44(42):6854–6859
62. Rodolfa KT, Bruckbauer A, Zhou D, Shevchuk AI, Korchev YE, Klenerman D (2006) Nano Lett 6(2):252–257

63. Hu H, Xie S, Meng X, Jing P, Zhang M, Shen L, Zhu Z, Li M, Zhuang Q, Shao Y (2006) Anal Chem 78(19):7034–7039
64. Kim K-H, Ke C, Moldovan N, Espinosa HD (2003) In: Proceedings of the 4th international symposium on MEMS and nanotechnology. The 2003 SEM annual conference and exposition on experimental and applied mechanics, Charlotte, NC, 2–4 June, pp 235–238
65. In: Materials Research Society symposium proceedings, fall MRS meeting, 2004, p A5.56.1
66. Espinosa HD, Moldovan N, Kim K-H (2007) In: Bhushan B, Fuchs H (eds) Applied scanning probe methods, vol VII. Springer, Berlin, pp 77–134
67. Kim K-H, Sanedrin RG, Ho AM, Lee SW, Moldovan N, Mirkin CA, Espinosa HD (2007) Direct Delivery and Submicron Patterning of DNA by Nanofountain Probe. In press in Advanced Materials, 2007
68. Deladi S, Tas NR, Berenschot JW, Krijnen GJM, de Boer MJ, de Boer JH, Peter M, Elwenspoek MC (2004) Appl Phys Lett 85(22):5361–5363
69. Deladi S, Berenschot JW, Tas NR, Krijnen GJM, de Boer JH, de Boer MJ, Elwenspoek MC (2005) J Micromech Microeng 15:528–534
70. Xu S, Miller S, Laibinis PE, Liu G-Y (1999) Langmuir 15:7244–7251
71. Liu G-Y, Amro NA (2002) Proc Natl Acad Sci USA 99(8):5165–5170
72. Naujoks N, Stemmer A (2004) Colloids Surf A 249(1–3):69–72
73. Agarwal G, Naik RR, Stone MO (2003) J Am Chem Soc 125:7408–7412
74. Jang J, Hong S, Schatz GC, Ratner MA (2001) J Chem Phys 115(6):2721–2729
75. Sheehan PE, Whitman LJ (2002) Phys Rev Lett 88(15):156104
76. Cho N, Ryu S, Kim B, Schatz GC, Hong S (2006) J Chem Phys 124(2):024714
77. Weeks BL, Vaughn MW, DeYoreo JJ (2005) Langmuir 21:8096–8098
78. Rozhok S, Sun P, Piner R, Lieberman M, Mirkin CA (2004) J Phys Chem B 108(23):7814–7819
79. Weeks BL, Noy A, Miller AE, De Yoreo JJ (2002) Phys Rev Lett 88(25):255505
80. Zou J, Wang X, Bullen D, Liu C, Mirkin C (2006) Proc SPIE 6223:62230N-10
81. Hampton JR, Dameron AA, Weiss PS (2005) J Phys Chem B 109(49):23118–23120
82. Weeks BL, DeYoreo JJ (2006) J Phys Chem B 110(21):10231–10233
83. Schwartz PV (2002) Langmuir 18(10):4041–4046
84. Peterson EJ, Weeks BL, DeYoreo JJ, Schwartz PV (2004) J Phys Chem B 108(39):15206–15210
85. Nafday OA, Vaughn MW, Weeks BL (2006) J Chem Phys 125:144703
86. Manandhar P, Jang J, Schatz GC, Ratner MA, Hong S (2003) Phys Rev Lett 90(11):115505
87. Lee N-K, Hong S (2006) J Chem Phys 124(11):114711
88. Xu J, Lynch M, Nettikadan S, Mosher C, Vegasandra S, Henderson E (2006) Sens Actuators B 113(2):1034–1041
89. Lee SW, Oh BK, Sanedrin RG, Salaita K, Fujigaya T, Mirkin CA (2006) Adv Mater 18(9):1133–1136
90. Vettiger P, Despont M, Drechsler U, Durig U, Haberle W, Lutwyche MI, Rothuizen HE, Stutz R, Widmer R, Binnig GK (2000) IBM J Res Dev 44(3):323–340
91. Vettiger P, Cross G, Despont M, Drechsler U, Durig U, Gotsmann B, Haberle W, Lantz MA, Rothuizen HE, Stutz R, Binnig GK (2002) IEEE Trans Nanotechnol 1(1):39–55
92. Salaita K, Wang Y, Fragala J, Vega RA, Liu C, Mirkin CA (2006) Angew Chem Int Ed Engl 45(43):7220–7223
93. Bullen D, Chung S-W, Wang X, Zou J, Mirkin CA, Liu C (2004) Appl Phys Lett 84(5):789–791
94. Wang X, Bullen DA, Zou J, Liu C, Mirkin CA (2004) J Vac Sci Technol B 22(6):2563–2567
95. Bullen D, Liu C (2006) Sens Actuators A 125(2):504–511
96. Zhu H, Miao J, Chen B, Wang Z, Zhu W (2005) Microsyst Technol 11:1121–1126

97. Northwestern University (2007) Espinosa research group.
    http://clifton.mech.northwestern.edu/~nfp/
98. Rothuizen H, Drechsler U, Genolet G, Haberle W, Lutwyche M, Stutz R, Widmer R,
    Vettiger P (2000) Microelectron Eng 53(1–4):509–512
99. Kim CH, Jeong HM, Jeon JU, Kim YK (2003) J Microelectromech Syst 12(4):470–478
100. Lutwyche M, Andreoli C, Binnig G, Brugger J, Drechsler U, Haberle W, Rohrer H,
     Rothuizen H, Vettiger P, Yaralioglu G, Quate C (1999) Sens Actuators A 73(1–2):89–94
101. Ying L, Bruckbauer A, Zhou D, Gorelik J, Shevchuk A, Lab M, Korchev Y, Klenerman D
     (2005) Phys Chem Chem Phys 7:2859–2866
102. Ziegler C (2004) Anal Bioanal Chem 379:946–959
103. Grogan C, Raiteri R, O'Connor GM, Glynn TJ, Cunningham V, Kane M, Charlton M,
     Leech D (2002) Biosens Bioelectron 17(3):201–207
104. Gupta A, Akin D, Bashir R (2004) Appl Phys Lett 84(11):1976–1978

6 Self-Sensing Cantilever Sensor for Bioscience

Hayato Sone · Sumio Hosaka

Abstract. A simple and high-sensitivity detection system is desired in the fields of biotechnology and medical science. In order to develop the system, one of the techniques is the use of a micro-cantilever mass sensor using a harmonic vibration with a resonance frequency. In this chapter, we describe a harmonic vibration-type self-sensing cantilever sensor in bioscience applications. Firstly, we introduce the cantilever mass sensor and its vibrations using theoretical analysis of cantilever motion and finite element method simulation. Then, we explain details of the self-sensing system using a piezoresistive cantilever. Finally, we demonstrate two application studies to achieve femtogram sensitivity, one for water molecule detection in air and the other for the biomolecular reaction between an antigen and an antibody in water.

Key words: Biosensor, Piezoresistive cantilever, Mass detection, Femtogram Sensitivity, Biomolecule

6.1 Introduction

As medical technologies advance, it is revealed that various kinds of chemical and biological materials affect our health. Of particular interest to the public are environmental hormone-disrupting chemicals, allergenic substances, and genetic materials. Analytical instrumentation techniques such as gas chromatography–mass spectrometry (GCMS) or liquid chromatography—mass spectrometry (LCMS) [1–3], immunoassay [4–6], the quartz crystal microbalance (QCM) method [7–9], and surface plasmon resonance (SPR) [10–12] have been developed to detect these materials. Although these techniques have some advantages, they also have a few disadvantages, as follows. GCMS and LCMS have a high mass resolution of about 1 pg, but the preparation and measurement of samples take a long time. QCM is a simplified measurement method, but it has a low sensitivity of about 30 pg/Hz. SPR has a high sensitivity equivalent to a few picograms, but the measurement instrument is complex and expensive. One of the solutions to these problems is a microcantilever-based mass sensor using the detection technique of vicinal deflection.

The cantilever-based mass sensor was introduced by Berger et al. [13, 14]. They proposed various physical and chemical sensor applications, including as a mass sensor, a force sensor [15], a temperature sensor [16], a calorimeter [17], and a thermo-gravimetric sensor [18]. Water adsorption in the nanogram range has been measured using a cantilever with a zeolite sensor [13]. Applications of biomolecular detection for bovine serum albumin and deoxyribonucleic acid hybridization using the nanomechanical responses of cantilever bending have been reported [19–21]. The binding of thiol molecules with a total mass of a few femtograms has been monitored using the resonance frequency shifts of cantilevers [22]. Nanometer-scale mass sensors with subattogram sensitivity have been fabricated using nanoelectromechanical oscillators [23]. Cantilever harmonic-vibration-type mass sensors with picogram mass sensitivity using a laser beam deflection detection system have been reported in experiments of molecular adsorption on the cantilever [24, 25]. These reported works, however, used a laser beam deflection detection system for the detection of the resonance frequency. This method involves technical difficulties such as heavy equipment and complex adjustment. Furthermore, in the case of biomolecule detection, the cantilever mass sensor has to operate in water. In order to solve these problems and to respond to bioscience needs more effectively, a self-sensing system has been proposed to replace the laser beam deflection detection system. As the self-sensing system, several researchers have used a piezoresistive strain sensor embedded in the cantilever [26–33]. The cantilever has been used to detect the deflection of the cantilever by measuring the piezoresistance in an atomic force microscope cantilever [34–36]. In thischapter, we will discuss the role of harmonic-vibration-type self-sensing piezoresistive sensors in bioscience applications.

6.2
Basics of the Cantilever Mass Sensor

When we vibrate a microcantilever using an exciting piezoactuator which supports the cantilever base, as shown in Fig. 6.1a, the cantilever equation of motion is given by

$$M\frac{d^2z}{dt^2} + a\frac{dz}{dt} + kz = F_0\sin(\omega t) , \tag{6.1}$$

where M is the effective mass of the cantilever, z is the deflection of the cantilever top, a is the viscous resistance modulus, k is the spring constant, and $F_0\sin(\omega t)$ is the external vibration force from the exciting piezoactuator. The general solution is obtained using the differential equation of (6.1), as follows:

$$z(t) = C_1\exp\left[\left(-\frac{a}{2M} + \sqrt{\frac{a^2}{4M^2} - \omega_0^2}\right)t\right]$$
$$+ C_2\exp\left[\left(-\frac{a}{2M} - \sqrt{\frac{a^2}{4M^2} - \omega_0^2}\right)t\right]$$
$$+ \frac{F_0}{M\left[\left(\omega_0^2 - \omega^2\right)^2 + \left(\frac{a\omega}{M}\right)^2\right]}\sin(\omega t + \delta) , \tag{6.2}$$

where C_1 and C_2 are constants, $\omega_0^2 = k/M$ and $\delta = \tan^{-1}\left(-\frac{a\omega}{M\left(\omega_0^2 - \omega^2\right)}\right)$. When the cantilever is vibrated at the resonance frequency and given sufficient time, we only

Fig. 6.1. The harmonic-vibration-type microcantilever sensor using a piezoactuator. **a** Resonance frequency, ω. **b** Resonance frequency with adsorbed molecules on the cantilever, $\omega - \Delta\omega$

(a)

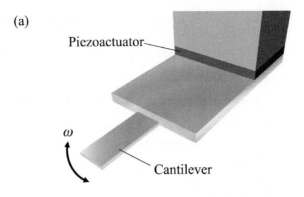

Piezoactuator

ω

Cantilever

(b)

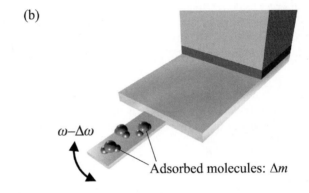

$\omega - \Delta\omega$

Adsorbed molecules: Δm

need to consider the third term in (6.2). Since the vibration amplitude reaches its maximum value at the resonance frequency, ω is given by

$$\omega = \sqrt{\frac{k}{M} - \frac{a^2}{2M^2}} \; . \tag{6.3}$$

Additionally, the sensitivity of the sensor is derived from the total differential of (6.3), as follows:

$$\Delta\omega = \frac{1}{2}\left(\frac{\frac{a^2}{M^3} - \frac{k}{M^2}}{\sqrt{\frac{k}{M} - \frac{a^2}{2M^2}}}\Delta M + \frac{1}{M\sqrt{\frac{k}{M} - \frac{a^2}{2M^2}}}\Delta k\right) \; . \tag{6.4}$$

When the spring constant does not change, we can substitute the differential of the spring constant, $\Delta k = 0$:

$$\Delta\omega = \frac{1}{2}\frac{\frac{a^2}{M^3} - \frac{k}{M^2}}{\sqrt{\frac{k}{M} - \frac{a^2}{2M^2}}}\Delta M \; . \tag{6.5}$$

This result shows that the resonance frequency changes as the cantilever mass changes, and that the amount of the frequency change depends on the spring constant and the viscous resistance. When we vibrate the cantilever in a low-viscosity environment, a is negligibly small; therefore, $\Delta\omega$ is given by

$$\Delta\omega = -\frac{1}{2M}\sqrt{\frac{k}{M}}\Delta M = -\frac{\omega}{2M}\Delta M . \tag{6.6}$$

M of a rectangular-shaped cantilever is given by nm, where n is a proportional constant and m is the mass of the cantilever. Depending on the geometry, n is estimated to be 0.24 for a rectangular cantilever [37]. According to the material adsorption on the cantilever, as shown in Fig. 6.1b, the mass, m, changes to $m + \Delta m$ and the resonance frequency, ω, changes to $\omega - \Delta\omega$. From (6.6), the mass change, Δm, is given by

$$\Delta m = -2\frac{m}{\omega}\Delta\omega . \tag{6.7}$$

In order to detect small Δm values, it is necessary to use a cantilever with a small mass and a high resonance frequency and to detect small frequency changes with a fine resolution. The microcantilever satisfies the requirements of small mass and high resonance frequency. The spring constant of the cantilever is given by

$$k = \frac{Et^3 w}{4l^3} , \tag{6.8}$$

where E is Young's modulus, t is the thickness, w is the width, and l is the length. The resonance frequency of the cantilever is obtained from

$$\omega = \frac{1.02t}{l^2}\sqrt{\frac{E}{\rho}} , \tag{6.9}$$

where ρ is the density. Figure 6.2 shows the calculated results for k and the resonance frequency, f, of the rectangular silicon cantilever, where w is 10 μm. The calculated

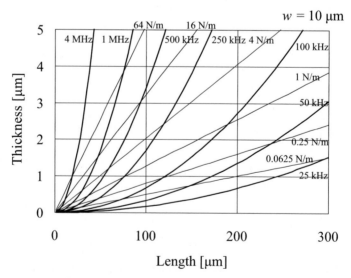

Fig. 6.2. Contour map of the resonance frequency and spring constant of the microcantilever at a cantilever width of $w = 10$ μm with changing length, l, and thickness, t

results show that k becomes larger with increasing t, and f becomes higher with increasing l. If we use a cantilever with a mass of 10 ng and a resonance frequency of 100 kHz, a mass sensitivity of $\Delta m / \Delta f = 200$ fg/Hz is obtained. This value is 100 times higher than that given by the QCM method.

6.3
Finite Element Method Simulation of the Cantilever Vibration

To realize highly sensitive mass detection using a piezoresistive cantilever, the resonance frequency and strain distribution were studied using a computer simulation and the commercially available software COMSOL Multiphysics (COMSOL). This software solves partial differential equations of physical parameters using finite-element modeling. In the simulation, the 3D structure of the cantilever is designed by the computer-aided design system, and the material properties, boundary conditions, and mesh size of the segment region are set in the designed 3D structures. Then, the software executes the calculations of the finite-element analysis together with adaptive meshing and error control using a variety of numerical solvers, and it can perform original frequency analysis or stationary-state analysis as well.

Figure 6.3a shows the strain distribution of the rectangular piezoresistive silicon cantilever. The length, width, and thickness of the cantilever are 110, 50, and 4 µm, respectively. The silicon base is fixed and the front end of the cantilever beam is pushed downward with an applied force of 40 nN. Consequently, the strain is widely distributed on the surface of the cantilever beam, and the maximum value of the strain, which concentrates in the base region of the cantilever beam, is approximately 0.17×10^{-6}. In addition, the original resonance frequency of the cantilever is about 473 kHz. In contrast, Fig. 6.3b shows the strain distribution on the cantilever base of the rectangular piezoresistive silicon cantilever with a narrow aperture 10 µm in length and 40 µm in width; i.e., the cantilever beam is connected with the silicon base via twin slim arms 10 µm in length and 5 µm in width. The length, width, and thickness of the cantilever and the applied force are same as for the previous rectangular cantilever. The results show that the strain is concentrated on the slim arms, and that the maximum value of the strain and the original resonance frequency are approximately 1.08×10^{-6} and 279 kHz, respectively. Therefore, it was found that the strain is concentrated in the slim arms of the cantilever base, and that it has the potential of generating a large resistance change, ΔR, unlike the cantilever without the narrow aperture.

The resonance frequency and strain of the rectangular piezoresistive cantilever with a narrow aperture for different cantilever lengths were calculated. Figure 6.4a shows the top view of the cantilever, and x_1 is the cantilever length. Figure 6.4b shows the calculated frequency and strain for cantilever lengths, x_1, of 20–110 µm. The frequency decreases from 5208 to 279 kHz with increasing length, while the strain increases linearly from 0.18×10^{-6} to 1.08×10^{-6} with increasing length. Although the higher resonance frequency and the smaller mass of the cantilever have the advantage of enabling the detection of small masses of adsorbed material, as shown by (6.7), the strain decreases with decreasing length. In addition, the detectable frequency is limited in the detection circuits. Under these conditions,

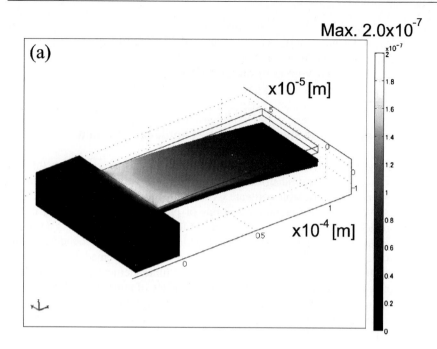

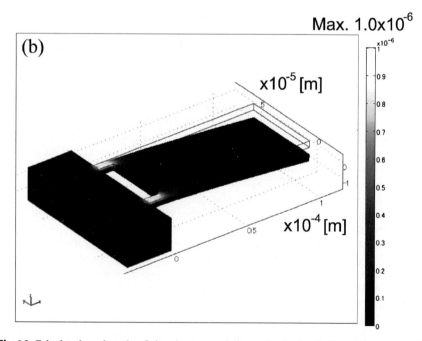

Fig. 6.3. Calculated results using finite-element modeling. **a** Strain distribution of the rectangular piezoresistive silicon cantilever. **b** Strain distribution of the rectangular piezoresistive silicon cantilever with a narrow aperture

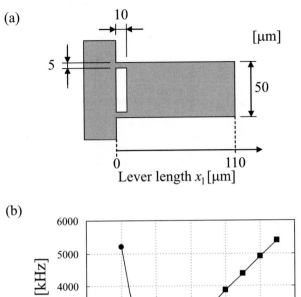

Fig. 6.4. a The piezoresistive cantilever with a narrow aperture. **b** Calculated results of the frequency and strain for cantilever lengths, x_1, from 20 to 110 μm

the optimum cantilever length (x_1) is considered to be about 80 μm. The optimum length changes depending on the cantilever width and the arm width.

Next, the resonance frequency and strain of the piezoresistive cantilever with an aperture for different aperture lengths were calculated. Figure 6.5a shows the top view of the cantilever, and x_a is the aperture length. Figure 6.5b shows the calculated results of the frequency and strain for aperture lengths, x_a, of 10–100 μm. The frequency decreases from 280 to 219 kHz in the x_a range from 10 to 50 μm, and increases from 219 to 312 kHz in the x_a range from 50 to 100 μm. The strain decreases rapidly from 1.08×10^{-6} to 0.93×10^{-6} in the x_a range from 10 to 30 μm, and increases gradually to 0.96×10^{-6} in the x_a range from 30 to 100 μm. As higher resonance frequency and higher strain have the advantage of enabling the detection of small mass changes, the optimum aperture length (x_a) is about 10 μm in the cantilever. This optimum length also changes depending on the cantilever width and the arm width.

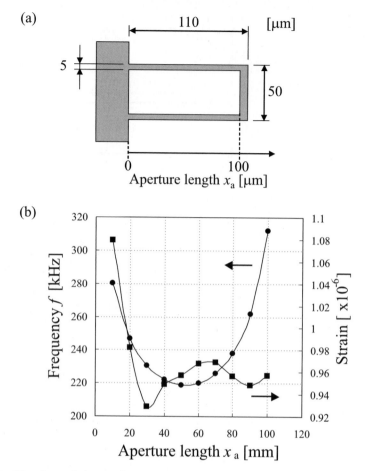

Fig. 6.5. a The piezoresistive cantilever with an aperture. **b** Calculated results of the frequency and strain for aperture lengths, x_a, from 10 to 100 μm

6.4
Detection of Cantilever Deflection

In order to evaluate the sensitivity of the cantilever sensor, we compared a position sensor using a laser beam deflection detection system with a piezoresistive sensor.

6.4.1
Using a Position Sensor

Laser beam deflection detection systems are widely used in atomic force microscope systems. A laser beam deflection detection system can magnify small position changes of the cantilever front end at a high optical magnification, from ×100 to ×1000. An incident laser beam irradiated from a laser diode is focused on the cantilever, and the position of the reflected laser beam is detected by a position-sensor

Fig. 6.6. The laser beam deflection detection system. *PSD* position-sensor diode

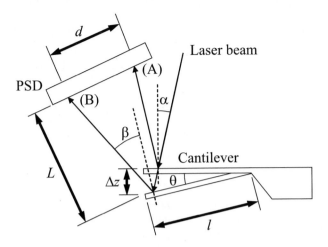

diode (PSD), as shown in Fig. 6.6. When the cantilever moves down with the angular change θ, the position of the laser beam reflected on the cantilever shifts from A to B on the PSD. The laser beams at A and B are reflected from the cantilever with the angles 2α and 2β, respectively. The position change of the cantilever top, Δz, is magnified to the distance d on the PSD. The magnification, $d/\Delta z$, of this system is given by

$$\frac{d}{\Delta z} = \frac{2L}{l} \, , \tag{6.10}$$

where l is the length of the cantilever and L is the distance between the cantilever and the PSD. A short cantilever should be used in order to obtain a large magnification; i.e., a system with a short cantilever has a high sensitivity for detecting the amplitude of the cantilever vibration. Here, we used a cantilever length, l, of 100 μm and an optical path length, L, of about 5 cm. Consequently, the magnification was 1000. As the PSD can detect a minimum beam position change of less than 0.1 μm, the minimum detectable amplitude of the cantilever vibration is about 0.1 nm; thus, this detection system can detect small cantilever vibrations.

6.4.2
Using a Piezoresistive Sensor

For biomolecule detection, the mass sensor must be operated in water. The reported detection method uses a laser beam deflection detection system for the detection of the resonance frequency, as discussed in Sect. 6.4.1. This method requires some technical adjustments of the laser path, focal point, etc. In order to respond to bioscience needs more effectively, a self-sensing system has been developed to replace the laser beam detection system. The self-sensing system is based on a piezoresistive sensor embedded in the cantilever [26–33].

The deflection of the piezoresistive cantilever is detected on the basis of the piezoresistance change generated by the bending strain. Figure 6.7 shows a cross-section diagram of a cantilever with thickness t and deflection length z. The length

Fig. 6.7. The bending behavior of the piezoresistive cantilever

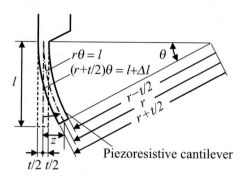

Piezoresistive cantilever

along the midline of the cantilever is l, which is the same as before bending. After bending the cantilever, when the angle of arc and the radius of the cantilever bending along the midline are θ (radians) and r, respectively, the length along the outer surface is given by

$$(r + \frac{t}{2})\theta = l + \Delta l ,\tag{6.11}$$

where Δl is the stretch of the cantilever. From $l = r\theta$, Δl is given by

$$\Delta l = \frac{t}{2}\theta .\tag{6.12}$$

Therefore, the strain, ε, in the bending cantilever is defined by

$$\varepsilon \equiv \frac{\Delta l}{l} = \frac{t}{2r} .\tag{6.13}$$

The strain can be detected on the basis of the resistance change, ΔR, of the piezoresistor, as follows:

$$\varepsilon \equiv \frac{1}{K}\frac{\Delta R}{R} ,\tag{6.14}$$

where K is the gage factor of the piezoresistor and R is the piezoresistance. Then, the deflection length, z, is given by

$$z = r(1 - \cos\theta) = r\left[1 - \cos\left(\frac{l}{r}\right)\right] .\tag{6.15}$$

From (6.14) and (6.15), we obtain

$$
\begin{aligned}
z &= \frac{t}{2\varepsilon}\left[1 - \cos\left(\frac{2\varepsilon l}{t}\right)\right] = \frac{tKR}{2\Delta R}\left[1 - \cos\left(\frac{2l\Delta R}{tKR}\right)\right] \\
&= \frac{tKR}{\Delta R}\sin^2\left(\frac{l\Delta R}{tKR}\right) .
\end{aligned}\tag{6.16}
$$

Since we use a piezoresistive cantilever, $l\Delta R$ is much smaller than tKR:

$$z \approx \frac{tKR}{\Delta R}\left(\frac{l\Delta R}{tKR}\right)^2 = \frac{l^2\Delta R}{tKR} .\tag{6.17}$$

Consequently, if we know the values of t, l, K, and R, the cantilever deflection can be obtained on the basis of the piezoresistance measurements. From (6.14), z is given by

$$z = \frac{l^2}{t}\varepsilon .$$ (6.18)

To detect the strain, ε, a strain meter with Wheatstone bridge circuits is used. The strain meter amplifies the strain signal generated by the cantilever deflection. For example, we obtain an output voltage from the strain meter of 1 V for a strain of 50×10^{-6}. As we can usually measure the voltage with a resolution of 10 mV, the minimum detectable strain is about 50×10^{-8}. If we use a piezoresistive cantilever length, l, of 100 μm and a thickness, t, of 4 μm, the minimum detectable deflection, z, is about 1 nm. This sensitivity is 10 times lower than that of the laser beam deflection detection system. But if we use small piezoresistive cantilevers or measure small voltage changes, it is possible to detect deflections about 1 nm.

Many kinds of piezoresistive cantilevers have been developed for atomic force microscope probes, and some of them are commercially available. We introduce here two types of piezoresistive cantilevers. One is the V-shaped piezoresistive silicon cantilever, shown in Fig. 6.8a, which is fabricated by a semiconductor microprocess. The cantilever has a length of 335 μm, a thickness of 3 μm, a surface area of about 5.8×10^{-4} cm^2 (total of both sides), a mass of about 200 ng, and a piezoresistance of about 1.3 kΩ. The piezoresistors R_1 and R_3 are formed by boron implantation on both beams of the cantilever, as shown in Fig. 6.9a. The other is a commercially available rectangular piezoresistive silicon cantilever (NPX1CTP004; SII Nanotechnology), shown in Fig. 6.8b. This cantilever has a length of 110 μm, a thickness of 4–5 μm, a surface area of about 9.8×10^{-5} cm^2 (total of both sides), a mass of about 46 ng, a spring constant of 40 N/m, and a piezoresistance of about 630 Ω.

The resistance change generated by the cantilever deflection was measured using a dynamic strain meter (e. g., DC-96A; Tokyo Sokki Kenkyujo) with Wheatstone bridge circuits, as shown in Fig. 6.9b. In the V-shaped piezoresistive cantilever shown in Fig. 6.8a, the bridge circuits comprise the piezoresistors R_1 and R_3 and the standard resistors R_2 and R_4. The voltage change, ΔV, is given by

$$\Delta V = \frac{R_1 R_3 - R_2 R_4}{(R_1 + R_2)(R_3 + R_4)} E ,$$ (6.19)

where E is the voltage applied to the bridge circuits. In the case of $R_1 = R_3 = R + \Delta R$, $R_2 = R_4 = R$, ΔV is given by

$$\Delta V = \frac{2R\Delta R + \Delta R^2}{(2R + \Delta R)^2} E ,$$ (6.20)

where ΔR is the piezoresistance change. As R is much larger than ΔR, ΔV is given by

$$\Delta V = \frac{\Delta R}{2R} E .$$ (6.21)

This equation substitutes for (6.17) and (6.18), as follows:

$$\varepsilon = \frac{t}{l^2}z = 2\frac{\Delta V}{KE} .$$ (6.22)

(a)

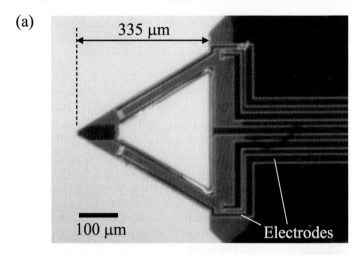

335 μm

100 μm

Electrodes

(b)

110 μm

20 μm

Fig. 6.8. Optical microscope image of the microcantilever. **a** V-shaped piezoresistive cantilever. **b** Commercially available rectangular piezoresistive cantilever

From this equation, the deflection of the cantilever can be detected on the basis of ΔV, and the resonance frequency of the cantilever is obtained from the change in the amplitude of the vibration signal.

The resistance, R, of the V-shaped piezoresistive cantilever is about 1.3 kΩ, and a voltage, E, of 2 V is applied to the bridge circuits. Consequently, the power induced in the piezoresistor is about 1.5 mW. In order to prevent the induced power from exerting any influence, we gave the bridge circuits ample time after supplying the power. Thus, we were able to measure the molecular adsorption on the cantilever under the thermal equilibrium condition.

The bridge circuits in Fig. 6.9b constitute a two-gage detection system that can calibrate the piezoresistance change with changing temperature. If we use a one-gage piezoresistive cantilever, as shown in Fig. 6.8b, which has the piezoresistance R_1,

ΔV is given by

$$\Delta V = \frac{\Delta R}{4R} E \ . \tag{6.23}$$

This equation substitutes for (6.17) and (6.18), as follows:

$$\varepsilon = \frac{t}{l^2} z = 4 \frac{\Delta V}{KE} \ . \tag{6.24}$$

The deflection of the cantilever can also be detected by a one-gage sensor on the basis of ΔV.

(a)

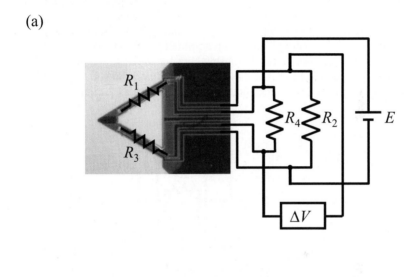

(b)

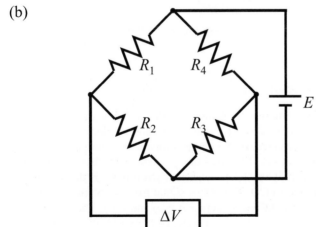

Fig. 6.9. a Piezoresistive cantilever and Wheatstone bridge circuits. **b** Equivalent circuits of **a**

6.5
Self-Sensing Systems

6.5.1
Vibration Systems

The cantilever mass sensor has the potential to achieve femtogram mass sensitivity using the harmonic-vibration method, as discussed previously. Figure 6.10 shows the setup of the piezoresistive cantilever mass sensor using the harmonic-vibration method. The mass sensor consists of a piezoresistive cantilever, a preamplifier, positive-feedback circuits, a piezoactuator, and phase-locked loop (PLL) circuits. To form an oscillator, the piezoresistive cantilever is set on a driver element using an exiting piezoactuator, and the output signal from the bridge circuits, ΔV, is input into the preamplifier and the positive-feedback circuits. Then, the output signal of the positive-feedback circuits is input into the piezoactuator to oscillate the cantilever. The oscillation signal output from the preamplifier is demodulated to the frequency signal by the PLL circuits. The relationship between the PLL output voltage and the frequency is calibrated in advance using a function generator. It is possible to use quadrature circuits or a frequency counter in place of the PLL circuits.

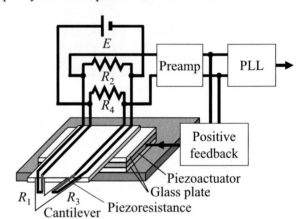

Fig. 6.10. Harmonic-vibration-type mass sensor system and detection circuits using a piezoresistive cantilever. *PLL* phase-locked loop

6.5.2
Vibration-Frequency Detection Systems

In order to detect the vibration frequency, the PLL circuits are usually used as the frequency modulation demodulator. Figure 6.11 shows a block diagram of the PLL system. The PLL circuits consist of a phase comparator, a loop filter, and a voltage-controlled oscillator (VCO). This system operates on the basis of the frequency and phase of the input signal, $v_i(t)$, consistent with those of the sine wave, $v_c(t)$, which is generated in the VCO. The output signal from the phase comparator, $v_e(t)$, corresponds to the phase difference between $v_i(t)$ and $v_c(t)$. The input signal, $v_i(t)$, is given by

$$v_i(t) = V_i \cos[\omega_i t + \theta_i(t)] . \tag{6.25}$$

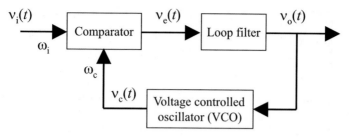

Fig. 6.11. The PLL circuits

The output signal from the VCO is given by

$$v_c(t) = V_c \sin[\omega_c t + \theta_c(t)] \, . \tag{6.26}$$

Since the phase comparator multiplies $v_i(t)$ by $v_c(t)$, the output signal from the comparator is given by

$$
\begin{aligned}
v_e(t) &= V_i V_c \cos[\omega_i t + \theta_i(t)] \sin[\omega_c t + \theta_c(t)] \\
&= \frac{V_i V_c}{2} \{\sin[(\omega_i + \omega_c)t + \theta_i(t) + \theta_c(t)] + \sin[(\omega_i - \omega_c)t + \theta_i(t) - \theta_c(t)]\}.
\end{aligned}
\tag{6.27}
$$

In the loop filter, the high-frequency component in the first term of $v_e(t)$ is cut off, which means that the output signal from the loop filter, $v_o(t)$, is given by

$$v_o(t) = \frac{V_i V_c}{2} \sin[(\omega_i - \omega_c)t + \theta_i(t) - \theta_c(t)] \, . \tag{6.28}$$

As this signal is fed back into the comparator through the VCO, the difference between ω_i and ω_c and the difference between $\theta_i(t)$ and $\theta_c(t)$ are gradually reduced. Finally, as the differences approach zero [$\omega_i = \omega_c$ and $\theta_i(t) = \theta_c(t)$], the $v_o(t)$ that corresponds to ω_i is obtained.

6.6
Applications

Application studies for high-sensitivity sensors using a self-sensing system with a piezoresistive cantilever have been demonstrated. In this chapter, we describe two application studies, one for water molecule detection in air [30,31] and the other for the biomolecular reaction between antigen and antibody in water [32,33].

6.6.1
Water Molecule Detection in Air

Figure 6.12 shows a schematic diagram of the measurement system for determining the water molecule adsorption on a cantilever. A piezoresistive cantilever and

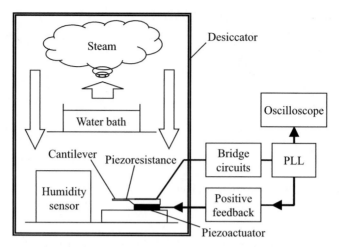

Fig. 6.12. The water adsorption measurement system using a piezoresistive cantilever sensor

a ceramic sensor (commercially obtained) were placed in a desiccator. The humidity inside the desiccator was reduced using silica gel before the experiment. First, we placed a water bath inside the desiccator. Then, we measured the PLL output voltage change with increasing humidity. Two types of piezoresistive cantilevers, shown in Fig. 6.8, were used in this experiment. The resonance frequency of the V-shaped piezoresistive cantilever is 174.2 kHz in air. Consequently, an oscillation frequency of about 174 kHz was obtained by the positive-feedback system, and the locked center frequency and PLL bandwidth were set at 173.5 and 1 kHz, respectively.

Figure 6.13a shows the resonance frequency, Δf, of the cantilever when the humidity increases from 19 to 57%. The humidity was measured by a ceramic humidity sensor. The resonance frequency decreases with increasing humidity. This fact shows that water molecules adsorb on the cantilever with increasing humidity.

Using (6.7), we can estimate the humidity dependence of the mass change due to the water molecule adsorption on the basis of the mass at the humidity of 19%, as shown in Fig. 6.13b. The total mass of the adsorbed water is about 550 pg for the humidity change from 19 to 57%. In this case, the resonance frequency shift is about 250 Hz, as shown in Fig. 6.13a. Consequently, the estimated mass detection sensitivity is about 2.2 pg/Hz, which is 10 times higher than the sensitivity of the quartz crystal oscillation method.

In Fig. 6.13b, the rate of mass change rapidly increases above the humidity of about 42%. The increased rates are about 10 pg/% and 21 pg/% in the humidity range under 42% and over 42%, respectively. It was considered that these differences are caused by the fact that the adsorbed water molecules covered the cantilever surface. The formation of water molecule islands on graphite, gold, and mica substrates has been observed by scanning force microscopy [38, 39]. The adsorbed water makes a number of water molecule islands on the cantilever surface at lower humidity. With increasing humidity, the height of each water island increases. Over the critical humidity, the islands join together, and the cantilever surface is covered by a continuous water film. This mechanism of water molecule adsorption on a cantilever was

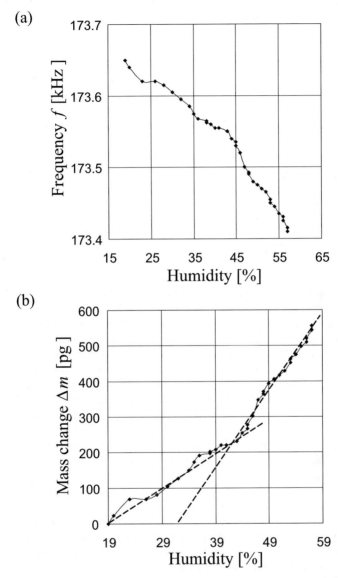

Fig. 6.13. Experimental results for the water molecule adsorption on the V-shaped piezoresistive cantilever. **a** Resonance frequency of the cantilever. **b** Humidity dependence of the mass change due to the water molecule adsorption

proposed in [25]. In our present experiment, it can be considered that water coverage expansion was achieved by an island growth mechanism in the humidity range under 42%. On the other hand, the coverage expansion shifted to a layer growth mechanism in the humidity range over 42%. In order to determine the amount of water adsorbed on the cantilever, we estimated the thickness of the water film at the critical humidity of 42%. Assuming that water adsorption occurs on the cantilever uniformly, the mass of a one-monolayer-thick water film, M_1, is given by

$$M_1 = \left(\frac{M_{H_2O}}{N_A} \right)^{1/3} S \,, \tag{6.29}$$

where M_{H_2O} is the molecular mass of water (18 g/mol), N_A is Avogadro's number (6.0×10^{23}), and S is the area of the cantilever surface ($5.8 \times 10^4 \, cm^2$). In this case, M_1 is about 18 pg. Because the mass of the adsorbed water is 230 pg at a humidity of 42%, the number of water layers is about 13 monolayers. This value is almost consistent with the number of 12 monolayers found in the experiment for water molecule detection using a laser beam deflection detection system [25].

The critical humidity of the water adsorption is reported to be about 68% in [13] and about 60% in [25]. These values are different from the critical humidity of 42% found in our present experiments. We consider that the critical humidity changes depending on the condition of the cantilever surface, which was covered by zeolite and gold film in the studies described in [13] and [25], respectively.

In order to obtain a higher mass sensitivity, we used a rectangular piezoresistive cantilever, as shown in Fig. 6.8b. The measurement system in this experiment was the same as in the previous experiment. The resonance frequency obtained for the cantilever was 286.7 kHz in air. The locked center frequency and the PLL circuit bandwidth were set at 285.9 kHz and 300 Hz, respectively.

Figure 6.14a shows the resonance frequency shift, Δf, of the cantilever when the humidity increases from 19 to 26%. The humidity was measured by a ceramic humidity sensor with a humidity resolution of 1%. The resonance frequency shift is about 114 Hz for the humidity change from 19 to 26%. Figure 6.14b shows the humidity dependence of the mass change due to the water molecule adsorption on the cantilever surface, which is based on the mass of adsorbed water at a humidity of 19%. The total mass of the adsorbed water is about 36 pg in the humidity range from 19 to 26%. The resonance frequency shift is about 114 Hz, as shown in Fig. 6.14a. Consequently, the estimated mass detection sensitivity is about 320 fg/Hz, which is 100 times higher than the sensitivity of the quartz crystal oscillation method. This result demonstrates that this technique has the potential to enable the realization of a practical and highly sensitive molecular sensor, if we can detect frequency shifts with a resolution of less than 1 Hz. In this experiment, a minimum resonance frequency shift of 3 Hz, which corresponds to the detectable minimum mass change of about 1 pg, was detected. In order to estimate the minimum detectable frequency shift, we discuss thermal noise of the cantilever [40]. The mean-square frequency modulation due to thermal noise is given by

$$\langle \delta \omega \rangle = \sqrt{ \frac{\omega_0 k_B T B}{k_L Q \langle z_{osc}^2 \rangle} } \,, \tag{6.30}$$

where ω_0 is the resonance frequency, k_B is the Boltzmann constant, T is the temperature, B is the bandwidth, k_L is the spring constant, Q is the quality factor, and $\langle z_{osc}^2 \rangle$ is the mean-square amplitude of the self-oscillating cantilever. In the case of the rectangular piezoresistive cantilever, ω_0 is 286.7 kHz, T is 300 K, B is 300 Hz, k_L is 40 N/m, Q is 350, and $\langle z_{osc}^2 \rangle$ is about 100 nm. This means that the mean-square frequency modulation is about 0.02 Hz. This result shows that we can detect frequency shifts of less than 1 Hz.

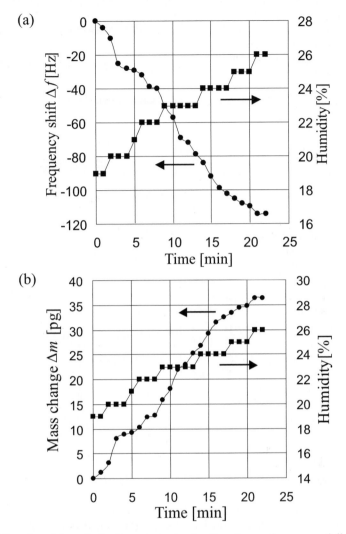

Fig. 6.14. Experimental results for the water molecule adsorption on the commercially available rectangular piezoresistive cantilever. **a** Resonance frequency shift (*circles*) of the cantilever with increasing humidity (*squares*). **b** Humidity (*squares*) dependence of the mass change (*circles*) due to the water molecule adsorption

To study the calibration of small-mass detection for this sensor, we prepared a cantilever on which a thin permalloy film was deposited by RF sputtering (MB96-1011, ULVAC), as shown in Fig. 6.15. Then, we measured the resonance frequency of the cantilever in air. The thickness of the deposited film was measured by atomic force microscopy (AFM). Figure 6.16 shows the frequency spectrums of the cantilever before and after the permalloy film deposition. The resonance frequency and quality (Q) factor were changed from 321.0 to 313.9 kHz and from 321 to 314 by the deposition, respectively. When a resonance frequency shift, Δf, occurs due to

Fig. 6.15. The preparation for the mass-detection check. **a** Cross-sectional scheme of the cantilever before permalloy film deposition. **b** After permalloy film deposition

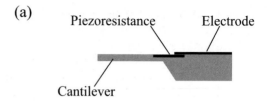

Fig. 6.16. Frequency spectrums of the cantilever before permalloy film deposition (*thin line*) and after deposition (*thick line*)

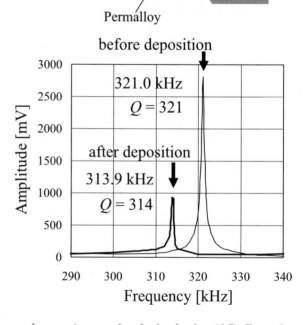

molecular adsorption, the mass change, Δm, can be obtained using (6.7). From the equation and the frequency shift of $\Delta f = 7.1\,\text{kHz}$ found in this experiment, we estimated the mass of the deposited film, Δm_e, to be about 2.02 ng. The thickness of the deposited film was about 90 nm, as measured by AFM. The mass, Δm_c, of the permalloy film, which had a thickness of 90 nm and a surface area of $4.9 \times 10^{-5}\,\text{cm}^2$ (single side of the cantilever), was about 3.83 ng. The Δm_c value almost agrees with the Δm_e value. The deviation between Δm_c and Δm_c can be considered by increasing the cantilever spring constant, which is suggested by decreasing the Q factor and decreasing the amplitude after deposition.

6.6.2
Antigen and Antibody Detection in Water

We measured the reaction between the antibody immunoglobulin G (22.5 μM) and the antigen egg albumen (440 μM). The molecular masses of the antibody and the antigen are 160 and 45 kDa, respectively. Figure 6.17 shows a schematic diagram of the measurement system for studying the reaction between the antigen and the antibody. The piezoresistive cantilever was submerged in 10 ml water. In the first experiment, we measured the resonance frequency shift of the cantilever upon supplying 10 μl (4.40 nmol) of the antigen using a micropipette, as shown in Fig. 6.18a. After the frequency shift saturated, we supplied 10 μl (0.225 nmol) of the antibody. In the second experiment, we supplied 10 μl of the antibody first and 10 μl of the antigen second, as the reverse of the first experiment, as shown in Fig. 6.18b.

Figure 6.19 shows optical images of the sensor system and the piezoresistive cantilever submerged in water, respectively. First, we measured the resonance frequency of the cantilever in air and in water, as shown in Fig. 6.20a and b, respectively. The resonance frequency and Q factor in air obtained were about 324.4 kHz and 368.6, respectively. Those in water were about 188.5 kHz and 11.8, respectively. In the water experiment, the resonance frequency shifted to a lower frequency, the Q factor decreased, and the amplitude decreased. These results suggest the influence of increased water viscosity. When we used a positive-feedback system, we obtained an oscillation frequency of about 190 kHz in water.

Figure 6.21a shows the experimental result for the frequency shift after supplying small volumes of the antigen and the antibody into the water bath. Seven minutes after supplying the antigen, the frequency shift became saturated. We assume that the cantilever surface was covered by the antigen. Next, we supplied a small volume of antibody into the water bath. After 18 min, the frequency shift became saturated again, which indicates that the reaction between the antigen and the antibody had finished. We estimated the mass change using (6.7), as shown in Fig. 6.21b. The mass change due to antigen adsorption was about 27.7 pg, and that due to antibody reaction was about 41.3 pg. Thus, a mass sensitivity of about 190 fg/Hz and a mass resolution of about 500 fg were obtained. This mass sensitivity is 100 times higher than that of

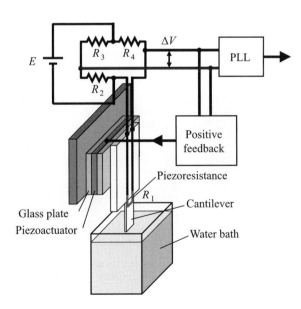

Fig. 6.17. Harmonic-vibration mass biosensor system using a piezoresistive cantilever in water

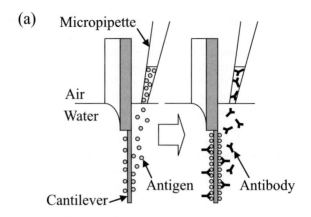

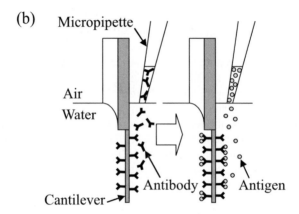

Fig. 6.18. The biochemical reactions for the adsorption of the antigen and the antibody. **a** Adsorption of the antigen on the cantilever followed by that of the antibody on the cantilever in water. **b** Adsorption of the antibody and antigen in the reverse process of that in **a**

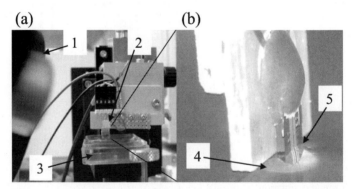

Fig. 6.19. Optical images of the piezoresistive cantilever biosensor. *a* Biosensor setup: *1* CCD camera, *2* cantilever sensor, *3* water bath. *b* Cantilever submerged in water: *4* water surface, *5* piezoresistive cantilever

(a)

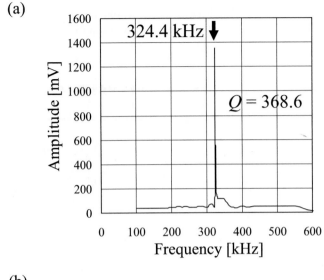

(b)

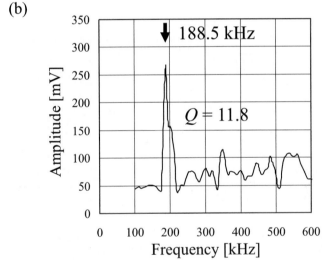

Fig. 6.20. Frequency spectrums of the cantilever **a** in air and **b** in water

the quartz crystal oscillation method. In this experiment, we used a commercially available piezoresistive cantilever designed for AFM measurement. If a cantilever with a small mass and a high resonance frequency is designed, a higher mass sensitivity of less than 190 fg/Hz can be achieved. In addition, we used a PLL with a locked center frequency and a detectable resonance frequency bandwidth of 190 and 1 kHz, respectively. We detected a minimum resonance frequency shift of about 2.5 Hz, which corresponds to the noise level of this sensor system. The mean-square frequency modulation due to thermal noise is less than 1 Hz [31, 40]. Therefore, if we optimize the electric circuits of the detection system, we can detect frequency

(a)

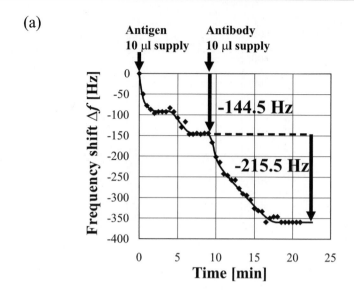

(b)

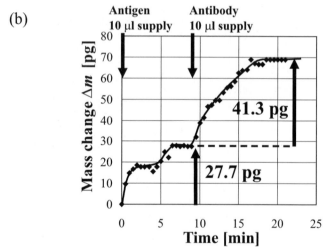

Fig. 6.21. Experimental results for the antigen–antibody reaction on the commercially available piezoresistive cantilever. **a** Frequency shift. **b** Mass change

shifts of less than 1 Hz. From the mass change, we were able to estimate the number of adsorbed molecules of the antigen and the antibody. The number of antigen molecules adsorbed on the cantilever was about 3.7×10^8, and that of the antibody reacted with the antigen was about 1.5×10^8. It has been reported in bioscience that an antibody has two binding sites for an antigen; therefore, perfect binding means that the ratio of antibody to antigen is 1:2. Our experimental results showed a ratio of 0.8:2.

Figure 6.22 shows the experimental results for the frequency shift and the mass change. The experiment was carried out using the reverse experimental procedure,

Fig. 6.22. Experimental results for the antibody–antigen reaction on the commercially available piezoresistive cantilever in the reverse process of that for Fig. 6.21. **a** Frequency shift. **b** Mass change

(a)

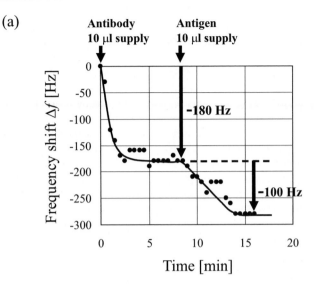

(b)

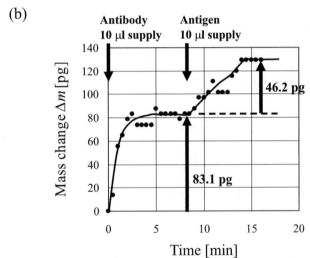

wherein we supplied the antibody first and the antigen second. We observed that the frequency shift saturated at 5 min owing to the adsorption of the antibody on the cantilever. At 13 min, the frequency shift saturated again, which indicates that the reaction between the antibody and the antigen had finished. The estimated mass changes of the adsorbed antibody and antigen were 83.1 and 46.2 pg, respectively. The estimated number of antibody molecules was about 3.1×10^8, and that of antigen molecules was about 6.2×10^8. These results show that the binding ratio for the antibody and the antigen is 1:2; therefore, we consider that the biosensor successfully monitored the reaction between the antibody and the antigen.

6.7
Prospective Applications

We conclude that the self-sensing mass sensor using a piezoresistive cantilever has the potential to enable femtogram-order mass detection in air as well as in water. Using this sensor system, we can monitor environmental changes in air and chemical or biological molecular reactions in water. It has a much broader range of application in the various fields of chemistry, biotechnology, and medical science, e. g., immunochemical assay, amino acid analysis, allergy testing, cancer testing, and genetic testing. Although some of the applications have already been studied, e. g., a cantilever array sensor system [14,41] and cantilever functionalization [19–21,42, 43], there are many obstacles for the commercialization of such products. In order to realize these applications, we must devise ways to modify the cantilever surface as well as time-efficient multiple-testing methods. Additionally, we must improve the sensitivity of the sensor to attain high reliability and high resolution.

Acknowledgements. The authors wish to express their special gratitude to H. Okano (Tokyo Sokki Kenkyujo Co.) and T. Izumi (Gunma University) for their valuable discussions and tremendous support. We would also like to thank Y. Fujinuma, T. Hieida, T. Chiyoma, and A. Ikeuchi for their great help and contribution. The authors gratefully acknowledge partial financial support from the Highland Kanto Liaison Organization.

References

1. James AT, Martin AJP (1952) Biochemistry 50:679
2. Liu R, Zhou JL, Wilding A (2004) J Chromatogr A 1022:179
3. Ma WT, Fu KK, Cai Z, Jiang GB (2003) Chemosphere 52:1627
4. Hamblin C, Barnett ITR, Hedger RS (1986) J Immunol Methods 93:115
5. Avremeas S (1969) Immunochemistry 6:43
6. Aozawa O, Ohta S, Nakano T, Miyata H, Nomura T (2001) Chemosphere 45:195
7. Janshoff A (2004) In: Baltes H (ed) Sensor update, vol 9. Wiley, New York, pp 313–354
8. Matsuno H, Niikura K, Okahata Y (2001) Chem Eur J 7:3305
9. Cooper MA (2004) J Mol Recognit 17:286
10. Liedberg B, Lundström I (1993) Sens Actuators B 11:63
11. Melendez J, Carr R, Bartholomew DU, Kukanskis K, Elkind J, Yee S, Furlong C, Woodbury R (1996) Sens Actuators B 35:212
12. Saenko E, Sarafanov A, Ananyeva N, Behre E, Shima M, Schwinn H, Josic D (2001) J Chromatogr A 921:49
13. Berger R, Gerber C, Lang HP, Gimzewski JK (1997) Microelectron Eng 35:373
14. Lang HP, Baller MK, Berger R, Gerber C, Gimzewski JK, Battiston FM, Fornaro P, Ramseyer JP, Meyer E, Güntherodt HJ (1999) Anal Chim Acta 393:59
15. Binnig G, Quate CF, Gerber C (1986) Phys Rev Lett 56:930
16. Gimzewski JK, Gerber C, Meyer E, Schlittler RR (1994) Chem Phys Lett 217:589
17. Berger R, Gerber C, Lang HP, Gimzewski JK, Meyer E, Güntherodt HJ (1996) Appl Phys Lett 69:40
18. Berger R, Lang HP, Gerber C, Gimzewski JK, Fabian JH, Scandella L, Meyer E, Güntherodt H-J (1998) Chem Phys Lett 294:363
19. Raiteri R, Nelles G, Butt H-J, Knoll W, Skladal P (1999) Sens Actuators B 61:213

20. Fritz J, Baller MK, Lang HP, Rothuizen H, Vettiger P, Meyer E, Güntherodt HJ, Gerber C, Gimzewski JK (2000) Science 288:316
21. Hansen KM, Ji H-F, Wu G, Datar R, Cote R, Majumdar A, Thundat T (2001) Anal Chem 73:1567
22. Lavrik NV, Datskos PG (2003) Appl Phys Lett 82:2697
23. Ilic B, Craighead HG, Krylov S, Senaratne W, Ober C, Neuzil P (2004) J Appl Phys 95:3694
24. Sone H, Fujinuma Y, Hieida T, Hosaka S (2003) Proc SICE Annu Conf 2985
25. Sone H, Fujinuma Y, Hosaka S (2004) Jpn J Appl Phys 43:3648
26. Boisen A, Thaysen J, Jensenius H, Hansen O (2000) Ultramicroscopy 82:11
27. Porter TL, Eastman MP, Pace DL, Bradley M (2001) Sens Actuators A 88:47
28. Zhou J, Li P, Zhang S, Huang Y, Yang P, Bao M, Ruan G (2003) Microelectron Eng 69:37
29. Tang Y, Fang J, Yan X, Ji H-F (2004) Sens Actuators B 97:109
30. Sone H, Fujinuma Y, Hieida T, Chiyoma T, Okano H, Hosaka S (2004) Proc SICE Annu Conf 1508
31. Sone H, Okano H, Hosaka S (2004) Jpn J Appl Phys 43:4663
32. Hosaka S, Chiyoma T, Ikeuchi A, Okano H, Sone H, Izumi T (2006) Curr Appl Phys 6:384
33. Sone H, Ikeuchi A, Izumi T, Okano H, Hosaka S (2006) Jpn J Appl Phys 45:2301
34. Tortonese M, Barrett RC, Quate CF (1993) Appl Phys Lett 62:834
35. Linnemann R, Gotszalk T, Hadjiiski L, Rangelow IW (1995) Thin Solid Films 264:159
36. Gotszalk T, Grabiecl P, Rangelow IW (2002) Ultramicroscopy 82:39
37. Chen GY, Warmack RJ, Thundat T, Allison DP, Huang A (1994) Rev Sci Instrum 65:2532
38. Luna M, Colchero J, Gil A, Gomez-Herrero J, Baro AM (2000) Appl Surf Sci 157:393
39. Gil A, Colchero J, Luna M, Gomez-Herrero J, Baro AM (2000) Langmuir 16:5086
40. Albrecht TR, Grütter P, Horne D, Rugar D (1991) J Appl Phys 69:668
41. Lang HP, Berger R, Andreoli C, Brugger J, Despont M, Vettiger P, Gerber C, Gimzewski JK, Ramseyer JP, Meyer E, Güntherodt H-J (1998) Appl Phys Lett 72:383
42. Wu G, Datar RH, Hansen KM, Thundat T, Cote RJ, Majumdar A (2001) Nat Biotechnol 19:856
43. Hyun S-J, Kim H-S, Kim Y-J, Jung H-I (2006) Sens Actuators B 117:415

7 AFM Sensors in Scanning Electron and Ion Microscopes: Tools for Nanomechanics, Nanoanalytics, and Nanofabrication

Vinzenz Friedli · Samuel Hoffmann · Johann Michler · Ivo Utke

Abstract. In this chapter the synergies upon the integration of atomic force microscope sensors in scanning electron and ion microscopes are outlined and applications are presented. Combining the capabilities of the standalone techniques opens the world to nanoscale measurements and process control. The high-resolution microscopy imaging provides direct visual feedback for the analysis of specific sample features and of individual nanostructures. Fundamental static and dynamic mechanics of cantilever beams are reviewed with an emphasis on the usage of the beams as force and mass sensors in a vacuum. Static force sensing is applied to probe the mechanical properties of nanowires in tensile, bending and compression experiments and dynamic force sensing is used for AFM in SEM applications. Cantilever-based dynamic sensing is discussed to measure the mass of material deposited or etched using the electron or ion beam inherent to the microscope.

Key words: Cantilever-based force/mass sensor, Piezoresistive cantilever, Vibration, Resonance, Scanning electron/ion microscope, Focused electron/ion beam induced deposition and etching

Symbols

α_{th}	Thermal expansion coefficient
A_c	Deflection amplitude at cantilever end
A_d	Excitation amplitude
A_s	Area of sample cross section
β_n	Vibration eigenmode of the nth mode
β_{th}	Temperature coefficient of Young's modulus
B	Measurement bandwidth
Δf	Frequency shift
Δf_{min}	Minimum resolvable frequency shift
Δm	Mass shift
Δm_{min}	Minimum resolvable mass shift
ΔW	Total energy lost per vibration cycle
δf	Frequency noise
δm	Mass noise
E	Young's modulus
f	Frequency
f_0	Resonance frequency (point-mass model)
f_d	Driving frequency
f_n	Resonance frequency of the nth mode
F_a	Applied force

h	Cantilever height (thickness)
I	Geometrical (area) moment of inertia
J	Molecular flux
k	Spring/force constant
l	Cantilever length
L	Sample length
m^*	Effective mass
m_c	Cantilever mass
n_e	Normalized effective mass
ϕ	Phase
Q	Quality factor
ρ	Mass density
R	Mass responsivity
σ	Strength
t	Time
T	Temperature
ε	Strain
V	Volume
w	Cantilever width
W_0	Stored vibrational energy
Z_n	Vibration shape of the nth mode

Abbreviations

AFM	Atomic force microscope
CNT	Carbon nanotube
CVD	Chemical vapor deposition
FEB	Focused electron beam
FIB	Focused ion beam
FWHM	Full width at half maximum
GIS	Gas injection system
NEMS	Nanoelectromechanical system
QCB	Quartz crystal microbalance
SEM	Scanning electron microscope
SIM	Scanning ion microscope
SW-CNT	Single-wall carbon nanotube
TEM	Transmission electron microscope

7.1
Introduction

In 1986 the Nobel Prize was given to Ruska for scanning electron microscopy and to Binning and Rohrer for scanning probe microscopy. Referring to the increasing number of publications in this field, the potential of combined or hybrid techniques is about to be explored. Atomic force microscopes (AFMs) with their nanomanipulation capabilities and their cantilever-based sensor derivatives are increasingly combined with scanning electron microscopes (SEMs) and scanning ion microscopes (SIMs). SEMs and SIMs evolved over the last two decades literally into "workshops" which can fabricate tailored 3D nanostructures using gas injection. With the aid of the focused electron beam (FEB) or the focused ion beam (FIB), deposition, etching, and milling can be performed at the nanoscale. Combining these powerful scanning probe observation techniques and their derivatives provides access to the measurement of individual nanostructures, in particular nanowires, nanotubes, and 3D structures in nanoelectromechanical systems (NEMS). Furthermore, FEB and FIB processing of nanostructures can be controlled quantitatively in terms of etching, milling, and deposition rates and can thus be optimized. Figure 7.1 shows the capabilities of SEMs, SIMs, and AFMs and their application in hybrid techniques.

The combination of techniques into hybrid systems is mainly driven by the investigation of *individual* nanostructures and of mechanisms during nanoscale growth and etching. This demands the manipulation of individual nanostructures to probe their physical and chemical properties. In other words, attachment, placement, and release actions must be performed under visual control at nanometer scale to move or to fabricate the nanostructure at the desired place.

We summarize our discussion of AFMs, SEMs, and SIMs in Table 7.1 showing a comparison of several specifications for their use in hybrid systems.

The techniques in Table 7.1 are well known and abundantly discussed in the literature: For information on SEMs the reader is referred to standard textbooks [64, 125]. SIMs are assembled much the same way as their electron counterparts but use other interaction mechanisms [116]. Atomic force microscopy is a very powerful and well-described technique [15,59,63] with derivative techniques—many of them described and discussed in this book series [13].

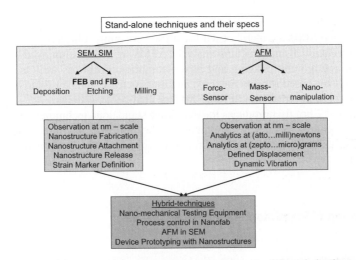

Fig. 7.1. Classic standalone scanning probe techniques and their standalone derivative techniques and tools. Combining their capabilities in observation, nanostructuring, and analytics into hybrid systems opens the world to nanoscale measurements and process control. Acronyms: *AFM* atomic force microscope, *SEM* scanning electron microscope, *SIM* scanning ion microscope, *FEB* focused electron beam, *FIB* focused ion beam

Table 7.1. Comparison of the scanning electron microscope (*SEM*), the scanning ion microscope (*SIM*), and the atomic force microscope (*AFM*)

	SEM	SIM	AFM
Probe	Focused electron beam	Focused ion beam (mostly Ga ions)	Probe tips (functionalized)
Lateral resolution	~1–10 nm	~10 nm	~0.1 nm
Depth resolution	Low (large depth of focus)	Low (large depth of focus)	High (~ subnanometer)
Environment	Vacuum: ~10^{-6} mbar Environmental, ~1 mbar	Vacuum: ~10^{-6} mbar Environmental, ~1 mbar	Ambient, liquids, vacuum
Analytics	Elemental analysis, crystal orientation, electric field contrast	Elemental analysis, grain orientation	Magnetic force microscopy, friction force microscopy, scanning capacitance microscopy, etc.
Limitations	Projected view	Projected view damage during imaging	Small lateral range
Structuring abilities	Lithography	Lithography, milling, implantation	Assembly of nano-objects via nanomanipulation
Derivative techniques	FEB-induced deposition and etching	FIB-induced deposition and etching	Haptic manipulation, force spectroscopy, mass sensing

FEB focused electron beam, *FIB* focused ion beam

SEMs and SIMs are mainly used for surface imaging at high resolution and for submicron chemical and structural analysis. AFMs are mainly employed for highest-resolution topography imaging and nanomanipulation. This as well as the use of their cantilever-based sensors for force and mass detection makes them particularly attractive for integration into SEMs, SIMs, or dual-beam systems.

Section 7.2 briefly introduces the derivative standalone techniques and their use in hybrid systems. Section 7.3 summarizes the fundamentals of cantilever-based sensors necessary to understand the three hybrid techniques presented in detail in Sect. 7.4.

7.2
Description of Standalone Techniques

7.2.1
FEB/FIB Nanofabrication

FEB/FIB maskless nanofabrication is a standalone technique with many applications in tailored device prototyping covering fields of nanoelectronics [9,37], nanophotonics [91,111,120], and functionalization of scanning probe sensors [10,51,152]. In

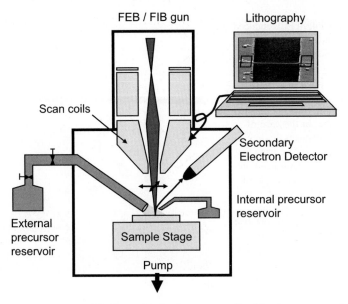

Fig. 7.2. Schematics of FIB and FEB nanofabrication systems. Gas injection systems with external and internal precursor reservoirs are shown. The laptop monitor shows (schematically) FEB- or FIB-written clamps to a nanowire

combination with nanomanipulation, FEB- and FIB-induced deposition is predominantly used as a technique for attachment to the sensor despite the fact that often the mechanical properties of the deposit are unknown. FIB milling by 30-keV Ga ions is predominantly employed as a specimen preparation technique which permits "cutting" nanostructures with dimensions down to sub-100-nm into the bulk. For electrical contacts to carbon nanotubes (CNTs) or nanowires, FEB-induced deposition is frequently used as a soldering technique [21, 54, 55]. The principle of both techniques is presented in Fig. 7.2

7.2.1.1
FEB-Induced Deposition and Etching

The principle of nanofabrication via FEB is shown in Fig. 7.3. Comprehensive overviews in this field are found in [18,20,31,124]. Basically, during deposition and etching surface-adsorbed gas molecules are decomposed by electron irradiation and form either a stable deposit and gaseous by-products or volatile reaction products with the substrates (Fig. 7.3). The molecules are supplied by a gas injection system (GIS). By switching between a deposition gas and an etch gas, one can use this technique as an attach and release tool to/from a nanomanipulator or sensor. Often hydrocarbon molecule sources are present in SEMs coming from oil vapors of the vacuum system and from the contamination on the sample itself. Under electron irradiation they lead to carbonaceous contamination deposits. Historically regarded as an unwanted side effect, they have proved very useful in attaching nanostructures

Fig.7.3.a Principle of FEB-induced deposition: Molecules adsorb at the surface and are dissociated under electron impact. Volatile fragments are pumped away and a deposit grows coaxially to the beam. Here molecules are injected by a microtube. In the case of contamination deposits, hydrocarbon molecules originate from the microscope backpressure and the substrate surface. **b** Principle of FEB-induced etching: The surface adsorbed molecules dissociate under electron impact into reactive species and form volatile compounds with the substrate material

a)

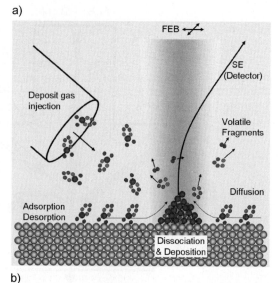

b)

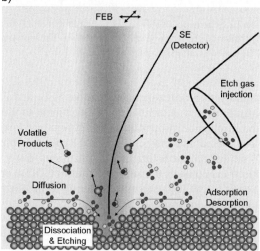

to cantilevers and as a marker technique for surface strain quantification detection (Fig. 7.4b,c).

Obviously, reproducibility of "contamination" attachments will strongly depend on the contamination level of the sample and the microscope. Introducing organic precursors into the SEM chamber allows control of the deposit composition and deposition rate [19]. One advantage of FEB-induced deposition is that tuning between mechanical stability and functionality of deposits can be performed using adapted metal–organic precursor molecules (Fig. 7.4a,d). The resulting deposit is a nanocomposite of metal nanocrystals embedded in a stabilizing carbonaceous matrix. In such deposits the metal content varies according to the precursor chemistry and deposition conditions. Higher metal contents are deposited when beam heating effects oc-a)

b)

c)

d)

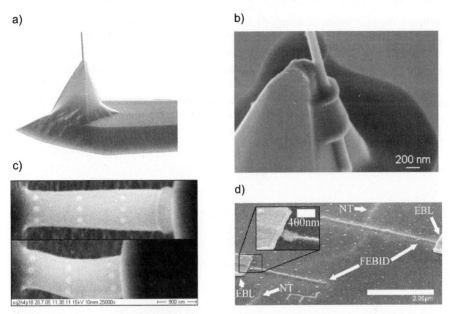

Fig. 7.4. Examples of FEB deposits. **a** AFM sensor functionalized with a magnetic tip deposit. **b** FEB clamp deposit to fix a nanowire to a cantilever tip from a paraffin precursor. **c** FEB marker deposits for strain detection during a compression experiment of micropillars from contamination. **d** FEB contacts to a carbon nanotube using a gold-containing precursor. (**a** From [152], **b** Reprinted with permission from [38]. Copyright (2005), American Institute of Physics, **c** from [105], **d** Reprinted with permission from [21].Copyright (2005), American Institute of Physics)

cur [147, 153]. Examples include FEB deposits from $(CH_3)_2–Au–C_5H_4F_3O_2$ [151], $Co_2(CO)_8$ [16], $W(CO)_6$ [88] which result in electrical resistivities about 100–1000 times higher than the corresponding metallic bulk value. Low-resistivity Au contacts comparable to standard Au liftoff techniques were achieved with an inorganic gold precursor $AuClPF_3$ [21] and a mixture of $(CH_3)_2–Au–C_5H_7O_2$ and H_2O gas [97]. Etching gases comprise H_2O, O_2 for carbon, and XeF_2 for Si and SiO_2 [124]. Water vapor also attacks the CNT under electron irradiation and can even be used for cutting CNTs [169].

Mechanical properties (strength, density, elastic modulus) of FEB (and FIB) deposits have been poorly investigated owing their small deposit volumes (less than $1\,\mu m^3$) and small masses (less than 10 pg) and require special force and mass sensors, as discussed later.

7.2.1.2
FIB-Induced Deposition, Etching, and Milling

Processing with FIB is mainly based on liquid gallium sources being available since the late 1980s. Besides imaging capabilities they provide the ability to remove or add locally material (metals or insulators) at sub-100-nm dimensions [101,110,127].

When a focused beam of energetic ions hits a surface, the energy of the incident particles is transferred to the substrate, resulting in ejection of neutral and ionized substrate atoms (sputtering), displacement of atoms (damage), and emission of electrons (imaging, charging) and phonons (heating); see Fig. 7.5a. A substantial number of the impinging ions is implanted into the substrate, leading to contamination problems, or can be used as material doping. The physical sputter process, also named "milling", is predominantly exploited in FIB nanostructuring. Injection of precursor gases leads to deposition or etching as already described for FEB. Major applica-

a)

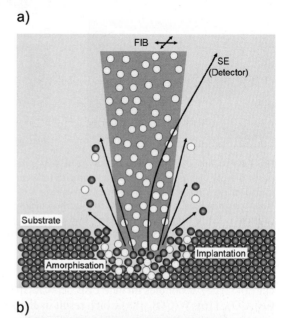

b)

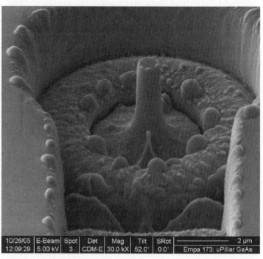

Fig. 7.5. a Principle of FIB milling (or sputtering). **b** GaAs micropillar for compression experiments machined with FIB

tions are device and circuit editing, lamellae preparation for transmission electron microscopy, and structure machining for mechanical tests (Fig. 7.5b). Owing to the ion damage, however, FIB is not well suited for nanostructure imaging and for direct contact writing to nanostructures.

7.2.2
Cantilever as a Static Force Sensor

In conventional AFM force spectroscopy the deflection of a cantilever during an approach–withdrawal cycle with a substrate is monitored and converted into force via the spring constant. Adhesion forces and molecule binding forces have been extensively studied using this technique [13, 26, 115]. Mechanical characterization of nanostructures like nanowires or nanotubes requires physical interaction with the object. With the AFM tip the object can be stretched or bent with nanometer positioning accuracy, while performing measurements with nanonewton to millinewton force resolution (Fig. 7.6). Brittle fracture of semiconductors, the yield point of metallic nanowires, and their Young's modulus can be measured by lying them down on a trench and bending them [14, 69, 119, 131]. The use of the cantilevers in a SEM together with nanomotion control is not only helpful for accurate positioning control, but also to gather additional visual information on the failure mechanism. Section 7.4.1 describes the use of a cantilever in a SEM to perform tensile and bending experiments.

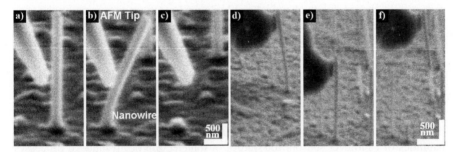

Fig. 7.6. SEM images of a bending and a tensile experiment. **a–c** Bending experiment: an AFM tip is used to manipulate the nanowire. **d–f** Tensile experiment: the freestanding nanowire is attached via a FEB-written bond to the AFM tip. The cantilever of the AFM tip is retracted and used as a force sensor

7.2.3
Cantilever as a Resonating Mass Sensor

Since the mid-1990s there has been significant development in the field of cantilever-based mass sensors for chemical-sensing and biosensing applications. These sensors are based on the fact that the resonance frequency of a vibrating structure depends on external stimuli, such as mass loading.

Using standard AFM cantilevers and equipment, the group of Thundat started to observe mass changes induced by adsorption of molecules on the cantilever

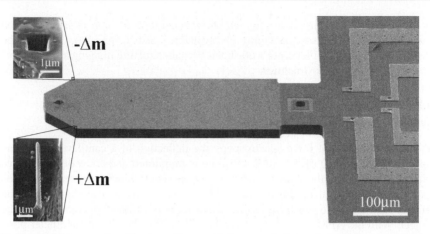

Fig. 7.7. SEM image of a silicon cantilever with integrated piezoresistive Wheatstone bridge (Nascatec, Germany) used as a resonating mass-sensing device in a SEM. The *insets* show FIB milling and FEB deposition on the cantilever surface

surfaces [29,142,156]. Further development has led to sensor arrays for the discrimination of volatile organic compounds [66, 89]. Cells of *Escherichia coli* bacteria have been selectively detected with antibody surface coated cantilever beams [76]. Another wide application field of cantilever-based mass sensors is process control, e. g., weight change due to chemical and biological reactions occurring on the cantilever surface. Berger et al. demonstrated in situ measurements of surface stress changes and kinetics during the formation of self-assembled monolayers [11] and thermally induced mass changes [12]. Applications of mass-sensitive resonating cantilever sensors for the detection of gas-phase analytes, liquid-phase analytes, and biological species are reviewed in [90]. Process control of additive or subtractive surface processing techniques based on cantilever sensors is a relatively new field. Sunden et al. [141] showed a weight change due to growth of CNTs on the cantilever surface. In Sect. 7.4.2 we discuss the exploitation of the analytical capabilities of cantilever sensors for studies of ion/electron beam induced processes (Fig. 7.7).

7.2.4
Nanomanipulation

Several kinds of AFM-based nanomanipulation systems have been developed [81, 126, 135] as well as virtual-reality interfaces to facilitate feedback during nanomanipulation [67, 92]. AFMs have been combined with different haptic devices, e. g., NanoMan [155], NanoManipulator™ [1], NanoFeel 300 manipulator [108], and the Omega haptic device [52, 109]. Haptic devices provide the operator with real-time force-feedback. The main drawback of the abovementioned devices is the lack of visual feedback of the manipulation process in real time. Integration of such devices into the SEM would compensate for the lack of real-time visual feedback. Presently nanomanipulation in the SEM is mainly performed with

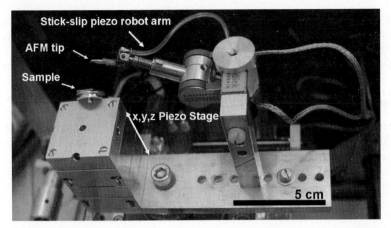

Fig. 7.8. Nanomanipulation setup mounted on a SEM stage comprising x, y, z stages with the sample and a robot arm like a nanomanipulator (Kleindiek MM3A) onto which a cantilever sensor is attached

nanopositioning stages [41, 139, 159, 166] or with robot-arm-like nanomanipulators [85, 86] (Fig. 7.8). In Sect. 7.4.3 AFM integration into a SEM environment is discussed.

7.3
Fundamentals of Cantilever-Based Sensors

7.3.1
Static Operation—Force Sensors

In static operation, the force F applied to the cantilever causing a vertical deflection Δz at the loading point is calculated by Hooke's law, $F = k_c \Delta z$. The static spring constant k_c of a cantilever (Fig. 7.9) for forces acting vertically on the cantilever end $x = l$ is

$$k_c = \frac{3EI}{l^3} = \frac{Ewh^3}{4l^3} ,\qquad(7.1)$$

where

$$I = \int dy \int z^2 \, dz ,\qquad(7.2)$$

Fig. 7.9. Cantilever with length l, width w, and thickness h. A rectangular cantilever generally fulfills $l \gg w \gg h$

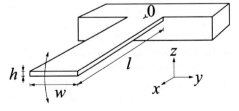

where the first relation in (7.1) is for general cantilever shapes in terms of the geometrical moment of inertia I, and the second relation is valid for rectangular cantilever beams. Equation 7.2 states the general formula of I for an arbitrary cross section; the origin of the coordinate system must lie on the neutral line of the beam. The cantilever deflection Δz can either be read out from the SEM image, or directly from the output of the deflection sensing system employed.

Finite-element simulations conducted by the authors revealed that taking into account the clamping of the cantilever to the chip makes the structure softer by roughly 10% than what the formula, which holds for a perfectly clamped beam, predicts. Also, for special shapes as in certain piezoresistive cantilevers (Fig. 7.7) (7.1) does not hold. For quantitative measurements the cantilever has to be calibrated; see Sect. 7.3.4.

7.3.2
Dynamic Operation—Mass Sensors

The dynamic properties of cantilevers are reviewed specifically for their use as mass-sensing devices in high vacuum and ultrahigh vacuum conditions. The discussion is based on the flexural vibrations of cantilevers.

7.3.2.1
Modal Analysis

The equation of motion for the flexural vibrations of a freely vibrating and undamped cantilever can be approximated by the Euler–Bernoulli equation for small vibration amplitudes [158]:

$$EI\frac{\partial^4 z(x,t)}{\partial x^4} + m_c/l\frac{\partial^2 z(x,t)}{\partial t^2} = 0 \,, \tag{7.3}$$

where z is the cantilever displacement at position $x \in [0,l]$ along the cantilever at time t, E is the Young's modulus, and m_c/l is the cantilever mass per length l. In the case of a clamped–free cantilever beam, the boundary conditions imposed are zero displacement and slope at the clamped end, $Z(x=0) = Z'(x=0) = 0$, and vanishing external torque and shear forces at the free end, $Z''(x=l) = Z'''(x=l) = 0$. They lead to the stationary modal shapes given by

$$Z_n(x) = \frac{A_c}{2}\left([\cos(\beta_n x) - \cosh(\beta_n x)] \right.$$
$$\left. -\frac{\cos(\beta_n l) + \cosh(\beta_n l)}{\sin(\beta_n l) + \sinh(\beta_n l)} [\sin(\beta_n x) - \sinh(\beta_n x)] \right) \,, \tag{7.4}$$

where $Z_n(x)$ is the nth eigenmode of (7.3) and $A_c = Z_n(l)$ is the vibration amplitude at the free cantilever end. In Sect. 7.4.1.4 these stationary solutions will be compared with the experimentally excited vibration modes in nanowires and nanotubes.

The boundary conditions further result in the characteristic equation $\cos(\beta_n l) \cosh(\beta_n l) = -1$, wherefrom the dimensionless eigenvalue $\beta_n l$ of the nth flexural resonance mode can be calculated numerically . The eigenvalues of the first three modes are $\beta_1 l = 1.87510$, $\beta_2 l = 4.69409$, and $\beta_3 l = 7.85476$. For modes $n > 3$: $\beta_n l \approx (n - 1/2)\pi$.

The resonance frequencies for flexural vibrations in the cantilever are obtained from (7.3) using the stationary solutions described by (7.4):

$$f_n = \frac{(\beta_n l)^2}{2\pi} \sqrt{\frac{EI}{l^3 m_c}} . \tag{7.5}$$

7.3.2.2
The Point-Mass Model

The complexities of the cantilever vibrations have given rise to models that simplify the dynamics considerably. In the point-mass model (also called the first-mode approximation), the cantilever is approximated by a one-degree-of-freedom mass-spring model, such that the higher-order flexural modes are neglected. The distributed cantilever mass is replaced by an effective point mass m^* attached to a massless spring with stiffness k_c (Fig. 7.10). m^* is chosen such that the fundamental resonance frequency f_1 of the free cantilever equals the point-mass resonance frequency [123]:

$$f_0 = 1/2\pi\sqrt{k_c/m^*} \tag{7.6}$$

from the combination of (7.5) and (7.6). For rectangular beams the normalized effective mass is obtained where $m^* = n_c m_c$ with n_c being the normalized effective mass by setting $n = 1$: $n_e = 3/(\beta_1 l)^4 = 0.2427$. It is noted that $n_e \approx 33/140 = 0.2357$ is sometimes used in the literature. This approximation is derived by assuming a static cantilever deflection curvature and results in a systematic error of 2.9%.

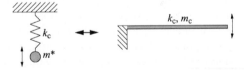

Fig. 7.10. The point mass attached to a spring (*left*) models the clamped–free cantilever with distributed mass (here without damping)

7.3.2.3
Dissipation—Quality Factor

As we will see later, dissipation has a crucial influence on the ability to resolve small shifts in resonance frequency. Operating the cantilever sensors inside the vacuum chamber of a SEM/SIM results in low damping working conditions, which are favorable for high-resolution mass sensing.

In the model of the harmonic oscillator, dissipation can be included in the system by introducing the dimensionless quality factor Q of the resonator which is a description of the total damping. The quality factor of a resonator is inversely proportional to the damping coefficient [99] and is defined as

$$Q = 2\pi \frac{W_0}{\Delta W} , \tag{7.7}$$

where W_0 is the stored vibrational energy and ΔW is the total energy lost per oscillation cycle. The dissipation channels that contribute to ΔW can be divided into internal damping due to the physical structure of the micromechanical resonator – dissipation via coupling to the support structure (clamping loss) and internal friction [46, 164] – and external damping caused by molecular, viscous, and turbulent flow of the surrounding media or acoustic radiation. In high and ultrahigh vacuum conditions external damping is negligible.

For low damping, in a vacuum, the modal shape solutions $Z_n(x)$ are the same as for zero damping (see (7.4)), but the dispersion relation giving the damped eigenfrequencies f_0' in terms of the undamped frequencies f_0 is

$$f_0' = f_0 \sqrt{1 - \frac{1}{2Q^2}} . \tag{7.8}$$

In high and ultrahigh vacuum conditions damping induced shifts are negligible since the quality factors are typically greater than 10,000.

The steady-state solution of the equation of motion for the driven damped harmonic oscillator leads to the following expressions for amplitude A_c and phase ϕ of the oscillations as a function of the drive amplitude A_d and the excitation frequency ω_d:

$$A_c = \frac{A_d \omega_0^2}{\sqrt{(\omega_0 \omega_d / Q)^2 + (\omega_0^2 - \omega_d^2)^2}} , \tag{7.9}$$

$$\phi = \arctan\left(\frac{\omega_0 \omega_d}{Q\left(\omega_0^2 - \omega_d^2\right)} \right) . \tag{7.10}$$

It is clear from (7.9) and (7.10) that if $Q \gg 1$, the amplitude maximum is reached at the frequency where the oscillation has a 90° phase lag relative to the excitation. This is, however, not generally true at higher damping conditions (in air and water).

In practice the quality factor is either determined from the full width at half maximum (FWHM) of the squared amplitude peak or from the phase variation $d\phi/df$ at resonance (see (7.10)) according to

$$Q \approx \frac{f_0}{\text{FWHM}} = \frac{f_0}{2} \frac{d\phi}{df} . \tag{7.11}$$

7.3.2.4
Mass Loading

In general, resonant mass sensing is performed by observing a shift in the resonance frequency Δf due to the added mass Δm. We will consider two extreme cases (Fig. 7.11): homogeneous surface coverage loading, $f_0 - \Delta f = (2\pi)^{-1}\sqrt{k_c/[n_e(m_c + \Delta m)]}$; and point-mass loading, $f_0 - \Delta f = (2\pi)^{-1} \cdot \sqrt{k_c/(n_e m_c + \Delta m)}$.

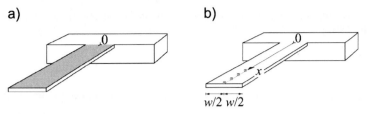

a)   b)

Fig. 7.11. a Homogeneous surface loading. **b** Point-mass loading

Homogeneous Mass Loading

Assuming that the added mass Δm is a small fraction of the cantilever mass m_c, we can write the frequency response of the fundamental mode upon homogeneous surface loading as

$$R = \frac{\Delta f}{\Delta m} = -2\pi^2 n_e \frac{f_0^3}{k_c} = -\frac{1}{2}\frac{f_0}{m_c} \,. \tag{7.12}$$

This expression assumes that the cantilever stiffness EI, surface stress, and damping are not significantly affected by the added mass. This still holds if the added species form a film, which has insignificant intra cross-linking and is not rigidly bonded to the cantilever surface. In practice, for physisorption (van der Waals interaction) of molecules on the sensor surface these assumptions are valid. It is noted that (7.12) is analogous to the Sauerbrey equation [133] but here is written in terms of the absolute mass rather than the mass density of the added species. Hence, it is noted that the frequency response given by (7.12) for a homogeneously loaded mass is dependent on the active sensor area.

Point-Mass Loading

Particle beam induced processes modify the cantilever surface at the microme-ter/nanometer length scale with the positioning accuracy inherent to the SEM/SIM. The imaging capabilities further allow the determination of the location of locally added or removed mass on the cantilever, e. g., from loaded nanoparticles or wires. For these reasons we consider here the response of the cantilever sensor upon load-ing of mass at a specific location on the beam in a small area wherein the mass responsivity can be considered as constant. Upon loading of such a point mass at the free cantilever end, the responsivity of the fundamental mode becomes

$$R = \frac{\Delta f}{\Delta m} = -2\pi^2 \frac{f_0^3}{k_c} = -\frac{1}{2n_e}\frac{f_0}{m_c} \,. \tag{7.13}$$

Using the approach of equality of the kinetic energy of the system with distributed mass $m(x)$ and with effective mass m^*, we can derive a position-dependent formu-lation of mass responsivity along the x-axis of a rectangular cantilever. An integral formulation for the effective mass in terms of the transverse mode shape $Z_1(x)$ is given by [102]

$$m^* = \frac{1}{l} \int\limits_0^l m_c(x) \left(\frac{Z_1(x)}{A_c} \right)^2 dx .$$

(7.14)

As can be seen from this relation the mass $m_c(x)$ along the cantilever contributes to the effective mass m^* weighted by a factor $[Z_1(x)/A_c]^2$. Accordingly, a small mass Δm added at the position x is also weighted by the same factor [138], and thus

$$R(x) = R \left(\frac{Z_1(x)}{A_c} \right)^2 .$$

(7.15)

A useful approximative relation $R(x) = R(x/l)^3$ was proposed by Sader et al. [129], which replaces the trigonometric and hyperbolic functions in the modal shape expression $Z_1(x)$ (7.4). The third-order position dependency deviates less than 2% from (7.15) for $x > \frac{2}{3}l$. Equations (7.13) and (7.15) were both obtained by assuming that the load position lies in the axis of symmetry of the cantilever. For cantilevers with large length-to-width ratio, the application of an off-axis load only produces a small deviation from the on-axis value (less than 2%) [129].

The position-dependent responsivity of the fundamental mode given in (7.15) was experimentally verified as shown in Fig. 7.12. In these experiments SEM imaging allows the determination of the mass loading position and the microparticle volume with very high precision.

Comparison with finite-element simulations and experimental values [40] shows that (7.15) can be extended to higher-order modes, $n > 1$, by replacing $Z_1(x)$ by $Z_n(x)$ given in (7.4). Mass sensing at higher-order modes increases the mass responsivity owing to increased resonance frequency as seen from (7.5): $f_n/f_1 = (\beta_n l/\beta_1 l)^2$. This corresponds to a gain in responsivity of a factor of 6.3 and 17.5 for

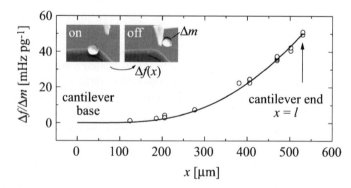

Fig. 7.12. Experimental measurement of the point mass loading response of a nearly rectangular silicon cantilever (see Fig. 7.7). The responsivity, $R(x) = \Delta f(x)/\Delta m$, was measured by adding and removing micrometer- sized copper spheres with mass $\Delta m = V\rho_{Cu}$ at position x along the cantilever (*inset*). The experimental data (*circles*) were fitted using (7.15) (*line*)

the second and the third mode, respectively. It is noted that the positioning sensitivity in point-mass loading is stronger for higher-order modes.

An analytic distributed-mass model was formulated in [43, 123], wherefrom the dynamic response to various tip–sample contact mechanisms and mass loading at variable positions along the beam can be derived including all higher modes.

7.3.2.5
Detection Limits

The fundamental limits of resonating cantilever sensors are determined by the ratio of their responsivity to the level of intrinsic noise. Accordingly, the minimum resolvable frequency shift δf_{min} of the measuring system determines the minimum resolvable mass change as $\delta m_{min} = \delta f_{min} R^{-1}$. In any actual implementation noise and drift sources are imposed by the environment, such as the temperature instability (Sect. 7.3.5), as well as by the measuring system (Sect. 7.3.6). From an engineering point of view these noise and drift contributions can be minimized in a carefully designed measurement setup. Here we restrict the discussion on the intrinsic noise mechanisms which determine the ultimate fundamental limits set by the thermomechanical noise. The minimal detectable relative frequency shift is then given by [90]

$$\frac{\delta f}{f_0} = \frac{1}{A_c} \sqrt{\frac{2\pi k_B TB}{f_0 k_c Q}} \, , \tag{7.16}$$

where $k_B T$ is the thermal energy and B the measurement bandwidth. From (7.16) it is seen that thermomechanical noise is dependent on the dissipation Q^{-1}, which favors the high-resolution operation in a vacuum. The measurement bandwidth is determined by trading off the minimum noise level and the maximum available measurement response. Another implication of (7.16) is that the thermomechanical noise can be minimized when driving the cantilever sensor at the maximum acceptable deflection amplitude. Further intrinsic noise sources which come into play for nanometer-scale structures are discussed elsewhere [32, 33].

7.3.3
Sensor Scaling

Pushing down the detection limit by design considerations of rectangular cantilever sensors is discussed in terms of size and material. In Table 7.2 the dependence on dimensions and material properties of the spring constant, resonance frequency, responsivity, and the minimum detectable mass at the thermomechanical noise limit are compiled.

If the size of a cantilever is reduced proportionally in all dimensions by the scaling factor a ($a < 1$), according to Table 7.2 the responsivity increases by a factor a^{-4} and the minimum detectable mass at the thermomechanical noise limit decreases by a factor of a^{-3}. Furthermore, in the case of a reduction of the cantilever length to bl ($b < 1$) and an increase of its height to cl ($c > 1$), the responsivity increases by a factor b^{-3} and the minimum detectable mass at the thermomechanical

Table 7.2. Scaling laws of rectangular cantilever mass sensors for point-mass loading. For homogeneous loading, the responsivity R must be divided by the active area

Spring constant (7.1)	$k_c \propto l^{-3} w h^3 E$
Resonance frequency (7.5)	$f_0 \propto l^{-2} h \sqrt{E \rho^{-1}}$
Responsivity ((7.12), (7.13))	$R \propto f_0 m_c^{-1} \propto l^{-3} w^{-1} \sqrt{E \rho^{-3}}$
Minimum detectable mass at the thermomechanical noise limit (7.16)	$\delta m = \delta f R^{-1} \propto \sqrt{l^7 w h^{-2}} \sqrt{E^{-3/2} \rho^{5/2}}$

noise limit decreases by a factor $\sqrt{b^7 c^{-2}}$. These examples illustrate that besides overall size reduction the l/h ratio should be minimized to obtain the best sensing capabilities [32]. At the same time, a proportional size reduction and a reduction of the l/h ratio increases the resonance frequency by a factor a^{-1} and $b^{-2} c$, respectively.

For homogeneous mass loading the conclusion regarding size reduction and l/h ratio minimization is similar to the case of point-mass loading.

Besides cantilever size reduction, the material choice influences the sensing capabilities. A comparison of conventional materials used in microfabrication reveals that monocrystalline silicon and polysilicon have an outstandingly high stiffness-to-density ratio, making them ideal materials for high-resolution cantilever mass sensors.

Experimental results show that other issues, such as the quality factor, additional intrinsic noise sources in nanoscale devices [33, 48], the integration of a sensitive readout system, and the device fabrication, must be taken into account for the optimization of high-resolution mass-sensing devices. Development of advanced surface machining techniques and the integration of nanowires and nanotubes into NEMS allow for the realization of resonators with minuscule masses in conjunction with excellent mechanical properties. This translates into opportunities for unprecedented mass resolution at the zeptogram scale as demonstrated in recent experiments [36, 47, 163].

7.3.4
Cantilever Calibration

Any quantitative force and mass measurement using cantilever-based sensors relies on the precise knowledge of the cantilever spring constant (7.1). Evaluation of cantilever spring constants can be performed using a number of complementary approaches. The first approach relies on theoretical calculations [112, 129] enabling a priori results based on the knowledge of the cantilever geometry and material properties. However, experiments show that theoretically calculated spring constants often differ significantly from experimentally measured ones [25, 129]. Practical shortcomings such as the strong dependence of the spring constant on a precise measurement of the cantilever thickness, $k_c \propto h^3$, and variable stoichiometry found in microfabricated cantilever materials [128, 157] restrict the applicability of these theoretical calculations.

Several experimental methods have been proposed to measure cantilever spring constants directly. Static methods rely on the application of a known force to the cantilever, which can be supplied by a small added mass [137], a reference lever [60, 143], hydrodynamic drag [98], reference artifacts [35], or an indentation device [73]. Dynamic methods utilize the resonance response of the vibrating cantilever [17, 24, 34, 61, 75, 129, 130]. In the widely used "dynamic add-mass" approach proposed by Cleveland et al. [34], the shift Δf in resonance frequency due to the loading of a known mass Δm reveals the spring constant using

$$k_c = \frac{4\pi^2 \Delta m}{1/(f_0 - \Delta f)^2 - 1/f_0^2} . \tag{7.17}$$

The add-mass is typically a spherical particle of gold or tungsten adhered to the end of the cantilever. The particle mass is calculated by measuring its radius and using the bulk density of the material. To adhere the particle usually no glue is necessary, the adhesion is sufficient and the method is nondestructive. As outlined in Sect. 7.3.2.4 and presented experimentally in Fig. 7.12 the accuracy of this calibration is very sensitive to the loading position of the particles. Spring constant values from off-end loaded particles can be corrected using $k_c = k'_c(x)[Z_1(x)/A_c]^2$, where $k'_c(x)$ is the spring constant found from (7.17) with the mass Δm loaded at position x along the beam. In the literature the precision of this method is claimed to be less than 10–20% [62, 128]. Performing such experiments inside the SEM using a nanomanipulator allows for a very precise control of the load position and determination of the particle size. Comparing with experiments conventionally performed under an optical microscope, the measurement precision achieved in the SEM relaxes the influence of these two main error sources and a precision of less than 5% can be achieved. This reduced error allows for precise measurements in analytics at the nanoscale, which will be discussed in Sect. 7.4.

For a critical discussion of spring constant calibration methods in terms of their simplicity, reliability, and precision, the reader is referred to the reviews by Burnham et al. [23], Gibson et al. [62], and Sader [128].

Besides the knowledge of the spring constant, quantitative static force measurements require the cantilever output signal to be related to the corresponding deflection. This is achieved by pushing the cantilever load point against a rigid surface and reading out the sensor output together with the displacement Δz from microscopy images or from a calibrated nanomanipulator stage [65].

7.3.5
Temperature Stability

Operation of cantilever sensors inside the vacuum chamber of a SEM/SIM is favorable in terms of short-term frequency stability owing to the achievement of high quality factors. However, such experiments suffer from long-term drifts due to thermal stabilization of the setup over a long time period (over a time scale of hours). This is mainly due to the fact that in vacuum conditions heat evacuation is dominated by thermal conductance through the microscope stage and electrical connections wired to the setup, while thermal convection is negligibly small compared with that in ambient conditions. Power dissipation in electrical components, e. g., piezoresistive sensors as discussed in Sect. 7.3.6, and environmental temperature fluctuations are significant sources of temperature instability.

A change in temperature influences the resonance frequency of a cantilever made from a single material in two ways. The geometrical dimensions and the stiffness are changed according to the material's thermal expansion and the temperature dependence of Young's modulus, respectively. The resonance frequency of the cantilever beam is accordingly modeled by [12]

$$f_0 = \frac{1}{2\pi} \sqrt{\frac{E(T)w(T)h(T)^3}{4l(T)^3 m^*}} . \qquad (7.18)$$

At the temperature T' (the prime is used for values at T') the relative change of resonance frequency with temperature results in

$$\frac{1}{f_0'}\frac{\partial f_0}{\partial T} = \frac{1}{2}\left(3\frac{1}{h'}\frac{\partial h}{\partial T} - 3\frac{1}{l'}\frac{\partial l}{\partial T} + \frac{1}{w'}\frac{\partial w}{\partial T} + \frac{1}{E'}\frac{\partial E}{\partial T}\right) = \frac{1}{2}(\alpha_{th} + \beta_{th}) , \quad (7.19)$$

where α_{th} is the thermal expansion coefficient assumed to be isotropic along l, w, and h, and β_{th} is the temperature coefficient of Young's modulus. For silicon at room temperature these coefficients are approximately $\alpha_{Si} = 2.6 \times 10^{-6}\,\mathrm{K}^{-1}$ and $\beta_{Si} = -90 \times 10^{-6}\,\mathrm{K}^{-1}$ [96]; thus, the temperature dependence of the resonance frequency of a silicon cantilever is dominated by the variation of Young's modulus. The expected relative frequency temperature dependence of $-43.7\,\mathrm{ppm\,K}^{-1}$ translates to a frequency shift of $-2.2\,\mathrm{Hz\,K}^{-1}$ for a typical fundamental resonance frequency of $f_0 = 50\,\mathrm{kHz}$. For a silicon cantilever with approximately rectangular shape such as the one shown in Fig. 7.7 a relative temperature-dependent frequency of $-23\,\mathrm{ppm\,K}^{-1}$ of the first flexural mode was measured by the authors. This value is clearly smaller than the expected value derived above from the material properties. The deviation might be mainly attributed to the oxide layer of several hundred nanometer thickness which protects the piezoresistive strain sensors at the bases extending to one sixth of the cantilever length. Even though this layer is not covering the cantilever completely, it has a large impact since the mechanical properties in the transition from the cantilever to the chip determine disproportionately the cantilever dynamic properties owing to maximum cantilever curvature in this region. The presence of this double-layer structure indicates that stress-stiffening effects as well as the positive temperature coefficient of the Young's modulus of thermally grown SiO_2 [132] compensate partly for the temperature dependence of the Young's modulus of silicon.

For the mass-loading responsivity $R(T)$ the equivalent relative temperature dependence holds as compared to the resonance frequency. As a consequence, to improve the frequency stability the temperature of the setup operated in a vacuum must be kept constant by a closed-loop temperature control to avoid long-term drifts. An alternative technique to account for thermal effects is to use two cantilever sensors, one of which acts as a reference sensor to monitor temperature shifts [89].

A similar development of the temperature dependence as for the resonance frequency can be made for the spring constant of a rectangular cantilever (7.1) and leads to: $1/k'_c \partial k_c / \partial T = \alpha_{th} + \beta_{th}$. For a silicon cantilever the relative temperature dependence of k_c is then -87 ppm K^{-1}, which translates to negligible errors in force measurements as for the ones described in Sect. 7.4.1.

7.3.6
Piezoresistive Detection

The most common cantilever deflection readout system is the well-known optical lever [103, 104]. This technique uses a collimated light beam emitted from a laser diode which is reflected off the cantilever and is projected onto a segmented photodiode which detects sensitively changes in the cantilever slope and thus in its deflection. A typical sensing head incorporating such a system has dimensions of several centimeters in width and length. This is clearly a disadvantage when a cantilever sensor is integrated in a SEM or a SIM, because in many of these instruments severe size constraints are imposed by the available space between the objective and the microscope sample stage (Fig. 7.2). In the case of a cantilever force sensor being moveable in at least three degrees of freedom (including tilt or rotation), which is a requirement in most applications, the complete sensing head must be manipulated with nanometer precision, which is not an easy task to achieve. For these reasons, for its ease of operation with no alignment required (as for all optical detection systems), and no disturbance of the ion/electron beam owing to the large electric fields applied (as for capacitive detection systems), an integrated piezoresistive deflection readout is favorable inside the SEM/SIM.

Piezoresistive deflection readout of cantilevers is accomplished by integrated piezoresistive elements measuring the variation in film resistance with respect to cantilever deflection as a consequence of a surface-stress change. Piezoresistivity is a material property where the bulk resistivity of a material is influenced by the mechanical stress applied to the material [87]. For piezoresistivity to be observable, the electrical conductivity along the thickness of the cantilever has to be asymmetric, which is often accomplished by a differential doping [96] of the material.

Cantilever sensors with piezoresistive deflection readout have mainly been implemented in two ways: by integrating a thin piezoresistive layer on one side of a cantilever which is typically U-shaped or by placing piezoresistors in the zone of maximum surface stress close to the clamped end of the cantilever. The first approach was proposed and implemented for an AFM with atomic resolution by Tortonese et al. [144]. For accomplishment of the resistivity measurement in the detection system, four resistors, of which one or two are for sensing deflection and the others are passive, are connected to form a Wheatstone bridge. The alternative design, depicted in Fig. 7.7, employs a single cantilever containing four piezoresistors connected in a Wheatstone bridge arrangement [94]. To improve the sensitivity of this design, the cantilever may be perforated at the center of the Wheatstone bridge, which increases the mechanical strain in the material next to the hole where the resistors are located [95]. The sensitivity and noise of piezoresistive cantilever sensors can be tailored by optimizing the dimensions of the integrated piezoresistors [68]. However, for a cantilever with piezoresistive detection the minimum detectable frequency shift is in most cases limited by the Johnson noise and the $1/f$ noise of the piezoresistors [168].

A symmetric Wheatstone bridge arrangement compensates to the first order for the influence of thermal drifts on the detection system. However, the power dissipated in piezoresistive devices tends to heat up the cantilever and the setup at a rate depending on the input power and the thermal resistance of the device [68]. As a consequence, the resonance frequency is subject to thermally induced drifts as discussed in Sect. 7.3.5.

7.4
Analytics at the Nanoscale

7.4.1
Nanomechanics

Macroscopic mechanical properties are dominated by a statistical distribution of defects with various characteristic length scales: the size and distribution of flaws in brittle fracture (100 μm), size and interdistance of persistent glide bands in metal fatigue (10 μm), subgrain boundary spacing in creep (1 μm), and dislocation inter-distance in metal plasticity (100 nm). At the macroscopic scale, the Weibull statistic governs the strength of brittle materials [160]. It is based on the fact that in small samples the probability to encounter a large defect that leads to failure is smaller than in larger samples, which makes them statistically stronger (Fig. 7.13). As struc-ture dimensions are further scaled down below the characteristic length, material

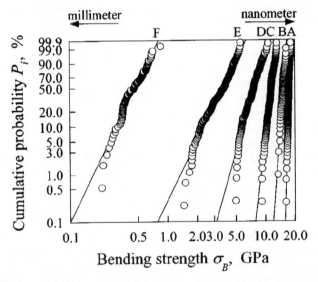

Fig. 7.13. Strength distribution from bending experiments on silicon beams. Specimens A and F have diameters of approximately 250 nm and 1.5 mm respectively. P_i is the cumulative probability with which the samples will fracture at the stress σ_B. For decreasing diameters the strength increases and the scatter in strength decreases. (Reprinted with permission from [107]. © 2004 IEEE)

properties become controlled by geometrical constraints. This includes device dimensions (physical size effect) as well as microstructure length scales like grain size (microstructure/nanostructure size effect). Whereas microstructure size effects have been used for decades for hardening of metals, physical size effects have been studied only recently, pushed by the ongoing miniaturization of micromechanical and nanomechanical systems. Physical size effects appear in particular if the characteristic length scale is comparable to the device dimensions. In this case the mechanical behavior of the material is governed by the interaction of a few defects, i.e., the behavior cannot be described by the statistical treatment of a large number of defects nor by their absence. The understanding of these size effects is a prerequisite for efficient materials processing and high reliability of nanodevices like NEMS or nanowires for sensor and electronic applications. A comprehensive description of nanomechanics and micromechanics can be found in [122]. The majority of experiments can be arranged in tensile, bending, and resonance configurations. For all of them the sample must have a high aspect-ratio geometry.

7.4.1.1
Tensile Experiment

In a tensile experiment the fracture strain, fracture stress, and Young's modulus are measured. It is preferred over the bending test because the stress distribution is homogeneous throughout the sample. The sample is attached perpendicular to a rigid support and a cantilever force sensor is attached on the free end by FEB/FIB-induced deposition of carbonaceous (or some precursor molecule) material inside the SEM, as sketched in Fig. 7.14a. The sample is then subjected to a tensile stress by either displacing the rigid support or the cantilever. A first approximation of the cantilever force F_c necessary to force failure of a perfect, brittle nanostructure in the tensile experiment can be estimated by

$$F_c = \sigma_c A_s \leq \frac{E_s}{10} A_s , \qquad (7.20)$$

where, A_s denotes the samples cross section and E_s its Young's modulus. Since E_s for a nanostructure is not known a priori it is approximated by the known bulk modulus. For a silicon nanowire having a diameter of 100 nm and $E_{Si(111)} = 188$ GPa, the critical force is $F_c \approx 0.1$ mN. For a single-wall CNT (SW-CNT) with an outer diameter of 2 nm, a wall thickness of 0.34 nm, and $E_s \approx 600$ GPa (diamond-like carbon) the surface becomes $A_s \approx 3$ nm^2 and the critical force is $F_c \approx 200$ nN. The cantilever's spring constant must be tuned to this range in accordance with the Δz increments of the nanopositioning system and the resolution of the deflection measurement. Let us assume $\Delta z = 100$ nm for such an increment, as for a stick–slip piezo device. Then we want at least ten data points for our stress–strain curve before failure, which gives a total cantilever deflection $\Delta z_{tot} = 10 \times \Delta z = 1$ μm at failure. The required cantilever spring constant k_c follows from the force balance $F_c = k_c \Delta z_{tot}$; hence, for the SW-CNT $k_c = 0.1$ N m^{-1} and for the Si nanowire $k_c = 100$ N m^{-1}. If the nanomanipulation system is capable of a better displacement resolution, a stiffer cantilever can be used. This prevents the cantilever from bending too much and therefore helps to minimize offset angle problems discussed later in this section (Fig. 7.14).

The measured force and sample elongation provide the stress–strain data for the calculation of the Young's modulus as well as for the fracture strength. The equations for stress and strain and their corresponding relative error are

$$\sigma_s = \frac{F_a}{A_s} = \frac{k_c \Delta z}{A_s} , \tag{7.21a}$$

$$\frac{\Delta \sigma_s}{\sigma_s} = \left| \frac{\Delta k_c}{k_c} \right| + \left| \frac{\Delta(\Delta z)}{\Delta z} \right| + \left| \frac{\Delta A_s}{A_s} \right| , \tag{7.21b}$$

$$\varepsilon_s = \frac{L_0 - L}{L_0} ; , \tag{7.22a}$$

$$\frac{\Delta \varepsilon_s}{\varepsilon_s} = \left| \frac{\Delta L_0}{L_0} \right| + \left| \frac{\Delta(L_0 - L)}{(L_0 - L)} \right| , \tag{7.22b}$$

where F_a is the force applied on the sample and L_0 and L are its initial and its actual length. The Young's modulus is obtained from $E_s = \sigma_s / \varepsilon_s$ in the linear, elastic region of the stress–strain curve (Fig. 7.14b). If the sample undergoes plastic or nonlinear elastic deformation, the Young's modulus can only be measured if enough data points are available at small deformation.

Equations (7.21b) and (7.22b) state the systematic errors. The largest error source in the stress measurement is typically the measurement of the cross section of the sample as, for example, the diameter of a circular nanowire goes as the square. If the deflection of the cantilever Δz_c is extracted from the SEM image, its accuracy is strongly dependent on the image quality. Manually a resolution of ± 1 to ± 5 pixels can be achieved, whereas with an image analysis tool based on a cross-correlation algorithm the resolution can be enhanced by a factor of 10–50. Δz_c may also be measured from the deflection signal of the cantilever (e. g., piezoresistive or optical readout), if provided. The accuracy of k_c depends on its calibration and is discussed in Sect. 7.3.4.

For the strain, the accuracy-limiting parameter is the change in length $L_0 - L$. It can be calculated either from the difference of the known displacements of the substrate with the sample (e. g., with a calibrated piezostage) and the deflection of the cantilever (e. g., through piezo resistive sensors), or from the SEM image. If it is extracted from the SEM image, suppose the whole sample spans 500 pixels. With the abovementioned pixel resolution, an absolute accuracy in strain of $\pm 0.2 - 1\%$ or $0.02 - 0.1\%$ can be achieved manually or with an image analysis tool, respectively. This yields a considerable relative error if the strain to be measured is on the order of 1%.

In (7.21b) and (7.22b) the errors arising from the offset angles are not taken into account. An offset between the sample and the tip can lead to bending of samples with low aspect ratios; these errors are discussed in [93]. In the following we consider the errors that follow from the projection of forces and distances. The total force F_a applied to a sample can be derived from Fig. 7.14a and c. First, we consider a tilt of the cantilever with respect to its movement Δz_c (Fig. 7.14a). In this case an effective spring constant $k'_c = k_c / \cos \gamma$ can be introduced [70], meaning that under tilt γ the cantilever displacement Δz will exert a force $F_a = (k_c / \cos \gamma) \Delta z_c$ along the z-direction. Second, from Fig. 7.14c it follows that the total applied force

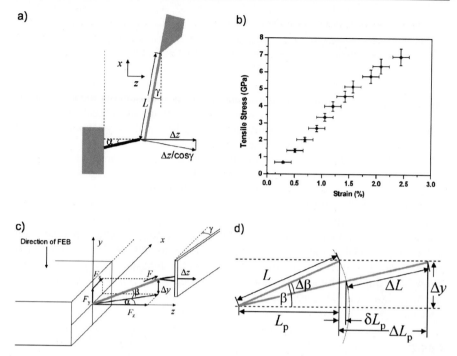

Fig. 7.14. A general cantilever–sample with offset angles in a tensile experiment. **a** Top view as seen in the SEM. The cantilever is mounted with a tilt γ and is retracted along the z-axis. **b** Stress–strain curve of a boron nanowire under tension. **c** 3D view: the nanowire tilt β not seen in the top-view SEM image and related force components. **d** View in the β-plane: Relation between true and observed elongation. (**b** Reprinted with permission from [39]. Copyright (2006), Elsevier)

is $F_a = (F_x^2 + F_y^2 + F_z^2)^{1/2} = F_z(1 + \tan^2\alpha + \tan^2\beta + \tan^2\alpha\tan^2\beta)^{1/2}$. Along the z-axis the force balance is $F_a = (k_c/\cos\gamma)\Delta z_c = F_z$. Hence, the total applied force is

$$F_a = \frac{k_c\Delta z_c}{\cos\gamma}\left(1 + \tan^2\alpha + \tan^2\beta + \tan^2\alpha\tan^2\beta\right)^{1/2}. \tag{7.23}$$

Here we assume that the AFM tip height is very small compared to the cantilever length, $H_c \ll L_c$, which is generally the case. If not the denominator becomes $(\cos\gamma - H_c\tan\alpha/L_c)$. Whereas the tilts α and γ can be corrected in real time as they can be observed by the SEM, the tilt β can pass unnoticed because of the projected SEM view. As an example, setting $\alpha = \gamma = 0$, we obtain the ratio $F_a(\beta = 20°)/F_a(\beta = 0) = 1.13$, i.e., a 13% systematic error in applied force.

The projected view also overestimates the strain, as depicted in Fig. 7.14d. The true length change ΔL of the sample while pulling along z at constant Δy is perceived as ΔL_p in projection. However, owing to the fact that β changes to $(\beta - \Delta\beta)$ during tensile loading, the true projected length change corresponding to ΔL is $(\Delta L_p - \delta L_p)$. The error in a strain measurement defined as the ratio $\delta L_p/\Delta L_p$

neglecting a β tilt follows from Fig. 7.14c:

$$\frac{\delta L_p}{\Delta L_p} = \frac{\cos(\beta - \Delta\beta) - \cos\beta}{(1 + \Delta L/L)\cos(\beta - \Delta\beta) - \cos\beta} , \tag{7.24}$$

where $\Delta\beta = \beta - \arcsin[\sin\beta/(1 + \Delta L/L)]$. Here we assume that the angle α changes negligibly keeping $\Delta\beta$ in the same plane as β throughout the experiment. Taking $\varepsilon_s = \Delta L/L = 10\%$ corresponding to the maximum theoretical strain before failure and keeping $\beta = 20°$ results in $\Delta\beta = 1.9°$ and a systematic error of $\delta L_p/\Delta L_p = 10\%$ in strain measurement at failure.

From the considerations of systematic errors above it becomes clear that 3D information is necessary for precise results. A 3D reconstruction can be performed by taking two projection images of the sample at different angles [79].

Results including Weibull statistics on the tensile strength were reported for polyacrylonitrile carbon fibers [171] and CNTs [7, 162]. ZnO [71] (see Fig. 7.6d–f) and boron [39] nanowires were subjected to tensile testing to extract their strength and Young's modulus. Further the strength of tungsten- and cobalt-containing FEB joints [149], the spring constant of FIB-deposited tungsten springs [106], and Young's modulus of biological samples like spruce wood cell wall material [117] and hairs of insects [118] were tested using SEM compatibel cantilever-based force sensors.

7.4.1.2
Bending Experiment

In such an experiment the sample is bent by a force sensor to measure the bending strength and Young's modulus. The sample is not stressed uniformly as in the tensile configuration, there is a tensile and a compressive zone at the sample root. For a bending experiment the force balance is $F_a = k_c\Delta z_c = k_s\Delta z_s$, where F_a is the force component acting perpendicular to the sample axis. From elastic beam theory [165], the maximum induced stress in the sample is

$$\sigma_{max} = \frac{k_c\Delta z_c L_s D_s}{2I_s} , \tag{7.25a}$$

$$\frac{\Delta\sigma_{max}}{\sigma_{max}} = \left|\frac{\Delta k_c}{k_c}\right| + \left|\frac{\Delta L_s}{L_s}\right| + \left|\frac{\Delta D_s}{D_s}\right| + \left|\frac{\Delta(\Delta z_c)}{\Delta z_c}\right| + \left|\frac{\Delta I_s}{I_s}\right| , \tag{7.25b}$$

where $k_c\Delta z$ is the force applied by the cantilever and D_s is the diameter of the sample in the direction of the applied force. For cylinders $I_{cyl} = \pi R^4/4$ and for hollow cylinders $I_{tube} = \pi(R^4 - r^4)/4$. The geometrical moments of other cross sections can be found in standard textbooks.

The strain can be measured without information on the applied force since it can be calculated from the deformed shape of the sample. Using $k_s = 3EI_s/L_s^3$, $\sigma_{max} = E_s\varepsilon_{max}$, (7.25a), and the force balance, one gets

$$\varepsilon_{max} = \frac{3D_s\Delta z_s}{2L_s^2} , \tag{7.26a}$$

$$\frac{\Delta\varepsilon_{max}}{\varepsilon_{max}} = \left|\frac{\Delta D_s}{D_s}\right| + \left|\frac{\Delta(\Delta z_s)}{\Delta z_s}\right| + 2\left|\frac{\Delta L_s}{L_s}\right| . \tag{7.26b}$$

Young's modulus is obtained by calculating $E_s = \sigma_{max}/\varepsilon_{max}$,

$$E_s = \frac{L_s^3}{3I_s}\frac{\Delta z_c}{\Delta z_s}k_c \,, \tag{7.27a}$$

$$\frac{\Delta E_s}{E_s} = 3\left|\frac{\Delta L_s}{L_s}\right| + \left|\frac{\Delta I_s}{I_s}\right| + \left|\frac{\Delta(\Delta z_s)}{\Delta z_s}\right| + \left|\frac{\Delta(\Delta z_c)}{\Delta z_c}\right| + \left|\frac{\Delta k_c}{k_c}\right| \,. \tag{7.27b}$$

The same considerations regarding accuracy as in the tensile experiment apply. The diameter of the sample must be measured accurately as it goes with the power of 4 in the geometrical moment of inertia for the calculation of stress and Young's modulus. If all the variables except the cantilever stiffness are read out from the SEM, the systematic errors originating from its calibration partly cancel out, for the strain even totally. Chen et al. [30] have investigated the errors originating from the way the sample is attached to the substrate.

If the sample undergoes plastic deformation, i.e., it does not return to its initial position when the force is released (Fig. 7.15), the above formulas are not valid as the plastic deformation releases the sample from stress. However, for small deflections the deformation is normally elastic, so the question is if the resolution of

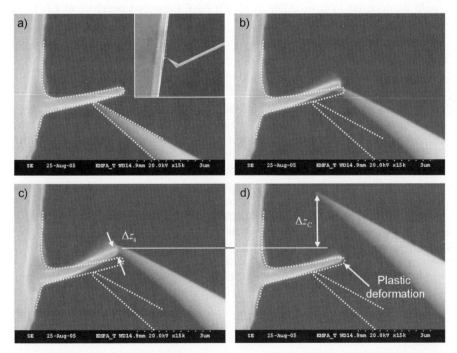

Fig. 7.15. SEM images of measurement of a bending experiment. **a** Initial unstrained contact position of the cantilever tip with the sample. The *inset* shows the arrangement of the cantilever and the sample. **b** Cantilever tip is pushed against the sample. Initial positions are marked with the *dotted lines*. **c** Positions just before loss of contact with the cantilever tip. **d** Cantilever and sample take their unstrained position

the equipment is good enough to measure stress and strain at small deflections. Then not only Young's modulus, but also the yield stress (the onset of plasticity) may be measured.

In situ bending experiments have been performed on CNTs [50], Si nano-wires [72], ZnO nanowires [71], and FIB deposits [57,77].

7.4.1.3
Compression Experiments

In compression experiments one can extract the compression strength as well as Young's modulus. The sample buckles if the force is not applied perfectly along its axis, and it cannot be predicted into which plane buckling occurs, so the projected view might in the extreme case not detect any buckling at all (buckling plane is parallel to the FEB incidence) and give the impression of a compression strength experiment because of a sudden jump in the stress–strain curve. A slight off-axis load breaks the rotational symmetry and can gives rise to a preferred buckling plane that leads to premature failure. The equations are basically the same as those for the tensile experiments. Owing to the difficulties of a well-aligned force application at this scale only few reliable compression experiments with an AFM could be done up to date. Akita et al. [2] have measured buckling of CNTs in a SEM.

7.4.1.4
Resonance Experiment

Another way of measuring Young's modulus is the resonance experiment in which the sample is excited at its resonance frequency by thermal, piezo, or electrostatic excitation. Vibration amplitude detection is possible with the imaging capabilities of the SEM: the overall modal shape is visualized by normal image scanning and a frequency spectrum is taken by secondary electron detection with a stationary beam [22]. If the electron beam is positioned at the maximum amplitude position, a secondary electron peak signal is detected at resonance while sweeping the excitation frequency (Fig. 7.16). If the electron beam is positioned at zero amplitude position the resonance manifests itself as a negative secondary electron peak owing to the decreasing dwell time of the vibrating sample. A frequency sweep can be achieved with a piezo actuator or an alternating electrostatic excitation with a counter electrode [121]. The electrostatic force acting on the sample is

$$F(t) = c\left[\Delta V + V_{dc} + V_{ac}\cos(\omega t)\right]^2$$
$$= c\left[2a\left(\Delta V + V_{dc}\right)V_{ac}\cos(\omega t) + 0.5V_{ac}^2\cos(2\omega t)\right] + cV'^2 , \qquad (7.28)$$

where ΔV is the bias voltage for the work function difference of the two electrodes, c is a nanostructure-specific proportionality constant, and $V'^2 = (\Delta V + V_{dc})^2 + 0.5V_{ac}^2$. It can be seen that a linear V_{ac} term at the driving frequency and a quadratic V_{ac} term at twice the driving frequency act on the sample. Correct assignment of the sample resonance frequency must thus be ensured by checking at double and half frequency.

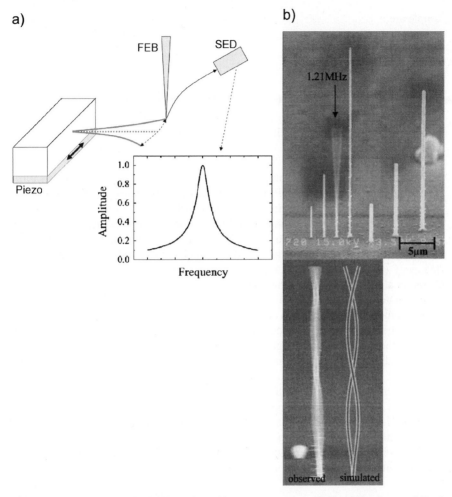

Fig. 7.16. a Measurement principle in a piezo-driven resonance experiment. **b** Fundamental (first) and third vibration mode of a FIB-deposited carbon nanopillar visualized by in situ scanning elecron microscopy (**b** reprinted with permission from [56]. Copyright (2001), American Institute of Physics)

The excitation strength of the electrostatic field is independent of the frequency, but the stronger the electric field needed for a detectable amplitude, the worse becomes the image quality. Here the use of piezoactuators is more advantageous. However, piezoactuators are specified for operation frequency windows; outside these windows the excitation amplitude can change dramatically with frequency and even drop to zero.

The modal shape of these flexural vibration modes is given by (7.4). The resonance frequencies depend on the sample material via $(E_s/\rho)^{1/2}$, the sound velocity in the material, and the sample's geometry via the geometric moment of inertia I_s and the cross section A_s. From (7.5) the frequencies for the flexural modes of the sample are

$$f_n = \frac{(\beta_n l_s)^2}{2\pi l_s^2} \sqrt{\frac{I_s}{A_s}} \sqrt{\frac{E_s}{\rho}} \; . \tag{7.29}$$

For cylindrical nanowires $(I_s/A_s)^{1/2} = D_s/4$, where D_s is the diameter, for nanotubes $(I_s/A_s)^{1/2} = \sqrt{D_o^2 + D_i^2}/4$, where D_o and D_i are the outer and inner diameter, respectively. For core–shell nanowires with a shell (subscript L) of different material the expression $(E_s I_s/A_s \rho)^{1/2}$ in (7.29) becomes $[E_s I_s + E_L I_L/(\rho_s A_s + \rho_L A_L)]^{1/2}$ [39].

Theory assumes perfect clamping of one end of the elongated nanostructure. In practice adhesion forces or FEB contamination depositions are widely used as clamps. Changes of up to 25% in resonance frequency have been reported between originally "adhesion-clamped" and then FEB-clamped nanostructures [39]. During FEB clamping care should also be taken that the remaining nanowire is not exposed to contamination deposition as this will add mass and shift the resonance frequency (Sect. 7.3.2.4). This is especially true for thin samples like nanotubes. Calibri et al. [27] investigated the influence of the presence of intrinsic curvature on the calculated Young's modulus.

From the resonance method only the ratio E_s/ρ can be extracted. Young's modulus and the density can be decoupled by the aid of an additional experiment determining either the density through mass detection (Sect. 7.4.2) or Young's modulus with a bending or tensile experiment (Sects. 7.4.1.1, 7.4.1.2). Often the bulk density is assumed and Young's modulus reported.

Resonance experiments using the thermal energy as excitation have been performed on CNTs in a transmission electron microscope (TEM) [145] and in a SEM [6]. An ac electric field in a SEM has been used to excite ZnO nanobelts [28] and ZnO nanowires [167]. Poncharal et al. [121] have observed elastic wavelike distortions in strongly bent nanotubes and made that effect responsible for the dependence of Young's modulus on the nanotube diameter.

7.4.2
Cantilever-Based Gravimetry

In this section we outline the extended range of applications of cantilever-based mass sensors when integrating them in FEB and FIB nanostructuring systems. We show results and the potentials of this field which started to be explored only very recently.

7.4.2.1
Process Control of FEB/FIB Chemical Vapor Deposition

Mass sensing inside the SEM/SIM can be applied for process control of chemical vapor deposition (CVD) by FEB and FIB and for the determination of material densities of deposits fabricated therewith. This is of interest since for the optimization of FEB/FIB-induced processes for reliable fabrication and applications in microdevices and nanodevices further understanding of the fundamental physics underlying the process is required.

Before microcantilever-based sensors were readily available, in situ process control of uniform CVD within square millimetres by defocused ion beams inside a SIM was performed using quartz crystal micro balances (QCMs) [42]. Results from these experiments allowed a growth model to be developed on the basis of the measurement of the deposition rates and yields and the corresponding adsorbed precursor surface coverage. However, the mass resolution of the QCM is insufficient to measure the growth of nanostructures by a FEB/FIB. This becomes necessary when noting that FEB/FIB CVD allows for the fabrication of features with dimensions at the micrometer scale and ultimately at the nanometer scale [154], where mass changes are on the order of picograms to zeptograms. Cantilever-based sensors including nanoscale resonators based on nanotubes and nanowires have the potential to resolve such small masses.

Recently, the group of Nakayama described a nanotube resonator for in situ mass sensing (Fig. 7.17a) which relies on the principle described in Sect. 7.4.1.4. They measured the mass of material grown at the tube extremity by FEB-induced deposition from hydrocarbon and $W(CO)_6$ precursor molecules [113, 114, 134] . While this approach offers a high mass resolution at the subattogram scale, quantitative mass sensing relies on a measurement of the sensor response as pointed out in Sect. 7.3.4. However, this is challenging for the nanotube-based sensor.

An alternative approach based on the integration of a silicon microcantilever with piezoresistive deflection readout for mass sensing inside a SEM/SIM has been proposed by the authors of this chapter [53]. In the following we discuss the process control of FEB/FIB CVD based on experiments using this technique. This device allows the measurement of FEB/FIB-induced processes in more conventional and well-understood nanofabrication conditions in terms of substrate material and geometry. The arrangement of the cantilever mass sensor and the GIS which locally supplies the volatile precursor into the vacuum chamber of the SEM/SIM is shown in Fig. 7.17b. Note that for illustrative convenience the microtube GIS is scaled down in this figure. A SEM image of the piezoresistive cantilever used is depicted in Fig. 7.7. All processing is restricted to a small area of constant responsivity at the extremity of the cantilever where the responsivity is maximal.

Calibration of the mass sensor's spring constant was performed by a direct in situ measurement of the mass responsivity of the fundamental mode R_1 based on Cleveland's method as discussed in Sect. 7.3.4. We will see later in this section that instead of attaching individual microparticles to the cantilever by nanomanipulation, a considerably simplified approach removing material by FIB milling can be employed. The typical mass responsivity of the fundamental mode of the piezoresistive cantilever shown in Fig. 7.7 is $50\,Hz\,pg^{-1}\,cm^{-2}$ for homogeneous mass loading on the cantilever topside and $0.1\,Hz\,pg^{-1}$ for point-mass loading at the cantilever end.

The mass sensor device is mounted on a Peltier-element heat sink which stabilizes the temperature of the cantilever at a constant value close to room temperature. Therefore, the heat due to dissipation of the applied bias power in the bridge is compensated for and long-term temperature fluctuations in the setup, i.e., environmental perturbation via conductive heat transfer through the microscope stage and the electrical connections, are avoided.

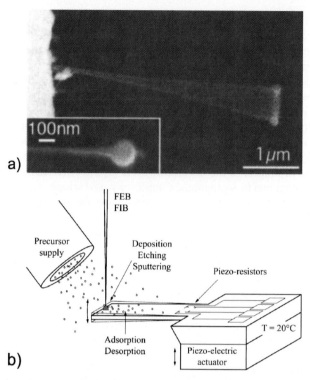

Fig. 7.17. Gravimetric sensors for in situ monitoring of FEB/FIB processes. **a** SEM image of a carbon nanotube resonator for high-resolution mass detection. The *inset* shows a deposit created by FEB-induced deposition. **b** A cantilever mass sensor with piezoresistive readout which allows the measurement of precursor adsorption/desorption, FEB/FIB deposition and etching, and FIB milling (the precursor supply tube is not to scale) (**a** Reprinted with permission from [113]. Copyright (2005), American Institute of Physics. **b** Reprinted with permission from [53]. Copyright (2007), American Institute of Physics)

One of the simplest methods to continuously sense resonance frequency shifts, which was originally employed for atomic force microscopy, is to excite the cantilever at a fixed frequency close to its resonance and to measure its amplitude and phase response (see (7.9) and (7.10)). Upon a change in resonance frequency the oscillation amplitude settles with a time constant of $\tau = 2Q/\omega$ to the new steady-state value. Operation in vacuum conditions with high Q thus limits the available bandwidth $B = \tau^{-1}$ and the dynamic linear sensing range [59]. A closed-loop implementation where the cantilever acts as a resonator in an active feedback circuit makes Q and B independent of each other and enables the separation of the response from conservative stimuli (e. g., mass loading) and dissipative interactions (e. g., damping) [4,44]. The cantilever is continuously driven at its momentary resonance frequency. This is accomplished with a phase-locked-loop implementation where the bandwidth B can be adjusted by the demodulation characteristics [45]. It is seen from (7.16) that it is essential to limit the bandwidth of the measurement system to minimize the noise level from thermomechanical noise and from other contributions, such as the noise of the piezoresistors. The amplitude of the driving signal A_d is closed-loop-controlled to maintain the deflection amplitude A_c constant at the maximal tolerable value which gives the best noise performances revealed by (7.16). Thus, the drive amplitude A_d is a direct measure of relative changes in the quality factor, since from (7.9), $A_d/A_c = Q^{-1}$ at $\omega_d = \omega_0$.

Running the phase-locked loop at a demodulation bandwidth of approximately 1 Hz was an acceptable trade-off between signal-to-noise ratio and response time to follow the mass changes in FEB/FIB CVD. Quality factors of the piezoresistive cantilevers at 10^{-5} mbar were typically greater than 10,000 and cantilever deflection amplitudes were between 100 and 500 nm. A short-term stability on the order of 1-mHz root mean square frequency noise was shown in [53] which allowed resolution of mass changes on the order of 10 fg.

Adsorption

For FEB/FIB CVD the microtube GIS depicted in Fig. 7.17 locally supplies a constant flux J of precursor molecules on the order of 10^{16}–10^{18} molecules per square centimeter per second depending on the vapor pressure of the gas molecules to the processing site [148]. The mass change due to adsorption of precursor molecules which impinge homogeneously on the topside of the cantilever mass sensor is detected as a frequency reduction and can be deduced from (7.12), $\Delta m = \Delta f R^{-1}$. Since the background chamber pressure remains at high vacuum level (typically below 10^{-5} mbar), the molecular flux to the backside of the sensor is negligible. Impinging molecules physisorb on the cantilever with a given sticking probability and diffuse on its surface before they eventually desorb after an average residence time. Knowledge of the mass of the time-averaged number of adsorbates and the impinging molecule flux allows the molecule surface coverage and the molecule residence time to be deduced, which enables the direct measurement of molecule dissociation efficiencies by ion or electron impact.

Nondissociative Langmuir adsorption predicts the mass distribution on the cantilever to be linearly proportional to the impinging flux distribution at low submonolayer coverage. In the case of a nonuniform flux, the product of the mass distribution $\Delta m(x)$ and the position-dependent responsivity $R(x)$ (7.15) can be integrated along the cantilever to find the induced frequency shift according to [53]

$$\Delta f = \frac{1}{l} \int_0^l \Delta m(x) R(x) \, dx \ . \tag{7.30}$$

This relation was applied based on the process control experiment shown in Fig. 7.18a during FIB-induced deposition from a platinum precursor. Opening the precursor supply leads to a negative frequency shift due to added mass from adsorption of precursor molecules on the cantilever surface (part A). After several minutes, the GIS and the vacuum chamber achieve their equilibrium pressure. The surface coverage was determined to be about 10% of a monolayer and the residence time was found to be on the order of tens of microseconds. Reversibility of the mass loading by desorption of molecules upon closing of the precursor supply is observed (part C).

Damping effects due to precursor gas flow were measured below 1% from the excitation amplitude. This means that no additional surface stress is generated owing to physisorbed molecules which can desorb freely. On the other hand, cantilevers functionalized with organic or inorganic adhesion coatings favor uptake of adsorbates, inducing surface stress response of several orders of magnitude. Simultaneous measurement of the resonance frequency and cantilever static bending allows differentiation of the response from mass loading and surface stress [8, 100].

Owing to their small size cantilever mass sensors can map spatial distributions of precursor flux when moving them with respect to the GIS. Similarly the FIB profile can be mapped via its sputtering action and a deconvolution of the finite sensor size. This is very helpful for optimization of FEB- and FIB-based nanofabrication.

Rate and Yield

Upon FEB/FIB-induced deposition, material is added and the resonance frequency of the cantilever sensor decreases (part B in Fig. 7.18a). The mass evolution of the FIB deposition in a $1 \times 1\,\mu m^2$ scan area shown in Fig. 7.18b was derived from the frequency measurement using (7.13). Its slope gives the deposition rate ($23\,fg\,s^{-1}$) and the deposition yield after normalization by the irradiation dose ($4.7\,pg\,nC^{-1}$).

Figure 7.19a shows the evolution of deposited mass during a 5-min FEB irradiation in continuous stationary spot mode. The deposition rate, shown in Fig. 7.19b, saturates after an initial increase. In terms of deposit height and volume, such be-

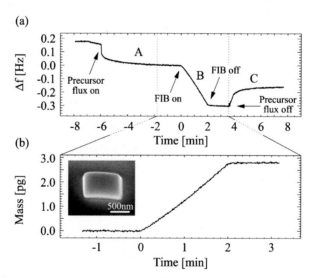

Fig. 7.18. a FIB-induced deposition experiment from the precursor $(CH_3)_3PtCpCH_3$: A mass loading due to adsorption and C mass loss due to desorption; B FIB exposure of a $1 \times 1\,\mu m^2$ rectangle. **b** Evolution of the FIB (30 kV, 5 pA) deposited mass corresponding to part B in **a**. *Inset*: SEM tilt view (52°) of the FIB deposit (**b** Reprinted with permission from [53]. Copyright (2007), American Institute of Physics)

a) 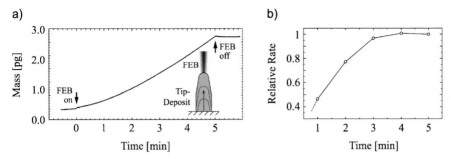 b)

Fig. 7.19. a FEB (5 kV, 200 pA) induced tip deposition from the precursor $Cu(C_5O_2HF_6)_2$ in stationary spot mode irradiation. The *inset* shows the corresponding tip growth evolution. **b** The corresponding relative rate change with ongoing deposition time

havior is well documented [19, 136]; however, with a correlated mass measurement density evolution during irradiation and eventually dose effects will become detectable.

The small resonance frequency jumps of several millihertz at the start and end of FEB deposition in Fig. 7.19a are attributed to electron-beam heating of the cantilever. An estimation based on a 1D linear thermal conductance model [74, 125] predicts a temperature rise on the order of a few millikelvin within a few nanoseconds. This translates well into the observed millihertz frequency jumps (Sect. 7.3.5).

The removal of cantilever mass by FIB milling of a $1 \times 1\,\mu m^2$ pit is shown in Fig. 7.20. A silicon milling rate of two atoms per ion was measured, which compares well with reported values. The constant sputter rate indicates that redeposition of ejected silicon atoms on the pit sidewalls is insignificant up to the final pit aspect ratio of 0.4 and less explored in this experiment. Since the mass of removed silicon can be deduced from the pit volume known from SEM imaging, this approach is an elegant way to perform a calibration of the cantilever sensor according to Cleveland's method (Sect. 7.3.4). Instead of a time-consuming manipulation and positioning of microparticles on the cantilever, a defined volume of the silicon cantilever is removed using FIB milling at a perfectly defined location. Together with the corresponding frequency shift, the mass responsivity and the spring constant of the cantilever can be determined. It is noted that the mass due to implanted Ga atoms at the pit bottom can be accounted for [53].

Fig. 7.20. Evolution of the cantilever resonance frequency during FIB (30 kV, 50 pA) milling of a $1 \times 1\,\mu m^2$ pit. *Inset*: SEM tilt view (45°) of the sputtered pit. (Reprinted with permission from [53]. Copyright (2007), American Institute of Physics)

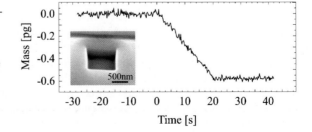

7.4.2.2
Deposit Density

Density determination of nanoobjects via mass and volume is straightforward, $\rho_s = \Delta m / V_s$ provided the sample geometry is well known. For porous and inhomogeneous samples an average density is measured.

A cantilever mass sensor in a conventional AFM setup was used to determine ex situ the density of FEB-deposited nanowires grown in various conditions [150]. Relations between the chemical deposit composition and deposit density were established and irradiation dose effects measured (Fig. 7.21).

A combination of a bending experiment (Sect. 7.4.1.2) and a resonance experiment (Sect. 7.4.1.4) was employed to determine both unknowns, the density and Young's modulus of vertically FIB deposited nanowires [57,77,78].

From the mass-sensing experiments using the piezoresistive cantilever mass sensor described herein, the deposit density of FEB/FIB-grown deposits could be determined with an error on the order of 10% [53].

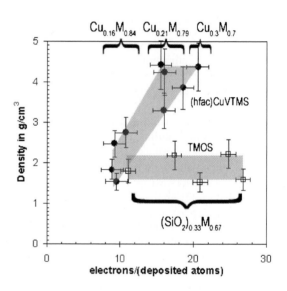

Fig. 7.21. Deposit density versus number of electrons per deposited atom. Deposit compositions are given in generalized form, where M stands for the carbonaceous matrix. The two precursors used were tetramethoxysilane (*TMOS*) and hexafluoroacetylacetonato–copper(I)–vinyltrimethylsilane [(*hfac*)*CuVTMS*]. (Reprinted with permission from [150]. Copyright (2006), American Institute of Physics)

7.4.3
Atomic Force Microscopy in a SEM

Conventional atomic force microscopy inside a SEM allows a sample to be observed using the electron beam and the local topography to be investigated by atomic force microscopy. The advantages of the standalone techniques are combined: the SEM's field of view of up to several square millimeters provides greater operating convenience than the relatively small one of the AFM. This advantage, coupled with the SEM's high scan rate—providing images at TV rate (50 Hz)—and large depth of field (several tens of micrometers), simplifies the survey process required to locate specific surface features for analysis by the AFM.

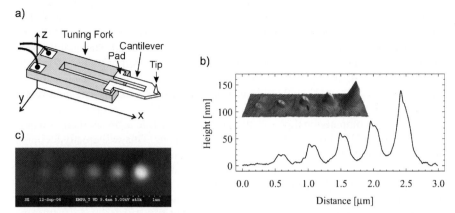

Fig. 7.22. a The self-sensing and self-actuating probe based on a tuning fork and a microfabricated cantilever. **b** Profile and 3D shape (*inset*) of FEB-grown dot deposits (precursor Cu(hfac)$_2$) for 2, 4, 6, 8, and 12 s with 500-nm pitch measured by frequency modulationatomic force microscopy in the SEM. **c** The same deposits as observed in a top-view SEM image (thermionic tungsten filament, 5 kV, 200 pA) (**a** Reprinted with permission from [3]. Copyright (2003), American Institute of Physics)

AFM–SEM systems have been developed in research [49, 58, 80, 83, 140, 146]. Three commercial AFMs for integration into a SEM are or were available on the market [87, 154]. Williams et al. [161] reported on the use of an AFM–SEM combination for the controlled placement of an individual CNT on a microfabricated test structure in a SEM.

Its simplest application, the topography imaging with the AFM of a surface area identified by the SEM, is especially useful in FEB/FIB processing. The system employed by the authors here consists of a self-sensing and self-actuating cantilever probe based on a quartz tuning fork, shown in Fig. 7.22a, which is simple to integrate into the vacuum chamber of a SEM/SIM. These sensors combine the advantages of the tuning fork's extremely stable frequency driving the tip vibration with the silicon cantilever's softer spring constant and, thus, higher sensitivity. Frequency modulation atomic force microscopy was performed inside a SEM to investigate FEB-grown dot deposits with aspect ratios less than 1. The results show the benefit of the 3D image capabilities of the AFM (Fig. 7.22b) when compared with the secondary electron image (Fig. 7.22c). The accurate topography data are useful to extract precisely the volume of the grown material. This is of interest for the determination of process rates and yields at the initial phase of growth and for material density measurements when correlated to mass measurements as discussed in Sect. 7.4.2.

7.5
Perspectives and Outlook

Among the newest developments, in situ high-resolution TEM nanomechanical experiments have the potential to image defects responsible for deformation mechanisms at the atomic scale such as grain effects, dislocation sources, twins, and stacking faults [84]. The development of microelectromechanical system or NEMS devices compatible with SEMs and high-resolution TEMs measuring directly and independently load and displacement goes along with this trend. Successful placement of nanowires and stress–strain measurements were demonstrated in a SEM and a TEM [170]. In situ characterization of NEMS including measurements of electrical properties and bistable deformation states [82] is an emerging field where the combined use of techniques discussed in this chapter has great potential.

Increasing computational power allows for faster image analysis such as object recognition and 3D reconstruction from projected views. This will greatly facilitate automatization of in situ nanostructure and microstructure assembly [5] for tailored devices as well as microscale and nanoscale testing. With their nanofabrication capabilities FEB and FIB techniques will become more and more important for device prototyping applications and as assistance tools in nanomechanics offering a solution for attach and release mechanisms in AFM-based nanomanipulation.

Ultimate high-resolution gravimetry is developing towards optimized NEMS designs with mass sensitivities having the potential to blur the traditional distinction between inertial mass sensing and mass spectrometry [163]. Such devices offer access to a parameter space for sensing and fundamental measurements that is intriguing and unprecedented.

References

1. 3rd Tech, USA, North Carolina (2007) http://www.3rdtech.com
2. Akita SM, Nishio et al (2006) Jpn J Appl Phys Part 1 45(6B):5586–5589
3. Akiyama TU, Staufer et al (2003) Rev Sci Instrum 74(1):112–117
4. Albrecht TR, Grutter P et al (1991) J Appl Phys 69(2):668–673
5. Aoki K, Miyazaki HT et al (2003) Nat Mater 2(2):117–121
6. Babic B, Furer J et al (2003) Nano Lett 3(11):1577–1580
7. Barber AH, Andrews R et al (2005) Appl Phys Lett 87(20):203106
8. Battiston FM, Ramseyer JP et al (2001) Sens Actuators B 77(1–2):122–131
9. Bauerdick S, Burkhardt C et al (2004) J Vac Sci Technol B 22(6):3539–3542
10. Becker M, Sivakov V et al (2007) Nano Lett 7(1):75–80
11. Berger R, Delamarche E et al (1997) Science 276(5321):2021–2024
12. Berger R, Lang HP et al (1998) Chem Phys Lett 294(4–5):363–369
13. Bhushan B (2004) Applied scanning probe methods. Springer, Berlin
14. Bhushan B (2004) In: Springer handbook of nanotechnology. Springer, Berlin, chap 21
15. Binnig G, Quate CF et al (1986) Phys Rev Lett 56(9):930–933
16. Boero G, Utke I et al (2005) Appl Phys Lett 86(4):042503
17. Bonaccurso E, Butt HJ (2005) J Phys Chem B 109(1):253–263
18. Bret T (2005) PhD thesis. Physico-chemical study of the focused electron beam induced deposition process. EPF Lausanne
19. Bret T, Utke I et al (2004) J Vac Sci Technol B 22(5):2504–2510
20. Bret T, Utke N et al (2006) Microelectron Eng 83(4–9):1482–1486
21. Brintlinger T, Fuhrer MS et al (2005) J Vac Sci Technol B 23(6):3174–3177
22. Buks E, Roukes ML (2001) Phys Rev B 63(3):033402
23. Burnham NA, Chen X et al (2003) Nanotechnology 14(1): 1–6
24. Butt HJ, Jaschke M (1995) Nanotechnology 6(1):1–7
25. Butt HJ, Siedle P et al (1993) J Microsc 169:75–84
26. Butt HJ, Cappella B et al (2005) Surf Sci Rep 59(1–6):1–152
27. Calabri L, Pugno N et al (2006) J Phys Condens Matter 18(33):S2175–S2183

28. Chen CQ, Shi Y et al (2006) Phys Rev Lett 96(7):075505
29. Chen GY, Thundat T et al (1995) J Appl Phys 77(8):3618–3622
30. Chen YX, Dorgan BL et al (2006) J Appl Phys 100(10):104301
31. Cividjian N (2002) Electron beam induced nanometer scale deposition. DUP Science, Delft
32. Cleland AN (2005) New J Phys 7:235
33. Cleland AN, Roukes ML (2002) J Appl Phys 92(5):2758–2769
34. Cleveland JP, Manne S et al (1993) Rev Sci Instrum 64(2):403–405
35. Cumpson PJ, Hedley J et al (2003) Nanotechnology 14(8):918–924
36. Davis ZJ, Boisen A (2005) Appl Phys Lett 87(1):013102
37. Dellaratta AD, Melngailis J et al (1993) J Vac Sci Technol B 11(6):2195–2199
38. Ding W, Dikin DA et al (2005) J Appl Phys 98(1):014905
39. Ding WQ, Calabri L et al (2006) Compos Sci Technol 66(9): 1112–1124
40. Dohn S, Sandberg R et al (2005) Appl Phys Lett 86(23):233501
41. Dong L, Arai F et al (2001) In: Proceedings of the IEEE international conference on robotics and automation, Seoul, pp 632–637
42. Dubner AD, Wagner A (1989) J Appl Phys 65(9):3636–3643
43. Dupas E, Gremaud G et al (2001) Rev Sci Instrum 72(10):3891–3897
44. Durig U, Gimzewski JK et al (1986) Phys Rev Lett 57(19):2403–2406
45. Durig U, Steinauer HR et al (1997) J Appl Phys 82(8):3641–3651
46. Ekinci KL, Roukes ML (2005) Rev Sci Instrum 76(6)
47. Ekinci KL, Huang XMH et al (2004) Appl Phys Lett 84(22):4469–4471
48. Ekinci KL, Yang YT et al (2004) J Appl Phys 95(5):2682–2689
49. Ermakov AV, Garfunkel EL (1994) Rev Sci Instrum 65(9):2853–2854
50. Falvo MR, Clary GJ et al (1997) Nature 389(6651):582–584
51. Folks L, Best ME et al (2000) Appl Phys Lett 76(7):909–911
52. Force Dimension, Switzerland (2007) http://www.forcedimension.com
53. Friedli V, Santschi C et al (2007) Appl Phys Lett 90(5):053106
54. Fritzsche W, Bohm K et al (1998) Nanotechnology 9(3):177–183
55. Fritzsche W, Kohler JM et al (1999) Nanotechnology 10(3):331–335
56. Fujita J, Ishida M et al (2001) J Vac Sci Technol B 19(6):2834–2837
57. Fujita J, Ishida M et al (2003) Nucl Instrum Methods Phys Res Sect B 206:472–477
58. Fukushima K, Kawai S et al (2002) Rev Sci Instrum 73(7):2647–2650
59. Garcia R, Perez R (2002) Surf Sci Rep 47(6–8):197–301
60. Gibson CT, Watson GS et al (1996) Nanotechnology 7(3):259–262
61. Gibson CT, Weeks BL et al (2003) Ultramicroscopy 97(1–4):113–118
62. Gibson CT, Smith DA et al (2005) Nanotechnology 16(2):234–238
63. Giessibl FJ (2003) Rev Mod Phys 75(3):949–983
64. Goldstein JI (2003) Scanning electron microscopy and X-ray microanalysis. Kluwer, New York
65. Gotszalk T, Grabiec P et al (2003) Ultramicroscopy 97(1–4):385–389
66. Hagleitner C, Hierlemann A et al (2001) Nature 414(6861):293–296
67. Hansen LT, Kuhle A et al (1998) Nanotechnology 9(4):337–342
68. Hansen O, Boisen A (1999) Nanotechnology 10(1):51–60
69. Heidelberg A, Ngo LT et al (2006) Nano Lett 6(6):1101–1106
70. Heim LO, Kappl M et al (2004) Langmuir 20(7):2760–2764
71. Hoffmann S, Östlund F et al (2007) Nanotechnology 18(20):205503
72. Hoffmann S, Utke I et al (2006) Nano Lett 6(4):622–625
73. Holbery JD, Eden VL (2000) J Micromech Microengin 10(1):85–92
74. Holman JP(2002) Heat transfer. McGraw-Hill, Boston
75. Hutter JL, Bechhoefer J (1993) Rev Sci Instrum 64(7):1868–1873

76. Ilic B, Czaplewski D et al (2001) J Vac Sci Technol B 19(6):2825–2828
77. Ishida M, Fujita J et al (2002) J Vac Sci Technol B 20(6):2784–2787
78. Ishida M, Fujita J et al (2003) J Vac Sci Technol B 21(6):2728–2731
79. Jähnisch M, Fatikow S (2007) Int J Optomechatron 1(1)4–26
80. Joachimsthaler I, Heiderhoff R et al (2003) Meas Sci Technol 14(1):87–96
81. Junno T, Deppert K et al (1995) Appl Phys Lett 66(26):3627–3629
82. Ke CH, Espinosa HD (2006) Small 2(12):1484–1489
83. Kikukawa A, Hosaka S et al (1993) J Vac Sci Technol A 11(6):3092–3098
84. Kizuka T, Takatani Y et al (2005) Phys Rev B 72(3):035333
85. Kleindiek Nanotechnik, Germany (2007) http://www.kleindiek.com
86. Klocke Nanotechnik, Germany (2007) http://www.nanomotor.de
87. Kloeck B (1994) In: Bau HH (ed) Mechanical sensors, vol XIII. VCH, Weinheim, p 674
88. Koops HWP, Weiel R et al (1988) J Vac Sci Technol B 6(1):477–481
89. Lang HP, Berger R et al (1998) Appl Phys Lett 72(3):383–385
90. Lavrik NV, Sepaniak MJ et al (2004) Rev Sci Instrum 75(7):2229–2253
91. Lezec HJ, Degiron A et al (2002) Science 297(5582):820–822
92. Li GY, Xi N et al (2004) IEEE-ASME Trans Mechatron 9(2):358–365
93. Li XD, Wang XN et al (2005) Rev Sci Instrum 76(3):033904
94. Linnemann R, Gotszalk T et al (1995) Thin Solid Films 264(2):159–164
95. Linnemann R, Gotszalk T et al (1996) J Vac Sci Technol B 14(2):856–860
96. Madou MJ (2002) Fundamentals of microfabrication the science of miniaturization. CRC, Boca Raton
97. Madsen DN, Molhave K et al (2003) Nano Lett 3(1):47–49
98. Maeda N, Senden TJ (2000) Langmuir 16(24):9282–9286
99. Majorana E, Ogawa Y (1997) Phys Lett A 233(3):162–168
100. McFarland AW, Poggi MA et al (2005) Appl Phys Lett 87(5):053505
101. Melngailis J, Musil CR et al (1986) J Vac Sci Technol B 4(1):176–180
102. Merhaut J (1981) Theory of electroacoustics. McGraw-Hill, New York
103. Meyer G (1988) Appl Phys Lett 53(24):2400–2400
104. Meyer G, Amer NM (1988) Appl Phys Lett 53(12):1045–1047
105. Moser B, Wasmer K et al (2007) J Mater Res 22:1004–1011
106. Nakamatsu K, Igaki J et al (2006) Microelectron Eng 83(4–9):808–810
107. Namazu T, Isono Y et al (2000) J Microelectromech Syst 9(4):450–459
108. NanoFeel, Switzerland (2007) http://www.nanofeel.com
109. Nanonis GmbH, Switzerland (2007) http://www.nanonis.com
110. Nellen PM, Bronnimann R (2006) Meas Sci Technol 17(5):943–948
111. Nellen PM, Callegari V et al (2006) Microelectron Eng 83(4–9):1805–1808
112. Neumeister JM, Ducker WA (1994) Rev Sci Instrum 65(8):2527–2531
113. Nishio M, Sawaya S et al (2005) Appl Phys Lett 86(13):133111
114. Nishio M, Sawaya S et al (2005) J Vac Sci Technol B 23(5):1975–1979
115. Noy A, Vezenov DV et al (1997) Annu Rev Mater Sci 27:381–421
116. Orloff J (1997) Handbook of charged particle optics. CRC, Boca Raton
117. Orso S, Wegst UGK et al (2006) J Mater Sci 41(16):5122–5126
118. Orso S, Wegst UGK et al (2006) Adv Mater 18(7):874
119. Palaci I, Fedrigo S et al (2005) Phys Rev Lett 94(17):175502
120. Perentes A, Bachmann A et al (2004) Microelectron Eng 73–74:412–416
121. Poncharal P, Wang ZL et al (1999) Science 283(5407):1513–1516
122. Prorok B, Zhu Y et al (2004) In: Nalwa HS (ed) Encyclopedia of Nanoscience and Nanotechnology. American Scientific, Valencia, CA
123. Rabe U, Turner J et al (1998) Appl Phys A 66:S277–S282
124. Randolph SJ, Fowlkes JD et al (2006) Crit Rev Solid State Mater Sci 31(3):55–89

125. Reimer L (1998) Scanning electron microscopy physics of image formation and micro-analysis. Springer, Berlin
126. Resch R, Baur C et al (1998) Langmuir 14(23):6613–6616
127. Reyntjens S, Puers R (2001) J Micromechan Microeng 11(4):287–300
128. Sader JE (2002) In: Hubbard AT (ed) Encyclopedia of surface and colloid science. Dekker, New York
129. Sader JE, Larson I et al (1995) Rev Sci Instrum 66(7):3789–3798
130. Sader JE, Chon JWM et al (1999) Rev Sci Instrum 70(10):3967–3969
131. Salvetat JP, Briggs GAD et al (1999) Phys Rev Lett 82(5):944–947
132. Sandberg R, Svendsen W et al (2005) J Micromechan Microeng 15(8):1454–1458
133. Sauerbrey G (1959) Z Phys 155(2):206–222
134. Sawaya S, Akita S et al (2006) Appl Phys Lett 89(19):193115
135. Schaefer DM, Reifenberger R et al (1995) Appl Phys Lett 66(8):1012–1014
136. Schiffmann KI (1993) Nanotechnology 4:163
137. Senden TJ, Ducker WA (1994) Langmuir 10(4):1003–1004
138. Shen ZY, Shih WY et al (2006) Rev Sci Instrum 77(6):065101
139. Sievers T, Fatikow S (2006) J Micromechatron 3(3):267–284
140. Stahl U, Yuan CW et al (1994) Appl Phys Lett 65(22):2878–2880
141. Sunden EO, Wright TL et al (2006) Appl Phys Lett 88(3):033107
142. Thundat T, Wachter EA et al (1995) Appl Phys Lett 66(13):1695–1697
143. Torii A, Sasaki M et al (1996) Meas Sci Technol 7(2):179–184
144. Tortonese M, Barrett RC et al (1993) Appl Phys Lett 62(8):834–836
145. Treacy MMJ, Ebbesen TW et al (1996) Nature 381(6584):678–680
146. Troyon M, Lei HN et al (1997) Microsc Microanal Microstruct 8(6):393–402
147. Utke I, Bret T et al (2004) Microelectron Eng 73–74:553–558
148. Utke I, Friedli V et al (2006) Microelectron Eng 83(4–9):1499–1502
149. Utke I, Friedli V et al (2006) Adv Eng Mater 8(3):155–157
150. Utke I, Friedli V et al (2006) Appl Phys Lett 88(3):031906
151. Utke I, Hoffmann P et al (2000) J Vac Sci Technol B 18(6):3168–3171
152. Utke I, Hoffmann P et al (2002) Appl Phys Lett 80(25):4792–4794
153. Utke I, Michler J et al (2005) Adv Eng Mater 7(5):323–331
154. van Dorp WF, B. van Someren et al (2005) Nano Lett 5(7):1303–1307
155. Veeco Instruments (2007) http://www.veeco.com
156. Wachter EA, Thundat T (1995) Rev Sci Instrum 66(6):3662–3667
157. Walters DA, Cleveland JP et al (1996) Rev Sci Instrum 67(10):3583–3590
158. Weaver W, Timoshenko SP et al (1990) Vibration problems in engineering. Wiley, New York
159. Weck M, Huemmler J et al (1997) Proc SPIE 3223:223–229
160. Weibull W (1951) J Appl Mech Trans ASME 18(3):293–297
161. Williams PA, Papadakis SJ et al (2002) Appl Phys Lett 80(14):2574–2576
162. Xiao T, Ren Y et al (2006) Appl Phys Lett 89(3):033116
163. Yang YT, Callegari C et al (2006) Nano Lett 6(4):583–586
164. Yasumura KY, Stowe TD et al (2000) J Microelectromech Syst 9(1):117–125
165. Young WC, Budynas RG (2002) Roark's formulas for stress and strain. McGraw-Hill, New York
166. Yu MF, Dyer MJ et al (1999) Nanotechnology 10(3):244–252
167. Yu MF, Wagner GJ et al (2002) Phys Rev B 66(7):073406
168. Yu XM, Thaysen J et al (2002) J Appl Phys 92(10):6296–6301
169. Yuzvinsky TD, Fennimore AM et al (2005) Appl Phys Lett 86(5):053109
170. Zhu Y, Espinosa HD (2005) Proc Natl Acad Sci USA 102(41):14503–14508
171. Zussman E, Chen X et al (2005) Carbon 43(10):2175–2185

8 Cantilever Spring-Constant Calibration in Atomic Force Microscopy

Peter J. Cumpson · Charles A. Clifford · Jose F. Portoles ·
James E. Johnstone · Martin Munz

Abstract. The measurement of small forces by atomic force microscopy (AFM) is of increasing importance in many applications. For example, in analytical applications where individual molecules are probed, or nanoindentation measurements as a source of information about materials properties on a nanometer scale. The fundamentals of AFM force measurement, and some of these applications, are briefly reviewed. In most cases absolute, not relative, measurements of forces are needed for valid comparisons with theory and other measurement techniques (such as optical tweezers). We review methods of AFM force calibration and the major uncertainties involved. The force range considered in this work is roughly from 10 pN to around 500 nN. We describe some issues of the repeatability of force measurements that can be important in common AFM instruments. In most cases the aspect that then limits the accuracy with which forces can be measured is the uncertainty in the stiffness (more specifically the normal force constant) of the atomic force microscope cantilever at the center of the instrument. It is known that commercially available microfabricated atomic force microscope cantilevers have a wide range of force constant, for cantilevers of nominally the same type and even the same production batch. Calibration is necessary, and many methods have been used over the years. We compare the accuracy that can be achieved and the ease of use of these different methods, including theoretical (dimensional), thermal, static and dynamic methods and their variants. A device developed at NPL should help overcome many of the problems of force constant calibration, at least for the most common AFM configurations. This is a microfabricated silicon device, which, because of its very small mass, is not susceptible to vibration as a larger device would be. A new calibration method based on electrical and Doppler measurements allows the calibration of the force constant of this device traceable to the SI newton. It can then be sent to AFM users for straightforward calibration of AFM force constants. We conclude with a brief discussion of the special problems of calibration of lateral forces, such as those obtained in frictional force measurements.

Key words: Atomic force microscopy, Cantilever spring constant, Force calibration, Force measurement, Force–distance curve, Microelectromechanical systems, Lateral force microscopy, Torsional spring constant

8.1
Introduction

The defining feature of atomic force microscopy (AFM) is that it is based on the measurement or maintenance of a small force, compared with traditional microscopies that are based on light or electron scattering by the sample. AFM was proposed [1] in 1986 as a way to extend the capabilities of scanning tunneling microscopy to imaging of nonconducting surfaces, and initially at least, the magnitude of the force involved was less important than the fact that it was kept constant during imaging. In the last two decades AFM has been continuously developed and has found a wide range of applications–many of them now requiring this force to be measured.

As a force sensing probe, AFM uses a flexible cantilever with a sharp tip at its free end. The tip is brought near or into contact with the surface of the sample, where the tip–surface interaction force causes the cantilever to bend. This bending gives a measurement of the force. The height and the lateral position of the tip over the surface are controlled by a piezo stage that enables positioning with nanometric precision. Initially AFM found applications in surface profiling in which the tip is raster-scanned over the surface in order to build up information on topography with subnanometric vertical resolution. Later on AFM was applied to force measurements where the tip is scanned vertically over a point on the surface while the cantilever bending is simultaneously recorded. A measurement of the force can be obtained by considering the cantilever as an elastic spring and applying Hooke's law:

$$F = k\delta , \tag{8.1}$$

where F is the force, k is the spring constant or normal stiffness of the cantilever and δ is the deflection of the cantilever at the position of the tip which works as a measurement of the cantilever bending. In this way, an accurate force measurement depends both on the possibility to measure δ and on our ability to obtain an accurate estimation of the cantilever's spring constant. Different methods have been developed in order to measure the cantilever deflection, including vacuum tunneling [2–4], optical interferometry [5–10], and electronic detection methods such as piezoresistive [11], piezoelectric and capacitance [12] measurement. Currently, though, the method of most extended application is the optical lever [13–15] which is implemented in most commercial atomic force microscopes. The optical lever is more sensitive in comparison with most of the electronic detection methods [16], and at the same time is easier to implement and integrate into the atomic force microscope than vacuum tunneling or interferometric detection methods. In this a laser beam is reflected on the back of the cantilever before reaching the center of a split-segment photodiode. The surface curvature of the cantilever under bending causes the laser spot at the photodiode to displace from its center, producing a proportional electronic deflection signal. In order to derive the actual deflection at the tip from the deflection signal it is necessary to calibrate the sensitivity of the optical lever, or in other words, to measure the photodiode deflection signal when the cantilever undergoes known small deflections. Sensitivity calibration is normally done by pressing the cantilever against a rigid surface by means of the Z-position piezo that controls the probe's height. Surface rigidity ensures that the vertical position of the tip is fixed during contact, while the vertical position of the cantilever base and consequently cantilever bending are controlled by the piezo. The Z-displacement is measured preferably with a built-in displacement sensor. The lever sensitivity depends on the cantilever geometry and on its relative positioning with respect to the laser beam and the photodiode; therefore, optical lever sensitivity must be calibrated every time a new cantilever is mounted or repositioned into the microscope. It must be emphasized that the validity of the sensitivity calibration is limited to those cases where the cantilever bends owing to a vertical load being applied at the tip. When the cantilever bends under different loading conditions, i.e., surface stress or free vibrations, further corrections become necessary in order to reconstruct the actual deflection from the photodiode signal [34].

8.2
Applications of AFM

AFM was initially used in nonconductive surface imaging applications probing subnanometric vertical resolution in a variety of surfaces [2–8,10,13–15], measuring atomic-scale friction for the first time [8] and nanomanipulation [9]. Demonstration that AFM images can be recorded in liquid environments [3], in particular in aqueous solution [15], and imaging of organic monolayers [4] opened the way for biological applications. Regarding surface imaging, we can distinguish two different categories of imaging modes in which the atomic force microscope can operate: contact and dynamic modes. In the contact mode the tip is in more-or-less continuous contact with the surface of the sample, and in order to keep a constant force on the tip the deflection signal is fed into a feedback loop that controls the Z (height) piezoelectric actuator as the tip raster-scans the sample surface. This allows one to obtain an accurate topography of the sample in terms of the Z position of the cantilever. The lateral resolution that can be achieved by this technique is the order of 1 nm and the vertical resolution is the order of 0.1 nm, for typical state-of-the-art commercial instruments.

The dynamic modes are based on the resonance properties of the cantilever. In intermittent contact mode or amplitude-modulation AFM [6] the cantilever is driven to resonance frequency through piezoelectric actuation. The tip–surface interaction produces a shift in the cantilever's oscillation amplitude and phase, which are recorded during the surface scanning. The cantilever amplitude of vibration provides the feedback signal to be kept constant during the scanning. This imaging mode is particularly useful for obtaining high-resolution images of biological materials in air or liquid. A similar method, noncontact AFM or frequency-modulation AFM [17], instead keeps constant the shift in resonance frequency across the surface scanned. This has found fundamental applications in surface measurements in ultrahigh vacuum. Details about dynamic methods can be found in García and Pérez [18].

AFM probes can be used to measure a wide range of physical interactions between the tip and the sample, including steric or contact forces, van der Waals forces [19, 33] and colloidal forces [20]. This capability finds applications also in the measurement of elastic properties of materials like Young's modulus, tribology and chemical force microscopy. Tips can be also functionalized with relevant biomolecules in order to investigate on a single-molecule basis properties of biological materials strongly linked to their function in living systems. An illustration of this approach is the identification and measurement of specific bonds strength (e.g., between complementary DNA strands [21–23], antibody–antigen interactions [24, 25] and ligand–receptor pair interactions [26, 27]), the measurement of elastic properties of single polymer molecules [28, 29] and the measurement of the forces that determine protein folding [30]. As mentioned above, however, accurate force measurements with the atomic force microscope rely on the ability to perform an accurate estimation of the cantilever's spring constant.

8.3
Force Measurements and Spring-Constant Calibration

Accurate calibration of spring constants of atomic force microscope cantilevers has often presented a considerable challenge. Part of the problem comes from their fabrication processes. The first atomic force microscopes used cantilevers made from a metal wire with a sharpened end as the tip. Soon it became evident that there was a need to produce cantilevers with simultaneously lower stiffness, higher resonance frequencies and shorter lengths [47]. This naturally led to wafer-scale microfabrication of the cantilevers by similar microlithography techniques as used for the manufacture of micro electromechanical systems (MEMS) [47]. These techniques have excellent lateral resolution, but the thickness of the microfabricated cantilevers has a substantial fractional uncertainty. Unfortunately the spring constant of a cantilever is strongly dependent on its thickness, increasing in this way the uncertainty in the determination of the spring constant. Additional problems related to the uniformity of material properties in those cantilevers made with silicon nitride [31] also contribute to variations in the cantilever spring constant between production batches, and even inside the same wafer [37]. This has led to the development of different methods in order to enable the calibration on a per cantilever basis. A wide variety of methods have been used to calibrate normal spring constants [32], including thermal vibrations [33–36], reference cantilevers of measured dimensions [37–39], radiation pressure [40, 41], use of a nanoindentation apparatus [42], capacitive sensors [43], structural analysis [45, 47–49] and cantilever dynamics models [56–62] and a differential pressure resulting from a know fluid flour rate [55]. Most of the existing methods fall in the following categories: theoretical methods, dynamic methods, thermal methods and methods based on force standards. There follows a discussion of the methods that are currently most widely used.

There are two main shapes of cantilever, the rectangular "diving board" shape and the triangular or V shape. The latter has proven to be the most popular design primarily owing to the belief that it has usefully high lateral stiffness for a given normal spring constant, although this has recently been questioned [44]. V-shaped cantilevers also have practical advantages in some cases (such as retaining a sphere near the apex to form a colloidal probe) so it seems likely they will continue to be used for many years. It should be noted that the V-shaped cantilevers vary slightly in basic shape and design between manufacturers and some are gold or aluminium coated to promote reflection as part of an optical lever. These factors all influence the spring constant of the atomic force microscope cantilever.

8.3.1
Theoretical Methods

So-called theoretical methods aim to calculate the spring constant of the cantilever from its geometry and material properties such as E (Young's modulus) or v (Poisson's ratio). This approach usually starts with the choice of a suitable model for the cantilever deflection, such as a Bernoulli beam for rectangular cantilevers or the clamped elastic plate for V-shaped cantilevers. The solution for the static deflection under end loading often involves numerical simulations, as exact analytical solutions are not often possible. The simulations for a specific cantilever geometry are typically repeated for a range of parameters (such as thickness and length) appropriate for that particular cantilever design geometry [45]. This helps build a picture of the dependence of the spring constant on the cantilever dimensional parameters and material properties [46]. Sometimes an analytical approximation to the model allows an exact solution to be obtained and in that case the error in the approximation can be assessed with respect to the more accurate numerical solutions. Rectangular cantilever bending can be modeled accurately by a Bernoulli beam, where the spring constant is given by

$$k_E = \frac{Ebt^3}{4L^3} \; ,$$

(8.2)

where k_E is the spring constant, E is the elastic modulus, b, t and L are the width, thickness and length of the beam, respectively. The main inconvenience presented by the theoretical methods is their dependence on the cantilever thickness and Young's modulus, which are both important sources of uncertainty. Furthermore, the dependence of the spring constant on the cantilever geometry makes it necessary to develop and solve entirely different models for different cantilever shapes.

8.3.2
V-Shaped Cantilevers and the Parallel-Beam Approximation

V-shaped cantilevers were initially thought to be properly described by the parallel-beam model that assumes the stiffness of the cantilever to be equal to that of two rectangular cantilever plates placed in parallel [47]. The accuracy of this approximation was later questioned [45], but a new version of the same approach [48] that provides values close to finite-element analysis (FEA) showed that the validity of the approximation depends on how the dimensions of the cantilever are defined. In another approach, the normal and lateral stiffnesses of V-shaped cantilevers are analyzed by adoption of a model that decomposes such cantilever geometry into two elements [49]: the first is a set of two prismatic beams that models the arms of the cantilever; the second is a triangular plate in order to model the cantilever end. The method proposed by Sader and White [45] develops a theoretical cantilever plate model for the V-shaped cantilever. Since the general equation of the cantilever plate does not allow an analytical solution, a zeroth-order approximation is proposed that assumes no inelastic curvature for the cantilever plate. Then, one compares this approximation with a full FEA carried out over a wide range of values for A (aspect ratio), d/b (width ratio) and v (Poisson's ratio). The FEA results are also compared with the parallel-plate model, showing an error of up to 80% depending on the exact geometry.

Hence, there are three main ways for calculating the spring constants of V-shaped cantilevers within the dimensional methods. These are, in increasing levels of complexity, (1) the parallel-beam approximation (PBA), where the V shape is modeled as two rectangular cantilevers in parallel, (2) detailed models that split the cantilevers into two or more shapes and (3) FEA.

Finite Element Analysis (FEA) should provide the most accurate solution. Hence, all the seven models available in the literature were compared to FEA [46] as a fuction of increasing the length of the legs of the cantilever, as shown in Fig. 8.1. It is shown that those of Neumeister and Ducker are closest in agreement with the FEA results, but that an important revision of their equation is necessary for the bending of the triangular portion of the V-shaped cantilever. This revision and the necessary equations are detailed in the paper by Clifford and Shea [46]. Figure 8.1 shows that as the length of the cantilever increases, the revised solution provides an excellent match to the FEA and never deviates by more than 0.8% in the region of typical V-shape lengths. Hence, for a V-shape cantilever, this modified Neumeister and Ducker Equation [46] gives the most accurate result. With careful measurement of the dimensions, the uncertainty in k_z can be around 11%.

Equations are also available [46] to account for the effect of a metallised layer on k_z of a cantilever and for the imaging tip not being at the cantilever apex.

Ultimately the estimation of cantilever spring constant via careful dimensional measurements is limited by the accuracy of those measurements and the cost of making them. The cubic dependence of the spring constant on thickness is typically the largest contributor to the overall error budget, even given access to special measurement equipment such as calibrated scanning electron microscopes. What is more, the material properties of the cantilever always need to be assumed, and are never checked, in any purely dimensional approach.

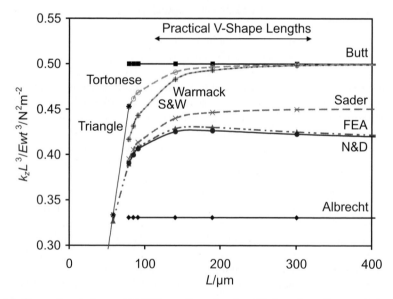

Fig. 8.1. Comparison between ($\triangle$) FEA results and various V-shaped cantilever equations for $k_z L^3 / E w t^3$ as the length of the legs, L, is varied for ($\bullet$) Neumeister and Ducker, (+) Warmack, ($\square$) Butt, (o) Tortonese, ($\blacksquare$) Sader and White, ($\times$) Sader and ($\blacklozenge$) Albrecht

8.3.3
Dynamic Experimental Methods

Dynamic methods combine a theoretical model of the cantilever's dynamics with an experimental measurement of a dynamic property such as resonance frequency. Sader et al. [56] modeled the cantilever as an elastic plate in order to derive its resonance properties. The problem is scaled on the basis of the form of the equations describing static and dynamic deflections of plates, defining the normalized effective mass of the cantilever as

$$M_e = \frac{k}{m\omega^2} , \tag{8.3}$$

where k, m and ω are, respectively, the spring constant, the mass and the resonance frequency of the cantilever. M_e is shown to be a length scale invariant [56] that can be determined via a FEA for cantilevers of different geometries and aspect ratios. In this way the spring constant can be determined from the dimensions (via FEA), mass and unloaded resonance frequency of the cantilever. This method has the advantage of being independent of the Young's modulus of the material, but it still depends on the cantilever thickness. Moreover it assumes that the cantilevers are in a vacuum. However it has been shown that damping introduces a shift in the resonance frequency [47, 56] experimentally estimated to be around 4% for V-shaped cantilevers and 2% for rectangular cantilevers for the resonance frequency in a vacuum respect to its value in air, for typical atomic force microscope cantilever dimensions. This corresponds to an increase between 4 and 8% in the predicted value of the spring constant [56]. Accurate treatment of atomic force microscope cantilever dynamics in viscous fluids offers a considerable challenge. Sader et al. [57, 58] developed a detailed mathematical description of the resonance response of prismatic cantilevers in viscous fluids under an arbitrary driving force. Such a description is based on the calculation of the hydrodynamic function of the cantilever $\Gamma(\omega)$, which depends on the geometry of the cantilever section and can be determined either analytically or numerically for every section shape of a prismatic cantilever. The cases for circular and rectangular sections have been developed analytically [57]. On the basis of this theory a simple method was proposed [59] that is valid for rectangular cantilevers in viscous fluids. It is summarized in the expression

$$k = 0.1906 \varrho_f b^2 L Q_f \Gamma_i (\omega_f) \omega_f^2 , \tag{8.4}$$

where ϱ_f is the density of the fluid, b and L are, respectively, the width and length of the cantilever, Q_f is the quality factor of the resonance in fluid, ω_f is the resonance frequency of the fundamental mode in the fluid and $\Gamma_i(\omega)$ is the imaginary component of the hydrodynamic function. Therefore, it provides a measurement of the spring constant independently of the density and thickness of the cantilever. The method can be indirectly extended to other cantilever shapes when a rectangular cantilever is present in the same chip [59] by assuming their thicknesses and material properties to be identical–an assumption that may or may not hold. A generalization of this method to the stiffness of an elastic body of arbitrary shape in a fluid was recently given by Sader et al. [60].

Cleveland et al. [61] proposed a method based on the measurement of the resonance frequency shift upon placement of tungsten spheres of different masses at the tip of the cantilever. The equation that determines the spring constants is as follows:

$$k = (2\pi)^2 \frac{M}{\left(\frac{1}{v_1^2}\right) - \left(\frac{1}{v_0^2}\right)} \, , \tag{8.5}$$

where M is the added mass and v_1 and v_0 are, respectively, the resonance frequency after and before the addition of the sphere. The method can also be applied for the consecutive addition of a series of masses by fitting the added masses and corresponding resonance frequency to the dependence

$$M = k(2\pi v)^{-2} - m^* \, , \tag{8.6}$$

where M is the added mass and v the resonance frequency. The fit will determine the values for k (spring constant) and m^* (the effective mass of the cantilever or the mass of the equivalent harmonic oscillator). This method has the advantage of being independent of cantilever-geometry, thickness and Young's modulus. However, the method is time-consuming and several problems limit its accuracy: the measurement of the added masses is proportional to the cube of the diameter of the tungsten spheres. This introduces an important source of uncertainty, especially for smaller spheres [61]. Moreover the accurate positioning of the tungsten spheres is difficult, which has been proposed as the main source of uncertainty for this method [56]. Additionally the method is potentially destructive since often glue is needed to attach the tungsten spheres to the tip [56, 62] with the consequent difficulty in restoring the tip surface to its original state. Cleveland et al. [61] propose also an approximate expression in order to estimate the spring constant from the unloaded or free resonance frequency of the cantilever. It is summarized in the following equation:

$$k = 2\pi^3 L^3 b \sqrt{\frac{\varrho^3}{E}} v_0^3 \, , \tag{8.7}$$

where L and b are, respectively, the length and width of the cantilever, ϱ and E are the mass density and Young's modulus of the material, and v_0 is the unloaded resonance frequency. This approximation though assumes a rectangular beam geometry or a double rectangular plate approximation for V-shaped cantilevers, which has been shown to be unsuitable for describing the stiffness of such cantilevers [45] unless the dimensions of the cantilever are redefined [48]. Gibson et al. [62] also propose a dynamic method involving mass addition; however, the mass here is added by depositing a gold layer on the back of the cantilever. Like the tungsten spheres of the method of Cleveland et al. the newly deposited gold layer produces no modification of the cantilever stiffness and its only contribution to the shift in resonance frequency is due to mass addition. This is true when the gold layer is thin enough as shown by Sader et al. [56]; however, this assumption becomes invalid above a certain thickness of deposited gold. This limiting gold thickness was established to be approximately 58 nm following a test over a set of different atomic force microscope cantilevers [62].

In contrast to the method of Cleveland et al., where the added mass is attached to the tip, the mass of the deposited gold layer is distributed across the whole area of the lever. This makes it necessary to consider the effective mass affecting the resonant vibrations of the lever instead of the actual added mass. Effective masses for different geometries of atomic force microscope cantilevers need to be calculated using numerical methods [56].

8.3.4
Thermal Methods

Thermal methods are based on models that link the spring constant of the cantilever to the thermal mechanical vibrations that the cantilever undergoes at thermal equilibrium. Such an approach was proposed initially by Hutter and Bechhoefer [33]. This method models the thermal vibrations of the cantilever with a simple harmonic oscillator at thermal equilibrium, which has a Hamiltonian of the form

$$H = \frac{p^2}{2m} + \frac{1}{2}m\omega_0^2 x^2 , \qquad (8.8)$$

where p is the oscillator momentum, x its displacement, m its mass and ω_0 its resonant angular frequency The Hamiltonian's quadratic form allows the application of the equipartition theorem [64], which states that at thermal equilibrium the average value of each quadratic term is given by $k_B T/2$, where k_B is Boltzmann's constant and T is the absolute temperature. Applying this to the potential term leads to the identity

$$\left\langle \frac{1}{2}m\omega_0^2 x^2 \right\rangle = \frac{1}{2}k_B T . \qquad (8.9)$$

As $\omega_0^2 = k/m$, the spring constant can be expressed as a function of the mean square displacement:

$$k = \frac{k_B T}{\langle x^2 \rangle} . \qquad (8.10)$$

Analyzing the displacement in the frequency domain, one can measure the mean-square displacement as the integral of the power spectrum under cantilever resonance (previously having subtracted the background noise corresponding to other noise sources–a process which is often difficult or subjective unless the best electronics and vibration isolation are available). Then the spring constant is given by

$$k = \frac{k_B T}{P} , \qquad (8.11)$$

where P is the area of the first resonance mode of the cantilever under the power spectral density curve. Hutter and Bechhoefer's method is based on the measurement of the fundamental mode of vibration, assuming that the contribution of higher modes to the power of the vibration is small in comparison. Butt and Jaschke [34] extended the calculation of the thermal noise to all the vibration modes of a rectangular cantilever, assuming also thermal equilibrium and using the equipartition theorem. They showed that the contribution of the higher modes cannot be ignored, yet their expressions are still valid if the total power spectrum of the cantilever is considered. It still remains possible to calculate the cantilever's spring constant from a single mode of vibration through a correction factor when the relative contribution of that mode to the total power of the thermal vibration is known. However, this introduces something of a problem for the method, since such relative contributions are dependent on cantilever geometry, limiting the validity of these results to rectangular cantilevers. In order to extend the method to another cantilever type, the correction factors must be determined for the particular geometry of that type.

Butt and Jaschke also introduced an important correction for the case of thermal vibrations being measured by an optical lever system. This new correction reflects two facts: first, the optical lever records deflection *angles* of the cantilever's surface instead of true linear displacement and, second, the cantilever bends into a different shape when it undergoes free vibrations from the shape it bends into when being statically bent during sensitivity calibration of the optical lever. This last part of their analysis is valid only for rectangular cantilevers, and again it needs to be reproduced for other cantilever geometries, despite their being less easy to deal with analytically. In this respect Stark et al. [65] determined the shape of the first ten modes of vibration for a particular type of V-shaped cantilever by means of a FEA, finding a systematic deviation of 7% in the spring constant from the assumption of rectangular shape [65].

Another thermal method was derived by Burnham et al. [36] by fitting the probability density distribution of a simple harmonic oscillator to the power spectrum of the first vibrating mode and again applying the assumptions of thermal equilibrium and the equipartition theorem. Instead of measuring the area under the resonance peak, this method relies on measurements of resonance frequency, resonance peak height and quality factor. It proved to be equivalent to Hutter and Bechhoefer's method but showed a smaller bandwidth limitation and fewer problems of noise substraction [36].

The main advantages of the thermal approach are its independence from the material properties of the cantilever as well as the possibility of being integrated into the atomic force microscope without additional equipment [22] in order to obtain a calibration in situ. In fact its use is currently available in most commercial AFM equipment, often as a routine within the operating software. Moreover this method can provide approximate values in liquid environments [36] with deviations due to viscosity and mass loading. The last feature is useful in measurements of biological samples with functionalized cantilevers that need to be kept in an aqueous environment. Another relevant feature of the thermal approach is that it relies on basic and well-established physics. Among the disadvantages, we can mention that the method depends on the optical lever sensitivity calibration, and this is an important source of uncertainty [36]. Dependence on the laser spot size and position has been also identified as a source of uncertainty [67]. It also appears to give systematic shifts in spring constant values for stiffer cantilevers in comparison with other methods [22,36], possibly because the thermal vibrations are smaller and might lead to overestimation of their contribution to the measured spectral power [22]. Another potential practical limitation for stiff cantilevers is the maximum sampling rate of the instrument. From the Nyquist theorem the sampling rate needs to be twice as high as the maximum frequency of the measured spectrum to allow a fast Fourier transform of the time domain data [66]. In practice, though, most signal-processing experts would recommend an even larger frequency range to avoid aliasing, and this may mean a wider frequency range is needed than in normal AFM operation and therefore beyond what the amplifier electronics has been designed to cope with. Also it must be pointed out that despite the thermal method initially being proposed as independent of the cantilever geometry, it has become clear that different corrections are needed that do depend on the geometry, and which require for their calculation the development of theoretical models of the cantilever dynamics [34, 65]. Considering all of these factors, number of estimates of the uncertainty of thermal methods have been published, lying in the range 10–20% [36, 67, 68].

8.4
Repeatability in AFM Force Measurements

We now consider issues of measurement repeatability as they apply to force measurements by AFM, in order to compare repeatability with the accuracy that can be achieved by the many methods of cantilever spring-constant calibration discussed in the previous sections.

The ability to perform highly repeatable quantitative measurements lies at the heart of a good instrument. By understanding the repeatabilities of different aspects of the measurement process, techniques can be refined and improved. By recording the repeatability of force measurements in the atomic force microscope, one can find experimental conditions where quantification is most accurate. Defining the repeatability of measurements in the atomic force microscope enables comparison with other force methods such as nanoindentation. It also allows judgements to be made as to the suitability of using such a technique for routine analysis, and has been the subject of AFM work at NPL in recent years [69].

Common measurements undertaken by AFM are force–displacement curves, the determination of the modulus for materials which generally have values less than 5 GPa and mapping of material stiffness. This range of materials includes most organic polymers [70] and soft biological samples, which hitherto have been very difficult to measure with traditional indentation equipment. The advantage in using AFM is the increased sensitivity with soft materials because of the availability of compliant microfabricated cantilevers and the ability to measure forces down to around 10 pN, i.e., below the force required to break a single chemical bond. This leads to the possibility of measuring bond-breaking forces in single molecules under certain conditions.

An essential concern for the AFM user is to consider how repeatabilities vary with time. A main issue is how the force measurement drifts over a few hours given the same experimental conditions and the same operator. Small drifts in the alignment of optical levers and drifts in the piezoelectric scanner can affect final measurements considerably; therefore, careful instrumental design is a critical factor. In studies at NPL the repeatability and quantification of noise levels and hysteresis of key measurable parameters have been measured [69]. The results show that the development of stable closed-loop feedback systems and careful design of instrument electronics are essential to ensure the best-quality force spectra.

Methods of normal force calibration are summarized and compared in Table 8.1.

8.4.1
z-Axis Displacement Repeatability

The typical translation stage used in an atomic force microscope is a piezoelectric tube scanner. The vertical displacement is controlled by an applied voltage to the appropriate piezoelectric segment. It is well known that piezoelectric materials exhibit creep, hysteresis and aging in their performance. This can result in large errors in measuring the z-displacement accurately. The development of capacitive, optical knife edge and linear variable differential transformer sensors has greatly improved closed-loop linearity. These have greater accuracy and linearity but at the cost of precision and are therefore not used when ultimate high-resolution scanning is required, typically below around 1 μm field of view. The repeatability of the open-loop displacement will vary according to the maximum extent of the displacement. Typical hysteresis can be up to 20%, translating into a repeatability of the same order.

8.4.2
Cantilever Deflection Repeatability

The cantilever deflection, δ, given as a voltage output can be measured with precision toward subangstrom displacements using optical beam and position-sensitive diode detection, as implemented in typical commercial atomic force microscopes. The measurement of cantilever deflection is prone, however, to instrument drift due to several effects. These include cantilever heating arising from imperfect reflection resulting in cantilever bending, and drift in laser alignment with time. To address the first problem, cantilevers can be coated with a thin layer of gold to increase the reflectivity. Unfortunately this overlayer forms part of an inconvenient bimetallic strip which will bend when heated because of the differing coefficients of expansion of the components, therefore aggravating the second problem. This causes the cantilever to bend downward, inducing a change in the position of the laser spot hitting the optical detector. This may tend to introduce an offset in the apparent cantilever deflection that is usually accommodated by fine adjustment of the detector by the user when setting up the atomic force microscope for imaging. Soft cantilevers are affected most by this because the heat-induced expansion stress must be balanced by a greater displacement of the cantilever. It may be expected that thermal equilibrium would be achieved rapidly owing to the small size of the cantilever, but in practice this is very dependent upon the design of the instrument. Thermal heating of the head assemblies and natural drift and settling of the alignment screws means that over the course of a few hours, the laser spot can, in some cases, move across the cantilever, affecting both the total intensity and the deflection constant. As an example, a typical stiff cantilever ($k_c = 3\,\mathrm{Nm}^{-1}$) was left for 2 h [69]. The laser spot moved across the cantilever by 15 μm during this time, causing the measured z-position to change by approximately 100 nm. This was the most critical drift component when assessing all the repeatabilities in force measurements from this cantilever.

Table 8.1. Summary of significant aspects of the published spring constant determination technique approaches

	Dimensional			Static experimental				Dynamic	Dynamic experimental	
	Butt [55]	Modified N&D*	Finite-element analysis	Static mass hanging	Cantilever on reference cantilever	Cantilever on reference spring	Nanoindenter	Dynamic mass attachment	Thermal noise analysis	Resonance frequency in air
Uncertainty of method (%)	25	11	10	15–25	10	5	8	15–25	15–20	15–20
Main uncertainty	Equation t and E and t	t and E	t and E	Mass attached	Reference cantilever	Contact mechanics of working lever on reference spring	Nanoindenter calibration	Mass attached	Interference from other noise sources	Reynolds number for the fluid-cantilever system
Potential damage to tip	Low	Low	Low	High	Medium	Medium	Low	High	Low	Low
User-friendliness	High	High	Medium	Poor	Medium	High	Medium	Poor	High once set up	Medium
Advantages	Simple equation	Simple equation and accurate	Accuracy	No geometry and E data needed	Simple idea, geometry- and coating-independent	Simple, potentially traceable to SI	Simple, quick	No geometry data needed	Only frequency and T data needed	Simple
Disadvantages	Not a good model, t and E needed	t and E needed	Finite-element analysis programs expensive, complex, t and E needed	Accurate placement of spheres, calibrated spheres, and calibrated deflection needed	Difficult to set one cantilever on other accurately. Need $k_{ref}k_{working}$	Contact method	Difficult to place indenter tip in correct position. Only for $k_z > 1\,Nm^{-1}$	Need accurate placement of masses. Calibrated spheres needed	Confusion over equation constant. Only for low-k_z cantilevers. Model validity unknown	Only for rectangular cantilevers

* Neumeister and Ducker [46]

The use of a silicon wafer as a test substrate is an excellent way to assess the key parameters of instrumental repeatability. One advantage is the good consistency across the surface coupled with a very low surface roughness. Care must be taken, however, to ensure that the sample is clean so that correct measurements can be made. With use of this substrate it has been shown that the force repeatability of a good commercial instrument is around 1.6% in the acquisition of force–distance curves and the is 25% in imaging [69]. Finally, when the instrument has been properly calibrated and characterized it is possible to perform nanoindentation experiments which have a repeatability comparable to that of traditional nanoindenters. In fact, these two techniques complement each other, with AFM providing sensitivity for soft materials and traditional indenters providing sensitivity for stiffer materials. The overlapping region for both AFM and nanoindentors covers most polymers.

8.5
Microfabricated Devices for AFM Force Calibration

Many of the advantages of AFM in precise force measurement stem from the small size –and indeed mass– of the atomic force microscope cantilever. In AFM imaging, a larger and more massive cantilever than those we use today would mean slower scanning, owing to the necessarily lower resonance frequency for a given force sensitivity. Therefore, when one comes to devise a calibration method whereby a small, known force is applied to calibrate an atomic fore microscope cantilever, the smaller the device the more suitable it is likely to be. One can, of course, build a large instrument capable of applying a small force; however, its size and mass would mean that it would be extremely sensitive to vibration, and indeed it may be a better seismometer than an AFM calibration device. Attention has therefore turned, among a number of researchers, toward devices for atomic force microscope cantilever calibration which, like the atomic force microscope cantilevers themselves, are microfabricated.

The first microfabricated devices for atomic force microscope cantilever calibration were reference cantilevers of carefully controlled dimensions produced by Tortonese and Kirk [37]. These were later made commercially available. They are rectangular cantilevers fabricated with a special etch-stop to control their thickness, minimizing the uncertainty in the calculation of their spring constants by dimensional methods. One potential source of uncertainty is accurate location of the position of the atomic force microscope tip along the length of the reference cantilever, which generally requires additional optics. From this point of view, ease of use can be improved by methods that make such optical measurement unnecessary, for example:

1. By fabrication of arrays of reference springs [38], each with a single, well-defined spring constant, or
2. By increasing the size of the carefully fabricated reference cantilever and providing a lithographic "ruler" for accurate measurement of the position of the atomic force microscope tip along its length [39].

An alternative approach has been taken more recently by some of the authors of this chapter. Rather than control the fabrication process to an exquisite level of accuracy, devices have been fabricated using standard methods, to a design that allows subsequent calibration traceable to the SI measurement system [71]. The calibration method measures the spring constant of these artifacts, by a combination of electrical measurements and a type of interferometry, Doppler velocimetry, that is particularly suited to traceable dynamic measurements well below 1 μm. The devices themselves are fabricated by silicon surface micromachining.

The principle on which the MEMS device is based is the balancing of a mechanical restorative force applied by microfabricated springs with an electrostatic force on a capacitor, which can be determined without knowledge of the capacitor dimensions. This avoids all of the difficult measurements of small dimensions critical to many other spring constant calibration methods. Indeed, no length or thickness measurements are required at all.

The static deflection of the platform is the result of the balance between the elastic restoring force applied by the folded springs and the electrostatic force from the comb drives. The stored electrostatic field energy, E, is

$$E = \frac{1}{2}CV_p^2 \,, \tag{8.12}$$

where C is the capacitance and V_p is the potential difference across it. The electrostatic force, F_{elec}, is

$$F_{elec} = \frac{1}{2}\frac{\partial C}{\partial z}V_p^2 \,, \tag{8.13}$$

which balances an elastic force, $F_{elastic}$, of

$$F_{elastic} = kz \,, \tag{8.14}$$

where z is the static deflection.

Now consider a small a.c. drive (in addition to the d.c. bias) applied to set the platform into resonance. The current to earth, $i(t)$, is

$$i(t) = \frac{d(CV_p)}{dt} \,. \tag{8.15}$$

We now separate the capacitance of the device into two parts:

1. The dynamic capacitance, $C(z)$, which changes as the platform is displaced.
2. The static or parasitic part, C_{para}. This is the capacitance between fixed parts of the device, for example, adjacent tracks and pads on the silicon die.

If we measure the response of the device over a narrow frequency interval around the mechanical resonance, we expect the static capacitance to be constant, but the dynamic capacitance will vary with the motion of the platform.

$$i(t) = \left[C(z) + C_{para}\right]\frac{dV_p(t)}{dt} + V_p(t)\frac{\partial C(z)}{\partial z}\frac{dz}{dt} \,. \tag{8.16}$$

We apply a d.c. potential of φ_0 to the stationary part of the comb drives, together with a small a.c. component $v(t)$, so that the total voltage applied at time t is $V_p(t)$, where

$$V_p(t) = \varphi_0 + v(t) \,. \tag{8.17}$$

The purpose of the small a.c. component is to apply a small mechanical drive to the device, which, if this drive voltage is close to its mechanical resonance frequency, will cause it to vibrate mechanically with small but measurable amplitude. Typically φ_0 is chosen in the range 1–4 V, and $v(t)$ is a sinusoid of amplitude v_0 chosen in the range from 250 µV to 2.5 mV:

$$v(t) = v_0 \sin(\omega t) . \tag{8.18}$$

At each instant we have a measurement of the velocity $V(t) = V_0 \cos(\omega t + \theta)$ of the platform via Doppler velocimetry. For a given amplitude of a.c. drive, both the amplitude V_0 and the phase with respect to that drive $(\theta - \pi/2)$ vary as the drive frequency passes through resonance. We identify the Doppler velocity with the velocity (dz/dt) that appears in (8.16), to give

$$i(t) = \left[C(z) + C_{\text{para}}\right] \frac{dv(t)}{dt} + \left[\varphi_0 + v_0 \sin(\omega t)\right] \frac{\partial C(z)}{\partial z} V(t) . \tag{8.19}$$

For a particular bias voltage φ_0, and an a.c. component amplitude v_0 sufficiently small that the capacitance $C(z)$ varies linearly over the range of mechanical vibration, we obtain

$$i(t) = \left[C(z) + C_{\text{para}}\right] v_0 \omega \cos(\omega t) + \varphi_0 \frac{\partial C(z)}{\partial z} V(t) . \tag{8.20}$$

The first term on the right-hand side of (8.20) represents a parasitic capacitive current that is constant in amplitude for frequencies near the mechanical resonance, and $\pi/2$ radians in advance of the a.c. drive signal. The second term is the interesting one, because it is proportional to the capacitance gradient we wish to measure. This term has the same phase as the velocity of the mirror platform (and comb drives). At low frequencies the mirror displacement is in phase with the drive signal, whereas far above the resonance it lags by π radians. Therefore, the velocity is $\pi/2$ radians in advance of the a.c. drive voltage far below the resonance,

$$i(t) = \left[C(z) + C_{\text{para}}\right] v_0 \omega \cos(\omega t) + \varphi_0 \frac{\partial C(z)}{\partial z} V_0 \cos(\omega t) , \quad \text{for} \quad \omega \ll \omega_{\text{r}} \tag{8.21}$$

and lags by $\pi/2$ radians far above it,

$$i(t) = \left[C(z) + C_{\text{para}}\right] v_0 \omega \cos(\omega t) - \varphi_0 \frac{\partial C(z)}{\partial z} V_0 \cos(\omega t) , \quad \text{for} \quad \omega \gg \omega_{\text{r}} , \tag{8.22}$$

where $\omega_{\text{r}} = 2\pi f_{\text{r}}$ is the angular frequency of the mechanical resonance. The sharp mechanical resonance allows us to measure the magnitude of the second term of (8.22) as this phase change occurs, since the first term is essentially constant over this narrow frequency interval.

A typical plot of the amplitude of current and velocity signals is shown in Fig. 8.2. A complete analysis has involved fitting data such as those presented in Fig. 8.2 to an electrical equivalent circuit model, for which a signal analyzer with four

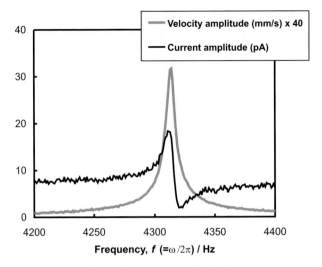

Fig. 8.2. Measured velocity amplitude, V_0, and current amplitude, i_0, for the microfabricated Watt balance device at a d.c. bias of $\varphi_0 = 3.0$ V

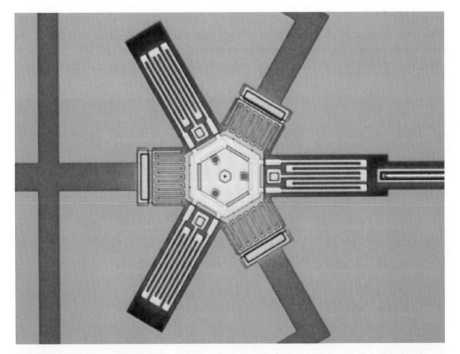

Fig. 8.3. Optical micrograph of a single tripod reference device. The device is fabricated using a three-layer polysilicon MEMS process. The central hexagonal stage (60 micrometres across) is surrounded by three electrostatic comb drives and three supporting legs. The spring constant of the reference device can be adjusted by varying the length of these three legs

channels would be necessary to achieve the most accurate results. However, since the resonance of this MEMS structure is so sharp, compared with the slow variation in parasitic capacitance with frequency, the transition from "in-phase" to "antiphase" addition of current offers us a good way to gain insight into the measurement of the spring constant. First, obtain an expression for the spring constant k by balancing electrostatic and elastic forces on the platform,

$$k = \frac{\varphi_0^2}{2\bar{z}} \frac{\partial C}{\partial z} \ , \tag{8.23}$$

where $\bar{z}$ is the time average of the displacement z. Then we obtain an expression for the gradient of capacitance $(\partial C / \partial z)$

$$\frac{\partial C}{\partial z} = \frac{i_0|_{\omega \gg \omega_r} - i_0|_{\omega \ll \omega_r}}{2\varphi_0 V_0} \ , \tag{8.24}$$

where the current amplitudes that appear at the top right-hand side of (8.24) represent the measured current amplitude above and below resonance, respectively. Using (8.24) to substitute for the capacitance gradient appearing in (8.23), we obtain

$$k = \frac{\varphi_0}{4\bar{z}V_0} \left(i_0|_{\omega \gg \omega_r} - i_0|_{\omega \ll \omega_r} \right) \ . \tag{8.25}$$

This has been applied to arrive at a traceable value of $0.193 \pm 0.01\,\mathrm{Nm^{-1}}$, in reasonable agreement with a more approximate value of $0.23 \pm 0.03\,\mathrm{Nm^{-1}}$ based on estimates of the mass and resonance frequency of the vibrating part of the device.

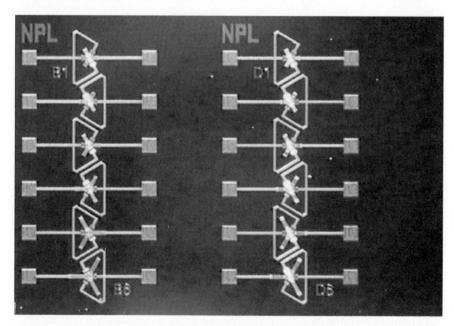

Fig. 8.4. Optical micrograph showing two arrays, each comprising six reference devices covering a range of spring constant

Early versions of the MEMS device were bipedal, i.e., have two supporting legs whose spring constant is measured using the procedure described above. More recently we have produced tripod versions of the same device, as shown in Fig. 8.3. Arrays of devices can be constructed on a single chip, as shown in Fig. 8.4, allowing the entire range of spring constants of commercial atomic force microscope cantilevers to be calibrated. The tripod design has the advantage of much reduced uncertainty resulting from of the atomic force microscope tip; the previous biped devices were prone to rotation around the axis of the two supporting legs unless the atomic force microscope tip was placed exactly on this axis (one example of the so-called corner loading error [72]). The newer tripod devices are much more forgiving of errors in tip placement; the atomic force microscope tip can be placed anywhere in the central hexagon of the device and will see essentially the same reference spring constant, as indicated by the measurements shown in Fig. 8.5.

A summary of the calibration method is shown schematically in Fig. 8.6. This emphasizes the fact that the first two steps of the calibration, although involving challenging electrical and interferometric measurements, do not take place in the atomic force microscope to be calibrated. Indeed, only the third step–comparison with the atomic force microscope cantilever in the atomic force microscope–need take place there. For the AFM user, steps 1 and 2 simply provide a calibrated reference spring, that is robust enough to arrive in their AFM lab through the post.

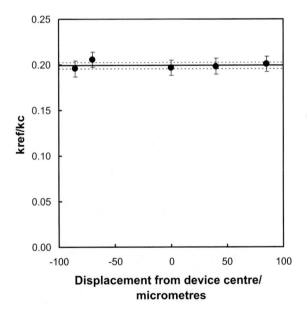

Fig. 8.5. Each data point represents the average of three force–distance curves. The *error bars* represent the average scatter of the sets of measurements on individual points on the surface. The *continuous line* represents the average of all measurements (indicating that the electrical nanobalance is 0.20 times the spring constant of the atomic force microscope cantilever) and the *dashed lines* represent the one standard deviation uncertainty limits on this value, given the measured scatter on the five individual data points. This is less than 2%

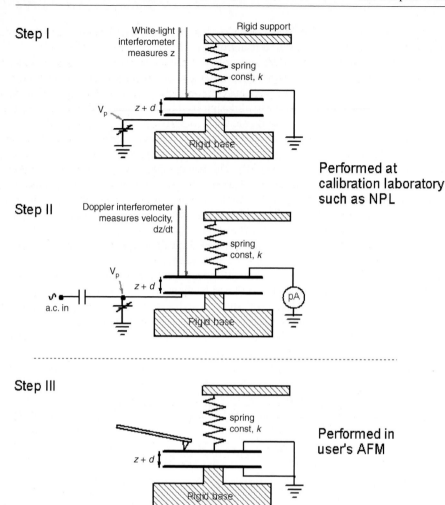

Fig. 8.6. The three-step cantilever calibration method described in this work. In steps I and II, static and dynamic measurements of the displacement of a moveable capacitor electrode, together with electrical measurements, allow the spring constant of the spring supporting that moveable electrode to be measured, potentially traceably to the SI. In step III this spring is then used as a reference spring within the atomic force microscope to calibrate the spring constant of the cantilever under test, without further electrical or other measurement. *AFM* atomic force microscope

8.6
Lateral Force Calibration

There is an increasing need for the accurate measurement of small lateral forces by AFM, in the mechanical analysis of contamination on semiconductor surfaces, polymer blends, functional thin films, recording media and measuring adhesion of nanoparticulates at surfaces [73]. The lateral force signal is useful for identifying surface composition where the materials are relatively flat but have significantly different friction characteristics [74,75]. When combined with the use of chemically functionalized atomic force microscope tips, lateral force imaging can reveal contrast between different surface species where none can be seen in any other scanned probe mode, and at a higher special resolution than that available from any other surface analytical technique. Many existing and future applications use the lateral force signal only to provide image contrast, but in many other applications the quantitative comparison of lateral force measurements is essential. This has been difficult so far, owing to the wide range of torsional constants seen in even supposedly similar cantilevers. Cantilever coatings, to improve reflectivity, or to chemically functionalize the tip, can have a significant effect on spring and torsional constants that are difficult to model. A calibration method is required.

Calibration of lateral forces is often more of a problem than calibration of normal forces [76–80]. Thermal vibrations can be useful [81], but there are few other

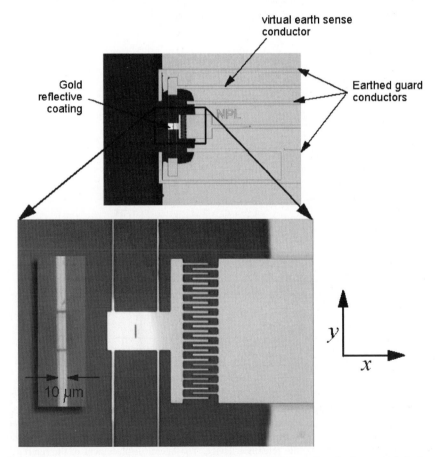

Fig. 8.7. Lateral electrical nanobalance for lateral force calibration. *Top*: In the expanded view, a side-on view of the 10 μm-thick structure

options. Many of the methods that have been tried for the purpose of normal force calibration have extensions [82] to allow the calibration of lateral forces, but some have no obvious method of being extended in this way, and are likely to be limited to the calibration of normal forces only. Those existing methods able to measure the torsional constant typically require accurate dimensional measurements (e.g., in an scanning electron microscope) or high frequency power spectrum measurement that is beyond the bandwidth of the signal amplifiers supplied as part of the atomic

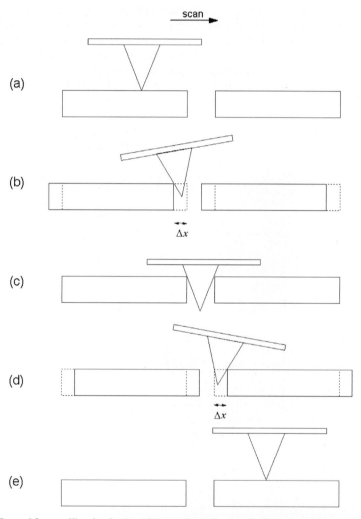

Fig. 8.8. Lateral force calibration by "continuous contact" using the lateral electrical nanobalance. In **a** the tip is being scanned across the surface of the reference device which has a calibrated lateral stiffness. In **b–d** the tip enters and leaves a narrow slit in this device, and thereby a lateral force is applied to the tip by the reference device. The lateral displacement of the reference device, Δx, is also visible in the AFM scanning data

force microscope electronics. In other words, these methods require additional facilities the AFM user may not have access to, and even if they are available, special training may be required to achieve the accuracy needed.

The most widely used lateral force calibration method, typically known as the "wedge" method, was reported by Ogletree et al. [77] around a decade ago. A torsional version [82] of the Sader method published more recently offers the prospect of eliminating any need to assume particular values for cantilever material properties or thickness. Cannara et al. [83] have developed the torsional Sader approach, and demonstrated agreement to within around 5% with the "wedge" method.

The MEMS device approach to normal force calibration described earlier in this chapter can readily be applied to direct lateral force calibration too [84]. By designing the MEMS platform to be constrained to move laterally, and placing the capacitive sensing in this plane (which is in fact a rather simpler and more linear approach than the normal force case), one can obtain calibrated springs free to deflect in the plane.

An example of one of these devices is shown in Fig. 8.7. In addition to electrical connections, this device has a narrow slit etched into the 10 μm-thick structural layer of highly doped single-crystal silicon. As shown in Fig. 8.8, an atomic force microscope tip can be scanned over this slit, and the lateral force on the tip can be used to calibrate the lateral force constant of the cantilever and the atomic force microscope optical lever in conjunction. A typical line scan is shown in Fig. 8.9, the

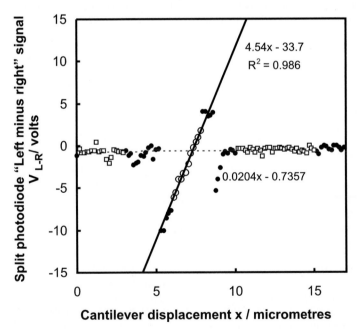

Fig. 8.9. Measured lateral force signal for continuous-contact lateral force calibration of anatomic force microscope cantilever. This should be viewed in conjunction with the diagram shown in Fig. 8.8

slope of which, together with the calibrated lateral force constant of the reference MEMS device, gives one a traceable calibration of subsequent (or indeed prior) lateral force measurements using the same cantilever.

Acknowledgements. This work forms part of the Valid Analytical Measurement Programme of the UK National Measurement System of the UK Department for Innovation, Universities and Skills (DIUS). J.F.P is grateful for PhD studentship funding from the NPL Strategic Research Programme.

References

1. Binnig G (1986) Phys Rev Lett 56:930–933
2. Albrecht TR, Quate CF (1987) J Appl Phys 62:2599–2602
3. Marti O, Drake B, Hansma PK (1987) Applied Phys Lett 51:484–486
4. Marti O, Ribi HO, Drake B, Albrecht TR, Quate CF, Hansma PK (1988) Science 239:50–51
5. McClelland GM, Erlandsson R, Chiang S (1987) Rev Prog Quant Non-destruc Eval 6:1307
6. Martin Y, Williams CC, Wickramasinghe HK (1987) J Appl Phys 61:4723–4729
7. Martin Y, Wickramasinghe HK (1987) Appl Phys Lett 50(20):1455
8. Mate CM, McClelland GM, Erlandsson R, Chiang S (1987) Phys Rev Lett 59:1942–1945
9. Stern JE, Terris BD, Mamin HJ, Rugar D (1988) Appl Phys Lett 53(26):2717–2719
10. Rugar D, Mamin HJ, Guethner P (1989) Appl Phys Lett 55(25):2588–2590
11. Tortonese M, Barrett RC, Quate CF (1993) Appl Phys Lett 62(8):834–836
12. Sarid D (1994) Scanning force microscopy. Oxford University Press, Oxford
13. Meyer G, Amer NM (1988) Appl Phys Lett 53(24):2400–2402
14. Alexander S, Hellemans L, Marti O, Schneir J, Elings V, Hansma PK, Longmire M, Gurley J (1989) J Appl Phys 65:164–167
15. Drake B, Prater CB, Weisenhorn AL, Gould SAC, Albrecht TR, Quate CF, Cannell DS, Hansma HG, Hansma PK (1989) Science 243:1586–1589
16. Hu Z, Seeley T, Kossek S, Thundat T (2004) Rev Sci Instrum 75(2):400–404
17. Albrecht TR, Grütter P, Horne D, Rugar D (1991) J Appl Phys 69(2):668–673
18. García R, Pérez R (2002) Surf Sci Rep 47:197–301
19. Hutter JL, Bechhoefer J (1994) J Vac Sci Technol 12(3):2251–2253
20. Ducker WA, Senden TJ, Pashley RM (1992) Langmuir 8:1831–1836
21. Lee GU, Chrisey LA, Colton RJ (1994) Science 266:771–773
22. Florin EL, Rief M, Lehmann H, Ludwig M, Dornmair C, Moy VT, Gaub HE (1995) Biosens Bioelectron 10:895–901
23. Strunz T, Oroszlan K, Schäfer R, Güntherodt HJ (1999) Proc Natl Acad Sci USA 96:11277–11282
24. Hinterdorfer P, Baumgartner W, Gruber HJ, Schilcher K, Schindler H (1996) Proc Natl Acad Sci USA 93:3477–3481
25. Allen S, Chen X, Davies J, Davies MC, Dawkes AC, Edwards JC, Roberts CJ, Sefton J, Tendler SJB, Williams PM (1997) Biochemistry 36:7457–7463
26. Lee GU, Kidwell DA, Colton RJ (1994) Langmuir 10:354–357
27. Florin EL, Moy VT, Gaub HE (1994) Science 264:415–417
28. Kienberger F, Pastushenko VP, Kada G, Gruber HJ, Riener C, Schindler H, Hinterdorfer P (2000) Single Mol 1(2):123–128
29. Oesterhelt F, Rief M, Gaub HE (1999) New J Phys 1:6.1–6.11
30. Rief M, Gautel M, Oesterhelt F, Fernandez JM, Gaub HE (1997) Science 276:1109–1112
31. Petersen K, Guarnieri CR (1979) J Appl Phys 50(11):6761–6766

32. Pratt JR, Smith DT, Newell DB, Kramar JA, Whitenton E (2004) J Mater Res 19:366–379
33. Hutter JL, Bechhoefer J (1993) Rev Sci Instrum 64:1868–73
34. Butt H-J, Jaschke M (1995) Nanotechnology 6:1–7
35. Lévy R, Maaloum M (2002) Nanotechnology 13:33–37
36. Burnham NA, Chen X, Hodges CS, Matei GA, Thoreson EJ, Roberts CJ, Davies MC, Tendler SJB (2003) Nanotechnology 14:1–6
37. Tortonese M, Kirk M (1997) SPIE Proc 3009:53–60
38. Cumpson PJ, Hedley J, Zhdan P (2003) Nanotechnology 14:918–924
39. Cumpson PJ, Zhdan P, Hedley J (2004) Ultramicroscopy 100:241–251
40. Sow C, Grier DG (1996) Abstract for the March meeting of the APS, 20–24 March, St Louis
41. Gallop JC (2002) UK Patent GB017812, 31 July
42. Holbery JD, Eden VL, Sarikaya M, Fisher RM (2000) Rev Sci Instrum 71(10):3769–3776
43. Scholl D, Everson MP, Jaclevic RC (1994) Rev Sci Instrum 65(7):2255–2257
44. Sader JE (2003) Rev Sci Instrum 74:2438
45. Sader JE, White L (1993) J Appl Phys 74(1):1–9
46. Clifford CA, Seah MP (2005) Nanotechnology 16(9):1666–1680
47. Albrecht TR, Akamine S, Carver TR, Quate CF (1990) J Vac Sci Technol A 8(4):3386–3396
48. Sader JE (1995) Rev Sci Instrum 66:4583–4587
49. Neumeister JM, Ducker WA (1994) Rev Sci Instrum 65:2527–2531
50. Albrecht TR, Akamine S, Carver TE, Quante CF (1990) J Vac Sci Technol A 8:3386
51. Sader JE, White E (1994) J Appl Phys 74:1
52. Sader JE (1995) Rev Sci Instrum 66:4583
53. Tortonese M (1997) IEE Eng Med Biol 16:28
54. Sarid D (1997) Exploring scanning probe microscopy with Mathematica. Wiley, New York
55. Butt H-J, Siedle P, Seifert K, Fendler K, Seeger T, Bamberg E, Wiesenhorn AL, Goldie K, Engel A (1992) J Microsc 169:75
56. Sader JE, Larson I, Mulvaney P, White LR (1995) Rev Sci Instrum 66(7):3789–3798
57. Sader JE (1998) J Appl Phys 84(1):64–76
58. Chon JEM, Mulvaney P, Sader JE (2000) J Appl Phys 87(8):3978–3988
59. Sader JE, Chon JWM, Mulvaney P (1999) Rev Sci Instrum 70(10):3967–3969
60. Sader JE, Pacifico J, Green JP, Mulvaney P (2005) J Appl Phys 97(124903):1–7
61. JP Cleveland, S Manne, D Bocek, PK Hansma (1993) Rev Sci Instrum 64(2):403–405
62. CT Gibson, BL Weeks, JR Lee, C Abell, T Rayment (2001) Rev Sci Instrum 72(5):2340
63. Lubarsky GV, Haehner G (2007) Review of Scientific Instruments 78:095102-1
64. Callen HB (1985) Thermodynamics and an introduction to thermostatistics, 2nd edn. Wiley, New York
65. Stark RW, Drobek T, Heckl WM (2001) Ultramicroscopy 86:207–215
66. Ramirez RW (1985) The FFT fundamentals and concepts. Prentice Hall, Englewood Cliffs
67. Proksch R, Schäffer TE, Cleveland JP, Callahan RC, Viani MB (2004) Nanotechnology 15:1344–1350
68. Matei GA, Thoreson EJ, Pratt JR, Newell DB, Burnham NA (2006) Rev Sci Instrum 77(083703):1–6
69. Johnstone JE, Clifford CA (2003) Force repeatability in imaging forces and force vs distance spectroscopy using an atomic force microscope, DQL report. NPL, Teddington, UK
70. Clifford CA, Seah MP (2005) Appl Surf Sci 252(5):1915–1933
71. Cumpson PJ, Hedley J (2003) Nanotechnology 14:1279–1288
72. Choi I, Kim M, Woo S, Kim SH (2004) Meas Sci Technol 15:237–243
73. Ecke S, Raiteri R, Bonaccurso E, Reiner C, Deiseroth HJ, Butt HJ (2001) Rev Sci Instrum 72:4164–4170
74. Dedkov GV (2000) Phys Status Solidi A 179:3–75
75. Carpick RW, Salmeron M (1997) Chem Rev 97:1163–1194

76. Carpick RW, Ogletree DF, Salmeron M (1997) Appl Phys Lett 70:1548–1550
77. Ogletree DF, Carpick RW, Salmeron M (1996) Rev Sci Instrum 67:3298–3306
78. Sader JE, Green CP (2004) Rev Sci Instrum 75:878–883
79. Pietrement O, Beaudoin JL, Troyon M (1999) Tribol Lett 7:213–220
80. Varenberg M, Etsion I, Halperin G (2003) Rev Sci Instrum 74:3362–3367
81. Drobek T, Stark RW, Heckl WM (2001) Phys Rev B 64:045401
82. Green CP, Lioe H, Cleveland JP, Proksch R, Mulvaney P, Sader JE (2004) Rev Sci Instrum 75:1988–1996
83. Cannara RJ, Eglin M, Carpick RW (2006) Rev Sci Instrum 77:053701
84. Cumpson PJ, Hedley J, Clifford CA (2005) J Vac Sci Technol B 23:1992–1997

9 Frequency Modulation Atomic Force Microscopy in Liquids

Suzanne P. Jarvis · John E. Sader · Takeshi Fukuma

Abstract. Frequency modulation atomic force microscopy is a sensitive and quantitative dynamic technique, which utilizes the change in resonance frequency of a cantilever to detect variations in the interaction force between the cantilever tip and the sample of interest. Although it has been used extensively in ultrahigh vacuum, it is rarely used in liquids. Here we explore the application of the technique in the liquid environment, covering various experimental implementations of the technique and its theoretical foundations. In addition, we describe a number of applications that demonstrate the potential of the technique in liquids and highlight future prospects.

Key words: Frequency modulation, FM-AFM, Dynamic AFM, Liquid environment, Physiological environment

Abbreviations

AFM   Atomic force microscopy
AGC   Automatic gain control
AM-AFM   Amplitude modulation atomic force microscopy
DPPC   Dipalmitoylphosphatidylcholine
FM-AFM   Frequency modulation atomic force microscopy
OBD   Optical beam deflection
PLL   Phase-locked loop
PSPD   Position-sensitive photodetector
RF   Radio frequency
OMCTS   Octamethylcyclotetrasiloxane
UHV   Ultrahigh vacuum

9.1
Introduction

Frequency modulation atomic force microscopy (FM-AFM) was first described by Albrecht et al. [1] as a highly sentitive dynamic feedback method for AFM operation. The exceptional capabilities of the method were first demonstrated in 1995 when Geissibl [2] applied FM-AFM to obtain the first true atomic resolution images with AFM in ultrahigh vacuum (UHV). Since this first demonstration of its true atomic resolution imaging capability, FM-AFM has become the method of choice in UHV [3–5]. However, with a few exceptions, it has not been widely applied in liquid environments. This may be due in part to the belief that the method only has clear benefits over alternative AFM techniques when combined with a high-Q system and thus has no apparent advantage in low-Q environments such as liquids. However, one major benefit over contact-mode AFM is that both stability and sensitivity can be retained in the near-contact region where the magnitude of the attractive force gradient will often exceed the stiffness of compliant levers used for highly sensitive static measurements [6]. Benefits also arise from using FM-AFM in liquids when compared with other dynamic techniques, the primary one being that that FM-AFM does not show a bistability like the widely used amplitude modulation AFM (AM-AFM) technique [7, 8]. When an extra feedback loop is added to FM-AFM to keep the cantilever oscillating at constant amplitude, the conservative and dissipative components of the interaction are decoupled. Further, when operation is in this constant-amplitude mode, both the frequency shift and the dissipation signals can be converted to quantitative conservative and dissipative forces in a manner which is independent of cantilever oscillation amplitude [9].

Unlike the many applications of FM-AFM in UHV, to date there are very few examples where FM-AFM has been applied in a liquid to obtain novel information which is not accessible with any alternative AFM technique. Instead the majority of work has focused on characterizing the technique itself or comparing it with other more familiar techniques. Ironically, one exception to this was the first application of FM-AFM to the liquid environment by Jarvis et al. [10] in 2000, where the technique was used to obtain a continuous and stable measurement of oscillatory forces in water in close proximity to a hydrophilic surface, a measurement not possible with alternative AFM techniques. To implement FM-AFM, they used magnetic activation of the cantilever [11] and combined this with a carbon nanotube probe [12] to reduce the long-range background interaction between the bulk of the atomic force microscope tip and the sample. With this setup they imaged the apparent striped phase of the hydrophilic self-assembled monolayer at high resolution as shown in Fig. 9.1.

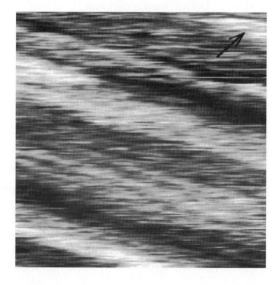

Fig. 9.1. $12 \times 12\,\mathrm{nm}^2$ area of a hydrophilic self-assembled monolayer showing faint periodicity and striped regions. The region indicated in the *top righthand corner* corresponds to the location of the water structure measurement shown in Fig. 9.2 (Reused with permission from Jarvis et al. [10]. Copyright 2000, American Chemical Society)

In the near-contact region they found evidence of structured water layers between the tip and the surface (Fig. 9.2). By utilizing FM-AFM, they could correlate highly sensitive force spectroscopy measurements of water structure obtained perpendicular to the surface with their nanoscale lateral characterization of the sample surface. These results indicated that FM-AFM could represent a powerful technique for probing the nature of aqueous environments as a function of precise lateral position on any surface, of particular relevance to biological materials.

Since this first measurement, FM-AFM has been used in liquid environments with inorganic, organic and biological samples and the modes of operation which have been explored have included imaging, force spectroscopy and force-extension. Recently the merits of FM-AFM over alternative dynamic techniques have come under increasing scrutiny owing to the enhanced imaging capabilities of FM-AFM in liquids when combined with a low-noise detection system [13]. This detection system made smaller oscillation amplitudes more readily accessible and controllable even on resonance, thus enhancing the sensitivity to short-range forces. Currently subangstrom resolution, the highest resolution to date under a liquid, has only been demonstrated using the FM-AFM technique. This was achieved on a lipid bilayer under physiological conditions [14]. The exceptional resolution attained on a *biological* sample is likely to trigger the wider application of FM-AFM to biological materials.

In this chapter we describe in detail the theoretical foundations of FM-AFM and explore a variety of implementation methods with examples. Not all issues regarding the relative merits of different implementations of the technique have been resolved; thus, we attempt to present an objective overview of the implementations currently in use. Table 9.1 presents the terminology used in the chapter with regard to the different implementations and subtle variations of dynamic techniques currently in use in liquids.

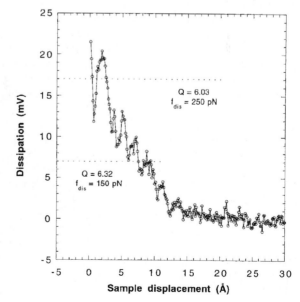

Fig. 9.2. Dissipation as a function of sample displacement away from the tip (zero on the x-axis is arbitrary). Oscillations in the dissipation signal are separated by approximately 2.2 Å and are associated with the displacement of individual water layers between the tip and the sample. (Reused with permission from Jarvis et al. [10]. Copyright 2000, American Chemical Society)

Table 9.1. Terminology used in the chapter with regard to the different implementations and subtle variations of dynamic techniques currently in use in liquids

Technique	Cantilever drive options	Constant	Imaging feedback setpoint
FM-AFM	Constant amplitude Constant excitation	Phase	Frequency (or sometimes dissipation)
AM-AFM	Constant excitation	Frequency	Amplitude
PM-AFM	Constant amplitude Constant excitation	Frequency	Phase

FM-AFM frequency modulation atomic force microscopy, *AM-AFM* amplitude modulation atomic force microscopy, *PM-AFM* phase modulation atomic force microscopy

9.2
Instrumentation

9.2.1
Basic Setup for FM-AFM

Figure 9.3 shows the experimental setup for FM-AFM [1]. In FM-AFM, the cantilever is mechanically oscillated at its resonance frequency using a self-excitation circuit, where the cantilever is used as a mechanical resonator. The cantilever vibration induced by an actuator is detected using a cantilever-deflection detector, which outputs a voltage signal proportional to the cantilever deflection. The deflection signal is fed back to the actuator through a phase shifter and an automatic gain control (AGC) circuit, forming a self-excitation loop. A phase shifter is used for adjusting the phase difference between the cantilever excitation signal and deflection signal to 90°. This ensures that the cantilever always oscillates at the resonance frequency. The deflection signal is also routed to a frequency detector that outputs a frequency shift (Δf) signal corresponding to the deviation from the free resonance frequency of the cantilever (f_0). A phase-locked-loop (PLL) circuit is most commonly used as a frequency detector. The detected Δf signal is fed into feedback electronics that control the vertical position of the sample scanner so that the Δf signal is kept constant. With this tip–sample distance regulation, the sample is raster-scanned across

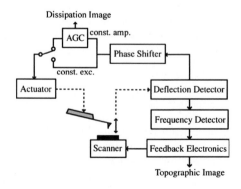

Fig. 9.3. Typical experimental setup for frequency modulation atomic force microscopy (FM-AFM). *AGC* automatic gain control

the surface and the output of the feedback electronics is recorded as a function of position to produce a topographic image. In some cases the PLL's oscillator signal is used to drive the cantilever as an alternative to the phase shifter. This has the advantage of greater spectral cleanliness, but having the PLL inside a self-excitation loop slows its time response.

FM-AFM can be operated either in constant-amplitude or in constant-excitation mode. The former utilizes an AGC circuit that adjusts the magnitude of the excitation signal such that the cantilever oscillation amplitude is kept constant. In this case, the dissipation of cantilever vibration energy due to the tip–sample interaction force results in an increase in the cantilever excitation signal. This signal can be recorded as a two-dimensional map to produce a dissipation image. In constant-excitation mode, the AGC circuit is bypassed and a fixed-amplitude excitation signal is used. In this case, the energy dissipation gives rise to an amplitude damping. Note that quantitative conservative and dissipative force measurements with FM-AFM require operation in constant-amplitude mode [9, 15, 16].

9.2.2
Cantilever Excitation in a Liquid

Dynamic-mode AFM [7] such as FM-AFM requires mechanical excitation around the resonance frequency. The various activation methods that have been combined with FM-AFM in a liquid include methods already widely used for dynamic AFM such as piezo activation [17] and magnetic activation [10] as well as the less common quartz tuning fork technique [18], chosen because of the high stiffness of the probe, which enabled smaller oscillation amplitudes of the order of 5 Å to be used [18]. The pulsed-laser technique [19, 20] is also likely to be combined with FM-AFM in a liquid because it does not require modification to the cantilever and can be used to activate cantilevers with very high resonance frequencies. In the following sections we discuss the two most common activation techniques in more detail.

9.2.2.1
Piezo Excitation

In this method, a small piece of piezoelectric actuator is placed adjacent to a cantilever holder. An ac voltage signal, which is referred to as the cantilever excitation signal, is applied across the actuator to induce its vibration. This vibration is transmitted through the cantilever holder and cantilever base to the cantilever itself. Since the actuator does not directly drive the cantilever end but instead drives it via other mechanical elements, the vibration of the actuator excites the resonances of all the elements mechanically coupled with the actuator. Consequently, the phase and the amplitude of the deflection signal can be influenced by the spurious resonances that do not represent the true cantilever vibration characteristics. Such influence can lead to an error in the quantification of force from frequency shift and dissipation measurements. It can also prevent stable cantilever vibration especially in FM-AFM, where the phase information of the deflection signal is used in the self-excitation circuit.

The influence of spurious resonances is not a serious problem in air and a vacuum, where the Q factor of a cantilever is much higher than that of the spurious resonances.

In such high-Q environments (typically $Q > 300$), the phase and the amplitude of the cantilever vibration show much stronger dependence on the excitation frequency than that of the spurious resonances. Thus, the influence from the spurious resonances is negligible in those applications. However, such influence is often a serious problem in a liquid. The Q factor in a liquid (typically $Q < 10$) is much lower than that in air and a vacuum, so the phase delay and the amplitude distortion induced by the spurious resonances are not negligible. In addition, the actuator is mechanically coupled with additional elements (e. g., the liquid cell) through the liquid. This forms a complicated mechanical transmission path through which the vibration energy of the actuator can excite a number of spurious resonances.

9.2.2.2
Magnetic Excitation

In order to avoid the influence from the spurious resonances, it is desirable to apply an excitation force directly to the cantilever end. Magnetic excitation [11] is one of the ways to realize this idea. With this method, a cantilever is modified to have magnetic sensitivity either by coating with a magnetic thin film or by attaching a magnetic particle on the backside of the cantilever. The film, or the particle, is magnetized by applying a strong magnetic field. An excitation signal is fed into a $V–I$ converter, which drives a coil placed close to the cantilever (typically underneath the sample), to produce an ac magnetic field. The coil is usually aligned such that the interaction between the ac magnetic field and the magnetic moment of the modified cantilever gives rise to a magnetic force perpendicular to the cantilever in order to excite the cantilever vibration normal to the surface.

Figure 9.4 shows amplitude and phase versus frequency curves obtained by driving a cantilever with piezo excitation and with magnetic excitation. The curves obtained with magnetic excitation show the typical amplitude and phase response of a simple harmonic oscillator, while those obtained with piezo excitation show a number of spurious peaks. Simple phase versus frequency characteristics are desirable for stable operation and quantitative measurements by FM-AFM, where the phase information is used for driving the cantilever. However, magnetic excitation requires modification of the cantilever, which limits the range of applicable cantilevers. In addition, the coated magnetic film or the attached magnetic particle may not be passive in all the solutions used in the experiment and hence may constitute a source of contamination. Therefore, both piezo and magnetic excitations have been used extensively in FM-AFM in liquids depending on the specific requirements of each application.

9.2.3
Cantilever-Deflection Detection

Various methods have been proposed for detecting the cantilever deflection for FM-AFM in liquids, including optical-beam deflection (OBD) [10] and homo-dyne [21] and heterodyne [22] interferometry. The main considerations when choosing a deflection-detection system for operation in a liquid include ease of implementation, accuracy of calibration and noise level. Interferometry-based systems usually

Fig. 9.4. Amplitude and phase versus frequency characteristics of a cantilever oscillated in a liquid with **a** piezo excitation and **b** magnetic excitation. The *insets* show fast fourier transform spectra of the cantilever deflection signals showing thermal Brownian vibration peaks. (Reused with permission from Higgins et al. [43]. Copyright 2005, Institute of Physics)

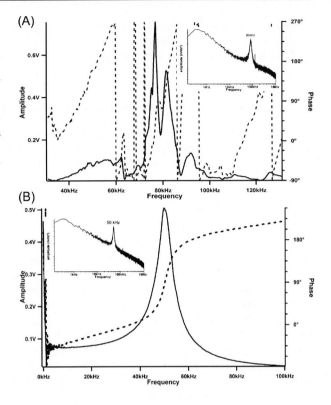

outperform the OBD system for the last two considerations but are not in widespread use owing to difficulties in implementation.

In FM-AFM, cantilever oscillation amplitudes of 0.1–50 nm are detected using the cantilever-deflection detection system. Low noise deflection detection is important in FM-AFM not only for obtaining optimal sensitivity in the frequency-detection signal but also for stable self-excitation of the cantilever. In order to access the optimum oscillation amplitudes for sensitivity to short-range forces ($A < 1$ nm) [23], stringent noise requirements are placed on the deflection measurement. Stable self-oscillation with such small amplitudes requires a low noise cantilever deflection signal [13].

The deflection signal generally contains two different noise components: noise from the cantilever-deflection measurement system and thermal Brownian motion of the cantilever. The signal-to-noise ratio of the force detection is determined by the noise in the deflection signal at the frequency range from $f_0 - B$ to $f_0 + B$, where B denotes the force detection bandwidth. Thus, in order to obtain the optimal force sensitivity, the deflection noise density arising from the deflection-detection system must be less than that from thermal Brownian motion (n_{zB}) given by [1]

$$n_{zB}(f) = \sqrt{\frac{2k_B T}{\pi f_0 k Q} \frac{1}{[1 - (f/f_0)^2]^2 + [f/(f_0 Q)]^2}} , \tag{9.1}$$

where k_B, T and k are the Boltzmann constant, absolute temperature and the spring constant of the cantilever, respectively.

Equation (9.1) shows that thermal noise around f_0 becomes smaller with increasing f_0 and k and decreasing Q. This suggests that achieving thermal-noise-limited performance becomes more difficult when using a higher resonance frequency and higher stiffness cantilever in a low-Q environment such as in a liquid. The use of a high-f_0 cantilever is desirable for high force sensitivity and a fast time response [1]. A stiff cantilever is required for oscillating the cantilever with small amplitude without suffering from instabilities such as tip adhesion. Therefore, a low noise cantilever deflection measurement is essential for high-resolution imaging with FM-AFM in a liquid.

9.2.3.1
OBD Detection

The OBD method [24] is the most widely used detection technique, especially in commercial AFM systems, because of the simple experimental setup and easy optical beam alignment. Figure 9.5 shows typical setups for the deflection-detection system using the OBD method in air and in a liquid. With this method, a focused laser beam is directed onto the back of the cantilever and the reflected beam is detected by a position-sensitive photodetector (PSPD). The cantilever deflection results in an angular deviation of the bounced laser beam, which in turn gives rise to a displacement of the laser spot on the PSPD. The PSPD produces a differential current signal in proportion to the laser spot displacement. The current signal is converted to a voltage signal with an I–V converter, which is referred to as the preamplifier, and is fed into a differential amplifier, generating a voltage signal proportional to the cantilever deflection. This signal is referred to as the cantilever-deflection signal.

In the case of a liquid environment, the optical path of the laser beam is influenced by the refraction at the liquid–air interface as shown in Fig. 9.5b. With this effect

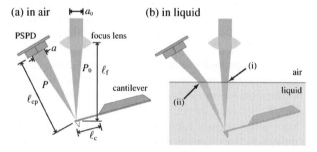

Fig. 9.5. Typical experimental setups for the cantilever-deflection measurement systems using the optical beam deflection method in **a** air and **b** a liquid. *PSPD* position-sensitive photodetector. (Reused with permission from Fukuma and Jarvis [25]. Copyright 2006, American Institute of Physics)

taken into account, the deflection sensitivity (S_z) is given by [25]

$$S_z = 6\chi\eta\alpha n_m R_{IV} A_{diff} \frac{P_0 \ell_f}{a_0 \ell_c} , \qquad (9.2)$$

where η is the efficiency of photocurrent conversion at the PSPD, α is the laser power attenuation in the optical path from the laser source to the PSPD and n_m is the refractive index of the medium of the operating environment. R_{IV} and A_{diff} are the transimpedance of the I–V converter and the gain of the differential amplifier, respectively. P_0 is the laser power measured at the output of the beam source. ℓ_f, ℓ_c and a_0 are the dimensions defined in Fig. 9.5a. χ is the correction factor for a Gaussian laser beam. If we define a_0 as the diameter of the beam cross section at $1/e$ (or $1/e^2$) of the maximum laser intensity, the coefficient χ will be 1.13 (or 1.60) [25].

9.2.3.2
Laser Noise in the OBD Method

There are three major noise sources in the OBD detection system: photodiode shot noise, transimpedance Johnson noise and laser beam noise. In most cases, the transimpedance Johnson noise can be suppressed to less than the other two by choosing a proper R_{IV} value, typically $10-100\,\mathrm{k}\Omega$, unless insufficient bandwidth of the photodetection system significantly deteriorates the deflection sensitivity [13]. Theoretically, the photodiode shot noise should determine the ultimate limit of the noise performance. However, in many cases, the laser beam noise dominates especially for AFM in the liquid environment. Thus, one of the most difficult parts of designing an OBD detector is reducing the laser noise to less than the photodiode shot noise.

In general, the deflection noise from the intensity fluctuation of the laser diode driven by an automatic power control circuit is negligible compared with those from the photodiode shot noise and the transimpedance Johnson noise [26]. This is because noise from the laser intensity fluctuation is mostly eliminated as a common mode noise at the differential amplifier if its common mode rejection ratio at the cantilever vibration frequency is sufficiently large. However, even if the laser intensity fluctuation is negligible, a small mode fluctuation in the laser diode beam could induce a drastic increase in the deflection noise [13]. This is because the laser mode fluctuation causes a fluctuation of the spatial distribution of the laser spot on the PSPD. Such a fluctuation of the laser spot pattern produces a differential mode noise that is magnified by a differential amplifier. Since the behavior of this noise component is strongly dependent on the characteristics of the laser diode and other optical components used in the OBD method, its quantitative estimation is practically very difficult.

A laser diode operated with a relatively small output power compared with its rating shows intensity fluctuations originating from spontaneous light emission. This noise is referred to as "quantum noise." As the laser power increases, the quantum noise becomes negligible compared with the averaged laser power. Instead, the intensity fluctuation from the hopping of the laser oscillation mode becomes evident in the high output power regime. This noise is referred to as "mode hop noise" and leads to a large deflection noise as described above. The magnitude of the mode hopping does not show a monotonic dependence on the output laser power but a steep increase at a threshold value [13]; thus, the laser power should not exceed the threshold for the onset of the laser mode hopping. Since high output laser power is desirable for reducing the noise from the photodiode shot noise, the laser power should be set at a value slightly below the threshold.

Besides the intrinsic noises of the laser diode such as quantum noise and mode hop noise, there are two other noise sources related to the laser beam. The reflection and scattering of the laser beam occasionally cause some part of the laser beam to go back into the optical resonator of the laser diode. This optical feedback generates additional laser oscillation modes, leading to an increase in mode hopping. This noise is referred to as "optical feedback noise." On the other hand, some of the reflected or scattered laser beam is incident on the PSPD. This part of the laser beam interferes with other parts of the laser beam that travel through different optical paths. Such unexpected optical paths are often unstable owing to the thermal and mechanical drifts in the system. Thus, the laser spot pattern on the PSPD shows a fluctuation over a relatively long time scale. This noise is referred to as "optical interference noise." These sources of noise are particularly evident in liquid-environment AFM [13]. In the setup for liquid-environment AFM, many obstacles such as a cover glass and a solution are inserted into the optical path of the laser beam. These obstacles induce reflection and scattering of the laser beam, which in turn increases the optical feedback noise and the optical interference noise.

These types of optical noise can be suppressed by modulating the laser power with a radio frequency (RF) signal whose frequency is typically 300–500 MHz [13]. The RF laser power modulation changes the laser oscillation mode from single mode to multimode. Since the mode hopping takes place owing to the competition among the possible laser oscillation modes in the optical resonator, multimode laser oscillation is much more insensitive to optical feedback than single-mode laser oscillation. Consequently, RF modulation considerably reduces the mode hopping induced by the optical feedback. In addition, the multimode laser beam has a lower coherence than the single-mode laser beam. Thus, RF modulation also works to suppress the optical interference noise.

Figure 9.6 shows a waveform of the frequency shift signal measured in water before and after turning off the RF laser power modulation. The waveform shows that the magnitude of the frequency noise suddenly increased after turning

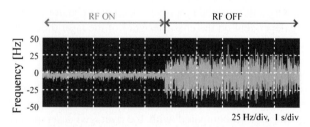

Fig. 9.6. Waveform of the frequency shift signal measured in water before and after turning off the radio frequency (*RF*) laser power modulation ($A = 5$ nm, $f_0 = 140$ kHz, $k = 42$ N/m). (Reused with permission from Fukuma et al. [13]. Copyright 2005, American Institute of Physics)

off the RF modulation. This demonstrates that the RF modulation is effective in reducing optical feedback noise and optical interference noise. Although the RF modulation also reduces the intrinsic mode hop noise, it is often difficult to completely suppress it. Thus, the average laser power should be set at a value below the threshold for the onset of mode hopping noise even with the RF laser power modulation.

9.2.3.3
Shot-Noise-Limited Noise Performance

Once the laser beam noise has been sufficiently reduced, the photodiode shot noise starts to dominate the deflection noise. The deflection noise density arising from the photodiode shot noise (n_{zs}) is given by [23]

$$n_{zs} = \frac{a_0 \ell_c}{6 \chi n_m \ell_f} \sqrt{\frac{2e}{\eta \alpha P_0}} , \qquad (9.3)$$

where e is the electron charge and other parameters are as defined for (9.2). This equation shows the theoretical limit of the noise performance attainable with the OBD method. The equation contains only independent parameters and hence gives us a direct guideline for reducing the deflection noise. Namely, increasing ℓ_f, α and P_0, and decreasing a_0 and ℓ_c will improve the noise performance of the deflection-detection system using the OBD method.

Figure 9.7 shows the frequency spectra of a cantilever-deflection signal measured in water with a low-noise OBD detector [25]. The experimentally measured spectra (solid lines) show good agreement with the thermal noise spectra (dashed lines) calculated with (9.3) around the resonance frequency. This is a necessary condition to obtain theoretically limited noise performance.

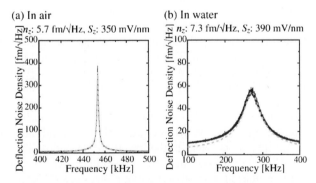

Fig. 9.7. Frequency spectra of cantilever thermal Brownian motion measured in **a** air and **b** water. The *solid lines* show experimentally measured values, while the *dotted lines* show theoretically calculated values for thermal Brownian motion obtained with (9.3). A Si cantilever (NCVH, Nanosensors) was used with a gold backside coating ($k = 21.5$ N/m, $f_0 = 271$ kHz, $Q = 6.6$). (Reused with permission from Fukuma and Jarvis [25]. Copyright 2006, American Institute of Physics)

9.3
Applications

Owing to the comparatively recent application of FM-AFM to the liquid environment, and uncertainties over its capabilities and advantages, many experiments utilizing FM-AFM in liquids have been performed with the aim of further characterizing or understanding the technique rather than applying it to obtain novel information about the sample or liquid. This has often involved working with well-characterized samples or liquids that have been extensively studied with contact-mode AFM or other dynamic techniques. Such experiments have included demonstrations of lateral imaging resolution, exploring sample deformation, measuring force resolution, directly comparing with other dynamic techniques and validating theoretical frequency to force conversion.

9.3.1
Nonbiological Systems

Application of FM-AFM in a liquid to nonbiological systems has been used extensively as a means of characterizing the FM-AFM technique itself, as nonbiological systems are often better characterized and more robust than biological samples. In particular, mica has frequently been used [21,25,27,28], as has Au(111) on mica [27, 29] and in the case of liquids, octamethylcyclotetrasiloxane (OMCTS) [18, 30–32]. In order to characterize the imaging capabilities of FM-AFM combined with a low-noise detection system, a single crystal of polydiacetylene [poly-PTS; 2,4-hexadine-1,6-diol bis(p-toluene sulfonate)] in water was initially used [33]. This experiment demonstrated that FM-AFM in water could achieve a lateral resolution of 0.25 nm and was followed soon afterwards by a demonstration of true atomic resolution on mica [28].

Figure 9.8 shows FM-AFM images of a muscovite mica surface taken in pure water. Interestingly the images show two different contrast patterns depending on the frequency shift setpoint: a honeycomb-like pattern (Fig. 9.8a) and triangular-like pattern (Fig. 9.8b). The honeycomb-like pattern consists of a number of hexagons repeating with a period of 0.52 nm. This contrast pattern is found in the FM-AFM images taken with relatively small frequency shifts (less than +200 Hz when $A = 0.20$ nm), namely, a relatively weak tip–sample interaction force. This pattern agrees well with the atomic-scale structure of a mica surface. The FM-AFM image shown in Fig. 9.8c highlights more detailed structure of the honeycomb-like pattern. The image shows some bright spots on the hexagons. The structural model of the mica suggests that the bright spots found in the image are located on the Si atom sites in the hexagons. These bright spots may correspond to the Al^{3+} ions because Al^{3+} ions are likely to be imaged as relatively large protrusions owing to the negative charges offered from the surrounding O^{2-} ions. The distance between the adjacent bright spots is about 0.3 nm, showing good agreement with the distance between the two adjacent Si atom sites. Such nonperiodic atomic-scale contrasts in FM-AFM images demonstrate the true atomic-resolution imaging capability of FM-AFM in liquids.

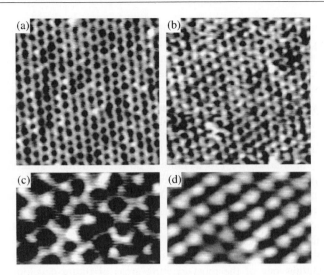

Fig. 9.8. FM-AFM images of the cleaved (001) surface of muscovite mica taken in water [28]. **a** 8 nm × 8 nm, $\Delta f = +54$ Hz, $A = 0.24$ nm, scanning speed 671 nm/s. **b** 8 nm × 8 nm, $\Delta f = +240$ Hz, $A = 0.20$ nm, scanning speed 1120 nm/s. **c** 4 nm × 2.5 nm, $\Delta f = +157$ Hz, $A = 0.16$ nm, scanning speed 934 nm/s. **d** 4 nm × 2.5 nm, $\Delta f = +682$ Hz, $A = 0.20$ nm, scanning speed 671 nm/s. The images were taken in constant-height mode. The cantilever used was an n-Si cantilever (NCH, Nanosensors) with a spring constant of 42 N/m and a resonance frequency of 136 kHz in water. (Reused with permission from Fukuma et al. [28]. Copyright 2005, American Institute of Physics)

FM-AFM images taken with a relatively large frequency shift (larger than $+200$ Hz when $A = 0.20$ nm), namely, a relatively large tip–sample interaction force, show a triangular-like pattern as shown in Fig. 9.8b. Under this condition, some of the bright dots show asymmetric contrast in the fast scanning direction (Fig. 9.8d), indicating some influence of lateral forces in the tip–sample interaction. The triangular-like pattern is similar to those seen in "lattice images" obtained by contact-mode AFM [34,35] where highly periodic images are produced by dragging a mica flake of long-range periodic order over a mica surface. However, FM-AFM images obtained in this experiment also show subnanometer-scale structural defects as shown in Fig. 9.8d. This difference from lattice images observed previously indicates that the image was formed through interactions acting over a subnanometer-scale tip–sample contact area.

9.3.2
Biological Systems

Applications of FM-AFM to biology have frequently involved imaging samples that have been studied by other AFM techniques so that comparisons can be drawn. Systems studied for comparative purposes have included living rat kidney cells (FM combined with Q control) [36], fibrilar tau proteins [37], DNA [38–40] and bacteriorhodopsin [21,41]. In addition, dynamic force-extension measurements have been made on single dextran molecules (FM combined with Q control) to measure effective viscosity [17] and with transverse excitation for comparison with dc AFM [42]. FM-AFM has also been applied to the measurement of ligand–receptor interactions where the well-studied biotin–avidin system was chosen [43]. Intermediates in the unravelling of the modular protein titin were also investigated as a function of oscillation amplitude in order to establish at what stage this becomes a limiting parameter in FM-AFM force-extension measurements [44].

The use of FM-AFM in liquids is now moving beyond method characterization and comparison studies of the technique to the investigation of specific biological problems where alternative AFM techniques do not seem to be suitable. One such experiment has been the application of FM-AFM to investigate hydration next to supported lipid bilayers as a function of bilayer fluidity [45]. In this case the FM technique was simply utilized for its sensitivity and stability in order to establish whether structured water layers were present and should be considered as an integral component in cell membrane theory.

The application of FM-AFM to biological systems under physiological conditions constitutes one of the most exciting areas of future development for the technique especially when high-resolution imaging and spectroscopy capabilities are combined, as shown recently [46].

Figure 9.9a shows a Δf versus distance curve measured on a mica-supported dipalmitoylphosphatidylcholine (DPPC) lipid bilayer in phosphate buffer solution [46]. The raw data (gray line) were smoothed (black line) and converted to a force versus distance curve (Fig. 9.9b) using the method reported by Sader and Jarvis [9]. The force curves obtained on the DPPC bilayer typically show an oscillatory force profile with one or two peaks. The averaged separation of the two peaks observed in some of the force curves is 0.28 ± 0.05 nm. This distance agrees with the size of the water molecule and hence suggests that the oscillatory force profile corresponds to the sequential removal of ordered water molecules between the tip and the sample as they approach.

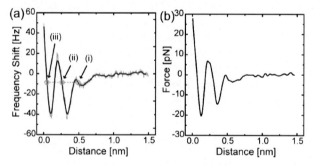

Fig. 9.9. a Δf versus distance curve measured on a dipalmitoylphosphatidylcholine (DPPC) bilayer in phosphate-buffered saline (PBS) solution, showing an oscillatory profile with two peaks. The smoothed line (*black line*) was obtained by averaging the raw data (*gray line*) over the distance range of ± 0.02 nm from each data point. **b** Force versus distance curve derived from **a** using the formula reported in [9]. (Reused with permission from Fukuma et al. [46]. Copyright 2007, Biophysical Society)

Repulsive and attractive tip–sample interaction forces, respectively, induce positive and negative Δf in FM-AFM. As a tip approaches a sample surface, Δf shows a steep increase owing to the short-range repulsive interaction potential near the surface. Thus, the tip–sample distance feedback regulation in the liquid-environment FM-AFM setup operates on the basis of the assumption that Δf increases with decreasing tip–sample separation such as in the force branches indicated by arrows i—iii in Fig. 9.9a. Therefore, if the Δf versus distance curve shows an oscillatory profile as shown in Fig. 9.9a, the feedback can operate at multiple tip positions corresponding to one setpoint as indicated by the circles in Fig. 9.9a. These equivalent positions will only be stable provided that the tip does not pass into a region where Δf decreases with decreasing separation. If it does, it will move continuously until it reaches the next equivalent position such as from i to ii in Fig. 9.9a. Thus, the tip can spontaneously jump between those positions even if the setpoint is unchanged.

Figure 9.10 shows an example of such spontaneous jumps during FM-AFM imaging of the DPPC bilayer in phosphate buffer solution. In the image shown

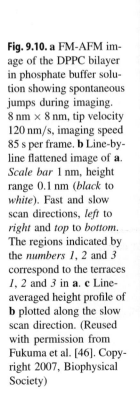

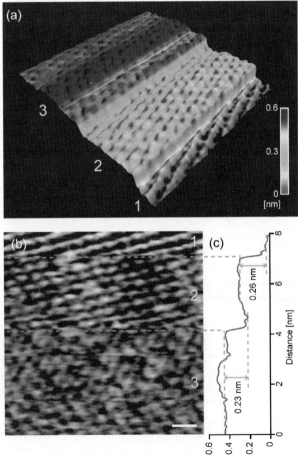

Fig. 9.10. a FM-AFM image of the DPPC bilayer in phosphate buffer solution showing spontaneous jumps during imaging. 8 nm × 8 nm, tip velocity 120 nm/s, imaging speed 85 s per frame. **b** Line-by-line flattened image of **a**. *Scale bar* 1 nm, height range 0.1 nm (*black* to *white*). Fast and slow scan directions, *left* to *right* and *top* to *bottom*. The regions indicated by the *numbers 1, 2* and *3* correspond to the terraces *1, 2* and *3* in **a**. **c** Line-averaged height profile of **b** plotted along the slow scan direction. (Reused with permission from Fukuma et al. [46]. Copyright 2007, Biophysical Society)

in Fig. 9.10, panel a, the tip is scanned from the lowest terrace (terrace 1). As imaging progresses the tip jumps spontaneously twice. Terrace 1 shows a highly ordered arrangement of bright spots separated by 0.50 ± 0.05 nm along the stripes. The first jump of 0.26 nm (terrace 1 to terrace 2) results in subtle changes to the image although discrete corrugations are still observed. The second jump of 0.23 nm (terrace 2 to terrace 3) results in a less-ordered contrast although imaging resolution is still high. The average height of the spontaneous jumps (0.29 ± 0.06 nm) agrees well with the size of a water molecule and the peak distance of the oscillatory force profile (0.28 ± 0.05 nm). This implies that the tip is jumping between water layers and the individual layers are imaged showing molecular-scale corrugations of the lipid headgroups.

Under physiological conditions, biological membranes are surrounded not only by water but also by various metal cations. The influence of these cations on membrane structure and stability has been studied intensively using model lipid bilayers. The addition of salts can trigger lipid bilayer phase separation [47], and vesicle aggregation and fusion processes [48]. The striking influence of the ions has highlighted the importance of lipid–ion interactions in biological processes. So far, a number of spectroscopy experiments have revealed that metal cations specifically interact with negatively charged moieties of the lipid headgroups [49–51]. These experiments, together with theoretical simulations [52], have led to the idea that individual ions may be interacting with multiple headgroups to form complex "lipid–ion networks." This idea has been used to account for the observed influence of the ions such as enhanced mechanical strength of membranes [53] and reduced mobility of the lipid molecules therein [54]. However, it has been a great challenge to experimentally access such lipid–ion networks owing to the lack of a method able to investigate local lipid–ion interactions with angstrom resolution. FM-AFM is an ideal tool for directly investigating these local lipid–ion networks formed at the water–lipid interface.

The membrane surface imaged by FM-AFM presents various structures depending on both time and location, reflecting the mobile nature of the ions interacting with the lipids [14]. FM-AFM images reveal the existence of at least two clearly different configurations as shown in Fig. 9.11, which are referred to as structure 1 and structure 2, respectively. Structure 1 shows pairs of two protrusions separated by 0.30 nm, corresponding to the phosphate and choline groups in the phosphatidylcholine headgroups without association with water molecules or ions. Structure 2 shows hexagonally arranged surface groups consisting of two oval-shaped subunits

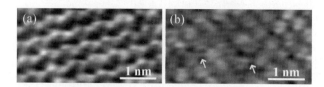

Fig. 9.11. FM-AFM images of the DPPC bilayer in PBS solution. Height scale 0.12 nm. **a** Structure 1. Tip velocity 146 nm/s. **b** Structure 2. Tip velocity 120 nm/s. (Reused with permission from Fukuma et al. [14]. Copyright 2007, American Physical Society)

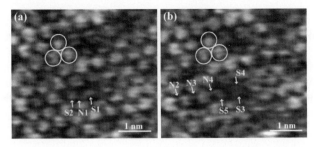

Fig. 9.12. Sequential FM-AFM images of the same area of the DPPC bilayer in PBS solution. Height range 0.1 nm. Tip velocity 120 nm/s. Imaging speed 85 s per image. (Reused with permission from Fukuma et al. [14]. Copyright 2007, American Physical Society)

with their longer axes parallel to each other. Detailed analyses of the FM-AFM images and theoretical simulations [14] have suggested that the angstrom-scale contrasts found in structure 2 represent the averaged position of mobile ions interacting with the lipid headgroups, namely, lipid–ion networks. The two subunits in structure 2 show the enhanced ion occupancies around the two negatively charged oxygen atoms in the phosphate group, which have been reported as the primary binding sites for the metal cations.

Sequential FM-AFM images taken on the same area (Fig. 9.12) reveal that some of the surface groups change their configuration upon the formation or disappearance of lipid–ion networks [14]. For example, subunit 1 is not pairing with another subunit in Fig. 9.12a, while a pair of subunits 3 and 4 appears in the next image (Fig. 9.12b), but with a darker contrast than other subunits, suggesting the lower height of this headgroup. It is likely that this particular headgroup was temporarily at an irregular tilt angle owing to the interaction between subunits 1 and 2 against the slight height difference when the first image was taken.

The observed structural changes due to the formation and disappearance of the lipid–ion networks indicate that the negatively charged phosphate groups are sharing the positive charge of cations, by which an attractive electrostatic force is exerted on all the headgroups involved in the network. The attractive interaction force mediated through such complex lipid–ion networks should bind the headgroups together and increase the global mechanical strength of the membrane. In fact, it is known that the addition of metal cations increases the mechanical strength of the DPPC bilayer [53]. The FM-AFM images shown here reveal the submolecular-scale origin of such an influence of ions on the mechanical properties of the biological membrane.

From the images it can be seen that experimentally FM-AFM in liquids has made significant progress in recent years to the extent that applications are moving away from characterization of the technique to exploring and understanding real biological problems with resolution and sensitivity not yet demonstrated by any of the alternative AFM techniques. In the following sections we move from experiment to cover the theoretical foundations of the technique.

9.4
Theoretical Framework for Quantitative FM-AFM Force Measurements

We now derive the theoretical framework for interpreting FM-AFM force measurements, noting that in such measurements the cantilever is normally excited at its fundamental resonance frequency ω_{res} only. Consequently, to describe the cantilever motion, it suffices to consider a damped harmonic oscillator with a single degree of freedom,

$$m \frac{d^2 w}{dt^2} + b \frac{dw}{dt} + m \omega_{res}^2 w = F_{int} + F_{drive} , \tag{9.4}$$

where w is the displacement of the cantilever tip from its unperturbed position, m is the effective mass of the cantilever, F_{int} is the interaction force experienced by the tip, F_{drive} is the driving force that excites the cantilever and b and ω_{res} are the damping coefficient and the resonance frequency of the cantilever in the absence of an interaction force, respectively.

The interaction force is assumed to be sufficiently weak so that changes in effective stiffness of the cantilever are small, leading to commensurately small changes in the resonance frequency. The motion of the cantilever is considered to be (approximately) harmonic, irrespective of the nonlinear nature of the interaction force. Importantly, these fundamental assumptions are satisfied in practice under standard operating conditions, and are normally assumed in analysis of force measurements using FM-AFM [3].

By employing a self-excitation mechanism, one fixes the phase φ of the cantilever tip displacement w relative to the driving force F_{drive}. Noting that the cantilever motion is (approximately) harmonic then enables one to express the displacement and the driving force as

$$w = a \sin (\omega t - \varphi) , \quad F_{drive} = F_0 \sin \omega t , \tag{9.5}$$

where a is the amplitude of oscillation and ω is the driving frequency. Typically, FM-AFM measurements are performed using a phase shift $\varphi = 90°$. Since the driving force is 90° out of phase with the cantilever tip displacement, it then follows that the only permissible oscillation frequency is the fundamental resonance frequency of the cantilever in the presence of an interaction force.

Next, we examine the nature of the interaction force F_{int}. Throughout Sects 9.4.1 and 9.4.2, we only consider the case where the cantilever is oscillating at a *fixed distance from the sample*, i.e., the distance of closest approach z between the tip and the sample is held constant.

9.4.1
Decomposition of Interaction Force

Importantly, F_{int} is a nonlinear function of the tip–sample separation in general, and can differ on the approach and retract portions of the oscillation cycle (Fig. 9.13). We therefore decompose the periodic force experienced by the tip into the sum of an "even" and an "odd" force: the even component is the average of the approach and retract force curves, whereas the odd component is obtained by taking the difference between the approach and retract curves and dividing by 2 (Fig. 9.14). We emphasize that this decomposition is completely general, irrespective of the nature of the force.

The physical significance of these "even" and "odd" forces can be interpreted by considering the work done W per oscillation cycle of the tip,

$$W = \oint \mathbf{F} \cdot \mathbf{ds} , \tag{9.6}$$

where $\mathbf{F}$ is the vector force experienced by the tip, $\mathbf{ds}$ is the elemental position vector in the direction of motion, and the integration is performed over a complete cycle of oscillation.

Even component of force: The work done on approach of the tip to the sample is equal in magnitude but opposite in sign to the work done on retraction, since the vector force is identical on approach and retraction, but the elemental position vector changes sign. Consequently, the *net* work done per oscillation cycle is zero, and the "even" force can be formally connected to the "conservative" component of the force *per cycle*.

Odd component of force: The work done on approach of the tip to the sample is identical to that done on retraction, since both the vector force and the elemental position vector are equal in magnitude but opposite in sign on approach and retraction. Consequently, the *net* work done per oscillation cycle is not zero in general, and the "odd" force can be formally connected to the "dissipative" component of the force *per cycle*.

We emphasize, however, that such connections to "conservative" and "dissipative" forces, although common and appealing [3], can be misleading and ambiguous since the origin of the forces cannot be determined solely by analyzing the harmonic motion of the tip in isolation. Indeed, it is entirely possible that the "conservative" force results from a combination of conservative, dissipative and energy-gaining processes over different parts of the oscillation cycle, even though the *net* energy lost or gained is zero over a complete cycle. Such a situation can occur, for example, when the sample and/or tip surfaces are deformable, and the interaction involves both truly conservative and dissipative forces. As an illustrative example, consider the interaction between two charged deformable surfaces with fluid imbedded between them. Here, the hydrodynamic (dissipative) force can affect the separation between the surfaces and, hence, the electrostatic (conservative) force. This results in the "even" and "odd" components of the force not being identical to the "conservative" (electrostatic) and "dissipative" (hydrodynamic) forces, respectively. Thus, even though the net energy lost *per cycle* by the "even" force is zero, referring to it as the "conservative" force is misleading and invalid. Importantly, a formal connection to conservative and dissipative forces can only be made if the precise nature and origin of the forces involved is known.

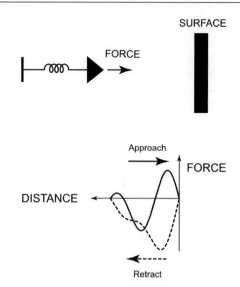

Fig. 9.13. Interaction force experienced by the tip during one oscillation cycle. (Reused with permission from Sader et al. [16]. Copyright 2005, Institute of Physics)

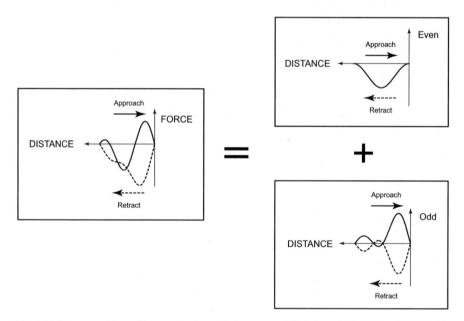

Fig. 9.14. Decomposition of interaction force during one oscillation cycle into 'even' and 'odd' components. Retract and approach curves for the "even" force are identical. Reused with permission from Sader et al. [16]. Copyright 2005, Institute of Physics

9.4.2
Governing Equations

The governing equations for the "even" and "odd" forces are obtained by first decomposing the interaction force as specified before,

$$F_{int} = F_{even} + F_{odd} . \tag{9.7}$$

Substituting (9.7) into (9.4) and performing a Fourier analysis then gives the required results. The Fourier cosine series gives the governing equation for the "even" force F_{even},

$$\frac{\Delta\omega}{\omega_{res}} = -\left(\frac{1}{ak}\right)\frac{1}{T}\int_0^T F_{even}(z + a + w(t))\cos(\omega t)\,dt , \tag{9.8}$$

whereas the Fourier sine series gives the governing equation for the "odd" force F_{odd},

$$F_0 + ab\omega = \frac{2}{T}\int_0^T F_{odd}(z + a + w(t))\sin(\omega t)\,dt , \tag{9.9}$$

where $\Delta\omega$ is the change in resonance frequency such that $\omega = \omega_{res} + \Delta\omega$, T is the oscillation period, z is the distance of closest approach between the tip and the sample, k is the dynamic spring constant of the cantilever [55] and F_0 is the magnitude of the driving force. We note that (9.8) was originally derived by Giessibl [6] using an alternative but equivalent approach, whereas the well-known relation for the average power dissipated per cycle [56] can be obtained directly from (9.9).

To enable the interaction force (both even and odd) to be determined, all forces must be uniquely defined for a given tip–sample separation. This condition is satisfied by F_{even}, but not by F_{odd} (Fig. 9.14). This limitation can be overcome by further decomposing the "odd" force into the product of an additional odd and even function. For the additional odd function, we choose the velocity of the cantilever tip $\dot{w}(t)$, as illustrated in Fig. 9.15. This generates what is commonly referred to as a "generalized damping coefficient" $\Gamma(z, a, \omega, w(t))$ as the corresponding even function. Importantly, this decomposition is formally exact and rigorous, irrespective of the nature of F_{odd}, and leads to the following exact expression

$$F_{odd} = \Gamma(z, a, \omega, w(t))\dot{w}(t) , \tag{9.10}$$

with the generalized damping coefficient being uniquely defined for a given tip–sample separation, provided the distance of closest approach z is held fixed, as has been assumed in this section. Note that the generalized damping coefficient can also depend on both the oscillation frequency and the amplitude.

From (9.8), (9.9) and (9.10), and using an appropriate change of integration variable, then gives the required governing equations for both the "even" force F_{even} and the generalized damping coefficient Γ:

$$\frac{\Delta\omega}{\omega_{res}} = -\frac{1}{\pi ak}\int_{-1}^1 F_{even}(z + a(1 + u))\frac{u}{\sqrt{1 - u^2}}\,du \tag{9.11a}$$

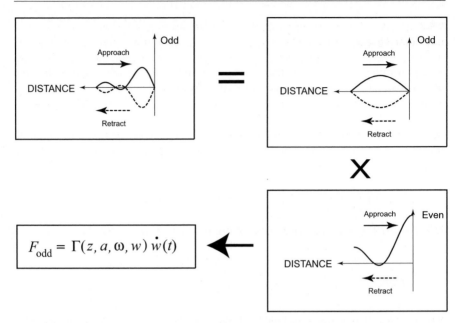

Fig. 9.15. Decomposition of "odd" force into the product of velocity and generalized damping coefficient. Retract and approach curves for generalized damping coefficient are identical. (From Sader et al. [16])

and

$$\frac{\Delta F_0}{\overline{F}_0} - \frac{\Delta \omega}{\omega_{\mathrm{res}}} = \frac{2}{\pi b} \int_{-1}^{1} \Delta \Gamma(z + a(1 + u))\sqrt{1 - u^2}\, du \,, \qquad (9.11b)$$

where $\overline{F}_0$ is the driving force in the absence of an interaction force, $\Gamma = b + \Delta\Gamma$ and $F_0 = \overline{F}_0 + \Delta F_0$. Note that ΔF_0 and $\Delta\Gamma$ are the changes in the driving force and generalized damping coefficient, respectively, resulting from the interaction.

9.4.3
Fundamental Conditions on Interaction Force

FM-AFM force measurements are usually performed by varying the distance of closest approach z and monitoring the change in driving force and resonance frequency. To determine the interaction force giving rise to these changes, (9.11a) and (9.11b) must be inverted. However, this is only possible if F_{even} and $\Delta\Gamma$ are *unique functions of the absolute tip–sample separation* $[z + a + w(t)]$, irrespective of the distance of closest approach z. In cases where this condition is violated, then inverting (9.11) to determine the interaction force is not justified. This essential condition is often ignored in measurements, and can lead to erroneous results. A specific case where this condition is violated occurs when bonds are created and broken at different tip–sample separations, e.g., adhesion measurements. In such a case, the breaking of bonds can lead to a reduced force as the distance of closest approach z is increased, even though the same absolute tip–sample separation is probed. Conversely, a case where this fundamental condition can be satisfied occurs in the measurement of long-range forces, e.g., van der Waals and hydrodynamic forces.

Also underlying the inversion of (9.11) is the requirement that the interaction force is zero when the distance of closest approach is infinite; see later. This condition is normally satisfied, but there do exist situations where it does not hold when interpreting measurements, requiring appropriate modification of the theory, e.g., see [43].

9.4.4
Determination of Forces

In this section, we derive analytical formulas that enable the "even" force F_{even} and the change in the generalized damping coefficient $\Delta\Gamma$ to be easily and explicitly evaluated from the measured change in drive force and resonance frequency [9, 16]. This is achieved by formally inverting (9.11a) and (9.11b), which provide the fundamental connection between the measured observables and the interaction force. Importantly, the explicit formulas presented in this section for both "even" and "odd" forces are valid for any oscillation amplitude.

To begin, we note that (9.11b) can be transformed into an identical form to (9.11a) using integrating by parts. This leads to the following equivalent formula for (9.11b):

$$\Theta = -\frac{1}{\pi ab} \int_{-1}^{1} B(z + a(1 + u)) \frac{u}{\sqrt{1 - u^2}} \, du , \tag{9.12}$$

where

$$\Theta = \frac{\Delta F_0}{F_0} - \frac{\Delta\omega}{\omega_{\text{res}}} \tag{9.13a}$$

and

$$B(x) = 2 \int_{x}^{\infty} \Delta\Gamma(\hat{x}) \, d\hat{x} . \tag{9.13b}$$

We therefore focus on the general solution of (9.12), since this also gives the solution to (9.11a). We first express the unknown function B as [9]

$$B(z) = \int_{0}^{\infty} A(\lambda) \exp(-\lambda z) \, d\lambda , \tag{9.14}$$

where $A(\lambda)$ is formally the inverse Laplace transform of $B(z)$. This implicitly requires that $B(z)$ approach zero as $z \to \infty$, which is a fundamental condition of the solution (as discussed above), and places no other restriction on $B(z)$.

Substituting (9.14) into (9.12) then gives

$$\Theta(z) = \frac{1}{a b} \int_{0}^{\infty} A(\lambda) T(\lambda a) \exp(-\lambda z) \, d\lambda , \tag{9.15}$$

where $T(x) = I_1(x) \exp(-x)$, and $I_n(x)$ is the modified Bessel function of the first kind of order n [57]. Importantly, the governing equations for $B(z)$ and $\Theta(z)$ presented in (9.14) and (9.15) differ only by the function $T(\lambda a)$ in their respective integrands. This feature immediately enables inversion of (9.12), leading to the following exact analytical solution for $B(z)$:

$$B(z) = \frac{ba}{2\pi i} \int_0^\infty \int_{c-i\infty}^{c+i\infty} \frac{\Theta(\hat{z})}{T(\lambda a)} \exp(-\lambda(z - \hat{z})) \, d\hat{z} \, d\lambda \,, \qquad (9.16)$$

where c is a real constant. However, (9.16) is of little value in practice, since it requires numerical evaluation of the inverse Laplace transform, which can pose a formidable challenge. We therefore construct an approximate representation for $T(x)$ by examining its asymptotic limits as $x \to 0$ and $x \to \infty$, and using a Padé approximant to connect these two limits:

$$T(x) \cong \frac{x}{2}\left(1 + \frac{1}{8}\sqrt{x} + \sqrt{\frac{\pi}{2}}x^{\frac{3}{2}}\right)^{-1} . \qquad (9.17)$$

This approximate equation gives an excellent representation for $T(x)$, see Fig. 9.16, with the error exhibited by (9.10) being less than 5% for all values of x. Note that (9.17) is exact in the limits as $x \to 0$ and $x \to \infty$.

Substituting (9.17) into (9.16) enables one to evaluate (9.16) explicitly [9], leading to the following result:

$$B(z) = 2b \int_z^\infty \left(1 + \frac{a^{\frac{1}{2}}}{8\sqrt{\pi(t - z)}}\right) \Theta(t) - \frac{a^{\frac{3}{2}}}{\sqrt{2(t - z)}} \frac{d\Theta(t)}{dt} \, dt \,. \qquad (9.18)$$

Substituting (9.13b) into (9.18), and applying (9.18) to (9.13a), then gives the required explicit formulas for the "even" force and the change in the generalized damping coefficient:

$$F_{\text{even}}(z) = 2k \int_z^\infty \left(1 + \frac{a^{\frac{1}{2}}}{8\sqrt{\pi(t - z)}}\right) \Omega(t) - \frac{a^{\frac{3}{2}}}{\sqrt{2(t - z)}} \frac{d\Omega(t)}{dt} \, dt \qquad (9.19a)$$

and

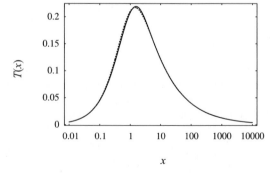

Fig. 9.16. Comparison of the function $T(x)$ (*solid line*) and its approximation, (9.17) (*dashed line*). (Reused with permission from Sader and Jarvis. [9]. Copyright 2004, American Institute of Physics)

$$\Delta \Gamma(z) = -b \frac{\partial}{\partial z} \int_z^\infty \left(1 + \frac{a^{\frac{1}{2}}}{8\sqrt{\pi(t-z)}} \right) \Theta(t) - \frac{a^{\frac{3}{2}}}{\sqrt{2(t-z)}} \frac{d\Theta(t)}{dt} \, dt \, ,$$

$$(9.19b)$$

where

$$\Omega(z) = \frac{\Delta\omega(z)}{\omega_{\text{res}}} \tag{9.20a}$$

and

$$\Theta(z) = \frac{\Delta F_0(z)}{F_0} - \frac{\Delta\omega(z)}{\omega_{\text{res}}} \, . \tag{9.20b}$$

Equations (9.19a) and (9.19b) are the results we seek, enabling the "even" force F_{even} and the generalized damping coefficient Γ to be evaluated explicitly from the measured frequency shift and change in driving force. We emphasize that these formulas are valid for all amplitudes of oscillation, provided the fundamental requirements detailed in Sect. 9.4.3 are satisfied.

9.4.5
Validation of Formulas

We now demonstrate the accuracy and validity of these formulas by presenting results for a simulated experiment. Note that (9.19a) and (9.19b) must exhibit similar accuracy, since they are derived using the same formalism. Hence, we focus our discussion on (9.19a) for the "even" force, noting that an identical conclusion holds for (9.19b). The reader is referred to [9] for a more detailed comparison and discussion.

To assess (9.19a), we first specify a force law, evaluate the resulting frequency shift using (9.11a), and then recover the force law using (9.19a). The accuracy and validity of (9.19a) can then be examined by comparing the original and recovered force laws. We choose a Lennard-Jones force law [9], consisting of a long-range attractive component and a short-ranged repulsive force:

$$F_{\text{even}}(z) = G \left(\frac{\ell^4}{3z^6} - \frac{1}{z^2} \right) , \tag{9.21}$$

where G is a constant and ℓ is the separation where the attractive force is maximum, which sets a natural length scale for the force law. Comparison of the original and recovered force laws is given in Fig. 9.17

Results are presented covering the entire spectrum spanning the case where the oscillation amplitude is small in comparison with the length scale ℓ, to the situation where the amplitude is large compared with ℓ. From Fig. 9.17 it is evident that (9.19a) gives excellent accuracy, with the original and recovered force laws being virtually identical, irrespective of the amplitude chosen. Note that the maximum error of approximately 5% occurs at intermediate amplitudes, which is expected, since the accuracy of (9.19a) (and (9.19b)) is dictated solely by the approximation for the function $T(x)$. Consequently, (9.19a) (and (9.19b)) can be used with confidence

Fig. 9.17. Actual (*solid line*) and recovered (*dashed line*) Lennard-Jones force laws using (9.19a). Amplitudes of oscillation used $a/\ell = 0.1, 0.3, 1, 3, 10$. (Reused with permission from Sader and Jarvis [9]. Copyright 2004, American Institute of Physics)

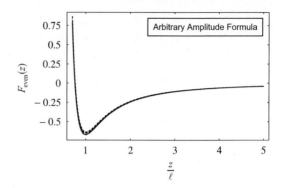

to determine the "even" force F_{even} and the generalized damping coefficient Γ, regardless of their nature and the oscillation amplitude used. Importantly, these formulas present a significant advance on other theoretical approaches for force extraction [3,15,56,58], by permitting the force to be evaluated simply and accurately for any amplitude of oscillation.

9.5
Operation in a Fluid

We now examine the application of the theoretical formalism presented in Sect. 9.4 to measurements conducted in fluid environments.

In the previous analysis, it was implicitly assumed that the effective mass of the cantilever is independent of the interaction force. While this is certainly true for operation in a vacuum, measurements performed in air or liquid environments require careful consideration, since hydrodynamic forces imposed on an oscillating cantilever contain both dissipative and inertial components [59]. Hydrodynamic dissipative forces pose no problem, since they contribute only to the generalized damping coefficient. However, inertial forces add directly to the effective mass of the cantilever and modify its resonance frequency. Therefore, unless accounted for, such inertial loading can lead to discrepancies in the measured interaction force, since the frequency shift is interpreted solely as a change in effective cantilever stiffness in the previous theoretical formalism.

9.5.1
Governing Equations and Resonance Frequency in a Fluid

It is well known that the resonance frequency of a cantilever immersed in a fluid can differ significantly from its value in a vacuum [59, 60]. For a cantilever far from a surface, calculations and measurements [59,60] show that for typical atomic force microscope cantilevers immersed in air, the resonance frequency is reduced by several percent, whereas the quality factor is reduced by several orders of magnitude from the corresponding values in a vacuum. The effect is greatly enhanced in a liquid, with the resonance frequency and quality factor decreasing by a further order of

Fig. 9.18. Measurement of the change in resonance frequency of a cantilever immersed in water as a function of separation. **a** Frequency shift over the last 14 μm before contact. **b** Frequency shift over the last 200 nm before contact. (Reused with permission from Sader et al. [16]. Copyright 2005, Institute of Physics)

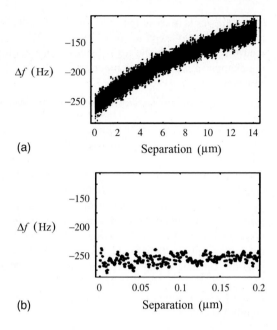

magnitude in some cases [60] from their values in air. However, the critical issue for FM-AFM force measurements is the change in inertial loading as the cantilever approaches the surface, since changes in dissipation are inherently accounted for in the theoretical formalism, as discussed earlier.

For the formalism presented in Sect. 9.4 to be valid, the following conditions must be satisfied: (1) the unperturbed resonance frequency must be independent of the tip–sample separation; (2) inertial loading must be independent of frequency, otherwise a change in frequency due to an interaction force will result in a change in effective mass.

First, we consider condition 1. We note that the hydrodynamic length scale for an atomic force microscope cantilever is typically given by its width [59]. Consequently, as a cantilever approaches the surface its inertial hydrodynamic loading will only be significantly affected if the cantilever–sample separation becomes comparable to the cantilever width. This property is demonstrated in the Fig. 9.18a, which shows the change in resonance frequency as a cantilever approaches a surface measured using FM-AFM. As expected, the length scale for the change in resonance frequency is given by the cantilever width. A magnetic particle is glued onto the backside of the cantilever to enable magnetic activation [11] and the resonance frequency of the cantilever (Nanosensors electrostatic force microscopy cantilever) in water is 13,120 Hz far from the surface. The cantilever dimensions were length 225 μm, width 28 μm and tip height 13 μm. Maximum travel by the piezo was 14 μm. The frequency shift presented is relative to the value when the cantilever is far from the surface.

Consequently, if the tip–sample separation is much smaller than the cantilever width, then a change in separation of the order of the tip–sample distance will have little effect on the resonance frequency. This feature is demonstrated in Fig. 9.18b, which presents results for the resonance frequency as a function of separation when the cantilever is in close proximity to the surface. Note that the resonance frequency is independent of separation in this case. Therefore, for practical FM-AFM force measurements where the tip is by necessity in close proximity to the surface, and the change in separation is much smaller than the cantilever width, condition 1 is satisfied.

Next, we consider condition 2. We note that the inertial hydrodynamic load is frequency-dependent [59]. This frequency dependence, however, is weak [61] and varies approximately as the square root of the oscillation frequency. Therefore, this variation can be neglected to leading order since it introduces an uncertainty of approximately 10% (calculated using the results for an unbounded fluid [59]) which is comparable to other uncertainties, e.g., errors in spring constant calibration. Condition 2 is therefore also satisfied.

Hence, for the theoretical formalism in Sect. 9.4 to be applicable to fluid measurements, the specified resonance frequency ω_{res} in (9.4) is the value when there is no interaction force F_{int} between the tip and the sample, and the cantilever is in close proximity to the surface, i.e., when the tip–sample separation is much smaller than the tip height and cantilever width, e.g., the resonance frequency shown in Fig. 9.18b. If the resonance frequency far from the surface or in a different fluid medium (such as air) is used for the unperturbed value ω_{res}, then this can lead to significant errors.

9.5.2
Validation of FM-AFM Force Measurements in a Liquid

To demonstrate the validity of using FM-AFM for quantitative measurements in a liquid, we present results of measurements obtained in OMCTS, which exhibits short-range ordering, and hence forces, on confinement. Importantly, the force is expected to be a unique function of the tip–sample separation in this system, since no bonds are formed or broken throughout the oscillation cycle of the cantilever, thus satisfying the fundamental conditions stipulated in Sect. 9.4.3. The reader is referred to [31] for a more complete description and comparison of these results. Measurements were performed against a freshly cleaved highly oriented pyrolytic graphite surface. To minimize hydrodynamic effects of the bulk standard atomic force microscope tip, and hence the required driving force, a multiwalled carbon nanotube was used as the probing tip. The carbon nanotube tip was made by attaching a nanotube onto a standard atomic force microscope cantilever tip (Nanosensors electrostatic force microscopy cantilever) with a spring constant of 3 N/m. A magnetic particle was glued onto the backside of the cantilever to enable magnetic activation [11]. The resonance frequency of the cantilever (with magnetic particle attached) was 28.19 kHz in air. When the tip was immersed in OMCTS and in close proximity to the surface (50-nm tip–sample separation), a resonance frequency of 19.54 kHz was measured, and this was used in (9.4), in accordance with the discussion in Sect. 9.5.1. Interestingly, this system shows little variation in the driving force as a function of separation [31], and hence we only focus on changes in resonance frequency as a function of separation.

Figure 9.19a shows frequency shift measurements for a range of oscillation amplitudes. Note that the frequency shift decreases as the oscillation amplitude in-

Fig. 9.19. Force measurements in OMCTS. **a** Measured relative frequency shifts as a function of separation and oscillation amplitude. **b** Recovered force versus separation curve using the data in **a** as a function of separation and oscillation amplitude. $a = 2.0$ nm (*solid line*), $a = 3.9$ nm (*dashed line*), $a = 7.2$ nm (*dotted line*). (Reused with permission from Uchihashi et al. [31]. Copyright 2004, American Institute of Physics)

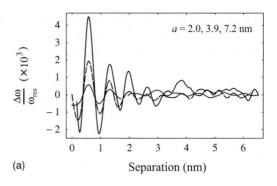

(a)

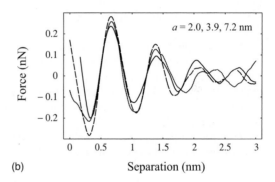

(b)

creases. This is expected, since the cantilever tip experiences a force over a smaller fraction of its oscillation cycle as the amplitude increases [3], hence decreasing the change in effective stiffness of the cantilever. Nonetheless, the force curves obtained from each of these individual frequency shift measurements using (9.16a) are in good agreement (Fig. 9.19b). A loss in sensitivity for the highest amplitude measurement ($a = 7.9$ nm) for separations greater than 3 nm is responsible for the apparent discrepancy with measurements at lower amplitudes. When the results given in Fig. 9.19b are scaled by the radius of the tip, force gradient values of the order of 10 mN/m are obtained, which are also in excellent agreement with previous independent measurements. These results demonstrate the validity of the FM-AFM technique for quantitative force measurements in a liquid, and when taken with the known capabilities of the method in a vacuum [3] establish FM-AFM as a universal technique for quantitative force measurements regardless of the environment.

9.6
Phase Detuning in FM-AFM

In this section, we examine the effects of varying the phase shift between the tip and the driving force from $\varphi = 90°$ on the coupling of the "even" and "odd" forces [62]; commonly referred to as "conservative" and "dissipative" forces, see Sect. 9.4.1.

This is essential in assessing the robustness of the FM-AFM technique and in interpreting the resulting force measurements. We also reexamine the experimental approach commonly used for achieving a 90° phase shift in FM-AFM measurements, which involves adjusting the phase to minimize the excitation amplitude far from the sample [63]. Importantly, this minimum does not formally coincide with the 90° phase point and can thus lead to significant phase detuning, particularly in liquid environments.

9.6.1
Governing Equations for Arbitrary Phase Shift

We define ω_{set} to be the oscillation frequency of the cantilever in the absence of an interaction force, i.e., far from the sample. Thus, if $\varphi = 90°$, ω_{set} coincides with ω_{res}, which is the resonance frequency of the cantilever in the absence of an interaction force.

Since the phase shift in FM-AFM measurements is fixed constant far from the sample, the phase shift φ between the cantilever-tip displacement and the driving force remains constant as the tip approaches and interacts with the sample. Consequently, all FM-AFM measurements of the change in frequency and excitation force are taken in reference to the frequency ω_{set} and the drive force $\overline{F}_0$, respectively, far from the sample.

Importantly, the relative difference in oscillation frequency ω with respect to the actual unperturbed resonance frequency ω_{res} is normally not accessible, since ω_{res} is inferred by adjusting the phase shift in the absence of an interaction force. As such, any phase error in the calibration procedure will be reflected directly in the measurement of ω_{res}. We therefore define

$$\omega = \omega_{\text{set}} + \Delta\omega , \quad F_0 = \overline{F}_0 + \Delta F_0 , \tag{9.22}$$

where $\Delta\omega$ and ΔF_0 are the frequency shift and the change in driving force, respectively, resulting from the tip–sample interaction. Substituting (9.5) and (9.22) into (9.4) and performing a Fourier analysis with respect to the phase-shifted functions $\sin(\omega t - \varphi)$ and $\cos(\omega t - \varphi)$ then gives the required results:

$$\frac{\Delta\omega}{\omega_{\text{set}}}\left(\frac{\omega_{\text{set}}}{\omega_{\text{res}}}\right)^2 + \frac{1}{2}\left[1 - \left(\frac{\omega_{\text{set}}}{\omega_{\text{res}}}\right)^2\right]\frac{\Delta F_0}{\overline{F}_0} = I_{\text{cons}} , \quad \frac{\Delta F_0}{\overline{F}_0} - \frac{\Delta\omega}{\omega_{\text{set}}} = I_{\text{diss}} ,$$

$$\tag{9.23}$$

where

$$I_{\text{cons}} = -\frac{1}{\pi k a}\int_{-1}^{1} F_{\text{even}}[z + a(1 + u)]\frac{u}{\sqrt{1 - u^2}}\,du \tag{9.24a}$$

and

$$I_{\text{diss}} = \frac{2}{\pi b}\int_{-1}^{1} \Delta\Gamma[z + a(1 + u)]\sqrt{1 - u^2}\,du , \tag{9.24b}$$

where we have used the property that the Fourier sine series only probes the even component of the force F_{even}, which is commonly referred to as the "conservative" component, and that the odd component of the force can be formally expressed in terms of a generalized damping coefficient $F_{odd} = \Gamma(z, a, \omega, w(t))\dot{w}(t)$ for a fixed minimum tip–sample separation z, where $\Delta\Gamma$ is the change in the generalized damping coefficient resulting from the interaction, i.e., $\Gamma = b + \Delta\Gamma$, see Sect. 9.4.1. Equation (9.22) can be solved for the measured observables $\Delta\omega/\omega_{set}$ and $\Delta F_0/\overline{F}_0$,

$$\frac{\Delta\omega}{\omega_{set}} = I_{cons} + \frac{\omega_{res}^2 - \omega_{set}^2}{\omega_{res}^2 + \omega_{set}^2} \left(I_{cons} - I_{diss}\right) , \tag{9.25a}$$

$$\frac{\Delta F_0}{\overline{F}_0} = I_{cons} + I_{diss} + \frac{\omega_{res}^2 - \omega_{set}^2}{\omega_{res}^2 + \omega_{set}^2} \left(I_{cons} - I_{diss}\right) , \tag{9.25b}$$

which gives the explicit coupling between the "conservative" I_{cons} and "dissipative" I_{diss} interaction force contributions.

9.6.2
Coupling of Conservative and Dissipative Forces

Several important features of the coupling between conservative and dissipative forces on the measured relative frequency shift $\Delta\omega/\omega_{set}$ and relative driving force $\Delta F_0/\overline{F}_0$ can be deduced directly from (9.25a). Importantly, we note that this coupling primarily depends on the relative difference between the unperturbed drive frequency ω_{set} and unperturbed resonance frequency ω_{res}, rather than the actual phase shift φ. This is particularly significant for cantilevers with high quality factors Q, such as encountered in UHV measurements [4], as we shall now discuss.

We first consider the effects of such coupling on the relative frequency shift $\Delta\omega/\omega_{set}$. In situations with high quality factors, large deviations of the phase from $90°$ may result in only minute changes in the oscillation frequency ω_{set} from the resonance frequency ω_{res}. Equation (9.25a) then directly indicates that the resulting coupling between conservative and dissipative forces will be commensurately small. As an example, consider a typical cantilever operating in UHV conditions with a resonance frequency of $f_{res} = \omega_{res}/(2\pi) = 150\,\text{kHz}$ and quality factor $Q = 10,000$ [63]. If the phase shift is set to $\varphi = 20°$, the corresponding unperturbed oscillation frequency will be $149.979\,\text{kHz}$, giving

$$\frac{\omega_{res}^2 - \omega_{set}^2}{\omega_{res}^2 + \omega_{set}^2} \sim 10^{-4} .$$

Assuming conservative and dissipative forces contribute equally for the purpose of discussion, i.e., I_{cons} and I_{diss} are of equal order (but not necessarily equal), (9.25) a establishes that the relative contribution of the dissipative force to the frequency shift $\Delta\omega/\omega_{set}$ will be 4 orders of magnitude smaller than that of the conservative force, despite the large phase offset. Clearly, as the quality factor is increased further, for example, by the use of quartz oscillators [4], this coupling becomes weaker for a given phase shift φ.

It is clear from (9.25b) that the relative driving force $\Delta F_0/\overline{F}_0$ always involves coupling between conservative and dissipative forces when operating on resonance. The change in this degree of coupling again only varies with the relative difference between the unperturbed drive frequency ω_{set} and the unperturbed resonance frequency ω_{res}. Consequently, rather than the phase being the key variable controlling this coupling, it is the relative frequency $\omega_{\text{set}}/\omega_{\text{res}}$; this can be very small even for large phase detuning. These findings allow for the operation of FM-AFM measurements away from the resonance peak while minimizing the effect on the measured observables $\Delta\omega/\omega_{\text{set}}$ and $\Delta F_0/\overline{F}_0$.

It is also interesting to note that the quantity $\Delta F_0/\overline{F}_0 - \Delta\omega/\omega_{\text{set}}$ in (9.23) is totally independent of the phase shift φ and thus always gives the relative energy dissipated in the interaction I_{diss}, as defined by (9.24b). Consequently, $\Delta F_0/\overline{F}_0 - \Delta\omega/\omega_{\text{set}}$ can be used to probe the frequency-dependent nature of the energy dissipated in the interaction, provided the frequency dependence of the cantilever damping coefficient b is known.

9.6.3
Operation of FM-AFM Away From the Resonance Frequency

We now examine the application of FM-AFM in quantitative force measurements away from the resonance frequency, i.e., $\varphi = 90°$. In such cases, (9.23) can be used to determine the resulting conservative and dissipative interaction forces. This is achieved using established inversion algorithms such as those presented in Sect. 9.4.4 and [9, 15, 16, 58] that enable conversion of the measured relative frequency shift and change in excitation force to conservative and dissipative forces. By using the following replacements in the current algorithms

$$\frac{\Delta\omega}{\omega_{\text{res}}} \rightarrow \frac{\Delta\omega}{\omega_{\text{set}}}\left(\frac{\omega_{\text{set}}}{\omega_{\text{res}}}\right)^2 + \frac{1}{2}\left[1 - \left(\frac{\omega_{\text{set}}}{\omega_{\text{res}}}\right)^2\right]\frac{\Delta F_0}{\overline{F}_0} \tag{9.26a}$$

and

$$\frac{\Delta F_0}{\overline{F}_0} - \frac{\Delta\omega}{\omega_{\text{res}}} \rightarrow \frac{\Delta F_0}{\overline{F}_0} - \frac{\Delta\omega}{\omega_{\text{set}}}, \tag{9.26b}$$

these methods [9, 15, 16, 58] are rigorously applicable, and facilitate measurement of conservative and dissipative forces as a function of frequency using the FM-AFM technique. This approach enables the FM-AFM technique to be used in a quantitative capacity in variable-frequency measurements using a single cantilever, which may be particularly useful for investigating systems where the interaction force is frequency-dependent [64].

9.6.4
Calibration of 90° Phase Shift

In this section, we reexamine the procedure commonly used to determine the $\varphi = 90°$ point in FM-AFM measurements. This point is typically achieved by varying the phase shift until the excitation force/amplitude is minimized, while maintaining constant tip amplitude [63]. This method implicitly assumes that the quality factor greatly exceeds unity. However, it is commonly used to set the 90° phase shift, regardless of the quality factor of the cantilever. Importantly, this method will not necessarily give the resonance condition $\varphi = 90°$, since the frequency where the peak in the amplitude resonance curve lies does not formally coincide with the resonance frequency of the cantilever. To understand this connection, we present the relation between the phase φ, quality factor Q, resonance frequency ω_{res} and drive frequency ω_{set}, in the absence of a tip–sample interaction force:

$$\cot \varphi = Q \left(\frac{\omega_{res}}{\omega_{set}} - \frac{\omega_{set}}{\omega_{res}} \right) . \tag{9.27}$$

The frequency where the peak in the amplitude resonance curve occurs is given by

$$\omega_{peak} = \omega_{res} \sqrt{1 - \frac{1}{2Q^2}} . \tag{9.28}$$

(A quality factor of $Q \leq 1/\sqrt{2}$ gives a peak frequency of $\omega_{peak} = 0$; therefore, this calibration procedure pertains only to $Q > 1/\sqrt{2}$ as required for dynamic operation.)

From these basic equations, we find that if the peak frequency ω_{peak} is used to set the 90° phase point, then the true phase shift is

$$\varphi = \tan^{-1} \left(2Q \sqrt{1 - \frac{1}{2Q^2}} \right) . \tag{9.29}$$

Consequently, provided $Q \gg 1$, the resulting phase shift φ will be very close to 90°. However, in ambient and liquid environments, where the quality factor can be small, the resulting phase shift obtained from this procedure can significantly deviate from 90°. This property is illustrated in Fig. 9.20, where the resulting phase shift φ obtained from (9.29) is plotted as a function of the quality factor Q. At high quality factors we recover the expected result of $\varphi = 90°$; however, for quality factors approaching unity, as is typical in liquid measurements [31,59], the resulting phase shift will be significantly lower than the expected 90°.

Since the coupling between conservative and dissipative forces depends only on the relative difference between the resonance frequency ω_{res} and the set frequency ω_{set} in the absence of an interaction force, the resulting coupling will only be significant provided the difference between these two frequencies is not small. For this phase-calibration procedure, (9.25a) and (9.25b) become

$$\frac{\Delta\omega}{\omega_{set}} = I_{cons} + \frac{I_{cons} - I_{diss}}{4Q^2 - 1} \tag{9.30a}$$

and

$$\frac{\Delta F_0}{F_0} = I_{cons} + I_{diss} + \frac{I_{cons} - I_{diss}}{4Q^2 - 1} . \tag{9.30b}$$

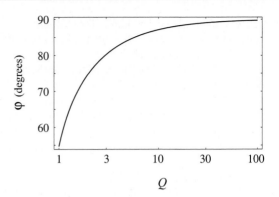

Fig. 9.20. The phase shift φ (degrees) as a function of the quality factor Q obtained by setting the drive frequency $\omega_{set} = \omega_{peak}$. (Reused with permission from Sader and Jarvis. [62]. Copyright 2006, American Physical Society)

While coupling is negligible for high quality factors $Q \gg 1$, significant coupling can occur for low quality factors, and alternative calibration of the 90° phase point may be required to avoid such coupling.

One approach to determine this 90° phase point is to measure the thermal noise spectrum of the cantilever. For liquid environments, this should be performed in close proximity to the surface, but sufficiently far from the surface so that no interaction forces are present, see Sect. 9.5.1. Fitting this noise power spectrum to the response of a damped harmonic oscillator, as in [60],

$$S(\omega) = \frac{A}{\left(\omega^2 - \omega_{res}^2\right)^2 + \frac{\omega^2 \omega_{res}^2}{Q^2}} , \tag{9.31}$$

where A is a constant, enables determination of the true resonance frequency ω_{res} and quality factor Q in a liquid. The phase in the self-excitation circuit can then be adjusted so as to ensure that the cantilever oscillates at its resonance frequency ω_{res}.

9.7
Future Prospects

FM-AFM is emerging as a robust technique for quantitative force measurement and imaging in liquids owing to the parallel development of instrumentation and theory. Recent applications of FM-AFM to biological systems have served to highlight the benefits of applying the technique to problems in biology and we anticipate that this area of application is likely to grow significantly in the near future. Many other areas of opportunity also exist involving the solid–fluid interface, for example, it would appear that no one has yet applied the technique to electrochemistry, or chemistry in general at the solid–fluid interface.

There are also some experimental challenges remaining as the technique still has some weaknesses, including the speed of scanning relative to methods with fewer feedback loops, such as phase modulation AFM, and difficulties associated with stability when imaging surfaces with high surface roughness. In addition, the intrinsic instabilities observed in Sect. 9.3.2, due to hopping of the tip between hydration layers, may require a careful choice of setpoint to avoid multiple equivalent feedback points when imaging in the presence of a liquid. However, they also indicate that FM-AFM can open up an additional avenue for understanding the role of water and ions in biological function.

It is refreshing to see that 20 years after the invention of AFM, instrumentation developments still have the capability of opening up major new areas of unexplored terrain.

References

1. Albrecht TR, Grütter P, Horne D, Ruger D (1991) J Appl Phys 69:668
2. Giessibl FJ (1995) Science 267:68
3. Garcia R, Perez R (2002) Surf Sci Rep 47:197
4. Giessibl FJ (2003) Rev Mod Phys 75:949
5. Sugimoto Y, Pou P, Abe M, Jelinek P, Perez R, Morita S, Custance O (2007) Nature 446:64
6. Giessibl FJ (1997) Phys Rev B 56:16010
7. Martin Y, Williams CC, Wickramasinghe HK (1987) J Appl Phys 61:4723
8. Lee M, Jhe W (2006) Phys Rev Lett 97:036104
9. Sader JE, Jarvis SP (2004) Appl Phys Lett 84:1801
10. Jarvis SP, Uchihashi T, Ishida T, Tokumoto H, Nakayama Y (2000) J Phys Chem B 26:6091
11. Jarvis SP, Oral A, Weihs TP, Pethica JB (1993) Rev Sci Instrum 64:3515
12. Dai H, Hafner JH, Rinzler AG, Colbert DT, Smalley RE (1996) Nature 384:147
13. Fukuma T, Kimura M, Kobayashi K, Matsushige K, Yamada H (2005) Rev Sci Instrum 76:053704
14. Fukuma T, Higgins MJ, Jarvis SP (2007) Phys Rev Lett 98:106101
15. Giessibl FJ (2001) Appl Phys Lett 78:123
16. Sader JE, Uchihashi T, Higgins MJ, Farrell A, Nakayama Y, Jarvis SP (2005) Nanotechnology 16:S94
17. Humphris ADL, Tamayo J, Miles MJ (2000) Langmuir 16:7891
18. Kageshima M, Jensenius H, Dienwiebel M, Nakayama Y, Tokumoto H, Jarvis SP, Oosterkamp TH (2002) Appl Surf Sci 188:440
19. Umeda N, Ishizaki S, Uwai H (1991) J Vac Sci Technol B 9:1318
20. Ratcliff GC, Erie DA, Superfine R (1998) Appl Phys Lett 72:1911
21. Hoogenboom BW, Hug HJ, Pellmont Y, Martin S, Frederix PLTM, Fotiadis D, Engel A (2006) Appl Phys Lett 88:193109
22. Kawakatsu H, Kawai S, Kobayashi D, Kitamura S, Meguro S (2006) e-J Surf Sci Nanotechnol 4:110
23. Giessibl FJ (1999) Appl Surf Sci 140:352
24. Meyer G, Amer NM (1988) Appl Phys Lett 53:1045
25. Fukuma T, Jarvis SP (2006) Rev Sci Instrum 77:043701
26. Sarid D (1994) Scanning force microscopy with applications to electronic, magnetic and atomic forces. Oxford University Press, Oxford
27. Okajima T, Sekiguchi H, Arakawa H, Ikai A (2003) Appl Surf Sci 210:68
28. Fukuma T, Kobayashi K, Matsushige K, Yamada H (2005) Appl Phys Lett 87:034101
29. Kobayashi K, Yamada H, Matsushige K (2002) Appl Surf Sci 188:430
30. Jarvis SP, Ishida T, Uchihashi T, Nakayama Y, Tokumoto H (2001) Appl Phys A 72:S129
31. Uchihashi T, Higgins MJ, Yasuda S, Jarvis SP, Akita S, Nakayama Y, Sader JE (2004) Appl Phys Lett 85:3575
32. Uchihashi T, Higgins MJ, Sader JE, Jarvis SP (2005) Nanotechnology 16:S40
33. Fukuma T, Kobayashi K, Matsushige K, Yamada H (2005) Appl Phys Lett 86:193108

34. Erlandsson R, Hadziioannou G, Mate CM, McClelland GM, Chiang S (1988) J Chem Phys 89:5190
35. Drake B, Prater CB, Weisenhorn AL, Gould SA, Albrecht TR, Quate CF, Cannell DS, Hansma HG, Hansma PK (1989) Science 243:1586
36. Tamayo J, Humphris ADL, Owen RJ, Miles MJ (2001) Biophys J 81:526
37. Sekiguchi H, Okajima T, Arakawa H, Maeda S, Takashima A, Ikai A (2003) Appl Surf Sci 210:61
38. Okajima T, Tokumoto H (2004) Jpn J Appl Phys 43:4634
39. Ebeling D, Hölscher H, Anczykowski B (2006) Appl Phys Lett 89:203511
40. Yang C-W, Hwang I-S, Chen YF, Chang CS, Tsai DP (2007) Nanotechnology 18:084009
41. Morita S, Yamada H, Ando T (2007) Nanotechnology 18:08401
42. Humphris ADL, Antognozzi M, McMaster TJ, Miles MJ (2002) Langmuir 18:1729
43. Higgins MJ, Riener CK, Uchihashi T, Sader JE, McKendry R, Jarvis SP (2005) Nanotechnology 16:S85
44. Higgins MJ, Sader JE, Jarvis SP (2006) Biophys J 90:640
45. Higgins MJ, Polcik M, Fukuma T, Sader JE, Nakayama Y, Jarvis SP (2006) Biophys J 91:2532
46. Fukuma T, Higgins MJ, Jarvis SP (2007) Biophys J 92:3603
47. Groves JT, Boxer SG, McConnell HM (2000) J Phys Chem B 104:11409
48. Ohki S, Arnold K (2000) Colloids Surf B 18:83
49. Herbette L, Napolitano CA, McDaniel RV (1984) Biophys J 46:677
50. Hermann TR, Jayaweera AR, Shamoo AE (1986) Biochemistry 25:5834
51. Binder H, Zschörnig (2002) Chem Phys Lipids 115:39
52. Berkowitz ML, Bostick DL, Pandit S (2006) 106:1527
53. Garcia-Manyes S, Oncins G, Sanz F (2005) Biophys J 89:1812
54. Böckmann RA, Hac A, Heimburg T, Grubmüller H (2003) Biophys J 85:1647
55. Sader JE (2002) In: Encyclopedia of surface and colloid science. Dekker, New York
56. Gotsmann B, Seidel C, Anczykowski B, Fuchs H (1999) Phys Rev B 60:11051
57. Abramowitz M, Stegun IA (1975) Handbook of mathematical functions. Dover, New York
58. Dürig U (2000) Appl Phys Lett 76:1203
59. Sader JE (1998) J Appl Phys 84:64
60. Chon JWM, Mulvaney P, Sader JE (2000) J Appl Phys 87:3978
61. Green CP, Sader JE (2005) Phys Fluids 17:073102
62. Sader JE, Jarvis SP (2006) Phys Rev B 74:195424
63. Holscher H, Gotsmann B, Allers W, Schwarz UD, Fuchs H, Wiesendanger R (2001) Phys Rev B 64:075402
64. Patil S, Matei G, Oral A, Hoffmann PM (2006) Langmuir 22:6485

10 Kelvin Probe Force Microscopy: Recent Advances and Applications

Yossi Rosenwaks · Oren Tal · Shimon Saraf · Alex Schwarzman ·
Eli Lepkifker · Amir Boag

Abstract. The Kelvin probe force microscopy technique is perhaps the most powerful tool for measuring the work function and the electric potential distribution with nanometer resolution. The work function is one of the most important values characterizing the property of a surface. Chemical and physical phenomena taking place at the surface are strongly affected by the work function. Although the work function is defined as a macroscopic concept, it is necessary to consider its microscopic local variations in understanding the behavior of semiconductor surfaces, interfaces and devices. In this chapter we describe and discuss recent applications of Kelvin probe force microscopy in the study of semiconductors. The method is introduced in the first section, and the second section examines the factors affecting the sensitivity and resolution of Kelvin probe force microscopy in general, and in semiconductor measurements in particular. An efficient numerical analysis of the electrostatic interaction between the measuring atomic force microscope tip and the semiconductor surface has allowed us to derive a point-spread function of the measuring tip and to restore the actual surface potential from measured images in almost real time. The third section describes the use of Kelvin probe microscopy to determine the density of surface and bulk states in inorganic and organic semiconductors, respectively. In inorganic semiconductors the method is based on scanning a cross-sectional pn junction; as the tip scans the junction, the position of the surface states relative to the Fermi level changes, thereby changing the surface potential. The energy distribution is then obtained by fitting the measured surface potential. The method is applied to various semiconductor (110) surfaces where a quantitative states distribution across most of the bandgap is obtained. In the case of organic semiconductors the density of states in obtained by injecting charge carriers into the channel of a bottom gate organic transistor. The measurement of the Fermi level shift together with the charge concentration allows us to derive the density of states of the highest occupied molecular orbital band.

Key words: Kelvin probe force microscopy, Electrostatic force microscopy, Semiconductor surface states, Density of states

10.1
Kelvin Probe Force Microscopy

Kelvin probe force microscopy (KPFM) is a method for measuring the surface potential distribution with nanometer resolution. Since its introduction in 1991 by Nonenmacher et al. and together with the fact that KPFM gives a direct quantitative measurement of the surface potential distribution, it has found many diverse applications in several fields. In materials research it has been used for work function mapping [1], ordering measurements in III–V compound semiconductors [2], local surface photovoltage and surface photovoltage spectroscopy, surface states and defects under different ambient conditions, domain characterization in ferroelectric materials, measurement of organics and self-assembled monolayers, and many more. KPFM has also proved to be a very important tool for potential mapping of passive and active semiconductor devices like pn junctions [3], resistors, and n–i–p–i heterostructures [4], high electron mobility transistors [5,6], light-emitting diodes [7,8], solar cells, and organic and polymer-based transistors. Recently several groups have reported high-resolution measurements of low-dimensional structures like 5-nm CdSe nanocrystal [9] single-quantum and multiquantum well structures. Although observations of atomic-scale features in KPFM were reported 5 years ago [10,11], their origin is poorly understood mainly owing to the long-range nature of the electrostatic forces, discussed in detail in the next section. Another important aspect of local potential measurements is related to quantitative two-dimensional dopant mapping [12]. Several groups have used KPFM for two-dimensional dopant profiling [13]; however, it was found that it is not the ideal tool for this purpose for two main reasons. One is the poor lateral resolution compared with the contact scanning capacitance and scanning spreading resistance methods, and second, and even more important, is that with its dynamic range of 1 V (typical bandgap energy) and sensitivity of a few millivolts it is practically impossible to quantify doping levels in the range required by the microelectronics industry.

The KPFM method is described briefly below. Sections 10.2.1–10.2.4 examine the factors affecting the sensitivity and resolution of KPFM in general, and in semiconductor measurements in particular. An efficient numerical analysis of the electrostatic interaction between the measuring atomic force microscope (AFM) tip and the semiconductor surface has allowed us to derive a point-spread function (PSF) of the measuring tip and to reconstruct the actual surface potential from KPFM images in almost real time.

The contact potential difference (CPD) between two materials, for example, between an AFM tip and a sample, is defined as

$$V_{CPD} = \frac{\phi_{sample}}{q} - \frac{\phi_{tip}}{q} \equiv \frac{\phi}{q} , \tag{10.1}$$

where ϕ_{tip} and ϕ_{sample} are the work functions of the tip and the sample, respectively, and q is the elementary charge. Therefore, if an AFM tip and a semiconductor with different work functions are held in close proximity to each other a force will develop between them, owing to the potential difference V_{CPD}; this is described schematically in Fig. 10.1. When the two materials are not connected, their local vacuum levels are aligned but there is a difference in their Fermi levels. Upon electrical connection, the Fermi levels will align by means of electron current, as shown in Fig. 10.1a. The two materials (electrodes) are now charged and there is a difference in their local vacuum levels. Owing to the charging of the tip and the sample, an electrostatic force develops as shown in Fig. 10.1b.

This force can be nullified by applying an external bias between the tip and the sample. The magnitude of this bias is the CPD and its sign depends on whether it is applied to the sample or to the tip.[1] A typical KPFM measurement is conducted

---

[1] The sign of the measured V_{CPD} will be positive (negative) if the nullifying voltage is applied to the sample (tip), respectively; see [7,8].

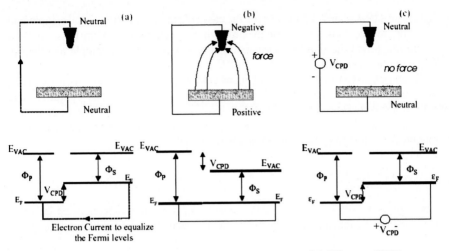

Fig. 10.1. Definition and basic measurement setup of contact potential difference (CPD)

in the following way. An AC bias at a frequency ω is applied between the tip and the sample. It is shown below [14] that the force component at this frequency is proportional to the CPD and, therefore, can be nullified using a feedback loop whose input is the component of the electrostatic force at a frequency ω. The most naïve way to derive this force is to treat the tip–sample system as a parallel-plate capacitor with one plate being the tip apex, and the other the sample underneath it. Under this assumption the force which is just the derivative of the electrostatic energy with respect to the tip–sample separation, z, is given by

$$F = -\frac{\partial U}{\partial Z} = -\frac{1}{2}\frac{\partial C}{\partial Z}V^2 \approx \frac{C}{d}V^2 \,, \tag{10.2}$$

where the electrostatic energy U is given for a parallel-plate capacitor configuration by $U = CV^2/2$, with C the tip–sample capacitance and V the potential difference between the AFM tip and the sample. Using the following expression for the potential difference: $V = V_{\mathrm{DC}} \pm V_{\mathrm{CPD}} + V_{\mathrm{AC}}\sin(\omega t)$, where V_{DC} is a nullifying voltage applied in order to measure the CPD, and inserting it in (10.2) gives

$$F \propto \frac{1}{2Z^2}\left[(V_{\mathrm{CPD}} - V_{\mathrm{DC}})^2 + 2(V_{\mathrm{CPD}} - V_{\mathrm{DC}})V_{\mathrm{AC}}\sin(\omega t) + V_{\mathrm{AC}}^2\sin^2(\omega t)\right] \tag{10.3}$$

and as expected the force at frequency ω is proportional to the CPD.

Hence, on the basis of (10.3) the sign of V_{DC} will be different if the nullifying voltage is applied to the tip or to the sample. The posteriori DC voltage difference is thus given for the two cases as

$$V_{\mathrm{CPD}}^{\mathrm{P}} = -\frac{\phi_{\mathrm{tip}}}{e} - \left(\frac{\phi_{\mathrm{sample}}}{-e} + V_{\mathrm{ext}}\right) = V_{\mathrm{CPD}} - V_{\mathrm{ext}} \,, \tag{10.4a}$$

$$V_{\mathrm{CPD}}^{\mathrm{P}} = -\frac{\phi_{\mathrm{tip}}}{e} - \frac{\phi_{\mathrm{sample}}}{-e} + V_{\mathrm{ext}} = V_{\mathrm{CPD}} + V_{\mathrm{ext}} \,. \tag{10.4b}$$

Equations (10.4a) and (10.4b) are for the cases of voltage applied to the sample and to the tip, respectively. Following the nullifying procedure, we obtain $V_{\text{ext}} = \pm V_{\text{CPD}}$ where the $+$ and the $-$ refer to the external bias applied to the sample and the tip, respectively.

Equation (10.3) although used by many authors is strictly correct only for a metallic sample and when the tip–sample system can be approximated as a parallel-plate capacitor configuration. Hudlet et al. [14] have presented a detailed one-dimensional analysis of the electrostatic force between a tip and a semiconductor, and have shown that the force at a frequency ω can be expressed as

$$F_\omega = \frac{Q_s}{\varepsilon_0} C_{\text{eff}} V_{\text{AC}} \sin(\omega t) \, , \tag{10.5}$$

where Q_s is the semiconductor surface charge, ε_0 is the dielectric constant, and C_{eff} is the effective capacitance of the electrode–air/vacuum–semiconductor system.

10.2
Sensitivity and Spatial Resolution in KPFM

10.2.1
Tip–Sample Electrostatic Interaction

It is well known that the finite tip size in scanning probe microscopes has a profound effect on the measured image. In electrostatic force based microscopies, the effect of the measuring tip is much larger because the measured forces have an infinite range. Tip effects in electrostatic force and KPFM were discussed and analyzed by several authors [15]. One of the simplest models was suggested by Hochwitz et al. [16], who modeled the tip by a series of (staircase) parallel-plate capacitors. Hudlet et al. [17] presented an analytical evaluation of the electrostatic force between a conductive tip and a metallic surface, while Belaidi et al. [18] calculated the forces and estimated the resolution in a similar system. Jacobs et al. [19] extended the calculations for the case of a semiconductor sample, by replacing its surface by a set of ideal conductors with mutual capacitances between them. In addition, they have shown that the KPFM images of infinitely large conducting surfaces are two-dimensional convolutions of the actual surface potential distribution with a transfer function, defined by the tip geometry [20].

To the best of our knowledge, in almost all the papers where semiconductor KPFM measurements have been analyzed, the semiconductor sample was replaced by a surface with a fixed or variable potential. This is only valid for the case of a weakly interacting tip–sample system, i.e., when there is no tip-induced band-bending phenomenon. As described above, a typical KPFM measurement is conducted by nullifying the electrostatic force at a frequency of ω, using an external bias V_{DC}. Thus, a calculation of the KPFM signal amounts to finding the voltage applied to the tip or to the sample, V_{DC}, that minimizes the total electrostatic force at the frequency ω. Below we calculate the static (DC) electrostatic force; i.e., it is assumed that the AC voltage has a negligible effect on the tip–sample forces. This is justified on the basis of the fact that the frequency ω used in the measurements (more than 300 kHz) is high and in addition the AC modulation amplitude applied to the tip is very low (100 mV).[2]

The electrostatic force was calculated as follows. First the electric field, E, for the semiconductor–air/vacuum–tip system is calculated using the Poisson equation with appropriate boundary conditions. Assuming that the tip is a perfect conductor, one can then calculate the electrostatic force by integrating the Maxwell stress over the entire tip surface:

$$F = \frac{1}{2\varepsilon_0} \int_{S^t} \sigma^2(r)\hat{n}\,dS \,, \tag{10.6}$$

where, σ is the tip surface charge density, dS is a tip surface element, and $\hat{n}$ is the outward normal unit vector. A similar approach for electrostatic force calculations was applied in the past for a metallic sample where the sample surface potential was assumed to be constant [18]. However, here we calculate whether the presence of the tip has any significant effect on the semiconductor surface potential, i. e., is there any tip–induced band bending at the semiconductor surface? Thus, our calculation takes into account the electrostatic energy present when the semiconductor energy bands are not flat owing to the presence of surface states and/or owing to tip-induced band bending. Hudlet et al. have shown that in the one-dimensional case, the tip–sample system can be modeled as two capacitors in series. Thus, when the distance between the sample and the tip is reduced, the sample capacitance cannot be neglected and it changes the force acting on the tip, and the measured CPD.

As an example, we have calculated the surface-induced band bending for a tip, having a potential of 0.1 V higher, then the sample surface ($n = 5 \times 10^{17}\,cm^{-3}$ GaP with no charged surface states, or zero band bending) and located 5 nm above it; the result is shown in Fig. 10.2. The figure shows that the surface band bending, expressed as $E_F - E_{is}$, where E_{is} is the surface Fermi level position, due to the presence of the biased tip, is zero everywhere except for a small region in the middle where the band bending is less than 6 mV. A similar calculation conducted for a tip–sample surface potential difference of 0.6 V resulted in an induced surface band bending of around 38 mV [21].

The small-induced band bending effect can be explained in the following way. The tip–vacuum–semiconductor system can be modeled as two capacitors, a tip–vacuum–semiconductor surface and the semiconductor space charge region (SCR) connected in series. Thus, an external voltage (in the present case it is the semiconductor–tip CPD) will drop mainly on the smaller of the two capacitors. The SCR capacitance is typically much larger (except for very low doped semiconductors), thus causing the voltage to drop mainly between the tip and the sample surface, and hence inducing a negligible band bending in the SCR. It must be reemphasized that in KPFM measurements the CPD between the tip and the sample is nullified, so typically the potential difference between any point on the tip, and on the sample

---

[2] We have studied extensively the effect of the AC modulation amplitude and frequency in the past; it was found [7, 8] that even for much larger V_{AC} amplitudes and at lower frequencies, the measured CPD is unchanged.

Fig. 10.2. Calculated local band bending, expressed as $E_F - E_{is}$ for GaP with no surface states, a tip–sample distance of 5 nm and an applied bias of $V_{tip} = 0.1$ V between the tip and the sample. The protrusion in the center is the tip-induced band bending at the GaP surface

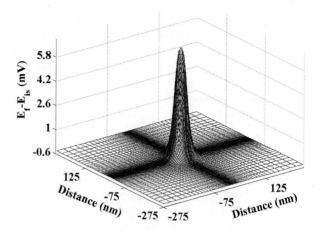

surface at a close distance to it, will be even lower than the 0.1 V used in calculating the potential in Fig. 10.2. This conclusion simplifies tremendously the simulation of semiconductor KPFM measurements. This is because it allows us to represent the whole semiconductor sample by a surface with a varying potential; the three-dimensional potential of the tip–sample system is then calculated using an integral equation based boundary element method (BEM), which is a much faster numerical process as shown in the following section.

10.2.2
A Fast Algorithm for Calculating the Electrostatic Force

Figure 10.3 shows schematics of the tip–sample system used for the BEM calculations. A perfectly conducting tip with a surface S^t is at a height d above a perfectly planar conducting sample surface S located at the $z = 0$ plane. The tip is scanning the sample and the vector $\rho_t = (x_t, y_t)$ represents the horizontal shift of the tip in the global coordinate system.

We use a notation where for any point, $r = (\rho, z)$, the two-dimensional vector $\rho = (x, y)$ represents the horizontal coordinates, whereas z denotes the vertical one. In addition, we define a local coordinate system attached to the tip. In the global coordinate system, a point on the tip surface $r \in S^t$ is represented as $r = \tilde{r} + \rho_t$, where $\tilde{r}$ are the coordinates of the point in the fixed coordinate system for the tip. In KPFM measurements, the CPD at each tip position, ρ_t, is the potential applied to the tip $V(\rho_t)$ to give minimum vertical (z-directed) force; this is equivalent to nullifying the first harmonic modulated force [17]. Let $\Phi(r|\rho_t)$ denote the electrostatic potential at point r when the tip is located at ρ_t. On the basis of the assumption that the presence of the tip is not modifying the surface potential, $\forall \rho_t \Phi(\rho, z = 0^+|\rho_t) = \Phi^s(\rho)$. The surface of the tip is assumed to be equipotential, namely, $\Phi(r|\rho_t)|_{r \in S^t} = V(\rho_t)$. Below we calculate the actual surface potential, $\Phi^s(\rho)$, on the basis of the tip potential measurements, $V(\rho_t)$.

The surface potential is represented by a dipole layer of density $\eta(\rho) = \Phi^s(\rho)/\varepsilon_0$ located on the top of a grounded plane (the sample) at $z = 0$. The total potential at

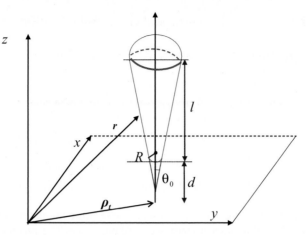

Fig. 10.3. The tip–sample system used for the electrostatic calculations. A perfectly conducting tip with a length l, half aperture angle θ_0, and spherical apex radius R is scanning the sample at a height d. The vectors ρ_t and r represent the position of the tip axis in the (x, y) plane and a point on the tip surface in the in the (x, y, z) space, respectively

a point r on the tip surface is given by the superposition

$$\Phi(r|\rho_t) = \Phi^t(r|\rho_t) + \Phi^d(r) , \tag{10.7}$$

where $\Phi^t(r|\rho_t)$ is the potential produced by the charges residing on the tip, and $\Phi^d(r)$ is the potential of the dipole layer. The potential of the tip charges and their images can be represented as

$$\Phi^t(r|\rho_t) = \int_{S^t} \left[G(r - r') - G(r - r'') \right] \sigma(r'|\rho_t) \, \mathrm{d}s' \equiv \mathcal{A}\sigma , \tag{10.8}$$

where $r'' = (\rho', -z')$ is the image point of $r' = (\rho', z')$, $G(r) = 1/(4\pi\varepsilon_0|r|)$ is the three-dimensional electrostatic Green function, and $\mathcal{A}$ is an operator yielding the contribution to the tip potential due to the charge density σ on the tip surface. Let $\Phi^d(\tilde{r}|\rho_t) = \Phi^d(r = \tilde{r} + \rho_t)$ represent the dipole layer contribution at a point $\tilde{r}$ on the tip located at ρ_t. We can express the dipole contribution to the potential as a convolution

$$\Phi^d(\tilde{r}|\rho_t) = D(\tilde{r}|\rho_t) * \Phi^s(\rho_t) , \tag{10.9}$$

where

$$D(\tilde{r}|x') = 2\varepsilon_0 \left. \frac{\partial G(r - r')}{\partial z'} \right|_{z'=0} \tag{10.10}$$

and the asterisk represents a two-dimensional convolution with respect to ρ_t.

Imposing the boundary condition $\Phi(r|\rho_t)|_{r \in S^t} = V(\rho_t)$ and substituting (10.8) in (10.7), we obtain an integral equation for the tip surface charge density:

$$\mathcal{A}\sigma = V(\rho_t) - \Phi^d(\tilde{r}|\rho_t) \quad \tilde{r} \in S^t . \tag{10.11}$$

Equation (10.11) can be approximated via the BEM and expressed in a matrix notation as

$$[A]\boldsymbol{\sigma} = V\mathbf{1} - \boldsymbol{\Phi}^{\mathrm{d}} , \tag{10.12}$$

where $[A]$ is an N by N matrix representing operator $\mathcal{A}$, and $\boldsymbol{\sigma}$, $\mathbf{1}$, and $\boldsymbol{\Phi}^{\mathrm{d}}$ are N-vectors representing the tip charge density, and the dipole layer potential, respectively. The system of linear equations represented by (10.12) can be formally solved as

$$\boldsymbol{\sigma} = V\boldsymbol{a} - [A]^{-1}\boldsymbol{\Phi}^{\mathrm{d}} , \tag{10.13}$$

where $[A]^{-1}$ is the inverse matrix of $[A]$ and $\boldsymbol{a} \equiv [A]^{-1}\mathbf{1}$ is a sum of all columns of $[A]^{-1}$.

The tip potential, V, is determined on the basis of the minimum force condition employed in the KPFM measurements described above. The z-component of the electrostatic force acting on the tip is given by

$$F_z = \frac{1}{2\varepsilon_0}\int_{S^{\mathrm{t}}} \sigma^2(\boldsymbol{r})(\hat{\boldsymbol{n}} \cdot \hat{\boldsymbol{z}})\,\mathrm{d}s \equiv B\sigma^2 , \tag{10.14}$$

where $\hat{\boldsymbol{n}}$ is the outward normal unit vector and we define B as a linear functional relating the vertical force to the charge density. In discrete form, the vertical force can be approximated as

$$F_z = \boldsymbol{\sigma}^{\mathrm{t}}[B]\boldsymbol{\sigma} , \tag{10.15}$$

where $[B]$ is an N by N diagonal matrix stemming from the BEM discretization of $\mathcal{B}$. The minimum-force condition $\partial F_z/\partial V = 0$ can be expressed via (10.15) as

$$\boldsymbol{a}^{\mathrm{t}}[B]\boldsymbol{\sigma} = 0 , \tag{10.16}$$

where, based on (10.13), we made a substitution $\partial\boldsymbol{\sigma}/\partial V = \boldsymbol{a}$. Substituting (10.13) in (10.16), we obtain an explicit expression for the measured tip potential in terms of the potential produced by the dipole layer:

$$V(\overline{\boldsymbol{x}}) = \frac{\boldsymbol{a}^{\mathrm{t}}[B][A]^{-1}\boldsymbol{\Phi}^{\mathrm{d}}(\overline{\boldsymbol{x}})}{\boldsymbol{a}^{\mathrm{t}}[B]\boldsymbol{a}} . \tag{10.17}$$

Finally, using (10.18), we can write the measured tip potential as a convolution of the actual sample surface potential $\Phi^{\mathrm{s}}(\boldsymbol{\rho}_{\mathrm{t}})$ with a continuous PSF as

$$V(\boldsymbol{\rho}_{\mathrm{t}}) = h_{\mathrm{c}}(\boldsymbol{\rho}_{\mathrm{t}}) * \Phi^{\mathrm{s}}(\boldsymbol{\rho}_{\mathrm{t}}) , \tag{10.18}$$

where

$$h_{\mathrm{c}}(\boldsymbol{\rho}_{\mathrm{t}}) \equiv \frac{\boldsymbol{a}^{\mathrm{t}}[B][A]^{-1}\boldsymbol{D}(\boldsymbol{\rho}_{\mathrm{t}})}{\boldsymbol{a}^{\mathrm{t}}[B]\boldsymbol{a}} \tag{10.19}$$

and $\boldsymbol{D}(\boldsymbol{\rho}_{\mathrm{t}})$ is an N-vector obtained via BEM discretization of $\boldsymbol{D}(\tilde{r}|\boldsymbol{\rho}_{\mathrm{t}})$.

In practice, scanning of the sample is conventionally performed with a discrete step Δ, such that $\rho_t = (m_x\Delta, m_y\Delta)$, where m_x and m_y are integers. By the same token, we can use a discrete approximation of the surface potential

$$\Phi^s(\rho) = \sum_{m_x, m_y} \Phi_d^s(m) b(\rho - m_x\Delta\hat{x} + m_y\Delta\hat{y}) , \tag{10.20}$$

where $\Phi_d^s(m)$ with $m = (m_x, m_y)$ are the discrete samples of the surface potential and $b(\rho)$ is a pixel interpolation function. Substituting $\rho_t = m_x\Delta\hat{x} + m_y\Delta\hat{y}$ and (10.20) in (10.18) and exchanging the order of summation and integration, we obtain the discrete counterpart of (10.18)

$$V(m) = h_d(m) * \Phi_d^s(m) , \tag{10.21}$$

where the asterisk denotes a two-dimensional discrete convolution with respect to m, and $h_d(m)$ is the discrete PSF computed by convolving the continuous PSF given by (10.19) with the pixel interpolation function. Equation (10.21) and its one-dimensional forms constitute the basis for the reconstruction of the actual surface potentials presented in Sect. 10.2.4. The convolutional nature of (10.21) is of major importance in terms of the numerical feasibility of the proposed approach.

10.2.3
Noise in KPFM Images

The electrostatic analysis presented in the previous sections did not include the noise inevitably present in any KPFM measurement. Therefore, (10.21) has to be modified in the presence of noise to

$$V(m) = h_d(m) * \Phi_d^s(m) + n(m) , \tag{10.22}$$

where n is the noise of the measured KPFM image. The noise in KPFM measurements is mainly due to the thermal noise, and other sources present in any AFM measurement, i.e., mechanical and electrical noise.

Performing many measurements has shown that in most cases the dominant noise factors are the image acquisition rate and the lock-in amplifier used for the measurement of the electrostatic force. Therefore changing these two parameters while scanning featureless samples gave noise spectra used for the image restoration.

Figure 10.4 demonstrates that the root mean square of the measured noise is smallest for a scanning rate of around 3–4 Hz. Such histograms have enabled us to construct the noise spectral distribution assuming Gaussian statistics, and to use it in the deconvolution of (10.23). Figure 10.5 shows a typical two-dimensional noise spectrum obtained by measuring a flat sample at a single point. Such a measurement performed at several scanning rates and with various lock-in amplifier time constants will give the dependence of the noise spectra on the above parameters.

10.2.4
Deconvolution of KPFM Images

The two-dimensional Fourier transform of the PSF h_d is concentrated at low spatial frequencies similar to the noise. In the case of additive noise, the Wiener filter [22] provides an optimal estimator in terms of minimum square error. The filter is defined in the spatial frequency domain as

$$W_{opt}(f) = \frac{S_{VV}(f) - S_{nn}(f)}{S_{VV}(f)} \frac{1}{H(f)}, \tag{10.23}$$

where S_{VV} and S_{nn} are the spectral power densities of the signal and noise respectively, and f is the two-dimensional spatial frequency. Also in (10.23), H is the

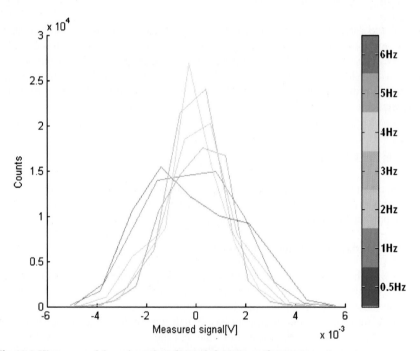

Fig. 10.4. Histogram of the noise values for each frequency of measurement

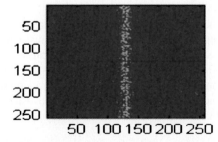

Fig. 10.5. Example of two-dimensional fast Fourier transform of a noise sample

two-dimensional Fourier transform of the PSF h_d. The actual surface potential is then restored as:

$$\hat{\Phi}_d^s(\boldsymbol{m}) = \mathcal{F}^{-1}\left[W_{\mathrm{opt}}\,\mathcal{F}\,(V)\right],\tag{10.24}$$

where $\mathcal{F}$ and $\mathcal{F}^{-1}$ stand for the fast fourier transform and its inverse, respectively.

Figure 10.6 shows a typical two-dimensional PSF calculated using (10.20) and the following tip parameters: $l = 4\,\mu m$, $d = 10\,nm$, $\theta_0 = 10°$, and $R = 10\,nm$. One-dimensional PSFs calculated for various tip parameters are presented in Fig. 10.7. The results show that the tip–sample distance, d, and the half aperture angle, θ_0, have a very large effect on the magnitude of the PSF, and thus on the sensitivity in KPFM measurements. For example, Fig. 10.7a shows that on increasing the tip–sample distance from 5 to 20–30 nm, the KPFM signal represented by the PSF decreases by about a factor of 3.

This implies that the sensitivity of ultrahigh vacuum (UHV) KPFM measurements, typically carried out at a distance $d = 5$, is about 3 times greater than that of air KPFM measurements, as was reported by us recently [15]. Similar results

Fig. 10.6. A typical two-dimensional point spread function (PSF) calculated for a tip with $l = 4\,\mu m$, $\theta_0 = 10°$, and $R = 10\,nm$ located at a height $d = 10\,nm$ above a sample

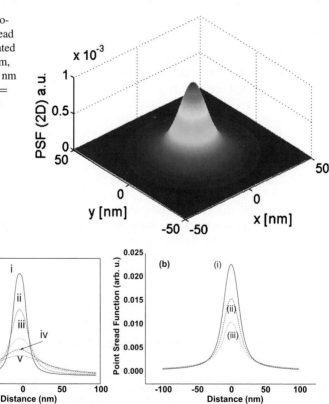

Fig. 10.7. One-dimensional PSFs calculated as a function of (**a**) tip–sample distances, d, of: (*i*) 5, (*ii*) 10, (*iii*) 20, (*iv*) 30, and (*v*) 40 nm, and (**b**) tip aperture angles, θ_0, of: (*i*) 10°, (*ii*) 20°, and (*iii*) 30°

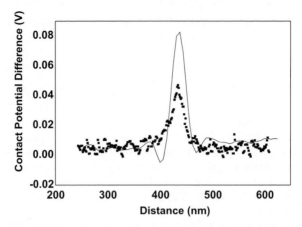

Fig. 10.8. Measured (*dots*) and restored (*solid*) surface potential of an atomic step on a GaP(110) surface. The Kelvin probe force microscopy (KPFM) measurement, described in detail in [15], was conducted under ultrahigh vacuum conditions where the tip–sample distance, d, was estimated as 5 nm. The sharp work function change at the step is due to localized surface charge

were reported by Kalinin et al. [23], who used carbon nanotubes as a tip calibration standard for electrostatic force microscopy. The effect of the tip length on the PSF shown in Fig. 10.7b is weak; this is due to the fact that at such small tip–sample separation (10 nm), the electrostatic force is mostly due to the tip apex as was also reported by others [17]. The tip apex radius, R, affects mainly the width of the PSF, in agreement with previous works [19].

Figure 10.8 shows measured and noise-filtered and restored surface potential line scans of an atomic step on a GaP(110) surface. The KPFM measurement was conducted under UHV conditions where the tip–sample distance, d, was estimated as 5 nm. The sharp work function change at the step is due to localized surface charge as discussed in detail in [15].

The potential restoration, based on the method developed here, shows that the reconstructed CPD increases by more than a factor of 2 (relative to the measured one) even at the small tip–sample separation used in UHV measurements. This result demonstrates the importance of the restoration in KPFM measurements of small features; in ambient KPFM measurements the effect is much larger, as discussed above.

10.3
Measurement of Semiconductor Surface States

10.3.1
Surface Charge and Band Bending Measurements

Probably the most important semiconductor surface electronic properties are the equilibrium surface band bending, V_S, and the concentration of the surface states within the semiconductor band gap. The former is a very important parameter, for example, in the formation of metal–semiconductor interfaces [24], oxidation, etc. In addition, in recent years, knowledge of the energy distribution (presented in the next section) and density of the surface states and of the surface band bending is essential in order to accurately quantify two-dimensional carrier profiling with high spatial resolution using scanning capacitance microscopy [25, 26] and scanning spreading resistance microscopy [27].

Surface photovoltage spectroscopy [28] is probably the only well-established technique for such measurements, but it has three main drawbacks: it relies on photoinduced transitions that make the determination of the equilibrium surface charge density possible only in very few cases; it does not work well for small and/or indirect band gap semiconductors like Si; and it does not have high spatial resolution. In this section we show that KPFM can measure the equilibrium surface band bending and surface charge in semiconductors. The method is based on cross-sectional surface potential measurements of very asymmetric $p^{++}n$ or $n^{++}p$ junctions. The measured built-in voltage on the surface of the junction is used to derive the spatial distribution of the diode surface band bending, and the total surface charge density on the low-doped side of the junction.

Figure 10.9a shows a two-dimensional numerical calculation of the potential distribution of a symmetric Si pn junction (dopant concentration of $1 \times 10^{17}\,\mathrm{cm}^{-3}$) with a density of $1 \times 10^{12}\,\mathrm{cm}^{-2}$ ($5 \times 10^{12}\,\mathrm{cm}^{-2}$) donor (acceptor) states located at an energy of $0.7\,\mathrm{eV}$ ($0.8\,\mathrm{eV}$) above the valence band maximum, E_V. The lower junction built-in potential on the surface, V_{bi}^s, compared with the built-in potential in the bulk, V_{bi}^b, observed at the "back" of the figure, is due to charged surface states; surface states trap holes (electrons) on the cleaved surface of the p (n) side of the junction, creating depletion-type band bending opposite in sign on each side of the junction. Thus, the bands will bend up (down) in the n (p) doped region, with the net result being a reduction of V_{bi}^s (relative to V_{bi}^b). In the case of the asymmetric diode with a density of $7.7 \times 10^{11}\,\mathrm{cm}^{-2}$ ($3.2 \times 10^{12}\,\mathrm{cm}^{-2}$) donor (acceptor) states located at an energy of $0.6\,\mathrm{eV}$ ($0.336\,\mathrm{eV}$) above E_V shown in Fig. 10.9b, the degenerate side of the junction serves as a potential reference since the band bending on this side is negligible. V_S is then determined as the difference between the calculated

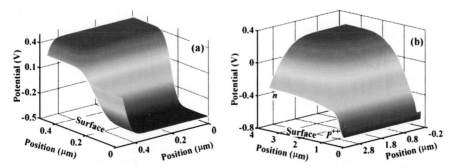

Fig. 10.9. Calculated two-dimensional potential distribution of (**a**) a symmetric Si pn junction and **b** a $p^{++}n$ diode showing the effect of the doping concentration on the surface band bending. Zero band bending is observed on the degenerate part of the junction (*front right* in (**b**))

bulk potential and the (simulated and fitted to the measured CPD) surface potential, calculated from a two-dimensional numerical solution of Poisson and Laplace equations for a semiconductor–air system. The surface charge and band bending were extracted from the following boundary condition at the semiconductor–air interface [29]: $\varepsilon_{SC}E_{SC} - \varepsilon_{air}E_{air} = Q_{SS}$, where ε is the dielectric constant, E is the electric field, and Q_{SS} is the surface charge and is a function of the surface band bending given by (for the case of a single acceptor state)

$$Q_{SS} = -qN_t \frac{1}{1 + \exp\left(\frac{(E_t - E_f)_b - qV_S}{kT}\right)} \, , \tag{10.25}$$

where N_t is the density of the surface state, E_t is the surface state level, the subscript b denotes bulk values, and the other symbols have their usual meaning.

The surface (110) band bending and charge of an air-cleaved n^{++}p Si diode (n-type, As-doped) implanted with a maximal dopant concentration of 2.1×10^{20} cm^{-3} on a p-type (B-doped) substrate with a dopant concentration of 1.8×10^{15} cm^{-3} resulting from the fitting procedure are shown in Fig. 10.10. The diode was cleaved using a specially designed cleaving machine (SELA); the doping was performed by ion implantation, and measured by secondary ion mass spectroscopy (SIMS) at IMEC, Belgium. Figure 10.10a shows the "deconvolved" CPD profile taken from the measured two-dimensional CPD image, the calculated (and fitted to the measurement) surface potential, and the surface charge. Figure 10.10b shows the calculated surface and the bulk potential and the surface band bending obtained from subtracting the surface from the bulk potential. As shown in Fig. 10.10b, the surface band bending is initially zero on the degenerate part of the junction as expected, then it increases owing to a decrease of the doping concentration, reaches a maximum, and then sharply decreases. This sharp change, which gives rise to the peak in the CPD curve, is due to a sharp decrease of the negative surface charge (marked with an arrow in Fig. 10.10a). This may result from a depopulation of donor surface states that move above the Fermi level as they follow the potential change along the pn junction. The surface band bending (surface charge) then decreases to zero in the vicinity of the metallurgical junction and increases again until it reaches a constant

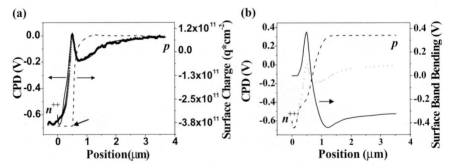

Fig. 10.10. a Deconvolved CPD profile (*triangles*), simulated surface potential (*solid line*), and surface charge (*dashed line*) of the n^{++}p Si diode. **b** Calculated surface (*dotted line*) and bulk (*dashed line*) potentials, and surface band bending (*solid line*)

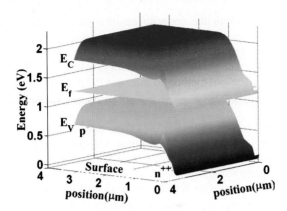

Fig. 10.11. Calculated two-dimensional band structure of an $n^{++}p$ Si diode based on the measured surface potential

value of 0.32 eV ($8.63 \times 10^{10}\, q\,\mathrm{cm}^{-2}$, where q is the elementary charge) in the neutral p region.

The error in the surface band bending is determined by the noise level of the measured CPD, typically $\pm 10\,\mathrm{mV}$ in our system, the error of the SIMS measurement to obtain the concentration of the dopants, and the error of the numerical calculation, which has the smallest effect. Therefore, we estimate that the accuracy of the surface band bending is $\pm 20\,\mathrm{meV}$. The error of the surface charge density depends on to the error of the surface band bending; for example, for a band bending of 0.32 eV the error is approximately $\pm 3.1 \times 10^{9}\, q\,\mathrm{cm}^{-2}$.

Knowledge of the junction surface potential allows to obtain the two-dimensional band structure of the $n^{++}p$ diode shown in Fig. 10.11. This is because once the surface charge has been determined, the potential at each point in the structure is known from the solution of the Poisson equation. Apart from the band bending, the figure also shows the spatial distribution and width of the surface SCR.

10.3.2
Measuring the Energy Distribution of the Surface States

The electronic properties of semiconductor surfaces are determined by the surface state energy distribution, $N_{SS}(E)$, within their band gap. This distribution is essential in determining the properties of semiconductor junctions [24], the phenomenon of Fermi level pinning, surface passivation [30], adsorption of molecules [31], leakage current in metal oxide semiconductor transistors [32], etc. The importance of contacts in molecular electronics [33] suggests that semiconductor electrodes may play an important role [34]; thus, knowledge of the distribution of surface states is crucial in determining the charge transfer processes and the current in such devices.

A widely used method to measure the energy distribution of surface states is photoemission spectroscopy, and angle-resolved photoemission spectroscopy. The energy distribution of interface states is measured using the capacitance–voltage

(*C–V*) method [35], where the capacitance of a metal–insulator–semiconductor structure is measured as a function of metal bias. The main disadvantage of the *C–V* method stems from the fact that it requires a top contact; thus measuring of bare surfaces (i.e., surface versus interface states) is impossible. Kronik et al. [28] have suggested a contactless method to measure the energy distribution of surface states using surface photovoltage spectroscopy with a tunable laser as the excitation source [36].

All the abovementioned methods (and others not mentioned here) share a common drawback: they have low spatial resolution. Scanning tunneling microscopy has been widely used in the last decade to measure local density of surface states [37], but it is limited mainly to highly conductive samples and very clean surfaces. We present here a method, based on KPFM, to measure the energy distribution of surface states on a local scale with very high sensitivity ($1 \times 10^9 \, \text{cm}^{-2} \, \text{eV}^{-1}$ or greater). The method is demonstrated by extracting the distribution of states on a polished oxidized Si(110) surface.

The method is based on measuring the surface potential on a cross-sectional pn junction using KPFM as shown schematically in Fig. 10.12. As the tip scans the junction surface from n- to p-side, the surface states depopulate, thereby changing the measured surface potential. The energy distribution is then obtained by equating the position derivative of the surface and the SCR charge (Q_{SS} and Q_{SC}, respectively), as described in detail in [38]. A key factor in this calculation is the knowledge of the absolute surface band bending, V_S, at each point on the surface [38, 39].

Figure 10.13a shows the measured surface potential of a polished and oxidized $p^{++}n$ Si diode (p-type, B-doped) implant with a maximum dopant concentration of approximately $1.75 \times 10^{20} \, \text{cm}^{-3}$ on an n-type (As-doped) substrate with a dopant concentration of approximately $2.9 \times 10^{14} \, \text{cm}^{-3}$. The KPFM measurements were conducted using a commercial atomic force microscope (Autoprobe CP, Veeco) operating in the noncontact mode. The KPFM measurements (based on a setup described previously [7]) were conducted inside a nitrogen-containing (less than 2-ppm relative humidity) glove box.

The distribution of surface states derived according to the procedure described in the Appendix is shown by the dashed line in Fig. 10.13b. This distribution is then used to calculate the surface potential; the result is shown by the dashed line in Fig. 10.13a. The large difference relative to the measured potential profile is because the position of $N_{SS}(E)$ does not change across the entire band gap as the tip scans

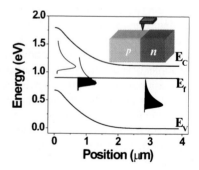

Fig. 10.12. Measurement of the energy distribution of surface states using KPFM. As the tip scans the junction surface (*inset*), the population of the surface states decreases, thereby changing the measured surface potential

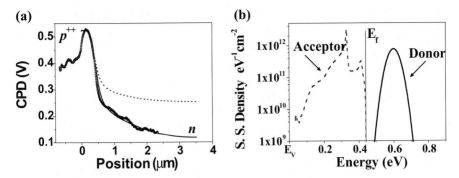

Fig. 10.13. (a) Measured (*triangles*) and initially calculated (*dashed line*) surface potential of a $p^{++}n$ diode; the *solid line* is a best fit based on a modified distribution. (b) Complete surface state distribution obtained from the fit (*solid line*) in (a). The acceptor part of the distribution was initially obtained according to the analysis described in the Appendix

the pn junction. In other words, the energy interval for which $N_{SS}(E)$ is calculated is less than the band gap and is further reduced by the surface band bending. In order to obtain the distribution in a wider energy range, we use an iterative process in which we modify the distribution of surface states and calculate the surface potential until a good fit like the one shown by the solid line in Fig. 10.13a is obtained; the distribution shown in Fig. 10.13b by the solid line is the part of $N_{SS}(E)$ resulting from this iterative procedure.

The assignment of the type (donor or acceptor) of surface state is based on the following argument. Since the low-doped n side of the junction has a depletion-type band bending (owing to the lower surface built-in voltage relative to the bulk), it must be due to negative surface charge resulting from acceptor-type states located below the surface Fermi level (0.44 eV above E_V). However, as shown above, calculating the surface potential using the initially obtained distribution of surface states (dashed line in Fig. 10.13b) resulted a lower surface potential relative to the measured one owing to an excess negative surface charge. Therefore, a Gaussian distribution of donor-type surface states located above the Fermi level (contributing a positive charge) was required to fit the measured surface potential. The final distribution, shown in Fig. 10.13, is composed of acceptor states (dashed line) with a peak density of $3.2 \times 10^{12}\,\text{cm}^{-2}$ at an energy of 0.336 eV above E_V, and donor states (solid line) with a peak density of $7.7 \times 10^{11}\,\text{cm}^{-2}$ at an energy of 0.6 eV and a full width at half maximum (FWHM) of 0.07 eV.

The accuracy and sensitivity of the last calculation is demonstrated in Fig. 10.14. The figure shows the surface potential calculated using $N_{SS}(E)$ with two different Gaussian donor-type surface state distributions. A change of the peak density of the surface states, N_t, from 7.7×10^{11} to $7 \times 10^{11}\,\text{cm}^{-2}$ changes the surface potential by approximately 25 mV (curve c), whereas changing the peak energy position, E_0, from 0.6 to 0.5 eV changes the surface potential by approximately 18 mV (curve b). On the other hand, changing the peak energy position from 0.6 to 0.7 eV changes the calculated surface potential only by approximately 1 mV, which shows that this calculation is not sensitive to the distribution of surface states far from the Fermi

Fig. 10.14. Calculated surface potential using different Gaussian donor distributions added to the initially calculated distribution. **a** and *solid line* $N_t = 7.7 \times 10^{11}$ cm^{-2}, $E_0 = 0.6$ eV (best fit); *b* and *dashed line* $N_t = 7.7 \times 10^{11}$ cm^{-2}, $E_0 = 0.5$ eV; *c* and *dotted line* $N_t = 7 \times 10^{11}$ cm^{-2}, $E_0 = 0.6$ eV. All energies are relative to the valance band maximum

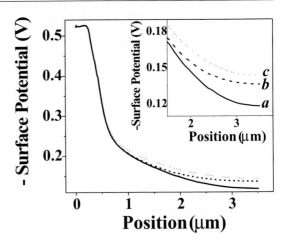

energy. This is the case only for the donor-type surface state distribution above E_f; the other part of the distribution, represented by the dashed line of Fig. 10.13b, is unambiguously obtained by our analysis described [38].

It is interesting to compare the distribution of surface states obtained here for oxidized Si(110) with previously reported data. Flietner [40] has reported on two distributions named P_L and P_H with FWHM of 0.12 and 0.08 eV located 0.4 and 0.7 eV above E_V, respectively. This is similar to our result apart from the fact that the P_L peak is a donor type. Fussel et al. [41] have attributed these distributions to dangling bond defect centers of the character Si$_{3-x}$O$_x$–Si. Poindexter et al. [42] have also reported on peaks of interface states located at similar band-gap energies. The well-known U-shaped distribution resulting from intrinsic surface states with donor-type (acceptor-type) states located at the lower (upper) half of the band gap cannot be observed in our measurements, because the method is insensitive to neutral states near the band edges.

10.3.3
Organic Semiconductors: Bulk Density of States

The field of semiconducting organic molecular films is developing very fast, driven by the enormous potential demonstrated by these materials for applications in opto-electronics and inexpensive electronics such as light-emitting diodes, field-effect transistors, and photovoltaic cells. As an example, organic thin-film transistors (OTFTs) present several possible advantages, such as low-cost processing, mechanical flexibility, and patterning for large-area applications. Improving the performance of OTFTs requires a deep understanding of charge carrier transport mechanisms through the organic layer. Although material technology advanced significantly at the last two decades, a fundamental understanding of the basic electronic and optical processes in these materials is still rather poor, and extensive research is necessary.

A good example of the above is the insertion of electronically active dopant molecules (i.e., electron donors or acceptors) into organic thin films, which has already been demonstrated by several groups. Dopants are introduced either intentionally to improve the film conductivity [43], modify charge injection barriers [44], add functionality (e.g., a pn junction [45]), and modify the optical properties of the film [46], or unintentionally as synthesis impurities [47], solution residuals [48], and chemically induced dopants [49] which deteriorate the performance of the organic material. In any case, an accurate understanding of the density of states (DOS) energy distribution and how it is affected by molecular doping is one of the keys to advancing basic research on, and technological applications of organic semiconductor films, since most electronic properties of organic semiconductors are closely related to the shape of the DOS [50, 51].

We review here the use of KPFM to measure several electronic properties of doped organic molecular thin films. KPFM has already been utilized for the study of charge mobility in OTFTs [52], and the potential drops at the source and drain contacts of operating OTFTs [53, 54]. Here, it is demonstrated that by using KPFM measurements across OTFTs with a relatively thin organic film (10 nm), it is possible to measure quantitatively the energy distribution of the electronic DOS and the Fermi level position.

10.3.3.1
DOS Measurements

A typical OTFT structure used in all measurements is schematically shown in Fig. 10.15. The OTFT substrates (fabricated by the Tessler group at the Technion-Israel Institute of Technology) consisted of a heavily doped p-type silicon gate

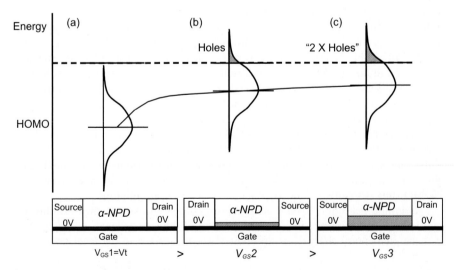

Fig. 10.15. *Top*: Qualitative scheme of the hole occupation of a Gaussian density of states at different positions with respect to E_f. *Bottom*: Cartoons of organic thin-film transistors (OTFTs) in which the relevant charge concentration is represented by a *shaded region*. **a** $V_{GS}1 = V_t$ i.e., zero gate induced hole concentration. **b** $V_{GS}2 < V_t$; thus, holes accumulate to form a conducting channel. **c** $V_{GS}3 < V_{GS}2$ such that the hole concentration is twice that in **b**. α-*NPD* N,N'-diphenyl-N,N'-bis(1-naphthyl)-1,1'-biphenyl-4,4'-diamine

electrode, a thermally grown 90-nm silicon oxide gate insulator, and 50-nm-thick gold strips evaporated on the oxide to form source and drain electrodes separated by 16 μm. A thin film (10 nm) of N,N'-diphenyl-N,N'-bis(1-naphthyl)-1,1′-biphenyl-4,4′-diamine was deposited on the substrate by sublimation in a UHV chamber (by the Kahn group at the Princeton University) and transported under a nitrogen atmosphere to a glove box (less than 2 ppm H_2O) in Tel Aviv University, where the KPFM instrument (Autoprobe CP, Veeco, with homemade Kelvin probe electronics) is located.

KPFM is used for high-resolution determination of the surface potential profile across the OTFT channel with a lateral resolution of tens of nanometers and an energy resolution of a few millielectronvolts. The transistors were scanned by the KPFM insrument while a semiconductor parameter analyzer (HP 4155C) was used to control the gate–source voltage (V_{GS}), with respect to the grounded source and drain electrodes, and to monitor the drain, source, and gate electrode currents. The contact resistance [55], leakage current through the gate insulator and to the periphery of the active area, and shifts of the threshold voltage [55] due to continuous voltage application were found to be negligible in these transistor structures.

We use the transistor structure to control the hole concentration by applying different V_{GS} while measuring the level shift with the KPFM instrument, as shown schematically in Fig. 10.15. Assuming a Gaussian distribution of the DOS related to the highest occupied molecular orbital (HOMO), when $V_{GS} = V_t$ (V_t is the threshold voltage defined as V_{GS} at the onset of the conduction channel) there is no induced charge in the organic film and the DOS is located below the Fermi energy level. However, when $V_{GS} < V_t$, holes are injected into the organic film from the grounded source and drain electrodes in order to screen the negative charge of the gate. The holes populate the tail states of the HOMO DOS, and the molecular energy levels shift towards the Fermi energy as shown (solely for the HOMO) in Fig. 10.15, part b. During the measurement, the Fermi energy is kept constant by the grounded source and drain contacts, and the molecular energy levels are shifted with respect to the Fermi energy by the gate-induced voltage. When $|V_{GS}|$ is further increased (Fig. 10.15, part c), such that twice as many holes are injected into the film, then in the former case, the HOMO energy levels are shifted less than before because there are more states available as the DOS is larger near the Fermi level. The above description illustrates the relation between the charge concentration and the HOMO shift to the specific shape of the DOS: a low state density results in a larger HOMO shift for a certain increase of hole concentration, while at higher state density the HOMO shift is smaller for the same increase in hole concentration.

Assuming negligible level bending (justified below), we can measure directly by the Kelvin probe force microscope the shift of the energy levels ($-qV_L$, where q is the elementary charge) for different V_{GS} with respect to the level position at $V_{GS} = V_t$:

$$V_L(x) \equiv CPD(x) - CPD_t(x) , \qquad (10.26)$$

where $CPD(x)$ is the CPD measured between the Kelvin probe force microscope tip and the sample at a given location (x) across the transistor, and $CPD_t(x)$ is measured at $V_{GS} = V_t$. Figure 10.16 shows a qualitative energy level scheme across an organic

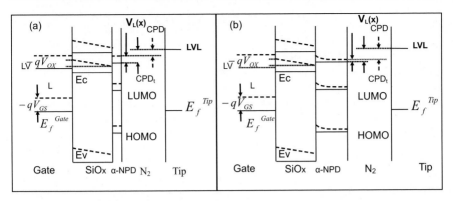

Fig. 10.16. Qualitative energy level scheme across a OTFT measured by KPFM far from the source and drain contacts for $V_{GS} = V_t$, $V_{DS} = 0$ V (*solid curves*), and $V_{GS} < V_t$, $V_{DS} = 0$ V (*dashed curves*) in the case of (**a**) negligible and (**b**) nonnegligible level bending. In the second case (**b**), the measured $V_L(x)$ is smaller because part of the applied gate voltage induces band bending in the organic layer. *HOMO* highest occupied molecular orbital, *LUMO* lowest unoccupied molecular orbital

field-effect transistor measured by a Kelvin probe force microscope far from the source and drain contacts for negligible (Fig. 10.16a) and nonnegligible (Fig. 10.16b) level bending. In Fig. 10.16a when the applied gate–source voltage $V_{GS} = V_t$, and the drain–source voltage $V_{DS} = 0$ V, there is no charge injection into the channel, and no level shift in the organic film, i.e: $V_L(x) = CPD(x) - CPD_t(x) = 0$. When $V_{GS} < V_t$ and $V_{DS} = 0$ V, holes are injected into the channel and $V_L(x) = CPD(x) - CPD_t(x) > 0$. In the case of nonnegligible level bending (Fig. 10.16b), the measured $V_L(x)$ is smaller because part of the applied gate voltage induces band bending in the organic layer.

Our method for DOS measurement is feasible when the energy level bending perpendicular to the gate can be neglected. According to our numerical calculations (O. Tal and N. Tessler, unpublished results) conducted for a 10 nm thick organic layer in an OFET structure and different Gaussian DOS distributions, the induced charge is homogeneously distributed across the organic film width and the energy level bending from the film surface to the film/gate-insulator interface is negligible up to a certain V_{GS} value. Above this given V_{GS} (O. Tal and N. Tessler, unpublished results), the conducting channel is squeezed towards the gate insulator and the level bending is not negligible; this behavior has already been reported elsewhere [52,53]. Under the negligible level bending condition, the shift of the energy levels (V_L) for different V_{GS} with respect to the level position at $V_{GS} = V_t$ can be measured directly by the Kelvin probe force microscope based on (10.26).

The hole concentration at any lateral location x in the channel is given by

$$p(x) = -\frac{C_{OX}}{q d_{org}}[V_{GS} - V_t - V_L(x)], \qquad (10.27)$$

where C_{OX} is the silicon oxide (insulating layer) capacitance per unit area, and d_{org} is the organic film thickness. Equation (10.27) is only valid when the charge

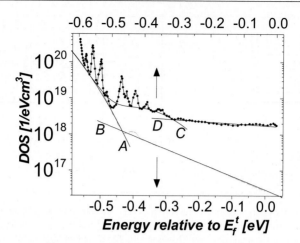

Fig. 10.17. Density of states (*DOS*) versus energy relative to E_f^t (E_f at $V_{GS} = V_t$) for undoped (*solid*) and doped (*circles*) samples. The *solid curves* are fits of a Gaussian function (*curve A*) and an exponential function (*curve B*) to the same ranges as in the DOS curve of the undoped sample, and an exponential fitting of (*curve C*) to the DOS curve of the doped sample

concentration is homogeneously distributed across the organic film depth. On the other hand, the concentration of holes in the channel is given by

$$p = \int_{-\infty}^{\infty} g(E) f_{FD}(E) \, dE \ , \tag{10.28}$$

where $g(E)$ is the DOS relevant for holes, and $f_{FD}(E)$ is the hole Fermi–Dirac distribution. Having the concentration and the energy level shift of the holes, we can calculate the hole DOS as described in [54] to give

$$g[qV_L(x)] = \frac{C_{OX}}{d_{org}q^2} \left(\frac{dV_{GS}}{dV_L(x)} - 1 \right) \ . \tag{10.29}$$

This analytic expression, which has no fitting parameters, is used to obtain the DOS from the measured V_L.

The DOS calculated using (10.29) is plotted versus qV_L in Fig. 10.17 for undoped and doped samples (doping concentration: $N_a = 1.4 \times 10^{18} \, \text{cm}^{-3}$ [54]). Each DOS curve is based on an average of 50 sets of V_L curves measured at different locations on the transistors to reduce the experimental uncertainty. The energy scale in Fig. 10.17 represents the energy relative to E_f^t defined as the E_f position at $V_{GS} = V_t$, and the negative sign denotes values below E_f^t. The sharp increase in the DOS near the high-energy end of the curves (left side of Fig. 10.17) reflects the termination of the level shift, possibly owing to the failure of the negligible level bending assumption at the corresponding V_{GS} value, or owing to the Fermi level pinning phenomenon. The rich DOS structure was discussed in detail by us recently [55]; in short, the sharp peaks observed in the doped sample may imply the presence of several doping-induced energy levels, and the differences in the general shape of the curves were ascribed to a broadening of the DOS owing to the presence of dopants.

10.3.3.2
Fermi Level Position

The Fermi energy position is subjected to two counteracting effects owing to the introduction of dopants into the molecular film. According to Bassler et al. [51], holes can be transferred from the lowest unoccupied molecular orbital (LUMO) of the dopant to the HOMO of the host even if the LUMO of the dopant is approximately

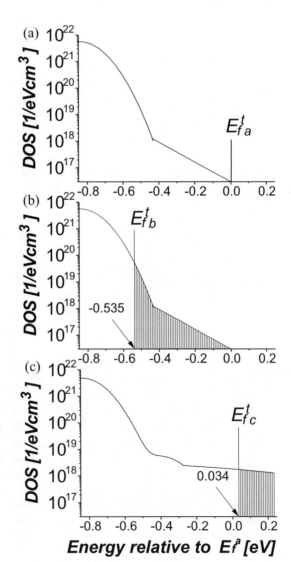

Fig. 10.18. The chemical potential and the position of the DOS edge for three cases: (**a**) undoped sample; where "0 eV" denotes the E_f^t position at the beginning of the DOS measurement (i.e., at $V_{GS} = V_t$); (**b**) hypothetical case of the undoped DOS populated by the dopant-induced holes (states occupied by holes appeared as the *lined region*); (**c**) doped sample ($V_{GS} = V_t$). The shape of the DOS is represented by the fits to the measured DOS (Fig. 10.17, curves A and B for the undoped case and curves C and D for the doped case, respectively)

0.8 eV below the HOMO of the host material. This is due to stabilization by Coulomb interactions between ionized dopants and released holes localized in nearby hopping sites. In our case $HOMO_{host} - LUMO_{dopant} \cong 0.28$ eV [55]; thus, all the dopants are most likely ionized under equilibrium conditions. As illustrated schematically by Fig. 10.18, plots a and b, respectively, the dopant-induced increase in the hole concentration shifts E_f^t from its original position (E_{fa}^t) in the undoped sample towards the HOMO center (position E_{fb}^t).

This shift (-0.535 eV) was calculated by finding the upper limit E_{fb}^t of an integral on the measured DOS in the undoped sample that equals the dopant-induced hole concentration (p_{doping}):

$$p_{doping} = N_a = 1.4 \times 10^{18}\,\mathrm{cm}^{-3} = \int_0^{E_{fb}} g(E)_{undoped} f_{FD}(E)\,\mathrm{d}E \ . \tag{10.30}$$

However, as measured and showed in the previous section, dopants induce a DOS broadening or conversion of shallow states into deep states in the HOMO–LUMO gap as shown schematically in Fig. 10.18, plot c. This effect should shift back E_f^t away from the HOMO center; based on the fitting of Gaussian functions to the measured DOS, the HOMO DOS centers of the doped and the undoped samples are located at $E_c = -0.88 \pm 0.04$ eV, and the E_f^t shift between the undoped sample to the doped sample is only 34 meV away from the HOMO center (E_{fc}^t position). Thus, in this case, the broadening of the HOMO level is compensating the Fermi level shift induced by the dopants, as shown in Fig. 10.18, plot c). This phenomenon is unique in comparison with crystalline inorganic semiconductors, where the DOS does not change upon doping and the introduction of p-type dopants causes E_f^t to shift towards E_c – the valance band edge.

From a technological point of view, any addition of dopants or the presence of impurities that act as dopants has a substantial effect on the DOS and consequently on the position of the Fermi level energy. This has important consequences for the physics and operation of all organic-based devices.

References

1. Nonenmacher M, O'Boyle MP, Wickramasing HK (1991) Appl Phys Lett 58:2921–2923
2. Leng Y, Williams CC, Su LC, Stringfellow GB (1995) Appl Phys Lett 66:1264–1267
3. Kikukawa A, Hosaka S, Imura R (1995) Appl Phys Lett 66:3510–3512
4. Chavez-Pirson B, Vatel O, Tanimoto M, Ando H, Iwamura H, Kanbe H (1995) Appl Phys Lett 67:3069–3071
5. Mizutani T, Arakawa M, Kishimoto S (1997) IEEE Electron Devices Lett 18:423-425
6. Arakawa M, Kishimoto S, Mizutani T (1997) Jpn J Appl Phys 36:1826–1829
7. Shikler R, Fried N, Meoded T, Rosenwaks Y (1999) Appl Phys Lett 74:2972–2974
8. Shikler R, Meoded T, Fried N, Mishori B, Rosenwaks Y (1999) J Appl Phys 86:107–113
9. Krauss TD, O'Brien S, Brus LE (2001) J Phys Chem B 105:1725
10. Kitamura S, Suzuki K, Iwatsuki M, Mooney CB (2000) Appl Surf Sci 157:222
11. Okamoto K, Sugawara Y, Morita S (2002) Appl Surf Sci 188:381

12. Duhayon N, Eyben P, Fouchier M, Clarysse T, Vandervorst W, Álvarez D, Schoemann S, Ciappa M, Stangoni M, Formanek P, Raineri V, Giannazzo F, Goghero D, Rosenwaks Y, Shikler R, Saraf S, Sadewasser S, Barreau N, Glatzel T, Verheijen M, Mentink SAM, Wiesendanger R, Von Sprekselen M, Maltezopoulos T, Hellemans L (2004) J Vac Sci Technol B 22:385
13. Henning K, Hochwitz T, Slinkman J, Never J, Hoffman S, Kaszuba P, Daghlian C (1995) J Appl Phys 77:1888–1896
14. Hudlet S, Jean MS, Roulet B, Berger J, Guthmann C (1995) J Appl Phys 59:3308
15. Rosenwaks Y, Glatzel T, Sadewasser S, Shikler R (2004) Phys Rev B 70:85320
16. Hochowitz T, Henning AK, Levey C, Daghlian C, Slinkman J (1996) J Vac Sci Technol B 14:457
17. Hudlet S, Saint Jean M, Guthmann C, Berger J (1998) Eur Phys J B 2:5
18. Belaidi S, Lebon F, Girard P, Leveque G, Pagano S (1998) Appl Phys A 66:S239
19. Jacobs HO, Leuchtmann P, Homan OJ, Stemmer A (1998) J Appl Phys 84:1168
20. Jacobs HO, Stemmer A (1999) Surf Interface Anal 27:361
21. Shikler R (2003) PhD thesis, Tel-Aviv University
22. Jain K (1989) Fundamentals of digital image processing. Prentice Hall, New York
23. Kalinin SV, Bonnell D, Freitag M, Johnson AT (2002) Appl Phys Lett 81:754
24. Mönch W (1993) Semiconductor surfaces and interfaces. Springer, Berlin
25. Zavyalov VV, McMurray JS, Williams CC (1999) J Appl Phys 85:7774
26. Yang J, Kong FC (2002) J Appl Phys Lett 81:4973
27. Eyben P, Xu M, Duhayon N, Clarysse T, Callewaert S, Vandervorst W (2002) J Vac Sci Technol B 20:471
28. Kronik L, Shapira Y (1999) Surf Sci Rep 37:1
29. Selberherr S (1984) Analysis and simulation of semiconductor devices. Springer, New York
30. Williams R (1962) J Phys Chem Solids 23:1057
31. Vilan A, Shanzer A, Cahen D (2000) Nature 404:166
32. Asuha AA, Maida O, Todokoro Y, Kobayashi H (2002) Appl Phys Lett 80:4552
33. Nitzan A, Ratner MA (2003) Science 300:1384
34. Guisinger NP, Greene ME, Basu R, Baluch AS, Hersam MC (2004) Nano Lett 4:55
35. Nicollian EH, Brews JR (1982) MOS physics and technology. Wiley, New York
36. Kronik L, Burstein L, Shapira Y (1993) Appl Phys Lett 63:60
37. Hamers RJ (1989) Annu Rev Phys Chem 40:531
38. Saraf S, Rosenwaks Y (2005) Surf Sci Lett 574:L35
39. Saraf S, Schwarzman A, Dvash Y, Cohen S, Ritter D, Rosenwaks Y (2006) Phys Rev B 73:35336–35342
40. Flietner H (1988) Surf Sci 200:463
41. Fussel W, Schmidt M, Angermann H, Mende G, Flietner H (1996) Nucl Instrum Methods Phys Res A 377:177
42. Poindexter EH, Gerardi GJ, Rueckel ME, Caplan PJ, Johnson NM, Biegelsen DK (1984) J Appl Phys 56:2844
43. Shen Y, Diest K, Man HW, Hsieh BR, Dunlap DH, Malliaras GG (2003) Phys Rev B 68:81204(R)
44. Gross M, Muller DC, Nothofer HG, Scherf U, Neher D, Brauchle C, Meerholz K (2000) Nature 405:661
45. Qibing P, Gang Y, Chi Z, Yang Y, Heeger AJ (1995) Science 269:1086
46. Welter S, Brunner L, Hofstraat JW, De Cola L (2003) Nature 421:54
47. Naka S, Okada H, Onnagawa H, Yamaguchi Y, Tsutsui T (2000) Synth Met 111:331
48. Sakai K, IkezakiK (2002) ISE11 Proc IEEE 151
49. Honhang F, Kachu L, Shukong S (2002) Jpn J Appl Phys 41:L1122

50. Arkhipov VI, Heremans P, Emelianova EV, Adriaenssens GJ, Bassler H (2002) J Phys Condens Matter 14:9899
51. Koehler M, Biaggio I (2003) Phys Rev B 68:752051
52. Horowitz G, Hajlaoui R, Delannoy P (1995) J Phys III 5:355
53. Horowitz G, Hajlaoui R, Bouchriha H, Bourguiga R, Hajlaoui M (1998) Adv Mater 10:923
54. Tal O, Preezant Y, Tessler N, Chan CK, Kahn A, Rosenwaks Y (2005) Phys Rev Lett 95:256405
55. Gao W, Kahn A (2003) J Appl Phys 94:359

11 Application of Scanning Capacitance Microscopy to Analysis at the Nanoscale

Štefan Lányi

Abstract. The scanning capacitance microscope is an instrument capable of imaging both conducting and insulator-covered surfaces, and hidden structures or other inhomogeneities in dielectrics and semiconductors. Since it has been identified as a promising tool for high-resolution and high-accuracy analysis of dopant concentration in semiconductor structures, in the last decade a great effort has been exerted to achieve specifically this goal. Though progress is inevitable, nanometre resolution and high accuracy at the same time remains a formidable task. Capacitance-based methods have been successfully used in the last 50 years to analyse and identify defects in dielectrics and semiconductors. Representatives of such methods are, for example, impedance spectroscopy, admittance spectroscopy and deep level transient spectroscopy. They yield information on the relaxation rate of defects and from its temperature dependence the activation enthalpies characterising them. Attempts to apply such methods on a microscopic scale have been rather scarce until now. The aim of this chapter is to review the recent developments and to sketch the prospects of this area.

Key words: Scanning capacitance microscopy, Impedance spectroscopy, Deep level transient spectroscopy, Isothermal charge-transient spectroscopy, Semiconductor analysis, Analysis of dielectric films

11.1 Introduction

The scanning capacitance microscope (SCM) is one of the first scanning probe instruments. The invention of the scanning tunnelling microscope (STM) by Binnig and Rohrer from the Zurich laboratory of IBM Research Division [1, 2] was announced on 12 August 1981, the same day as the production of the IBM PC started. Matey [3] of RCA filed a patent application on the SCM only 7 weeks later, on 30 September 1981. Yet the first paper by Matey and Blanc [4], describing the visualisation of the relief in a videodisc grove, read by means of a capacitive transducer—the CED VideoDisc pickup, appeared only years later. By that time the STM could already boast such fascinating results as the atomically resolved image of a 7×7 reconstructed surface of Si(111) [5]. Papers on true capacitance microscopy appeared by the end of 1980s [6–10].

A great impetus for the SCM became the definition of future needs of analytic methods for coming generations of ultra-large-scale integration (ULSI) [11]. The SCM has been identified as a prospective tool for dopant profile measurement for calibration of process design tools. The shrinking dimensions require increased spatial resolution and accuracy. The recent release of the International Technology

Roadmap for Semiconductors (2006 update) foresees 2.5 nm and 4% for 2008 and 1.5 nm and 2% for 2012 [12].

As a probe, the SCM uses a sharp conducting tip, very similar to a STM tip. It forms with the conducting surface, or a substrate covered by an insulating film, a small capacitor. The probe is frequently a conducting scanning force microscope (SFM) cantilever. It is raster-scanned in the vicinity of the surface or in contact with the insulating film. The signal proportional to the sensed capacitance or the control voltage needed to keep the capacitance at a preset value can be used to create a 2D image.

The capacitance of a capacitor depends, besides geometrical factors (like the distance of electrodes), on the properties of the dielectric material filling the space between the electrodes. Any of them may be the desired output to create an image. The problem is usually the separation of their contributions. Our interest will focus on the analysis of properties of the dielectric. In the following this term will be used also for semiconductor depletion layers, since they are free of mobile charges.

In a dielectric subjected to an electric field, a dielectric displacement field is built up. In an alternating field with frequency ω it can be expressed as

$$\boldsymbol{D}_0 e^{-i\omega t} = \hat{\varepsilon}(\omega)\boldsymbol{E}_0 e^{-i\omega t} , \tag{11.1}$$

where

$$\hat{\varepsilon}(\omega) = \varepsilon_0\hat{\varepsilon}_r(\omega) = \varepsilon_0 \left[\varepsilon'(\omega) + \varepsilon''(\omega)\right] \tag{11.2}$$

is the complex permittivity and ε_0 is the permittivity of a vacuum.

The material property formally characterising the dielectric is the relative permittivity ε_r. It is frequently called the dielectric constant, though as seen from (11.2), it is a complex function of frequency. If the material contains, for instance, aliovalent impurities and charge-compensating defects (e. g. vacancies) forming dipoles, these are normally oriented chaotically. In an electric field, they tend to be more ordered, to minimise the energy of the system. The reorientation contributes to the displacement current and thus to the permittivity. It takes place via thermally activated jumps, up to a frequency, above which they cannot follow the voltage changes. The capacitive current is shifted in phase with respect to an ac voltage by $-90°$. At a frequency where the reorientation lags behind it by $45°$, the imaginary part of ε_r attains a maximum. Its position depends on temperature. From its shift, the activation enthalpy of jumps, characteristic for the particular type of defects, can be estimated using Arrhenius plots. Other mechanisms, like space charge near grain boundaries, dislocations or electrodes also increase the capacitance. They are not a property of the material itself but rather a consequence of its defect structure, and as such may also carry useful information. Application of a square-wave or pulse voltage quickly charges the geometrical capacitance and settles the fast polarisation mechanisms. However, if the material contains any of the slower-responding defects, the fast, usually exponential transient is followed by slower ones that again carry information on these mechanisms, and on the concentration of entities responsible for them.

In semiconductors many defects form localised energy levels, traps, in the forbidden gap. If they are electron states below the Fermi energy (in an n-type semiconductor), they are filled, whereas above the Fermi level they are empty. If the band bending near metal/semiconductor and oxide/semiconductor interfaces or pn junctions is modified by an applied voltage, the charge state of the traps may change. Thus, normally empty traps in a depletion layer become charged by voltage pulses driving the semiconductor to accumulation. After the trailing edge of the pulse the depletion layer relaxes and the traps return to their position above the Fermi level; however, this time charged. Depending on the distance from the conduction band, the trapped electrons are released at a temperature-dependent rate. Thus, the observed charge transients may yield information on the kind and concentration of defects. The various technologically relevant defects may increase the capacitance from small fractions of a percent to orders of magnitude in the case of space charge near interfaces or electrodes.

In standard materials research the analysis can be performed on sufficiently large specimens. Then the information obtained from inhomogeneous materials is averaged and the true nature of defects eventually obscured. The microscopic analysis may show and disclose the local properties. High spatial resolution requires very local measurement, thus very small probed areas and volumes. The available sensitivity of the capacitance, current or charge sensors becomes one of the limiting issues. Another one, with impact on measurement accuracy, is the ability of the probe to sense sufficiently locally. We shall analyse the available options in detail. The questions may be posed as follows:

1. How can the capacitance be measured on the microscopic scale?
2. How sensitive are the available transducers?
3. How local is the discernible capacitance contribution?
4. How trustworthy is the image of an object obtained?
5. How should the optimal probe be chosen?

11.2
Capacitance Microscopes

The main components of a capacitance microscope are shown in Fig. 11.1. It contains a scanning mechanism and its control circuitry, a capacitance transducer consisting of a probe and electronics, and further hardware and software to collect the data and convert them into a virtual 3D object or a two-dimensional map. Most SCMs can parallelly produce an image of the topography of the surface.

The small capacitance that had to be measured by means of a SCM probe required a strategy different from those used before. The resolution and response time of available instruments were not satisfactory. It was a rare chance that at the time of emergence of probe microscopies a sensitive capacitive sensor, the RCA VideoDisc pickup, already existed. Though the videodisc was not a success and the company soon gave up the production of videodisc players, thanks to the SCM their pickup survived in later, not significant modifications. Though no new capacitance measurement principles have been invented since, with other dedicated capacitance transducers considerable performance has been achieved.

The capacitance measurement methods applied in the SCM can be divided into four major groups. The most widespread are the microscopes equipped with the RCA CED VideoDisc pickup or its clones. It contains a transmission line resonator,

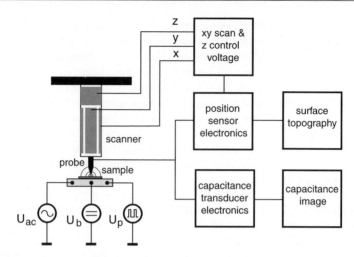

Fig. 11.1. A scanning capacitance microscope (SCM). The voltage sources may be connected to the sample depending on the requirements of the experiment performed. Others may be, e. g., a ramp or a triangular voltage source

coupled to an oscillator oscillating near 915 MHz, a frequency on the flank of its resonance curve. The frequency of resonance is fine-tuned by the tip/sample capacitance, giving rise to changes of the RF voltage amplitude in the resonator, and then demodulated [4, 7, 10, 13]. It is a simple FM demodulator used invertedly. In a radio receiver such a circuit would be tuned to a fixed frequency and the frequency-modulated input (intermediate-frequency) signal converted into an amplitude-modulated one, and then detected. A similar approach based on a lumped element circuit, operated at 90 MHz, was described by Bugg and King [6] and later by Dreyer and Wiesendanger [14]. Their instrument worked at 6 MHz. Figure 11.2 illustrates the principle of capacitance transducers utilising the resonance principle.

Martin et al. [8] have presented a principally different method based on measuring the electrostatic force f between the sample and a cantilever, to which a dc or an ac voltage V is connected:

$$f = \frac{1}{2}V^2\frac{\partial C}{\partial z} \ . \tag{11.3}$$

Modulation of the probe/sample spacing z produces a force component proportional to the derivative of the capacitance and to the square of the applied voltage. Since the method relies on force measurement, it is sometimes called scanning capacitance force microscopy [9].

Lányi et al. [15] introduced a low-frequency SCM. It performs admittance measurement based on Ohm's law—an ac voltage of known amplitude and frequency (2 MHz) is applied to the tip/sample capacitance and the current is measured phase-sensitively. This gives the possibility to separate the in-phase and quadrature components of the current, i. e. the real and imaginary (loss) parts of the capacitance. Figure 11.3 shows the possibilities of measurement of current flowing through the

Fig. 11.2. The principle of capacitance measurement by means of a simple FM demodulator. **a** Demodulation on the flank of the resonance curve and **b** a possible realisation using an LC circuit. C_p is the probe/sample capacitance that fine-tunes the resonator

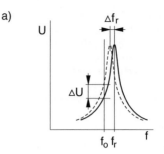

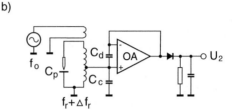

Fig. 11.3. Capacitance (admittance) measurement using **a** a high-input-impedance buffer and **b** a current-to-voltage converter. The output voltage U_2 is measured phase-sensitively by means of lock-in amplifiers that measure the in-phase and quadrature components. C_p is the probe/sample capacitance. The capacitances C_d and C_c denote the differential and common-mode capacitance of the operational amplifier. C_c affects the voltage transfer and the bandwidth. The high-frequency gain is C_p/C_c. C_i is the input capacitance of the amplifier and C_{fb} is the parasitic capacitance of the feedback resistor R_{fb}. At high frequencies the gain is $-C_p/C_{fb}$

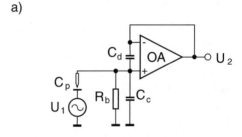

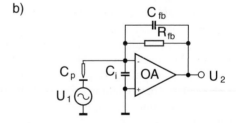

tip/sample capacitance. The last approach is the most traditional one—the use of capacitance bridges. Arakawa and Nishitani [16] have used a sensitive commercial bridge, albeit its resolution would hardly compete with capacitance transducers developed specially for application in microscopy. An electronic modification of the concept, a half-bridge compensating the current flowing through the large stray capacitance of the probe by subtraction of a current with opposite phase, was employed by Lee et al. [17]. At the low operating frequency of these instruments (1 and 5 kHz, respectively) the reactance of the probe/specimen capacitance attains very high values, which makes the measurement difficult. On the other hand, at low frequencies

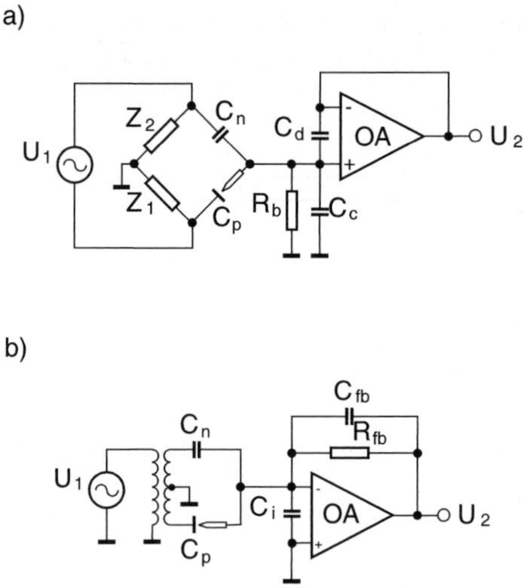

Fig. 11.4. Capacitance measurement by means of a capacitance bridge. **a** The two voltages with opposite phase are produced by impedances Z_1 and Z_2. **b** Transformer bridge. In **a** a buffer is shown as the balance indicator and in **b** a current-to-voltage converter, but either of them may be used in both cases. C_p is the probe/sample capacitance

a)

b)

it is easier to control the phase shifts, undesired resonances and other problems, and achieve higher accuracy and stability. Figure 11.4 shows examples of capacitance bridges.

11.2.1
Resolution of Capacitance Transducers

The resolution or sensitivity of the capacitance transducer determines the achievable spatial resolution. It is primarily a noise problem. Generally the noise can be efficiently suppressed by narrow-band signal amplification followed by demodulation and low-pass filtering, e.g. using lock-in amplifiers. However, the consequence of heavy filtering is slow response. In microscopy the measurement is made at many points. A response time of 1 s, a very good performance for single capacitance measurement, would mean an acquisition time of 65,536 s, i.e. more than 18 h for a 256 × 256 pixel frame, with corresponding requirements for stability, drift, etc. Therefore, it is reasonable to discuss the noise in a more practical bandwidth, e.g. 1 kHz. Such a bandwidth is achieved with low-pass-filter time constants from 0.16 to 1 ms, depending on the order of the filter (first order and fourth order, respectively) [18]. Such a response is of course beyond the possibilities of instruments measuring the capacitance at very low frequencies [16, 17].

The noise of most capacitance transducers used in SCMs was analysed in [19]. The output noise may have different origins. The low-frequency microscope in [15] measures the current with a unity gain buffer, with its approximately 40 fF input capacitance being connected in series with the probe/sample capacitance to form a voltage divider. The input stage of another recent design is a current-to-voltage converter, essentially an inverse-capacitance-to-voltage converter [20]. It has the advantage of low input impedance that enabled it to be configured to a combined SCM/STM using the same input stage. In both circuits a dc path is needed, through which the bias voltage is connected to the input field-effect transistor or operational amplifier, to define their operating point. This is realised by means of a large resistor. The thermal noise current of the resistor R

$$i_n = \sqrt{4kTB/R} \tag{11.4}$$

is one of the dominant sources of noise. B denotes the bandwidth.

The transducer noise is affected by the capacitance connected between the input of the operational amplifier and ground, e. g. of a coaxial cable. The signal-to-noise spectral-density ratio of the current-to-voltage converter is given by [21]

$$S/N_{sd} = U_1 / \sqrt{e_n^2 (1 + Z_1/R_2 + Z_1/X_i)^2 + i_n n Z_i^2 + 4kTZ_1^2/R_2} , \tag{11.5}$$

where Z_1 represents the impedance of the probe and R_2 is the feedback resistor. X_i is the reactance of the operational amplifier input capacitance plus eventual cable capacitance. e_n is the operational amplifier input voltage noise spectral density. The parasitic capacitance of the feedback resistor affects the bandwidth but not the signal-to-noise ratio; therefore, it is omitted for simplicity. A similar expression for a bootstrapped buffer can be found in [21]. Figure 11.5 shows the calculated noise and the signal-to-noise ratio of a fast low-noise operational amplifier in the current-to-voltage converter configuration and Fig. 11.6 shows the calculated noise and the signal-to-noise ratio of a buffer with bootstrapped supply nodes. While the

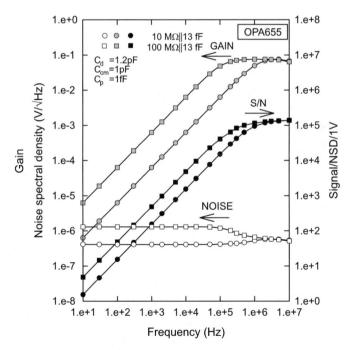

Fig. 11.5. The gain, noise and signal-to-noise ratio of a current-to-voltage converter connected through a 1-fF probe capacitance to a 1-V signal source, calculated for 10- and 100-MΩ feedback resistors

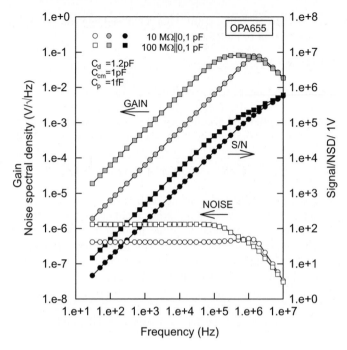

Fig. 11.6. The gain, noise and signal-to-noise ratio of a buffer connected through a 1-fF probe capacitance to a 1-V signal source, calculated for 10- and 100-MΩ biasing resistors

capacitance connected to the input reduces the transfer of the buffer at high frequencies without influencing the signal-to-noise ratio, it may increase the noise of the current-to-voltage converter without affecting much the bandwidth (Fig. 11.7). The increase is particularly high if the input capacitance of the bare amplifier is very small. The amplifier connected to a capacitance bridge as a balance indicator may be of either type, i.e. high impedance voltage or low impedance current amplifier, so from the point of view of noise it is similar to the previous cases. The increase of noise by a coaxial cable connection has been proved experimentally [17].

The transducers using a transmission line resonator or LC resonant circuit are a somewhat different case. The frequency noise of the oscillator is directly transferred to the output as a voltage noise. The real part of the resonator or resonant circuit impedance plays the same role as resistors in wide-band amplifiers. In resonance it is equal to the characteristic impedance multiplied by the quality factor Q. The impedance of a parallel LC circuit and the admittance of a series circuit scale in the same manner. Thus, in resonance the signal-to-noise ratio is proportional to the square root of Q. At a frequency on the flank of the resonant curve close to resonance, at which the signal is 6 dB below the maximum, the effective resistance R_{ef} of the parallel LC circuit is

$$R_{ef} = Re|Z_{LC} = R_0/(1 + Q^2 F^2) , \qquad (11.6)$$

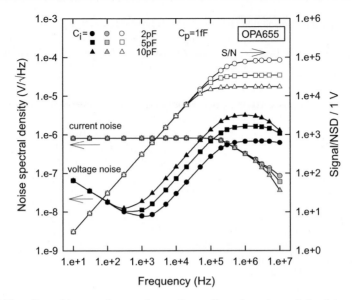

Fig. 11.7. The effect of input and external capacitance C_i on the noise and signal-to-noise ratio of the current-to-voltage converter

where R_0 is the impedance at the frequency of resonance f and $F = \Delta f / f_0$. Then the noise spectral density becomes

$$i_{nsd} = \sqrt{4kT/R_{ef}} = \sqrt{4kT(1 + Q^2 F^2)/R_0} \; . \tag{11.7}$$

This means that compared with resonance the signal-to-noise ratio is reduced. In a series circuit it is similar. The noise current appearing at the output is then the integral of (11.7) over the two low-frequency side bands around the operation frequency. Equation (11.7) defines a steeper function than the resonant curve, thus the farther is the operating frequency from resonance, the lower becomes the signal-to-noise ratio. The dependence of the noise and signal/noise ratio on detuning of the resonator has been studied experimentally in reference [22].

The results are summarised in Table 11.1. Since the resolution in all these cases is proportional to the applied voltage, they are normalised to 1 V. With an optimised transmission line resonator working in the range 1.8–2.0 GHz, Tran et al. [23] achieved a resolution of 3×10^{-20} F. This is a realistic value except for the stated peak-to-peak sense voltage of 0.3 V, which seems to be too small. Jaensch et al. [24] reported a sensitivity of 1.8×10^{-19} F for a coaxial resonator at 2.2 GHz, albeit without reference to the amplitude of the RF signal.

To achieve a high resolution to capacitance changes $\Delta C/C$ in the lumped element circuits, the circuit capacitance must be small. At a given resonant frequency this implies a relatively large inductance; however, coils also have parasitic capacitances. This limits the achievable resolution. The estimates given for the LC circuit at 50 and 200 MHz correspond to best compromise found.

Table 11.1. Capacitance resolution (F/V) in 1-kHz bandwidth

Frequency (MHz)	Buffer	Current-to-voltage converter	Lumped LC	Distributed LC
1	2.3×10^{-19}	6.9×10^{-19}		
3	1.3×10^{-19}	6.5×10^{-19}		
10	7.4×10^{-20}	7.1×10^{-19}		
50			1.1×10^{-19}[a]	
200			5.0×10^{-20}[a]	
250			2.5×10^{-19}[b]	
915				1.4×10^{-20}[a]
				2.5×10^{-20}[c]
				3.5×10^{-19}[d]

[a] Approximate noise limit
[b] Resolution of RCA lumped element pickup [13]
[c] Resolution of RCA CED VideoDisc pickup [13]
[d] Highest of reported sensitivities [10]

In electrostatic force capacitance sensing, with a SFM cantilever vibrating at its resonance frequency, the achievable resolution was estimated to be 2.5×10^{-20} F/V at 0.9 V ac [8]. In this case the resolution is proportional to V^{-2}.

Surprisingly, the resolution achieved with a current-to-voltage converter around 3 MHz was comparable with that of the VideoDisc pickup. High operating frequency is beneficial from the point of view of capacitance resolution. The reason why some solutions using lower frequencies than the transmission-line-based transducers may be competitive lies in the possibility to achieve higher impedance, by means of which currents are converted to voltage. While the transmission lines in resonance represent an impedance of about $10 \, k\Omega$, near 1 MHz values of about $10 \, M\Omega$ can be achieved.

A problem with resonator-based transducers is the amplitude of the voltage between the probe and the sample. It cannot be directly measured, only deduced from the length of the strip-line. It depends not only on the oscillator voltage but also on the properties of the sample. In the VideoDisc pickup its peak value was set to 5 V [13]. In microscopes, amplitudes as large as 10 V were suggested [10]. For analysis of nonlinear systems like semiconductor depletion layers the voltage should be on the order of kT/e (approximately 25 mV at room temperature) for the linearised theories to apply. In transducers measuring directly the current and in capacitance bridges the voltage can be set in a defined manner.

In a FM radio a limiter cares for more or less constant voltage that is then demodulated. In this way any AM or interference from electric discharges like lightnings are, in contrast to AM receivers, suppressed. This is not the case of demodulators used in microscopes. They are, in fact, amplitude-frequency demodulators; therefore, their sensitivity to capacitance changes is proportional to the voltage applied to the probe/sample capacitance.

A noisy voltage source introduces a noise proportional to the current circulating in the measured capacitance. In the cases analysed, except for the force-based

Fig. 11.8. The effect of stray capacitance on the generator noise transfer. **a** Resonator-based measurement. **b** Current to voltage converter based measurement

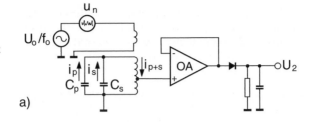

a)

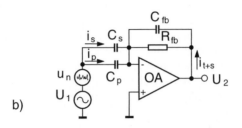

b)

measurement, the transducers are linear with respect to applied voltage. Though the force is proportional to the square of the voltage, for small voltage variations as the noise is, it can be also considered linear. This means that the noise of the ac voltage source is transferred to the sensor input proportionally to the probe/sample capacitance [20] (Fig. 11.8). For instance, the root mean square (rms) noise of a generator with 100-dB signal-to-noise ratio is $10\,\mu V$ at 1 V. For the generator connected to a 1 pF probe/sample capacitance, the rms noise would correspond to 10 aF, i.e. to a value more than 1 order of magnitude larger than the resolution of good capacitance transducers. Thus, the generator noise combined with large stray capacitance may significantly reduce the signal-to-noise ratio.[1] Estimates of resolution, based on scaling the applied voltage, may be misleading. We shall, therefore, look at it in more detail.

11.2.2
Stray Capacitance of the Probe

The transducers analysed operate with an ac voltage. However, the length scale of probes and their distance from the surface are negligibly small compared to the wavelength of the applied signal even at 2 GHz. This justifies the treatment of the electric field and interaction as static.

The electrostatic interaction is of long-range nature; therefore, the electrostatic field between two unshielded conductors is poorly localised. Let us assume a parallel-plate capacitor with circular plates of radius r, separated by d. Its capacitance in air is larger than given by the textbook formula $C = \varepsilon_0 \pi r^2 / d$, because the field near the perimeter of the plates is not homogeneous owing to fringing and even their backside contributes to a small extent to the capacitance. To force the capacitance to follow

---

[1] This statement is not exact in the case of force measurement, as we shall explain later.

exactly the formula, closely spaced guard rings are needed, surrounding the plates and held at their potentials, or at least around one of them if the size of the other is larger. This pushes the inhomogeneity of the field away from the active plate.

In microscopy a very small plate or sphere as the probe would be advantageous. A plate with 10-nm radius separated from the surface by 3 nm would give approximately 1 aF. To ensure enough rigidity, the probes usually have an approximately conical shape. The capacitance of such an unshielded body with respect to a plane becomes orders of magnitude larger. To reduce the stray capacitance, a loosely shielded probe has been used in the low-frequency SCM [15]. Figure 11.9 shows such tip with a schematically superimposed cross-section of the shield and the insulating tube. Though shielded pyramidal tips have been produced for millimetre-wave oscilloscopy [25], scanning Kelvin probe microscopy (SKPM) and electrostatic force microscopy [26], and a tungsten needle, insulated and then shielded by an evaporated film, has been produced for scanning potential microscopy [27], they have not yet been applied to capacitance microscopy. Moreover, they limit the spreading of the field but cannot make it homogeneous. To follow the same strategy as with macroscopic capacitors, i.e. shielding up to the face of the tip, could be difficult but is probably feasible. Asami [28] has built a microscope with a flat-faced probe with a robust guard, albeit the diameter of the working electrode was 76 μm. On the other hand, a homogeneous field would not always be advantageous.

Let us have a closer look at this point. As already mentioned, the capacitance depends on both the separation of electrodes and the properties of the material between them. Both a rough substrate and the surface of a dielectric film covering it lead to fluctuations of the film thickness, and hence to local variations of the capacitance. The same would result from film inhomogeneity. It is sometimes difficult to

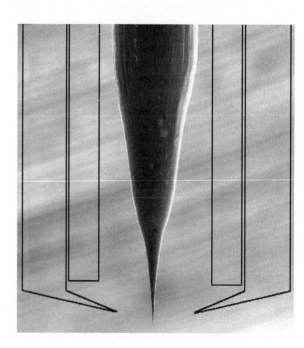

Fig. 11.9. Tungsten tip electrochemically etched in two steps, with the insulating quartz tube and the shield. The orifice diameter allows mounting and replacing of the core

distinguish the two. It is relatively easy to obtain independent information on the morphology of the surface by means of force microscope scans [29], since nowadays most SCMs are combined with SFMs. The substrate/dielectric interface is not accessible by such means. As we shall demonstrate later, the contrast produced by the capacitance probe is a sensitive function of distance—it decreases rapidly with increasing tip/substrate distance, as illustrated by Fig. 11.10. This is a consequence of field inhomogeneity, caused by the nonplanar face of the probe. The field strength is highest at small radii of curvature. This has two consequences. The diminishing contrast lets the distant electrode appear as a structureless background, and the information is obtained from the surface and the layer close to it in the insulator. Thus, in most cases the surface topography is sufficient as additional information.

The mutual capacitance of two conductors is the ratio of the accumulated charge and the potential difference between them. Its calculation requires solving the Laplace equation $\Delta V = 0$ in three dimensions. The surface charge density σ^ρ, induced in the electrodes by the applied voltage, is coupled with the electrostatic field E^ρ through the Poisson equation. ρ is the radial variable. In general, no closed-form analytic solution of this problem is known, except for some relatively simple cases. Configurations like sphere/plane, cone/plane, paraboloid/plane or hyperboloid/plane combinations can be solved in prolate spheroidal coordinates. For instance, the effect of an electron on the potential and tunnelling [30] and the electrostatic potential of a hyperbolic probe [31] were calculated using this method.

Kleinknecht et al. [7], assuming that the probe tip can be represented by a small sphere, have made use of the image charge method to estimate the resolution of their microscope. The mapping of the electrostatic field is used also for estimation of electrostatic force acting on the SFM probe [32], e. g. in [33] using the generalised image charge method, and for calculation of capacitance versus distance in [34].

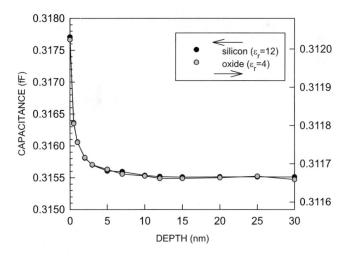

Fig. 11.10. The capacitance change resulting from a spherical metal inclusion in a 100-nm-thick oxide ($\varepsilon_r = 4$) or silicon ($\varepsilon_r = 12$) film on a conducting substrate, as a function of its distance from the surface. The inclusion diameter is 10 nm and the tip with radius 25 nm is placed 10 nm above the surface

Knowledge of the capacitance is useful in the interpretation of SKPM images [35,36]. However, there is an important difference between the two cases—force is a vector, whereas capacitance is a scalar quantity. So the forces acting on opposite sides of the probe or a topographic feature subtract, whereas fluxes, i.e. capacitance contributions, add.

A flexible tool for the calculation of complex geometries is the finite-element method [37]. It can be used for arbitrary tip shapes as well as surface structure. However, until now axially symmetric cases, sections of 3D problems in Cartesian coordinates [38, 39] and the combination of both have been analysed [40–43]. Such a procedure is illustrated by Fig. 11.11 and the result obtained simulating a constant height scan over a conducting plane with cylindrical holes filled with a dielectric is shown in Fig. 11.12.

The method was first applied to mapping the field and estimation of the capacitance of shielded probes [44]. Figure 11.13 shows the confinement of the stray field by a cylindrical shield, the same shield with insulation and a shield with a reduced

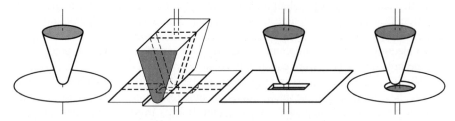

Fig. 11.11. The procedure to derive the approximate position-dependence of the capacitance of a probe scanning over a cylindrical hole. **a** The tip/plane capacitance is calculated with rotational symmetry. From a scan of a knife-edge-shaped body over a plane in Cartesian coordinates a slice with equal capacitance as of the conical tip/plane is cut. **b** The slice of the same width is scanned over a trench. This is assumed to be approximately equal to the scan of a conical tip over an oblong depression (**c**) and to the scan of the conical tip over a cylindrical depression (**d**)

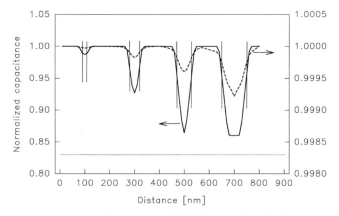

Fig. 11.12. Simulated scan over a conducting plane with dielectric-filled 20-nm-deep holes. Tip 10 nm above the surface in air (*dashed line*) and with a water bridge connecting the tip with the surface. The *dotted line* corresponds to the correct depth in air [40]

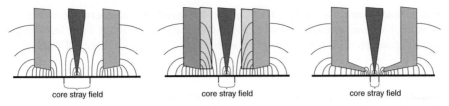

Fig. 11.13. The confinement of the stray field by shielding, shielding combined with an insulating tube and by reduction of the shield orifice diameter

orifice. It has been found that even with a large shield orifice of 125-μm radius it is possible to reduce the stray capacitance of a macroscopic electrochemically etched probe to approximately 1 fF and with 25-μm radius to less than 300 aF.

Figure 11.14 shows the capacitance versus tip radius of shielded and of bare conical tips. The height used to define the boundary conditions was 250 μm; the cone angle was 20°. The findings can be summarised as follows. For the stray capacitance to be low, the cone angle must be small and the tip, any lead connected to it or the cylindrical shank holding it, should be perpendicular to the surface. The length of the unshielded part of the probe should be as short as possible [45]. Taking such precautions, one can keep the stray capacitance of the probe etched from thin wire (approximately 50 μm) below 10 fF. With a thick (approximately 0.25 mm) unshielded sharpened wire the capacitance can easily exceed 100 fF.

Conducting SFM cantilevers, the standard probes of commercial SCMs, are thin narrow beams extending tenths of a millimetre from a silicon chip with approx-

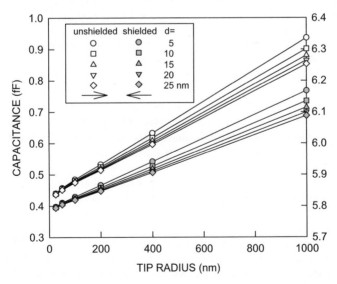

Fig. 11.14. The capacitance of shielded and unshielded tips with a hemispherical apex. The increase with tip radius of both is different, indicating the contribution of the conical part of the tip body. Cone angle 20°, height 0.25 mm

imate dimensions $2 \times 3\,\text{mm}^2$. They are held in position and contacted by spring clamps. The estimated capacitance of the whole fixture is about 1 pF. Such large stray capacitances are the probable cause of increased noise (M. LeFevre cited in [17]).

11.2.3
Sensitivity of the Probe

The resolution of capacitance transducers offers an idea of measurable capacitance changes. However, this is not sufficient to predict the performance of the microscope. The lateral resolution was usually inferred from resolved features in images. In the pioneering work of Matey and Blanc the resolution was determined by the metalised face of the sledge-shaped probe and the width of the videodisc grove—200 nm in the direction of movement and 2.5 μm in the transverse direction [4]. Later 2 μm [6] and 200 nm laterally and 5 nm normal to the surface was reported [7]. Williams et al. achieved a lateral resolution of 25 nm [10] and recently less than 10 nm [46]; Lányi et al. achieved 10 nm [15], later 5 nm [47]. Kleinknecht et al. [7] concluded that the theoretical lateral limit is 7 nm and 1 nm normal to the surface. However, they approximated the tip by a sphere, which underestimates the stray capacitance of the probe [44], and they assumed that the resolution would roughly correspond to the tip diameter.

Such estimates of lateral resolution are frequently not comparable. To image a single isolated surface feature and to resolve complicated surface roughness are completely different things. The features seen may be smaller than the radius of the tip apex [47]. Similarly as in optics, the resolution should be related to achieved contrast. The contrast of an image or a scene is usually defined as the ratio of the deviation of the brightness from its mean value, divided by the mean value $(B_{max} - B_{min})/(B_{max} + B_{min})$. In the case of capacitance imaging the brightness would represent the respective capacitance values. If the sensed capacitance responded to a sudden step in surface topography by a similar abrupt change, the contrast of the "scene" would be correctly reflected in the acquired capacitance image.

The contrast as a function of spatial frequency of rough conducting surfaces was derived using a local-height approximation and perturbation approach by Bruce et al. [41], although only in two dimensions. Another approach is based on the analysis of distortion of surface features [40]. A scan over 20-nm-deep cylindrical holes in a conducting plane, with diameters ranging from 20 to 100 nm was simulated in quasi-3D geometry. The holes were assumed filled with dielectric ($\varepsilon_r = 5$). Figure 11.12 shows that with a sharp tip the true depth of the holes could not be assessed. The smallest error occurs in the centre of the hole.

The contribution to the capacitance of the tip vicinity was analysed by Lányi et al. [48]. They found that a local enhancement of the flux occurs if the tip/surface separation is much smaller than the radius of curvature of the tip apex. In spite of such local improvement, the integral capacitance increases with radius, since the area involved is proportional to its square. In the case of an unshielded probe the extent of the field is badly defined—the reach of the field is large, albeit with vanishing intensity. The consequence for finite element method analysis is a dependence of the results on the choice of the boundary conditions. The sensitivity of the probe, the contribution of a small surface area (1 nm^2) to the capacitance, has been obtained using similar boundary conditions as in Fig. 11.15 with one modification—the planar electrode was replaced by concentric rings and the capacitance of each was divided by its mean length (Fig. 11.16). The tip/surface spacing plays a crucial role. In contrast to the STM, in the SCM the flow of a tunnelling current is undesirable; therefore, the minimum tip/surface distance is usually kept larger or a thin oxide layer is used as a barrier to prevent it.

Knowing the probe sensitivity, the achievable resolution and contrast can be simulated. This has been done by superposition of laterally shifted sensitivity curves (Fig. 11.17). It would correspond to the image of two small cylindrical protrusions or spheres [47]. It explains the formation of a high-resolution but low-contrast image with a tip radius larger than the distance of the surface features. A similar result was obtained by simulation in [16]. Figure 11.18 shows SCM images of a sputtered Pt film on a Si substrate, acquired in forward and backward scans. As an example, the enhanced squares show that features as small as 5 nm across can be correlated in the two images. The tip radius was approximately 30 nm.

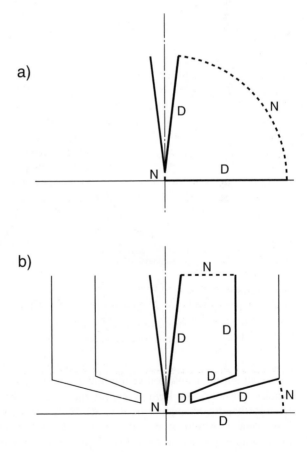

Fig. 11.15. Boundary conditions used in modelling the unshielded (**a**) and the shielded (**b**) tip. D and N denote Dirichlet and Neumann boundary conditions, respectively. In calculations of sensitivity the plane was divided into many concentric rings and the capacitance of each was calculated

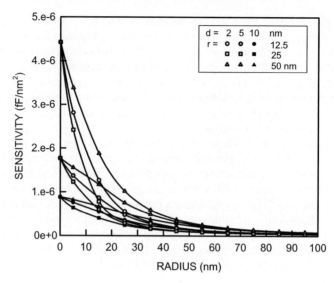

Fig. 11.16. The sensitivity, the contribution of 1-nm^2 area to the tip/surface capacitance, of probes with different tip radii r, separated by d from the surface

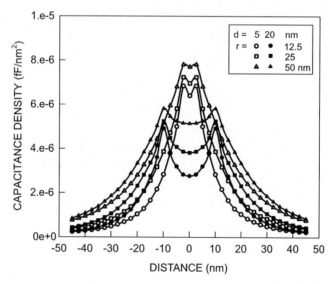

Fig. 11.17. Simulation of resolution of two cylindrical topographic features. Even with a relatively large tip radius they can be resolved but with low contrast

Assuming the resolution of the buffer-based sensor and allowing for a signal-to-noise ratio of 5, the limiting capacitance, measurable with 1-V applied voltage in 1-kHz bandwidth, would be 0.37 aF. With a tip radius of 25 nm at a distance $d = 5$ nm, it would be reached at $\rho \approx 10$ nm. If both the tip radius and the probe/sample distance were scaled down, the local contribution to the capacitance

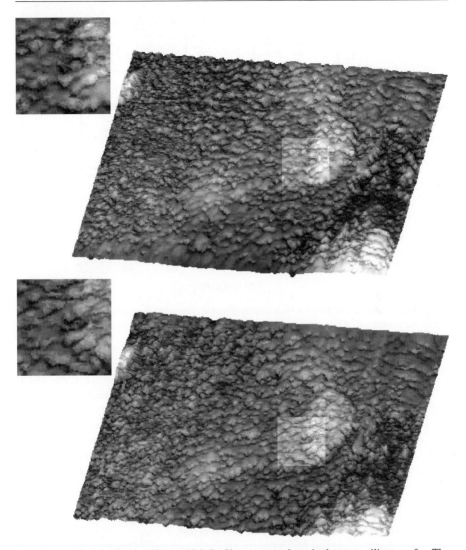

Fig. 11.18. SCM image of a 20-nm-thick Pt film vacuum-deposited onto a silicon wafer. The scanned area was $180 \times 180 \, nm^2$. The *highlighted square* shows that in **a** forward (from bottom up) and **b** backward (from top down) scans details approximately 5 nm in diameter can be correlated. The tip radius was approximately 30 nm. The images show that the resolution is not tip-radius-limited. (From [47])

would also decrease proportionally. Then, because of the finite resolution of the transducer, beyond an optimal value the lateral resolution would deteriorate, since the resolvable capacitance contribution would be reached on increasing ρ, as shown by the full lines in Fig. 11.19. The dotted lines correspond to constant d values. However, with large R, the sensitivity drop with increasing distance would be slower; hence the contrast would decrease, eventually diminishing completely for $R \to \infty$,

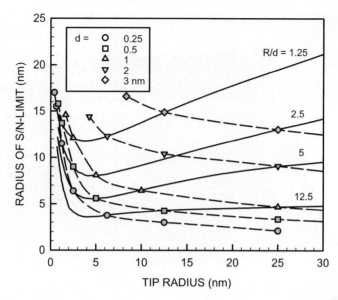

Fig. 11.19. The sample radius for which the capacitance contribution reaches a value corresponding to the resolution of transmission-line-based transducer at a signal-to-noise ratio of 5 (with 1-V measuring signal in 1 kHz-bandwidth)

i. e. the parallel-plate capacitor. There is an optimal tip radius of about 5 nm, which offers the best lateral resolution, though decreasing with probe/sample separation. For other transducer sensitivities or signal-to-noise ratios it would be different.

Until now a spherical tip apex has been assumed. This is a good approximation of electrochemically etched tungsten wires, seen in Fig. 11.20. However, a SCM probe used in contact mode undergoes wear, resulting in a more or less flat shape. This degradation may be very fast [49]. A simulation of the effect of flattening on the sensitivity is shown in Fig. 11.21. The assumed amount of worn material corresponds to the volume of the spherical apex with 25-nm radius. The sensitivity becomes approximately constant up to the radius of the face. The local contribution to the capacitance increases but the resolution is lost.

The achievable spatial resolution is an important issue. Another would be the ability to image correctly topographic artefacts or structural inhomogeneities. As already shown in Fig. 11.10, the inhomogeneity of the electrostatic field makes it possible to resolve inhomogeneities that are close to the surface of the dielectric but hardly those which are deeper, though for correct quantitative analysis this would be important too.

The error of capacitance measurement on surfaces with topographic artefacts has been analysed in more detail. The capacitance seen by the probe placed over the centre of cylindrical protrusions and depressions was simulated in [50]. Such artefacts can be simplified models of islands in semiconductors with higher (protrusions) or lower (depressions) doping level. The analysis has revealed three important facts. The measurement error, defined as a difference between the capacitances sensed over the artefact and over a plane, becomes small (approximately 10%) only at

Fig. 11.20. Scanning electron
microscope image of a tungsten tip with approximately
hemispherical apex

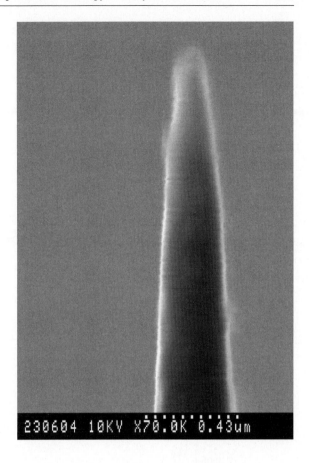

230604 10KV X70.0K 0.43um

radii on the order of micrometres. This confirmed the results of previous quasi-3D modelling [40]. More important, though not surprising, is the finding, seen also in Fig. 11.22, that the measurement error can be reduced using tips with a larger radius of curvature. They increase the otherwise very small contribution of the tip apex. Unexpected was the third finding—the resolution for small artefact radii is not dramatically ruined.

For quantitative analysis to be meaningful, the resolution of microscopes must be known. The sensitivity to the size of imaged artefacts complicates the calibration. The error introduced by the stray field over a trench of the finite width will be smaller than over a hole of the same diameter. A calibration standard should ideally contain features with the same size as and similar shape to the sample to be analysed. This is rarely possible. Therefore, to compare the performance of different instruments and data processing methods, usually dedicated structures are used.

Figure 11.16 shows that a change of probe/surface separation mostly affects the vicinity of the probe apex. By vibrating the probe normal to the surface and extracting the modulation, one can reduce the influence of stray capacitance [6, 7, 15, 51]. A similar effect is achieved by modulation of semiconductor depletion width by means of an ac bias voltage [22, 49, 52]. Such, in principle differential, measurement

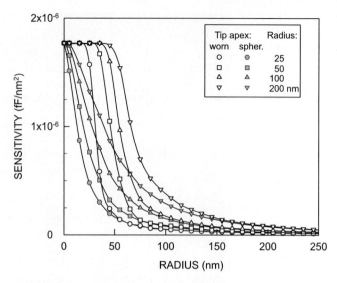

Fig. 11.21. Sensitivity of tips with a hemispherical apex and of worn tips. For each tip, the worn volume was equal to the hemispherical tip with 25-nm radius

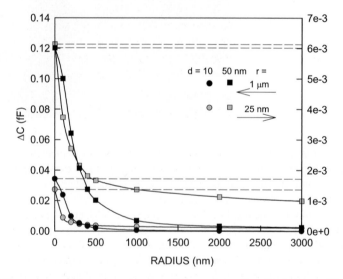

Fig. 11.22. The measurement error as a function of depression radius. The holes are 10 and 50 nm deep

has a side effect. It increases the noise. In the case of abrupt switching between two positions or depletion widths and conversion to rms voltage, the decrease of the signal-to-noise ratio would be at least 6 dB, since the output voltage would represent 50% of the difference of the two signal levels. Harmonic modulation would result in an output rms voltage equal to $1/(2\sqrt{2})$ of the peak-to-peak voltage, i.e. 9-dB reduction of the sensitivity. Figure 11.23 shows that mechanical vibration

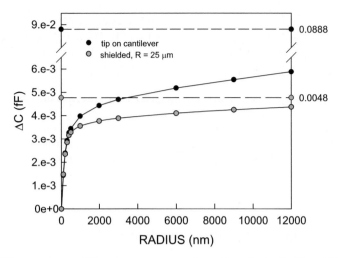

Fig. 11.23. The capacitance difference versus protrusion radius, obtained with a shielded probe and a conducting cantilever if the tip/surface distance is changed from 50 to 10 nm. The *dashed lines* correspond to limits reached for infinite radius, i. e. over a plane [50]

of a conducting cantilever is less efficient, because the contribution of the cantilever beam and fixture cannot be suppressed to a satisfactory level [50]. However, if the cantilever is vibrated in resonance, the amplitude of the cantilever end can be much larger than that of the fixture, thus improving the situation to some extent.

11.2.4
SCM Operation Modes

Similarly as in other probe microscopies, the SCM probe must be positioned over the surface in a defined way. In the STM, when atomic resolution is the goal, constant height scanning is preferred, since it can be done faster than in the constant-current mode, in which a feedback modifies the tip position. This is not suitable for large-area scans, since the scanning plane must be parallel within 1 nm or even less with the atomically smooth scanned surface. The lateral resolution of the SCM is orders of magnitude lower than for the STM and the typical scanned areas are on the order of micrometres. Though larger tip/surface distance fluctuations are acceptable, submilliradian tilts are still challenging. The solution is constant input quantity control.

The change of probe/surface capacitance with distance is on the order of attofarads per nanometre, depending on tip radius and distance. The large stray capacitance causes the tip capacitance versus separation dependence to be not steep enough to be used as an input quantity for control purposes.

Probe vibration, reducing the effect of stray capacitance, is one of the possible options [6, 15]. Another possibility, the most widespread, is force sensing and control [29]. Commercial SCMs are force microscopes equipped with a capacitance transducer. Their advantage is the ability to image independently also the surface topography. The major problem is the large stray capacitance of the probe. An interesting solution was found by Naitou and Ookuba [53]. They used shear-force control by means of a piezoelectric actuator with a sharpened tungsten wire glued to it perpendicularly to the sample. Their microscope electronics could produce, besides the topographic signal, dC/dV and dC/dx outputs. Probe vibration and signal detection on higher harmonics was recently attempted by Breitschopf et al. [54]. They used a quite large vibration amplitude of 50 nm. This gives enough capacitance modulation but less background suppression, which is anyway difficult with cantilevers. Using detection on second and third resonance, the contrast increased and they could resolve, in their not very high resolution images, details not seen on the first resonance but at the cost of 20- and 41-dB reduction of the signal-to-noise ratio, respectively.

11.3
Looking at the Invisible—Capacitance Contrast

As already mentioned, the capacitance measurement yields information on the area/distance of electrodes and the properties of the material filling the space between them. Of these nearly exclusively only the first option has been utilised hitherto. It was, for instance, the visualisation of the variation of oxide thickness on Si by Mang et al. [55]. The regions with different dopant concentration in Si, and deposited charges in $PbTiO_3$ ferroelectric film could be resolved in the work by Nakagiri et al. [56]. The position of the pn junction at zero bias has been imaged by the SCM and the surface potential by means of the Kelvin probe force microscope by Buh et al. [57]. Examples of IC failure analysis have been shown by Isenbart et al. in a SCM method oriented paper [58]. Similar problems and implant shape have been studied by Chao et al. [59] on cross-sectioned ICs. In both papers the cause of failure could be found.

Higher contrast with dC/dx than with dC/dV analysis of a Si memory device having an n-type epitaxial layer, a p-type well, metal wiring and capacitors was reported by Naitou and Ookuba [53]. A Si multilayer structure, consisting of 400-nm-thick layers with alternating high and stepwise decreasing lower dopant concentration was imaged by Basnar et al. [60]. They could confirm the increase of the SCM signal with decreasing dopant concentration from 9×10^{18} down to less than 10^{15} cm^{-3} and decreasing oxide thickness from 11 nm to 3 nm.

In a paper aiming at calibrated thickness measurement, Lee et al. [61] could resolve 1-nm variations of a dielectric film. Since they used a capacitance bridge [17] instead of the more frequently used transmission-line-based transducers that can measure only capacitance variations, their method can be applied also to dielectric films on metals. They also found the positive effect of the presence of a water bridge or meniscus between the tip and the imaged surface that improves the capacitance resolution by enhancing the contribution of the tip apex [40].

An interesting attempt to image nanocrystals embedded in SiO_2, thermally grown on Si, was made by Tallarida et al. [63]. Sn atoms were implanted and clusters formed by rapid thermal processing at 900 °C. Small depressions, seen in the force-based SCM image but not in topography, were assigned to the clusters. Their diameter, as seen in transmission electron micrographs, was about 20 nm.

   The SCM is not sensitive to fixed charges. In spite of that, charge deposited in ferroelectric film can be detected on semiconducting substrates [56]. This can be explained by local modification of the free carrier distribution in the semiconductor, creating a space charge that is mobile and can respond to the applied ac voltage.

   In the examples in [57–60] the contrast arose from differences in the conductivity of areas of the semiconductor substrate with different dopant concentration. At zero or small constant bias voltage different depletion widths [63] create wells of different depth in the boundary between the conducting bulk and the depleted volume, which represents the counter electrode to the probe. The capacitance profile sensed creates the capacitance image.

   The metal inclusions in dielectric film increase the local capacitance, which has a similar effect as a protrusion at the electrode/insulator interface or a hole in the surface. In all cases in which a semiconducting substrate has been used an ac bias was also applied.

11.4
Semiconductor Analysis

The first application of the SCM to analysis of semiconductors was an attempt to analyse the dopant concentration in silicon by Williams et al. [51]. They connected a low-frequency ramp voltage with ± 10 V to the sample. In this way the sample was driven between accumulation and deep depletion, which is a standard technique for MOS analysis. They achieved a lateral resolution of 200 nm. The high-frequency C/V dependences produced were smoothed by means of a boxcar averager. The depletion width

$$L_{\mathrm{d}} = \sqrt{2\varepsilon \Psi / qN} \tag{11.8}$$

is proportional to the square root of the concentration of free charge carriers N. q is the elementary charge and Ψ is the surface potential, i.e. the band bending between the surface and bulk [63]. Figure 11.24 illustrates the principle of high-frequency C/V measurement. The resulting capacitance variation should reflect the concentration of charge carriers. To take into account the inhomogeneity of the electrostatic field near the contact of the tip with oxide, the C/V data were interpreted using a 3D model, a hemispherical MOS structure [52]. Qualitative agreement with theoretical prediction was found.

   Maps created using isoconcentration contour lines have been provided by Neubauer et al. [64]. They imaged the cross-section of slowly varying dopant concentration in an implant in a Si sample. The measurements were made across a thin native oxide created during polishing of the sample. The concentration ranged from 10^{15} to 10^{20} cm^{-3}. The lateral resolution achieved was from 20 to 150 nm, depending on the concentration and the tip.

   The maximum depletion width at dopant concentrations of interest (approximately 10^{16}–10^{20} cm^{-3}) varies from 300 nm to approximately 5 nm. Driving the semiconductor to maximum depletion means a very large probed volume. This can

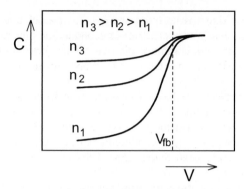

Fig. 11.24. In high-frequency C/V measurement the MOS capacitance is probed by a small ac voltage (usually at 100 kHz or 1 MHz) while a ramp bias drives it from accumulation to depletion and inversion (equilibrium) or deep depletion (nonequilibrium). The capacitance modulation depends on the oxide thickness (accumulation) and dopant concentration. In pn junctions and Schottky diodes the capacitance in accumulation is not limited as in MOS by the oxide

seriously limit the lateral resolution. The solution found rested on the accumulation-to-deep depletion measurement and replaced it by measuring the slope of $\Delta C/\Delta V$ at varying bias that keeps the capacitance modulation constant [65]. In this case only a small modulation of the depletion width takes place. It is conveniently measured by means of lock-in amplifiers.

The success of the first experiments drew attention to the SCM as a potential tool for a nondestructive, high-resolution dopant profiling technique, leading to cooperation with industry (two-dimensional semiconductor dopant profiling development project, funded by Digital Instruments/SEMATECH). Later studies comprising a series of experimental techniques confirmed that expectation [66]. Besides scanning capacitance microscopy, scanning spreading resistance microscopy (SSRM) was the second candidate. It measures the resistance of the sample between a point pressed to the surface and the back electrode. For a point contact with radius r it is given by

$$R_{SR} = \rho/4r , \tag{11.9}$$

where ρ is the resistivity, which is inversely proportional to the concentration of charge carriers. By placing the tip at desired positions, one can detect local conductivity differences. For stable performance a force on the order of micronewtons is needed, which creates imprints on the surface. Depth profiling is achieved and the vertical resolution is increased by grinding or etching the sample into the shape of a wedge. The small-angle-bevelled surface opens access to layers of interest and transforms the depth scale into a larger lateral scale.

Strictly speaking the SCM cannot image the dopant concentration, only that of the free charges [67]. On a large scale the concentration of electrons or holes and that of uncompensated ionised active dopant are the same, except in the depletion layer and near its edge. On the scale of nanometres the equality of the two is only approximate. The lower the dopant concentration, the greater the difference.

To test the feasibility of conversion of SCM data to dopant concentration, McMurray et al. [68, 69] compared SCM and secondary ion mass spectrometry

(SIMS) data measured on the cross-section of a gatelike structure, using a conversion algorithm, and later also with a process simulator [70]. For a slowly varying dopant profile they achieved an encouraging agreement with the conversion algorithm between 10^{16} to 10^{20} cm^{-3} but larger deviations from the prediction of the process simulator. It became clear that for reliable modelling high-quality data are required. The oxide in most early measurements was rather thick. This could cause stretching of the C/V curves. The thickness, as well as the interface charge causing also uncontrollable flatband shifts, had to be reduced. Since high-temperature oxidation to grow a device-quality oxide had to be avoided, the options were studied in detail. Ozone oxidation at 300 °C by means of a UV lamp has been found satisfactory [49]. It reduced the flatband shift from as high as 7 V to less than 1 V. The thickness of the film grown in this manner was about 3 nm, and 1 nm at room temperature [71], though in special reactors 10 and 2 nm were achieved.

In ULSI circuits with very small features steep concentration profiles must be created. To test the capability of the SCM to analyse them, experiments on abrupt dopant profiles were carried out. For this, epitaxially grown test structures, consisting of 50-nm-thick layers with defined dopant concentrations, have been used. The SCM data were compared with SIMS profiles. With use of sharp tips an approximately 25 nm resolution has been reported [73]. The results achieved on 2D dopant profiling by his group by the end of 1998 was reviewed by Williams [74].

The position of the pn junction is another important point for technology. The positions of electric and metallurgic junctions do not coincide, since the mobilities of electrons and holes are different. Owing to their opposite charge, the application of a bias voltage also has the opposite effect on n-type and p-type semiconductors. It drives one of them to depletion and the other to accumulation and vice versa; therefore, also the C/V curves are opposite, as well as the polarity of dC/dV. This is a convenient way to distinguish the sign of majority carriers. In the vicinity of a pn junction the net charge is smaller, eventually reaching zero at the junction. The dC/dV dependence should have sharp maxima.

The dopant profiles across junctions were analysed in [58, 72, 75–79]. Edwards et al. [76] obtained, by integrating the measured dC/dV, spatially resolved C/V curves. Zavyalov et al. [77] described the method of pn-junction delineation and presented SCM and SIMS data. They found that the dc offset of flatband is the cause of major concern, and that on shallow junctions the probe interacts with both sides and introduces uncertainty. Kopanski et al. [72] analysed the importance of working conditions of capacitance transducers, namely of the amplitude of high-frequency voltage, which has to be kept small, in order to produce data that can be compared with simulation. Large amplitude smears the dC/dV curves. Their fit produced a high-frequency voltage that seems to be unrealistically small for the resolution achieved and the known noise of transducers [10, 13]. Chim et al. [78] developed a method using inverse modelling and forward simulation of SCM measurement on a pn junction. Their model included effects like lateral spreading of the field in the oxide, and trapped charge in the oxide and at the interface, ignored in earlier models. The procedure is iterative and gives good results. O'Malley et al. [79] used electrical simulation of the pn junction and its interaction with the tip. They found that no general optimal bias exists but that imaging near midgap gives the best junction position.

In [80] the use of a semiconductor tip was reported and in some cases higher contrast and resolution was achieved. This is explained by the existence of a depletion layer in the tip; thus, with an n-type tip on a p-type semiconductor the resulting capacitance change is larger, whereas if the conductivity type of both semiconductors is the same, their effects subtract. The advantage is also a smaller radius compared with that of tips coated by a metal film. Such a tip had been used before by Goto and Hane [81] for imaging the electric field in a dielectric film on a metal and for reading stored charge in the dielectric. The resolution is said to depend entirely on the field in the tip apex and not on the stray field. However, other investigators consider such tips to be unsuitable for quantitative analysis [49]. Good analysis reproducibility and wear resistance have been reported for diamond-coated tips [82, 83].

In spite of evident progress, the improvement of achieved resolution in dopant profiling was slow. Therefore, the technique used to increase the resolution of spreading resistance analysis, namely angle bevelling, has been applied [84–87]. Giannazzo et al. [85] could resolve B-doped layers with a concentration of 10^{19} cm^{-3} with nominal thickness down to 0.5 nm, embedded in thicker layers with 6×10^{16} cm^{-3}, and 3-nm-wide layers with concentrations between 10^{17} and 10^{19} cm^{-3}. The magnification used was up to $\times 160$. Stangoni et al. [86] simulated the results of this study. By the same approach Giannazzo et al. [87] analysed quantum wells formed by 5-nm-thick B-doped $Si_{0.75}Ge_{0.25}$ strained between Si. Duhayon et al. [88] studied, using this technique, the movement of the pn junction under bias and the extent and effect of current spilling.

Crucial for resolution and accuracy is correct and efficient modelling. Kopanski et al. [89] and Marchiando et al. [90–92] attempted to simulate realistically the tip/insulator/semiconductor structure. The physical models used are rather simple. They take into account and simulate correctly the space charge in the semiconductor but important effects like those of interface charge and similar, introduce as corrections mostly on the experimental side by biasing the sample. The 2D and quasi-3D models are solved using the finite-element method in a dielectric and air and collocation of Gaussian points in the semiconductor [89]. Model databases have been computed and used for rapid conversion of capacitance to concentration [90, 91], and the limits of simulation, using a least-squares fitting approach, have been studied [92]. Ciampolini et al. [93] used a device simulator. They found that the resolution is little affected by the tip radius but with a sharp tip the signal-to-noise ratio suffers, in accord with the simulation in [50].

A recent review of the results of a round-robin analysis of test samples, undertaken within the European project HERCULAS, was provided by Duhayon et al. [94]. The methods used in participating laboratories comprised scanning capitance microscopy, SSRM, Kelvin probe force microscopy, electron holography and scanning electron microscopy. Structures with known dopant profiles and bipolar transistors have been analysed.

The confrontation of scanning capacitance microscopy or SSRM measured dopant profiles also has another side. It is the reliability of SIMS data used for calibration. SIMS provides information on chemical composition, i.e. on the presence of atoms of a distinct sort. The concentration of free charge carriers reflects the concentration of active dopant. Activation after the growth or implantation is achieved by annealing the material at high temperature. For shallow or very narrow profiles such high-temperature treatment must be reduced to a minimum to prevent diffusion. Then the SIMS concentration is not guarantied to be equal to the active dopant concentration.

11.5
Other Semiconductor Structures

The greatest amount of dopant profiling work has been performed hitherto on Si structures. In addition, there are a series of other materials and structures, promising future applications like SiC for high-temperature and high-power applications, various heterostructures for fast electronics or laser structures, which have also been addressed. The methods developed for Si proved to be universally applicable, though other materials may have specific problems. These may be, for example, of mechanical origin or connected with other defects, since few materials can be prepared in crystallographic quality competing with silicon. Bowallius et al. [95] investigated the dopant incorporation in the regrowth of n-type and p-type 4H-SiC around dry-etched mesas. They found a good correlation of doping level above 10^{17} cm^{-3} with SIMS data and improvement at lower concentrations after hydrofluoric acid treatment. Giannazzo et al. [96, 97] investigated the dopant concentration in epitaxially grown and N-implanted n-type 6H-SiC by means of scanning capacitance microscopy, transmission electron microscopy and Rutherford backscattering. They observed contrast created by damage caused by the implantation. The carrier profiling in SiC was reviewed by Calcano and Rainieri [98], with focus on C/V measurements, scanning capacitance microscopy and SSRM. They have found problems with SSRM. At low doping the probe could not create an ohmic contact with the sample.

Regrowth of InP mesas is used for current confinement and capacitance minimisation in optoelectronic devices. Hammar et al. [99] investigated the dopant distribution in hydride vapour epitaxy regrown areas by means of scanning capacitance microscopy. To overcome the lack of calibration data or standards, they grew layers with different dopant content directly on the mesas. They found the analysis complicated by the presence of slowly responding electrons in Fe-doped areas. Zhou et al. [100] studied In$_{0.15}$Ga$_{0.85}$N/GaN quantum-well structures by scanning capacitance microscopy. They could identify areas with increased In concentration, which are responsible for increased luminescence efficiency owing to the higher probability of radiative recombination. Douhéret et al. [101] studied similar problems on GaAs/AlGaAs buried heterostructure laser structures regrown with semi-insulating Fe-doped GaInP [99]. Ban et al. [102] found on multi-quantum-well buried heterostructure laser structures that SSRM gave bias-dependent results on n-type material, whereas scanning capacitance microscopy reflected correctly the doping level. On p-type material the error with SSRM was about 30%, while scanning capacitance microscopy did not yield quantitative results. Smith et al. [103] observed charging effects in Al$_x$Ga$_{1-x}$N/GaN heterostructures. The samples were grown by metal oxide chemical vapour deposition on 4H-SiC with excess Ga flux, at which Ga droplets form at the surface. These were etched away before the investigation. The observed charge contrast depended on location and could be correlated with variations of thickness and composition, being largest in the etch pits. Slow decrease of contrast and spreading of the trapped charge were observed.

11.6
Looking Deeper

The oxide represents a series capacitor C_{ox} connected between the probe and the semiconductor depletion capacitance C_D. Thick oxide makes the C/V transition from accumulation to depletion less steep and sets a limit on the measured capacitance $C_m = C_{ox}C_D/(C_{ox} + C_D)$ in accumulation. This complicates the analysis at high doping levels, since the capacitance modulation becomes very small. The reason is the charging and the voltage drop in the oxide. The consequence for an MOS transistor is the need for a larger gate voltage amplitude to switch it on or off. With extremely thin oxides the tunnelling probability becomes significant. The equilibrium of charge/potential balance assumed in the depletion layer model is not guaranteed. Electrons may be injected and trapped in the oxide, causing shifts of the flatband voltage [49]. At the interface of the semiconductor with a nonideally matched insulator, interface states exist that trap charge carriers [63]. Changes of gate voltage lead to changes in the amount of trapped charge. From the point of view of voltage transfer, the effect resembles that of thick oxide.

Silicon is probably the only element that can form an oxide with a low density of interface states. Standard MOS technology, using high-temperature oxidation (approximately 1150 °C), achieves a negligible level of such states. As already mentioned, the requirement of ultrashallow junctions and spatial limitations in ULSI complicates high-temperature treatments [104].

The relative permittivity of SiO_2 is only 3.9. The problems with leakage current through extremely thin oxides can be circumvented using materials with larger permittivity, called high-k oxides. With them the sufficiently large gate capacitance can be achieved at higher thickness; hence, there is less leakage. Unfortunately, they are not the ideal Si/SiO_2 combination anymore and it is difficult to achieve comparably low density of interface states.

Besides the oxide and oxide/semiconductor interface energy states in the semi-conductor gap may act as charge traps. They may be introduced by intentional doping, for instance to increase the recombination rate of injected minority carriers and thus the switching speed of bipolar junction transistors. In other cases they may be undesired impurities or intrinsic defects.

The defects in crystalline materials at not too high concentration form discrete energy levels. At high concentration they create narrow energy bands. In other cases continuous energy distributions are observed. Knowledge of them is useful in that it may help to identify the kind of defects responsible for them. The trapping is not permanent. The electrons/holes are emitted at a temperature-dependent rate. The emission probability depends upon the distance of the defect energy level from the conduction band edge (electron traps) or the valence band edge (hole traps) and their capture cross-section. The two represent fingerprints of the defects. In noncrystalline materials, owing to disorder, the energies are less discrete. The relaxation times of defects span many orders of magnitude and may exceed hours. In some cases, like metal nitride oxide semiconductor (MNOS) EPROMs and metal oxide nitride oxide semiconductor (MONOS) flash memories, ideally even permanent trapping is required.

The effect of interface states has been observed also in SCM analysis of semiconductors [73, 105–108] as a broadening of dC/dV curves and as a hysteresis appearing with dependence on the sign of bias voltage change.

In other materials, such as dielectrics or ionic conductors, many processes may lead to relatively slow relaxation. These may be grain boundaries, representing barriers that the charge carriers must overcome, or electrodes, at which space charge forms. They may adversely affect the performance of devices, or may be useful. The voltage-dependence of conductance of the many grain boundaries in a varistor makes it a nonlinear resistor. The space charge in dielectrics with artificially introduced interfaces is used to produce capacitors with considerable capacitance-to-volume ratio. Electrochemical processes at the electrode/solid electrolyte interface are utilised in solid-state batteries.

11.6.1
Impedance Spectroscopy

Many of the effects mentioned can be investigated using capacitance-based methods. Most frequently modifications of impedance spectroscopy are applied [109]. It is an impedance measurement in a wide frequency range, sometimes spanning from a few microhertz [110] to 100 MHz [111] or even beyond. The conductivities of the materials investigated may be very different. On high-conductivity materials measurements up to high frequencies are possible, whereas if the bulk resistance is large, such measurements make no sense, since they give only the geometric capacitance, which in the case of a conducting SFM tip is dominated by the stray capacitance. Measurements at low frequencies may be rather demanding. The impedance may attain very high values, with all the consequences like signal-to-noise ratio and phase resolution. Owing to cable and input capacitances of amplifiers the possibilities of small current measurement at high frequencies are limited. At very high frequencies incorrect impedance matching to a tip/sample impedance causes spurious inductive behaviour [112].

The impedance of a parallel RC circuit in a complex plain has the form of a semicircle. Though it is frequently observed in impedance measurements on various materials, often the arc is only a quarter-circle or something between a semicircle and a quarter-circle, or it is asymmetrically deformed. The deformations are explained by the distribution of relaxation times, e. g. caused by the distribution of the activation energies of the respective processes or interface inhomogeneity. The impedance diagrams are very instructive and help to deduce the possible nature of the processes. Unfortunately, one sees frequently the bad habit not to indicate the frequencies at which the data points or at least the vertex of the arc were acquired. Then the reporting of the spectrum is reduced to information on dc conductivity, which can be measured much more easily directly. Another possibility is to display the frequency dependence of capacitance and conductance in the form of Bode plots. In this case the information is complete, albeit sometimes less illustrative. The least readable, though still complete, plots are the plots of frequency dependence of the absolute value and the phase of the impedance. In real-world materials more effects can yield arc-shaped contributions to the impedance diagrams. These may be, for instance, two kinds of dipoles, with different relaxation times. Then their contribution to the permittivity ceases in different parts of the frequency spectrum. In an equivalent network an RC pair can be assigned to each.

Impedance spectra have been measured by Rodewald et al. [113] on polycrystalline Fe-doped $SrTiO_3$. They used 20-μm-diameter electrodes deposited on the surface. The large grain size allowed two such dots to be placed on a single grain; thus, the impedance within a single grain or across the grain boundary could be measured. In this material the grain boundaries form double Schottky barriers that reduce the conductance. In polycrystalline AgCl the conductance in grain boundary regions is higher than in the grains. Sever Škapin et al. [114] have measured impedance spectra from 10 mHz to 64 kHz. They contacted the grains with silver-coated tungsten tips. They found a coincidence of the grain conductivity activation energy with that known for single crystals, whereas a lower value was found in grain boundaries.

The abovementioned measurements were performed with relatively large electrodes. Layson et al. [111] investigated poly (ethylene oxide) (PEO)–lithium triflate and PEO-filled porous polycarbonate membranes using a SFM in an impressing frequency range from 400 Hz to 100 MHz. PEO is an ionic conductor used in solid-state batteries. They could well distinguish the nonconducting polycarbonate and the ionic-conductor-filled 100-nm-diameter pores.

The spreading resistance of the point contact may obscure processes taking place in the bulk. The situation was analysed by O'Hayre et al. [115]. They mapped the impedance of a gold/nitride test structure, ZnO varistor and Nafion, an ion-conducting polymer membrane. In [116] the Pt/Nafion contact was investigated in more detail. The impedance measurements were carried out in the frequency range from 100 mHz to 100 kHz. Electrochemical processes could be quantitatively analysed. Shao et al. [112] investigated polycrystalline ZnO in a wider frequency range from 40 Hz to 110 MHz. The goal was to measure the impedance of grain boundaries. They used two configurations, one across the sample between the SFM tip and the bottom of the sample, the other between the tip and a micropatterned electrode at the surface. They found two well-expressed arcs in impedance diagrams, both bias-dependent, and attributed them to the tip/surface Schottky contact and to grain boundary impedance.

11.6.2
Deep Level Transient Spectroscopy

On a macroscopic scale, deep level transient spectroscopy (DLTS) has become a well-established method for characterisation of defect states in semiconductors and their interfaces with insulators [117]. In this method periodic voltage pulses are applied to the sample biased to depletion. The pulses drive the sample to accumulation. Electrons from the conduction band (in n-type semiconductors) or holes from the valence band (in p-type semiconductors) become rapidly trapped by defects present in the depletion layer or at the semiconductor/insulator interface. Following the trailing edge of the pulse, the semiconductor returns to depletion and the trapped charge carriers are emitted to the conduction (valence) band and swept out by the built-in potential difference to the bulk of the sample. Figure 11.25 illustrates the sequence. The emission probability depends on the capture cross-section of the traps and the distance of its energy level from the respective band edge; thus, it is temperature-dependent. By heating the sample, the emission rate increases. The change of net charge in the depletion layer can be monitored by measuring the depletion layer capacitance. The capacitance transients are processed by a suitable correlator that transforms the monotonous temperature dependence of relaxation into a peak. Lang [117] used sampling at time instances t_1 and $t_2 = 10t_1$, which defined a rate window, and displayed their difference. At low temperature the emission is negligible; thus, the difference of the two samples is nil. At high temperature the charge is emitted already before the first sample; thus, the difference becomes

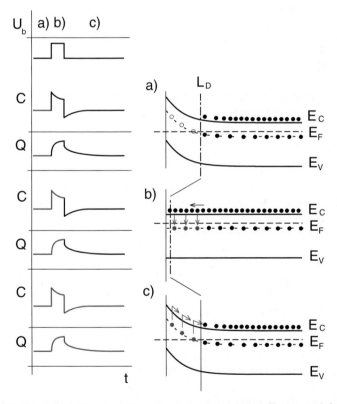

Fig. 11.25. The deep level transient spectroscopy measuring sequence. The empty defect level (**a**) is filled by a voltage pulse (**b**) and then allowed to relax (**c**). Electrons fill the previously empty conduction band in the depletion layer (**b**) and may fill the traps. After returning to the quiescent bias, electrons from the filled traps are emitted to the conduction band (**c**) and are swept out from the depletion layer. The state of the depletion layer can be followed by monitoring the capacitance, current or charge

negligible too. At a temperature between the two the difference attains a maximum. Its temperature is a function of the rate window. The position versus rate window dependence yields the energy of the defect level, and the peak height yields the concentration. The method is very sensitive and defect densities on the order of 10^{-5}–10^{-6} of the dopant concentration can be detected using electrodes of 0.1– 1 mm^2. The measurements are usually carried out in the temperature range from -190 to 25 or 125 °C.

For application in microscopy the sensitivity of capacitance transient detection must be extremely high. Let us assume a 1D case for simplicity. The capacitance of the depletion layer is $C_D = \varepsilon A/L_D$. The depletion layer width in the presence of ionised traps with concentration T is

$$L_D = \sqrt{2\varepsilon\Psi/q(N+T)} \qquad (11.10)$$

and the relative change of depletion layer capacitance is

$$\Delta C_D/C_D = T/2N . \qquad (11.11)$$

The rate window corresponds to relaxation time [117]

$$\tau_{max} = (t_1 - t_2)\left[\ln(t_1/t_2)\right]^{-1} . \qquad (11.12)$$

The peak maximum with the abovementioned sampling is $0.697\Delta C_d$ and the maximum appears at the temperature at which the relaxation time is $\tau_{max} = 3.91t_1$; thus, the peak height is $0.35T/N$.

The application of DLTS to MOS structures of very small gate area is a challenging task. It requires a many orders of magnitude increased sensitivity. Equation (11.11) allows an estimation of the achievable resolution to be made. Assuming a dopant concentration 10^{16} cm^{-3}, a tip radius of 50 nm and spherical symmetry, the capacitance change from depletion to accumulation is approximately 80 aF. With a transducer resolution of 10^{-2} aF, achievable in 1-Hz bandwidth using 1-V measuring signal and an acceptable peak height of 0.1 aF, the minimum detectable trap density would be 3.5×10^{13} cm^{-3}; for $N = 10^{18}$ cm^{-3} it would be 3.5×10^{14} cm^{-3}. In macroscopic measurements the 1-V applied voltage would be considered too high. With a sharper tip with $r = 25$ nm, the results would be 2.5×10^{14} and 2.5×10^{15} cm^{-3}.

It would be rather impractical in microscopy to perform such analysis under continual heating. The analysis would probably have been carried out at different points and thermal drifts would cause serious problems. Therefore, measurements should rather be isothermal, limited to a few temperatures. Such a possibility is offered by scanning capacitance transient spectroscopy, a derivative of DLTS. It uses instead of temperature scanning rate window scans. Tóth et al. [118] have investigated by this method gold-doped Si. They recorded also the capacitance transients using a commercial capacitance microscope. Though the transients were extremely small, they were measurable. The deep-level spectrometer connected to the microscope uses a lock-in amplifier as a correlator.

This approach was followed by Kim et al. [119]. They recorded capacitance transients in boron-doped silicon with a concentration 10^{15} cm^{-3}, with 8-nm-thick thermal oxide. The transients originated from hole traps. They used a force microscope equipped with an etched copper wire attached to a tuning fork. This arrangement reduced significantly the stray capacitance compared with SFM cantilevers. They could estimate the energy and apparent capture cross-section of the traps but encountered difficulties with estimation of trap concentration. The problem was the uncertainty of the volume scrutinised.

The original version of DLTS used high-frequency capacitance sensing to monitor the relaxation of the double layer. An alternative, comparably sensitive method is based on direct measurement of the transient charge [120]. Its advantage is that it can be applied also to low-conducting samples, like hydrogenated amorphous Si (a-Si:H) or dielectric films. The sensitivity problems in measurements on small-gate-area MOS structures have been solved recently by applying the excitation pulses to the gate and measuring the channel conductance [121] or drain current [122], thus utilising the gain of the transistor. Such solutions cannot be used in microscopy, since the analysis must be two-terminal, i. e. the sample must form a diode or a capacitor.

The first successful two-terminal measurements on small MOS structures were reported recently by Nádaždy et al. [123]. They applied isothermal charge-transient spectroscopy (IQTS) to thin-film transistors in a capacitor arrangement. The required sensitivity was achieved by means of a charge-to-voltage converter with resolution in the attocoulomb range [124]. The principle of the gated integrator employed for this purpose is seen in Fig. 11.26. The role of gating is to eliminate the large transient coming from charging the semiconductor depletion layer, to reset the integrator after each excitation pulse and to maintain its stability. The charge transients were sampled at time instants t_1, $2t_1$ and $4t_1$. The IQTS signal was created by combining the samples to $\Delta Q = Q(t_1) - 1.5Q(2t_1) + 0.5Q(4t_1)$ [125]. The rate window defined by such sampling is $1/t_1$. This filter yields higher selectivity than the standard double boxcar processing and completely removes the linear component of the response, which can be caused by a constant current appearing at the converter input, either leakage or operational amplifier bias current. The charge-transient spectra were obtained by sweeping t_1 from $2\,\mu s$ to $10\,ms$.

The transistors were fabricated in crystalline grains, created by laser crystallisation of amorphous Si films. References to the process can be found in [123]. It produces a regular array in which grain boundaries radiate from the centre, from

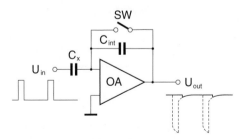

Fig. 11.26. The principle of charge-transient detection by means of a gated integrator. The large transient charge appearing during the filling pulse and shortly after it (*dashed lines*) does not contain interesting information and is avoided. During this time interval the integrator is reset. The remaining transient is then processed by means of a correlator

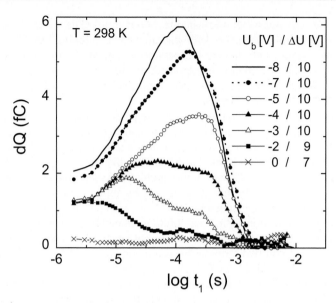

Fig. 11.27. Isothermal charge-transient spectroscopy (IQTS) spectra acquired on a $4 \times 4\ \mu m^2$ gate of thin-film transistor (TFT). At negative bias larger than -4 V the MOS capacitor is in inversion and charge generation takes place. Characteristic for the defects analysed are the spectra obtained with smaller bias [123]

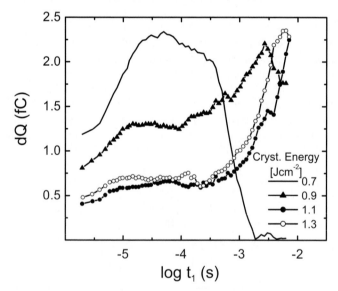

Fig. 11.28. IQTS spectra of TFTs in grains crystallised with different laser energy densities. gates with an area of $4 \times 4\ \mu m^2$ were positioned at the centre of the the grains [123]

where the crystallisation is initiated. Transistors with different gate sizes and orientations were placed on the grains. Figure 11.27 shows the bias dependence of IQTS spectra recorded on a centred gate with an area of $4 \times 4 \, \mu m^2$, produced in material crystallised by a laser energy density of $0.7 \, Jcm^{-2}$. In Fig. 11.28 the dependence of

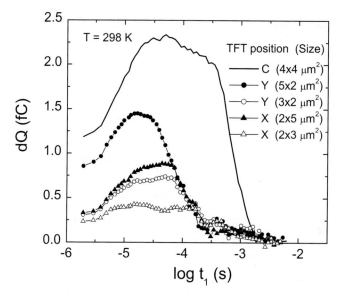

Fig. 11.29. The effect of channel size and position on the spectra from material crystallised with the lowest laser energy density of $0.7 \, Jcm^{-2}$ [123]

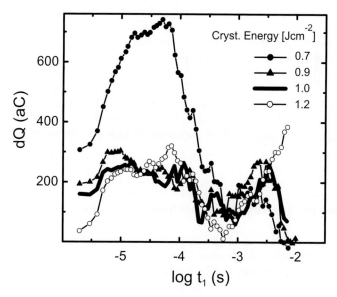

Fig. 11.30. The influence of laser energy density on IQTS spectra obtained on $2 \times 3 \, \mu m^2$ gates in y position

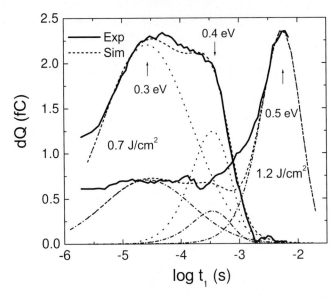

Fig. 11.31. Decomposition of IQTS spectra from Fig. 11.28 (energy densities 0.7 and 1.3 Jcm^{-2}) [125]

the spectra on laser energy and in Fig. 11.29 the dependence of the spectra on the position of gates are shown. Figure 11.30 shows a dependence of spectra recorded on smallest gates on crystallisation energy and Fig. 11.31 the decomposition of spectra corresponding to the lowest and highest laser energy density from Fig. 11.28. The temperature dependence of spectrum and the Arrhenius plots constructed from the position of peak maxima is seen in Fig. 11.32. They yielded energies of defect states 0.3 and 0.4 eV below the conduction band edge. The third identified energy level is 0.5 eV below the conduction band edge.

The resolution of the instrument used in [123, 124] is limited, on the one hand, by the sensitivity of the integrator, and on the other, by not well mastered mains and computer interference. The noise of the integrator has been found to be $2.5q = 4 \times 10^{-19}$ F/$\sqrt{\text{Hz}}$. The scatter of data points in the spectra is much larger. Thus, avoiding the interference offers an improvement in resolution of about 2 orders of magnitude.

11.7
Optimising the Experimental Conditions

The sensitivity and resolution of SCMs can be optimised by the choice of tip radius [50]. The lateral resolution in semiconductor structures is determined by the 3D depletion region width rather than by the microscope tip apex radius. Moreover, even in the case of high defect density the average separation of defect centres may be as high as tens to hundreds of nanometres. The choice of larger tip radius increases the local, i.e. relevant, contribution to capacitance, without reducing significantly

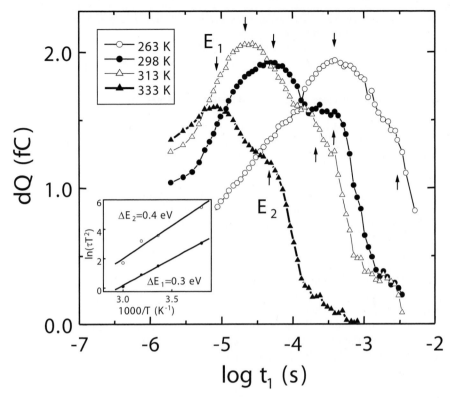

Fig. 11.32. Spectra recorded at different temperatures on a centred gate, and Arrhenius plots of the maxima of the peaks (*inset*), yielding the energies 0.3 and 0.4 eV

the lateral resolution. Table 11.2 contains some illustrative data on the role of the tip radius and shielding. In the first column are the probe geometry, the tip radius r, the cone angle and the radius of the shield orifice R. In the second column is the stray capacitance of the probe placed 5 nm over a conducting plane in air. The third column contains the capacitance change obtained by changing the tip/plane distance from 10 to 50 nm, and the last column contains the capacitance change if the probe were in the same position as in the second column but above a hole with 1-μm radius, and 50 and 10 nm deep, respectively. This would approximately correspond to imaging of low-doped wells in a semiconductor. For protrusions, corresponding to high-doped regions, the result would be similar, as demonstrated by Fig. 11.22.

From the above note a recommendation can be formulated. Since the measurement error with a large tip radius is improved without dramatic loss of lateral resolution, whenever high capacitance or charge resolution is aimed at, it is advantageous to choose a spherically terminated tip with a large radius of curvature. The capacitance resolution is enhanced also with flat tips, albeit the lateral resolution is seriously affected. On the other hand, for imaging or analysis of very thin films or properties of surface layers, sharper tips should enhance the resolution.

Table 11.2. The capacitance and its change over a plane and over a depression with 1-μm radius upon displacement of the probe [50]

Probe type	C_∞ (5 nm) (fF)	ΔC_∞ (10–50)[a] (aF)	$\Delta C_{1\,\mu m}$ (10–50)[b] (aF)
$r = 25$ nm/10° unshielded	4.2347	4.506	3.147
$r = 25$ nm/20° unshielded	5.6206	5.695	3.471
$r = 1$ μm/20° unshielded	6.0953	85.985	79.759
$r = 25$ nm/10°/shield $R = 25$ μm	0.3146	4.770	3.560
$r = 25$ nm/10°/shield $R = 1$ μm	0.0257	3.220	3.075
$r = 25$ nm/30°/microscopic tip	0.6002	3.960	3.510
$r = 25$ nm/30°/tip on cantilever	111.8173	88.779	3.982

[a]Capacitance difference for displacement from plane from 10 to 50 nm
[b]Capacitance difference for depression depth change from 10 to 50 nm

The tip/surface distance is usually controlled by the SFM sensor or sometimes with a tunnelling junction of the STM [126]. Their interaction is more local than that of a SCM probe. Small protrusions or depressions, which would hardly be visible under constant-capacitance control, may cause large capacitance changes. They may be easily misinterpreted. Scans on rough surfaces may make the impression of very high lateral resolution. In reality, they are only a different manifestation of surface topography, experienced by the capacitance sensor.

11.8
Conclusions

SCMs have become useful instruments for a variety of investigations. They are most frequently applied to dopant profiling in semiconductor structures—the need and demand for such analytic tools is evident and the SCM has contributed a lot to the interest in this kind of microscopy. In contrast to the scanning spreading resistance microscope, they do not need a well-defined electric contact that destroys the surface; they can work even contactless and achieve high lateral resolution.

The way in which the microscopes are used is far from optimal. It is first of all the application of only slightly modified SFM cantilevers that limits their performance. Reducing the stray capacitance by metal tips, narrow leads to them and isolation from the fixture would dramatically reduce the stray field and improve the signal-to-noise ratio and the contrast achieved [127]. However, the very best solution is shielding.

Applications other than dopant profiling are relatively rare but their frequency is growing. Of these, impedance spectroscopy and isothermal modifications of DLTS are promising. ICTS has the advantage of single-frequency measurement and thus the possibility to avoid RF or other interference. Its disadvantage is the limitation to semiconductors and other structures, in which a sufficiently large concentration of charge carriers, thus space charge, exists. Isothermal charge-transient microscopy can be used also on low-conductivity materials. Its problem is a larger propensity to interference, since it represents a wide time span, or wide-band measurement, combined with frequency-swept excitation, for which any periodic signal may be disturbing.

The combination of capacitance or charge and spatial resolution of the best sensors is close to the statistical limits. The noise spectral density of capacitance microscopes is lower than a few electron charges. At high spatial resolution the probability that enough charge carriers appear within the reach of the probe may be low. Thus, the analysis methods using conventional statistical physics may fail or would need to be modified, e. g. towards noise analysis.

Acknowledgements. Partial support of the Research and Development Agency (contract no. APVT-51-013904) and the VEGA Grant Agency (project no. 2/5097/25) is acknowledged.

References

1. Binnig G, Rohrer H, Gerber C, Weibel E (1982) Phys Rev Lett 49:57
2. Binnig G, Rohrer H (1982) Helv Phys Acta 55:726
3. Matey JR (1984) US Patent 4,481,616
4. Matey JR, Blanc J (1985) J Appl Phys 57:1437
5. Binnig G, Rohrer H, Gerber C, Weibel E (1983) Phys Rev Lett 50:120
6. Bugg CD, King PJ (1988) J Phys E 21:147
7. Kleinknecht HP, Sandercock JR, Meier H (1988) Scanning Microsc 2:1839
8. Martin Y, Abraham DW, Wickramasinghe HK (1988) Appl Phys Lett 52:1103
9. Kobayashi K, Yamada H, Matsushige K (2002) Appl Phys Lett 81:2629
10. Williams CC, Hough WP, Rishton SA (1989) Appl Phys Lett 55:203
11. The National Technology Roadmap for Semiconductors (1997) Semiconductor Industry Association, San Jose
12. International Technology Roadmap for Semiconductors (2006) ITRS 2006 update. http://www.itrs.net/Links/2006Update/2006UpdateFinal.htm
13. Palmer RC, Denlinger EJ, Kawamoto H (1982) RCA Rev 43:195
14. Dreyer M, Wiesendanger R (1995) Appl Phys A 61:357
15. Lányi Š, Török J, Řehůřek P (1994) Rev Sci Instrum 65:2258
16. Arakawa H, Nishitani R (2001) J Vac Sci Technol B 19:1150
17. Lee DT, Pelz JP, Bhushan B (2002) Rev Sci Instrum 73:3525
18. Horowitz P, Hill W (1980) The art of electronics. Cambridge University Press, Cambridge
19. Lányi Š (2002) Acta Phys Slovaca 52:55
20. Lányi Š, Hruškovic M (2002) Rev Sci Instrum 73:2923
21. Lányi Š (2001) Meas Sci Technol 12:1456
22. Zavyalov VV, McMurray JC, Williams CC (2000) J Vac Sci Technol B 18:1125
23. Tran T, Oliver DR, Thomson DJ, Bridges GE (2001) Rev Sci Instrum 72:2618
24. Jaensch S, Schmidt H, Grundmann M (2006) Physica B 913:376–377
25. van der Weide DW, Neuzil P (1996) J Vac Sci Technol B 14:4144
26. Pingue P, Piazza V, Baschieri P, Ascoli C, Menozzi C, Alessandrini A, Facci P (2006) Appl Phys Lett 88:043510
27. Clippendale AJ, Prance RJ, Clark TD, Brouers F (1994) J Phys D Appl Phys 27:2426
28. Asami K (1994) Meas Sci Technol 5:589
29. Barrett RC, Quate CF (1991) J Appl Phys 70:2725
30. Feenstra RM (2003) J Vac Sci Technol B 21:2080
31. Peridier VJ, Li-Hong Pan, Sullivan TE (1995) J Appl Phys 78:4888

32. Watanabe S, Hane K, Ohye T, Ito M, Goto T (1993) J Vac Sci Technol B 11:1774
33. Mesa G, Dobado-Fuentes E, Sáenz JJ (1996) J Appl Phys 79:39
34. Kurokawa S, Sakai A (1998) J Appl Phys 83:7416
35. Nonnenmacher M, O'Boyle MP, Wickramasinghe HK (1991) Appl Phys Lett 58:2921
36. Koley G, Spencer MG (2001) Appl Phys Lett 79:545
37. Zienkiewicz OC (1977) The finite element method. McGraw-Hill, New York
38. Marchiando JF, Kopanski JJ, Lowney JR (1998) J Vac Sci Technol B 16:463
39. Marchiando JF, Kopanski JJ (2004) J Vac Sci Technol B 11:411
40. Lányi Š (1999) Surf Interface Anal 27:348
41. Bruce NC, Garcia-Valenzuela A, Kouznetsov D (2000) J Phys D Appl Phys 33:2890
42. Ciampolini L, Ciappa M, Malberti P, Fichtner W (2001) Solid State Electron 46:445
43. Marchiando JF, Kopanski JJ (2002) J Appl Phys 92:5798
44. Lányi Š, Török J (1995) J Electr Eng 46:126
45. Tomiye H, Kawami H, Yao T (1997) Appl Surf Sci 117/118:166
46. Bussmann E, Williams CC (2004) Rev Sci Instrum 75:422
47. Lányi Š, Hruškovic M (2003) J Phys D Appl Phys 36:598
48. Lányi Š, Török J, Řehůřek P (1996) J Vac Sci Technol B 14:892
49. Zavyalov VV, McMurray JC, Williams CC (1999) Rev Sci Instrum 70:158
50. Lányi Š (2005) Ultramicroscopy 103:221
51. Williams CC, Slinkman J, Hough WP, Wickramasinghe HK (1989) Appl Phys Lett 55:1662
52. Huang Y, Williams CC (1994) J Vac Sci Technol B 12:369
53. Naitou Y, Ookuba N (2001) Appl Phys Lett 78:2955
54. Breitschopf P, Benstetter G, Knoll B, Frammelsberger W (2005) Microelectron Reliab
    45:1568
55. Mang KM, Khang Y, Park YJ, Kuk Y, Lee SM, Williams CC (1996) J Vac Sci Technol B
    14:1536
56. Nakagiri N, Yamamoto T, Suzuki Y, Miyashita M, Watanabe S (1997) Nanotechnology
    8:A32
57. Buh GH, Chung HJ, Kim CK, Yi JH, Yoon IT, Kuk Y (2000) Appl Phys Lett 77:106
58. Isenbart J, Born A, Wiesendanger R (2001) Appl Phys A 72:S243
59. Chao K-J, Kingsley JR, Plano RJ, Lu X, Ward I (2001) J Vac Sci Technol B 19:1154
60. Basnar B, Golka S, Gornik E, Harasek S, Bertagnolli E, Schatzmayr M, Smoliner J (2001)
    J Vac Sci Technol B 19:1808
61. Lee DT, Pelz JP, Bhushan B (2006) Nanotechnology 17:1484
62. Tallarida G, Spiga S, Fanciulli M (2003) Mater Res Soc Symp Proc 738:G5.1.1
63. Sze SM (1969) Physics of semiconductor devices. Wiley-Interscience, New York
64. Neubauer G, Erickson A, Williams CC, Kopanski JJ, Rodgers M, Adderton D (1996) J Vac
    Sci Technol B 14:426
65. Huang Y, Williams CC, Slinkman J (1995) Appl Phys Lett 66:344
66. De Wolf P, Stephenson R, Trenkler T, Clarysse T, Hantschel T, Vandervorst W (2000)
    J Vac Sci Technol B 18:361
67. Diebold AC, Rump MR, Kopanski JJ, Seiler DG (1996) J Vac Sci Technol B 14:196
68. McMurray JS, Kim J, Williams CC (1997) J Vac Sci Technol B 15:1011
69. McMurray JS, Kim J, Williams CC, Slinkman J (1998) J Vac Sci Technol B 16:344
70. Synopsys (1996) TSUPREM4. Technology Modeling Associates, Palo Alto
71. Born A, Wiesendanger R (2000) In: Proceedings of the 26th international symposium for
    testing and failure analysis, 12–16 November 2000, Bellevue, p 521
72. Kopanski JJ, Marchiando JF, Rennex BG (2002) J Vac Sci Technol B 20:2101
73. Huang Y, Williams CC, Wendman MA (1996) J Vac Sci Technol A 14:1168
74. Williams CC (1999) Annu Rev Mater Sci 29:471
75. Kopanski JJ, Marchiando JF, Lowney JR (1996) J Vac Sci Technol B 14:242

76. Edwards H, McGlothlin R, San Martin R, Gribelyuk EUM, Mahaffy R, Shih CK, List RS, Ukraintsev VA (1998) Appl Phys Lett 72:698
77. Zavyalov VV, McMurray JC, Williams CC (1999) J Appl Phys 85:7774
78. Chim WK, Wong KM, Teo YL, Lei Y, Yeow YT (2002) Appl Phys Lett 80:4837
79. O'Malley ML, Timp GL, Moccio SV, Garno JP, Kleiman RN (1999) Appl Phys Lett 74:272
80. O'Malley ML, Timp GL, Timp W, Moccio SV, Garno JP, Kleiman RN (1999) Appl Phys Lett 74:3672
81. Goto K, Hane K (1998) Appl Phys Lett 73:544
82. Yabuhara H, Ciappa M, Fichtner W (2001) Microel Reliab 41:1459
83. Goghero D, Giannazzo F, Raineri V (2003) Mater Sci Eng B 102:152
84. Ciampolini L, Giannazzo F, Ciappa M, Fichtner W, Raineri V (2001) Mater Sci Semicond Process 4:85
85. Giannazzo F, Goghero D, Raineri V, Mirabella S, Priolo F (2003) Appl Phys Lett 83:2659
86. Stangoni M, Ciappa M, Fichtner W (2004) J Vac Sci Technol B 22:406
87. Giannazzo F, Raineri V, La A Magna, Mirabella S, Impellizzeri G, Piro AM, Priolo F, Napolitani E, Liotta SF (2005) J Appl Phys 97:014302
88. Duhayon N, Clarysse T, Eyben P, Vandervorst W, Hellemans L (2002) J Vac Sci Technol B 20:741
89. Kopanski JJ, Marchiando JF, Lowney JR (1966) In: Seiler DG, Diebold AC (eds) Semiconductor characterization: present status and future. American Institute of Physics, New York, p 308
90. Marchiando JF, Kopanski JJ, Lowney JR (1998) J Vac Sci Technol B 16:463
91. Marchiando JF, Lowney JR, Kopanski JJ (1998) Scanning Microsc 12:205
92. Marchiando JF, Kopanski JJ, Albers J (2000) J Vac Sci Technol B 18:414
93. Ciampolini L, Ciappa M, Malberti P, Fichtner W (2002) Solid State Electron 46:445
94. Duhayon N, Eyben P, Fouchier M, Clarysse T, Vandervorst W, lvarez D, Schoemann S, Ciappa M, Stangoni M, Fichtner W, Formanek P, Kittler M, Raineri V, Giannazzo F, Goghero D, Rosenwaks Y, Shikler R, Saraf S, Sadewasser S, Barreau N, Glatzel T, Verheijen M, Mentink SAM, von Sprekelsen M, Maltezopoulos T, Wiesendanger R, Hellemans L (2004) J Vac Sci Technol B 22:385
95. Bowallius O, Anand S, Nordell N, Landgren G, Karlsson S (2001) Mater Sci Semicond Process 4:209
96. Giannazzo F, Musumeci P, Calcagno L, Makhtari A, Raineri V (2001) Mater Sci Semicond Process 4:195
97. Giannazzo F, Calcagno L, Roccaforte F, Musumeci P, La Via F, Raineri V (2001) Appl Surf Sci 184:183
98. Calcagno L, Raineri V (2002) Curr Opin Solid State Mater Sci 6:47
99. Hammar M, Rodriguez Messmer E, Luzuy M, Anand S, Lourdudoss S, Landgren G (1998) Appl Phys Lett 72:815
100. Zhou X, Yu ET, Florescu DI, Ramer JC, Lee DS, Ting SM, Armour EA (2005) Appl Phys Lett 86:202113
101. Douhéret O, Anand S, Angulo C Barrios, Lourdudoss S (2002) Appl Phys Lett 81:960
102. Ban D, Sargent EH, St. Dixon-Warren J, Grevatt T, Knight G, Pakulski G, Spring Thorpe AJ, Streater R, White JK (2002) J Vac Sci Technol B 20:2126
103. Smith KV, Dang XZ, Yu ET (2000) J Vac Sci Technol B 18:2304
104. Borland JO (2002) Mater Res Soc Symp Proc 717:C1.1.1
105. Chim WK, Wong KM, Yeow YT, Hong YD, Lei Y, Teo LW, Choi WK (2003) IEEE Electron Device Lett 24:667
106. Hong YD, Yeow YT, Chim W-K, Wong K-M, Kopanski JJ (2004) IEEE Trans Electron Devices 51:1496
107. Yang J, Kopanski JJ, Postula A, Bialkowski M (2005) Microelectrn Reliab 45:887

108. Schaadt DM, Yu ET (2001) Mater Res Soc Symp Proc 680:E5.1.
109. Macdonald JR (ed) (1987) Impedance spectroscopy: emphasizing solid materials and systems. Wiley, New York
110. Lányi Š, Tuček J (1987) Solid State Ionics 24:273
111. Layson A, Gadad SH, Teeters D (2003) Electrochim Acta 48:2207
112. Shao R, Kalinin SV, Bonnell DA (2003) Appl Phys Lett 82:1869
113. Rodewald S, Fleig J, Maier J (2001) J Am Ceram Soc 84:52
114. Sever Škapin A, Jamnik J, Pejovnik S (2000) Solid State Ionics 133:129
115. O'Hayre R, Lee M, Prinz FB (2004) J Appl Phys 95:8382
116. O'Hayre R, Feng G, Nix WD, Prinz FB (2004) J Appl Phys 96:3540
117. Lang DV (1974) J Appl Phys 45:3023
118. Tóth AL, Dózsa L, Gyulai J, Giannazzo F, Rainieri V (2001) Mater Sci Semicond Process 4:89
119. Kim CK, Yoon IT, Kuk Y, Lim H (2001) Appl Phys Lett 78:613
120. Kirov KI, Radev KB (1981) Phys Status Solidi A 63:711
121. Kolev PV, Deen MJ (1998) J Appl Phys 83:820
122. Exarchos M, Dieudonne F, Jomaah J, Papaioannou GJ, Balestra F (2004) Microelectron Reliab 44:1643
123. Nádaždy V, Rana V, Ishihara R, Lányi Š, Durný R, Metselaar JW, Beenakker CIM (2006) Mater Res Soc Symp Proc 910:0910-A19-02
124. Lányi Š, Nádaždy V (2007) Ultramicroscopy 107:963
125. Thurzo I, Gmucová K (1994) Rev Sci Instrum 65:2244
126. Imtiaz A, Anlage SM (2002) Ultramicroscopy 94:209
127. Yamamoto T, Suzuki Y, Miyashita M, Sugimura H, Nakagiri N (1997) J Vac Sci Technol B 15:1547

12 Probing Electrical Transport Properties at the Nanoscale by Current-Sensing Atomic Force Microscopy

Laura Fumagalli · Ignacio Casuso · Giorgio Ferrari · Gabriel Gomila

Abstract. In this chapter, we review the fundamentals and recent advances of current-sensing atomic force microscopy (CS-AFM) with particular emphasis on instrumental aspects. After discussing some generic aspects concerning the measurement of electrical currents at the nanoscale, we review the main CS-AFM techniques developed to probe the electrical transport properties at the nanoscale, namely, conductive atomic force microscopy, nanoscale impedance microscopy and electron noise microscopy. In each case we describe the electronic instrumentation implemented and the main applications of the technique to the fields of material science, electronics and biology. It is concluded that the measurement of direct and alternating currents and of current fluctuations with nanoscale spatial resolution provides an invaluable tool for an understanding of the spatially resolved electrical transport properties at the nanoscale.

Key words: electrical transport, conductance, impedance, noise, current-to-voltage amplifier, current-sensing atomic force microscopy, conductive-AFM, nanoscale impedance microscopy, nanoscale capacitance microscopy

Abbreviations

AFM	Atomic force microscopy
C-AFM	Conductive atomic force microscopy
CS-AFM	Current-sensing atomic force microscopy
DUT	Device under test
ENM	Electrical noise microscopy
NIM	Nanoscale impedance microscopy
SCM	Scanning capacitance microscopy
STM	Scanning tunneling microscopy

12.1
Introduction

The *nanoscale electrical transport properties* of materials, i.e., the properties that determine how the electric current flows through matter at the nanoscale, are of enormous interest in modern semiconductor device design [1,2] and in the emerging fields of molecular electronic devices [3–6] and single-molecule biosensors [7,8]. To address these properties, experimental setups capable of measuring electric currents with high sensitivity and nanoscale spatial resolution have become necessary [9].

Current-sensing atomic force microscopy (CS-AFM) is a scanning force microscopy method that precisely satisfies these requirements. By using a conductive probe with a sharp tip, it measures the current flowing through the sample with nanoscale spatial resolution, while simultaneously probing the surface topography under force feedback control. With this method, dc current, ac current and current fluctuations can be measured at the nanoscale, thus offering the extraordinary possibility of measuring the spatially resolved properties of electrical transport, such as nanoscale electrical resistance, impedance and noise. Furthermore, by obtaining a simultaneous map of surface topography, the correlation between electrical properties and morphological structure of material surfaces can be investigated with this technique.

Over the last decade a whole range of CS-AFM techniques have been developed to specifically measure dc currents (conductive AFM, C-AFM) [10–43], ac currents (ac CS-AFM and nanoscale impedance microscopy, NIM) [44–54] and electric current noise (here referred to as electrical noise microscopy, ENM) [55].

In C-AFM a stationary (dc) voltage bias is applied between the AFM probe and the sample and the corresponding dc electric current flowing through the sample is acquired, from which the *conductance* of the sample can be mapped locally. In addition, by sweeping the dc voltage between predefined values, one can measure the dependence of conductivity on the applied bias at specific locations of the sample. In some cases, a low-frequency voltage is applied instead of a dc voltage and the differential conductance (dI/dV) is measured.

In NIM a small alternating (ac) voltage signal is applied between the probe and the sample and the corresponding ac current is measured, from which the local *impedance* and hence the capacitance of the sample can be probed. In this case, by sweeping the frequency between predefined values or by superposing a dc voltage, one can probe the dependence of the impedance as a function of frequency or voltage respectively at specific points of the sample.

Finally, in ENM one measures the current fluctuations over a certain bandwidth with no need to apply any signal excitation, so that the *electrical noise* spectra inherent to the sample can be measured locally. Here again, if a dc voltage is applied and swept between predefined values, the dependence of the electrical noise on the applied voltage can be determined.

These techniques have been applied to the characterization of the electrical transport properties of large variety of samples, including thin dielectric films, self-assembled monolayers, single proteins, carbon nanotubes, nanoscale electronic devices and circuits. By shedding light on phenomena otherwise inaccessible to conventional techniques, CS-AFM is expected to become a standard tool for the electrical characterization of nanoscale systems. This chapter is intended to describe the basic principles of CS-AFM and report on the main results obtained so far, with special emphasis on instrumental aspects not so often reviewed in the literature.

By focusing on current-sensing techniques, we will not cover a number of other techniques used to probe electrical properties of materials at the nanoscale. In particular, this chapter will not review the main force-sensing techniques, namely, electrostatic force microscopy [56, 57], Kelvin probe force microscopy [58, 59] and scanning impedance microscopy [60, 61], with which one can obtain surface elec-

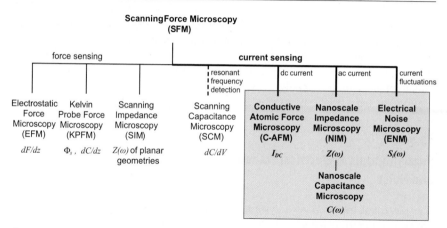

Fig. 12.1. Main scanning force microscopy (*SFM*) techniques used for nanoscale electrical measurements. Highlighted are the current-sensing techniques reviewed in this chapter. Note that scanning capacitance microscopy (*SCM*) has been commonly implemented with a resonant frequency detector, although it can be performed in a more general and quantitative approach using ac current sensing, which we refer to as nanoscale capacitance microscopy

trical properties such as dielectric constant, surface potential and surface charge by measuring the electrostatic interaction force between the probe and the sample. Some excellent overviews of these techniques are currently available [62–66]. We will not include either scanning capacitance microscopy (SCM) [67–70], which has traditionally employed a resonant frequency detector for dC/dV measurements on voltage-dependent semiconductor capacitance. We remark that this is a particular detection technique that was originally used as a feedback control quantity to mechanically scan the tip over the surface and hence the name of "scanning capacitance microscopy" [71]. Nanoscale capacitance measurement can be performed in a more general and quantitative way using an ac current-sensing approach, which we refer to here to as *nanoscale capacitance microscopy*.

Finally, we will not discuss either electrical measurements by the scanning tunneling microscopy (STM) technique [72], which in the last few decades has been adapted and used to perform most of the electrical transport measurements described in this chapter, including dc current measurements [73, 74], ac current measurements [75–77] and noise spectroscopy measurements [78–81]. In spite of many extraordinary results obtained by means of this technique, several advantages have imposed the use of CS-AFM for electrical transport measurements at the nanoscale, among others the ability of CS-AFM to simultaneously obtain the topographic and the electrical image *with independent detection methods*, which allows the electrical transport properties of insulating samples to be addressed, or the use of a force-sensing cantilever, which enables one to accurately control and quantify the tip–sample distance and the applied load on the sample surface, independently of the current level flowing through the tip.

This overview on CS-AFM will start in Sect. 12.2 by presenting briefly the main electrical transport properties probed by CS-AFM and the best-suited measuring strategies to address them. Then, we will describe in Sect. 12.3 the details of the experimental setups necessary to perform CS-AFM measurements, including a description of the required probe properties, measuring modes and electronic instrumentation. In the following sections, from Sect. 12.4 to Sect. 12.6, we will present an overview of the main results obtained by means of CS-AFM in its various implementations, namely, C-AFM, NIM and ENM. Finally, we will end with a summary and a brief outlook on the expected future developments of the field.

12.2
Fundamentals of Electrical Transport Properties: Resistance, Impedance and Noise

The electrical transport of a system can be described through three fundamental physical properties, namely, the electrical resistance, the small signal impedance and the electric noise spectrum. They can all be obtained by a measurement of the current flowing through the system followed by some signal postprocessing, and hence they are all accessible to CS-AFM. Note that in the case of CS-AFM, the system under study (here referred to as device under test, DUT) is composed of the sample, one contacting electrode (the conductive substrate or alternatively a lateral electrode) and the AFM probe, which acts as second electrode (Fig. 12.2).

The *electrical resistance R* gives an indication of the opposition of a system to the electric current flow. The resistance value of a linear system can be obtained as the ratio $R = V_{in}/I_{in}$ according to Ohm's law. In the case of nonlinear systems, in which the electrical resistance is a voltage-dependent quantity, it can be extracted as a differential value defined as $r = dV_{in}/dV_{in}$. Very often in nonlinear systems the whole of the current–voltage characteristics is taken as the property to be analyzed.

Measuring the resistance value simply requires a current meter to perform a stationary (dc) current measurement and a voltage source to apply a dc voltage across the system. In the case of analyzing the properties of insulating or semi-insulating materials, as is very often the case in CS-AFM, the current meter has to have high sensitivity and an extended measurement range over several current decades.

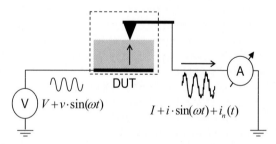

Fig. 12.2. Simplified diagram of electrical transport measurement with current-sensing atomic force microscopy (AFM). To perform current sensing, one applies across the electrode/device under test (*DUT*)/electrode system a dc voltage bias V superimposed on an ac small signal v and measures with a current detector at one electrode the electric current flowing through it, which, in the general case, is composed of a dc component I, an ac component i and an intrinsic current noise component i_n

The *electrical impedance* $Z(\omega)$ is the equivalent of the resistance extended to the frequency domain, i.e., it gives an indication of the opposition to the flow of a sinusoidal current at a given frequency f, where $f = \omega/2\pi$. It can be represented in complex notation with a real and an imaginary component as $Z(\omega) = R_{ac}(\omega) + jX(\omega)$, where R_{ac} is the ac resistance and X the reactance. While the real part is associated with the energy dissipation required to move charges through the material, the imaginary part is related to the ability of storing charges (capacitance) and sustaining magnetic fields (inductance). It can also be written in a vector notation as $Z(\omega) = |Z|e^{i\theta}$, where $|Z| = \sqrt{R^2 + X^2}$ is the impedance magnitude and $\theta = \arctan(R/X)$ is the phase, and is plotted in a polar graph as shown in Fig. 12.3a.

When tracing the impedance vector as a function of frequency, the so-called Nyquist plot is obtained. Measuring the Nyquist plot, that is, the real and imaginary parts of the impedance as a function of frequency, allows one to build an electrical equivalent model of the system under study made of resistors, capacitors and inductances [85]. Since in some cases the impedance can be voltage-dependent, Nyquist plots can be measured at different dc bias voltage conditions.

Impedance spectroscopy is widely applied to study the frequency response of a variety of systems. To cite a few applications, it is an essential tool in semiconductor process control/test to obtain characteristic parameters of devices such as dielectric constant, charge carrier density, junction capacitance and electron mobility [86–89], as well as in electrochemistry [90, 91] to study interface adsorption and charge transfer reaction.

In principle, the simplest procedure for measuring the impedance would only require a current-to-voltage amplifier to convert the output current into a voltage signal and an oscilloscope. The impedance can be directly obtained by applying an input sinusoidal voltage at a frequency f and using the oscilloscope to measure the amplitude and phase of the applied voltage and the output current (properly converted to voltage) as sketched in Fig. 12.3b. In practice, the output signal cannot be prevented from being corrupted by noise fluctuations of the measurement instruments or inherent to the sample under analysis. In addition, nonlinearities of the

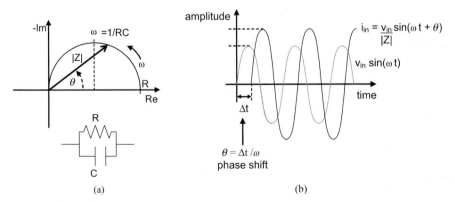

(a) (b)

Fig. 12.3. a Nyquist plot of impedance with $R//C$ equivalent model. **b** Input signal (ac applied voltage) and output signal (ac current) during impedance measurement

measurement instrumentation or the sample itself, as common for semi-insulating materials, determine the output signal distortion, which can be far from being an ideal sinusoid. Thus, the direct method is not sufficiently accurate and more complex measurement techniques are required to filter out noise and nonlinear harmonics.

The use of a current-to-voltage amplifier in a correlation analysis scheme of the input and output signals, typically referred to as *lock-in detection*, is the best approach for impedance measurements. By simply multiplying the ac output of the amplifier by the applied ac voltage signal and averaging the result for a certain time by means of a low-pass filter, the lock-in detector provides constant values proportional to the module and phase of the admittance $1/Z$, i.e., the inverse of impedance. Alternatively, the real and imaginary parts of the admittance can be separately obtained using a dual-phase lock-in scheme given in Fig. 12.4, based on two multiplying channels (one processing the signal in phase and the other in quadrature). The great advantage of this method is that the noise contribution decreases with increasing averaging time. Thus, noise fluctuations that affect the measured signal can be reduced to very low levels. Furthermore, the contribution of nonlinear harmonics that differ from the input frequency is properly rejected by the correlation technique.

Remarkably, in this scheme the impedance resolution (maximum detectable ac resistance and minimum detectable capacitance) is essentially determined by the current-to-voltage amplifier. The lock-in detection simply acts like a bandpass filter that follows the amplifier, recovering the single-frequency response of the system from noise fluctuations. For a given amplitude and frequency of the applied signal, the resolution only depends on two parameters, being directly proportional to the current noise power density $S_i(f)$ of the amplifier at the selected frequency and inversely proportional to the square root of the measurement time T. Once the noise floor given by the current-to voltage amplifier has been fixed, higher resolution can only be achieved by a longer acquisition time, thus paying the penalty of slowing down the measurement. This is why the use of a low-noise current-to-voltage amplifier is crucial, particularly when performing nanoscale measurement, and some guidelines for the correct choice/design of the amplifier will be given in Sect. 12.3.3.

The *electrical noise* is not simply a perturbation limiting the resolution of any measurement. It is a physical quantity given by the fluctuations of the measured

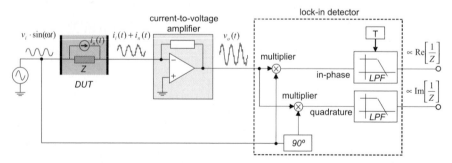

Fig. 12.4. Simplified diagram of dual-phase lock-in detection used for impedance measurement. The system (DUT) is represented here by its small signal equivalent circuit. *LPF* low-pass filter

current and is related to the electron transport mechanisms through the material. These fluctuations can be described in terms of a current noise power spectral density $S_i(f)$. Thus, considering the current noise fluctuations $i_n(t)$ in the time domain measured with a frequency bandwidth Δf centered on the frequency f, one can define the current noise power spectral density as

$$S_i(f) = \frac{\overline{i_n^2(t)}}{\Delta f} \, ,$$

where $\overline{i_n^2(t)}$ is the time average of the square of $i_n(t)$.

Electrical noise spectroscopy is an effective method for investigating fundamental properties of electron transport [92–95]. To cite a few examples, single carrier trapping and detrapping [96–98], diffusion of carriers in solids [99, 100], fractionally charged quasiparticles in quantum Hall systems [101, 102] and single electron spin resonance [103] have all found a clearer explanation through noise analysis.

The basic experimental setup for noise measurements is shown in Fig. 12.5a. Similarly to the impedance measurement setup, a current-to-voltage amplifier is required to convert the current fluctuations into voltage. The amplifier has to be followed by a bandpass filter selecting the frequency at which the spectral density is obtained (frequency selector), and then by a final block providing the mean squared value proportional to the current noise power density at the selected frequency. Also in this method the measurement resolution (minimum detectable noise) directly depends on the noise performance of the current-to-voltage amplifier. Indeed, only values of noise spectral density greater than that of the amplifier can be detected in this scheme.

The measurement resolution can be greatly improved using a cross-correlation technique [104–106], shown in Fig. 12.5b. In this configuration, generally referred to as *correlation spectrum analyzer*, the instrument is composed of two independent measurement channels simultaneously processing the current fluctuations of the sample. These are measured by two distinct input amplifiers and are processed in phase by two frequency selectors, thus resulting in a signal at the multiplier output with a mean value proportional to the current noise at the selected frequency. To a first approximation, the noise fluctuations produced by each amplifier are uncorrelated with each other. They are multiplied with a random phase and at the output of the multiplier they give a random signal with zero mean value. Therefore, similarly to the lock-in detection scheme, the final averaging stage reduces the instrumental fluctuations to very low values by increasing the measurement time, thus detecting the sample noise with extremely high precision. Note that in this configuration an external voltage source cannot be directly connected to the sample to apply a voltage bias to the sample. This is fixed by the input voltage of the two amplifiers, which are required to be tuneable by the user as shown in Fig. 12.5b. Frequency selection and correlation can be performed digitally or analogically using a superheterodyne scheme [107]. For noise spectroscopy up to tens of MHz a digital system is preferred, giving the possibility to process in parallel many frequencies of the spectrum. A few commercial instruments can be used to automatically perform the signal processing, but in any case the two external analog amplifiers are always necessary.

Here again, the ultimate performance of the instrument in terms of resolution is determined by the current-to-voltage amplifiers. The measurement resolution is

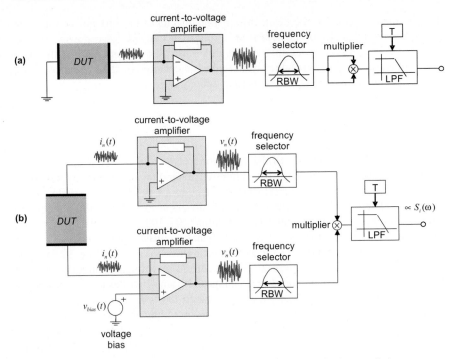

Fig. 12.5. Simplified diagram of (**a**) the noise spectrum analyzer and (**b**) the correlation spectrum analyzer for electrical noise measurement

given by the amount of correlated instrument noise, which cannot be distinguished from the noise fluctuations inherent to the sample and that are not suppressed by the instrument. The principle responsible for that noise is the input voltage noise $\overline{e_{OA}^2}$ of the two amplifiers applied across the sample, which produces in-phase current fluctuations on both channels. This is another reason that calls for an accurate and proper design of the amplifier as well as its careful connection to the sample under test, as will be discussed in Sect. 12.3.3.

To end this section, we note that the extraction of quantitative values for the electrical properties of the sample (e.g., resistivity, dielectric constant, charge trap density distribution, electron transition probabilities) from the electrical transport measurements (resistance, impedance and noise) requires a considerable theoretical effort to model the system geometry and contacts. The nanometric size of the AFM probe makes the well-known macroscopic expressions invalid, and in the nanoscale measurements new expressions have to be derived for every measuring situation. Moreover, in some cases (e.g., single-molecule measurements) the small size of the probe can also demand quantum mechanical transport models instead of semiclassical models. A revision of the various theoretical approaches followed to model the CS-AFM measurements lies outside the scope of the present chapter and will not be considered, but should be taken into account in the case of a quantitative interpretation of measurement data.

12.3
Experimental Setups for CS-AFM

In the previous section we reviewed the definition of the main electrical transport properties probed by CS-AFM and the best-suited measuring principles to address them. Here, we address the actual experimental CS-AFM setups implemented and the operating modes conventionally used.

Figure 12.6 shows the basic experimental setup required for CS-AFM. The local current is measured using the typical two-electrode configuration. The cantilever with a sharp tip covered with a conductive layer acts like a nano-electrode, while the second electrode is the conductive plate beneath the sample (or alternatively a lateral electrode deposited in any other part of the sample). Current images are obtained by scanning the tip over the surface, while a voltage is applied between the tip and the substrate, and a current-to-voltage amplifier measures the tip–substrate current. The measured current converted to a voltage value is acquired and properly processed to quantify the value of the electrical quantities as described in Sect. 12.2.

In order to implement successfully this measurement scheme an appropriate choice of the conductive probe, measuring mode and current preamplifier has to be made, as detailed in what follows.

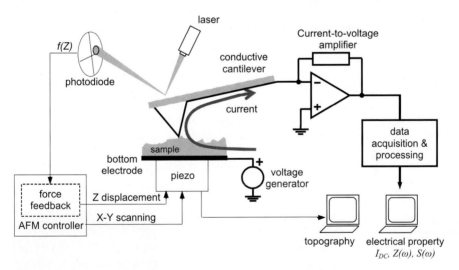

Fig. 12.6. Simplified diagram of the experimental setup for current-sensing AFM

12.3.1
Conductive Probes

Conductive probes used in CS-AFM are standard AFM probes typically made of silicon or silicon nitride coated with a thin conductive film, usually of gold, platinum or doped diamond. The main requirements that a conductive probe has to meet in CS-AFM are contamination and wear resistance (to avoid peeling off of the conductive coating during scanning). Normally, metal-coated probes do not satisfy these conditions as the metallic layer rapidly gets contaminated or peels off owing to tip–sample friction during tip–sample contact, and are thus unusable after a few contacts. Particularly tip contamination poses a severe problem, since, even in the case of careful preparation of the sample, a contamination layer and local debris are always present on the surface and they easily migrate on the tip surface during tip–surface contact. Typically one would use the AFM dynamic mode, in which the tip is oscillated in the proximity of the surface, in order to avoid tip–surface contact and hence tip contamination and wear. However, mechanical and electrical contact between the tip and the sample is a fundamental requisite of CS-AFM (see Sect. 12.3.2) and therefore dynamic mode cannot be used during current measurement. In this case, it is convenient for tip preservation to localize the target on the sample in dynamic mode, and then to switch to contact mode only when nanoscale current sensing is to be performed.

The best tips in terms of wear resistance currently available for CS-AFM are the doped diamond-coated probes [108–110], such as that shown in Fig. 12.7.

The extremely hard conductive coating of these probes can be used for days of tip–surface contact experiments under high shear forces, thus offering the possibility of reliable electrical measurements. It is worth noting that two drawbacks can limit the use of doped diamond-coated probes. On one hand, they display the typical granular structure of polycrystalline films, with approximately 50 nm grains as shown in Fig. 12.7c. With such a large surface roughness, the smallest tip radius that can be obtained is quite large, nominally on the order of 100–200 nm, which can tend to approximately 50 nm if a sharp edge of a grain is randomly obtained on the apex. Thus, the highest spatial resolution of images measured with these probes is rather limited. On the other hand, doped diamond-coated probes have a large intrinsic resistance on the order of several tens of $k\Omega$, which can be a limitation to conduction measurement on highly conductive samples such as carbon nanotubes. In these samples, also the intrinsic ac dispersion induced by the probes should be taken into account.

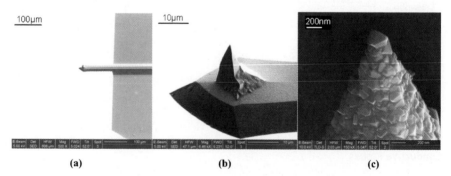

Fig. 12.7. Boron-doped diamond-coated probes that are commercially available (Nanosensors) and suited to current-sensing AFM. Scanning electron microscopy images with increasing magnification: **a** cantilever, **b** cone and **c** tip apex (nominal radius of 100–200 nm)

A main drawback of all the conductive probes described above is that the millimeter-scale probe assembly is completely covered with the conductive coating. This fact makes, on the one hand, the probe very sensitive to electrostatic forces given by tip–substrate voltage bias. As a result, the sample can be subjected to undesirable large voltage-bias-dependent loads applied by the probe, which can be a problem on soft materials, like biological samples. On the other hand, the conductive coating of the probe induces significant stray contributions in ac current measurements (impedance and electrical noise). As described in Sect. 12.5, stray contributions can be a serious limitation of NIM in some cases. To overcome these difficulties, innovative conductive probes are being developed by different groups, as probes with the conductive area limited to the tip apex or with an electrostatic shielding that properly screens the background electrostatic field lines [111–113].

12.3.2
Operating Modes of AFM

A main requisite in CS-AFM measurements is that the probe has to be in contact with the sample during the electrical measurement. To attain a stable tip–sample electrical contact, the atomic force microscope can be operated in contact AFM mode. This is a straightforward mode that maintains the tip in contact with the surface in the attractive van der Waals region by a constant force set point during scanning (Fig. 12.8a). In spite of its simplicity, contact mode has to be carefully used in CS-AFM, as in many cases the electric current measured depends on the load applied by the probe on the sample. Therefore, the dependence of the measured electrical property on the contact mode set point has to be investigated in order to provide reliable results. In addition, the scan velocity plays a key role, setting the measurement time per pixel of the image. Thus, it must be properly chosen to guarantee that the tip remains fixed in a location for enough time for the amplifier to probe the current and, in the case of impedance measurements, to achieve the desired electrical accuracy.

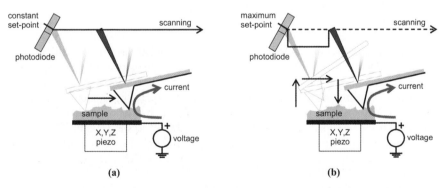

Fig. 12.8. Simplified diagrams of the AFM modes suited for current-sensing AFM. **a** Contact mode, in which the tip is maintained in contact with the surface by a constant-force set point, and **b** jumping mode, in which the tip probes the surface property in contact, but is moved laterally to the next pixel only after retraction

Contact mode, however, has the drawback of large shear forces during the lateral scanning, which becomes destructive for soft materials. As an alternative mode to electrically characterize biomaterials, such as biolayers and plasma membrane, jumping mode can be used [114]. In this case, as sketched in Fig. 12.8b, the tip is approached to contact until a maximum force set point is reached. After the electrical property in that position has been measured, the tip is retracted from contact and hence moved to the next point, thus suppressing the lateral forces acting on the sample. As a further advantage, jumping mode allows a precise control of the contact load applied by the tip, correcting the unavoidable drift in the zero force level [115]. This method has recently been combined with CS-AFM in the study of electrical conductivity of purple membrane [43].

Importantly, we remark here that the capability of controlling the tip deflection during the electrical measurement is one of the outstanding advantages of AFM. It offers the possibility of accurately tuning and monitoring the load applied onto the sample as well as the tip–substrate distance while the electric measurement is performed. Alternatively, the electrical transport properties can also be measured as a function of compression force and tip indentation into the sample [34, 36, 41].

12.3.3
Current Detection Instrumentation

For nanoscale electric current measurements, in which a very small signal or signal fluctuations are required to be detected, the proper design of the current-to-voltage amplifier is crucial.

Figure 12.9a shows the typical current-to-voltage amplifier scheme [116], generally used in C-AFM and available in many commercial atomic force microscopes. It consists of an operational amplifier in a transimpedance configuration. The feedback virtual ground (the inverting input of the operational amplifier) enables one to measure the current $i_i(t)$ flowing through the DUT, which is converted to voltage by the feedback resistance R_f. The output of the amplifier obtained is $v_o(t) = -R_f i_i(t)$. The choice of the resistor R_f is of fundamental importance for the circuit performance. In addition to setting the current–voltage gain, this resistor is responsible for the instrument noise and the settling time, i.e., the bandwidth, of the amplifier. Indeed, in this scheme the main source of noise at low frequency is the thermal noise of the feedback resistor R_f, equal to $4kT/R_f$. Therefore, to improve the current resolution, the feedback resistor must be chosen as large as possible. For example, to sense current levels ranging from nanoampere down below picoampere values, as in the case of insulating materials [13, 25], a feedback resistor of hundreds of MΩ is necessary. The main problem of this circuit resides in that, using such a high value resistor, a parasitic capacitance C_f shunts the resistor R_f, as shown in Fig. 12.9a, and hence reduces the frequency bandwidth to the pole $1/2\pi R_f C_f$. Although the amount of parasitic capacitance can be diminished by a careful feedback layout, it cannot be reduced below a few hundred femtofarads, therefore limiting the frequency bandwidth of the amplifier to a few kHz. In C-AFM, in which only the stationary current is measured, the transimpedance configuration is normally adequate. In some cases, when a large dynamic range of several decades is necessary, more sophisticated instrumentations are implemented such as those based on logarithmic amplifiers in

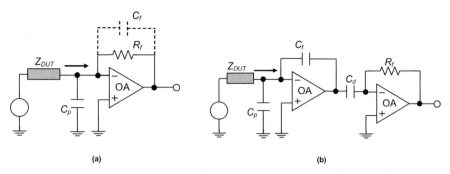

Fig. 12.9. Traditional current-to-voltage amplifiers: **a** transimpedance scheme, commonly used for current-sensing AFM; **b** integrator–differentiator scheme, used in patch-clamp measurement

scanning spreading resistance measurements [11] or those based on a semiconductor parameter analyzer [117].

Conversely, in ac CS-AFM (NIM and ENM), large bandwidth is a fundamental requirement to perform impedance/noise spectroscopy. To this end, using the transimpedance amplifier, one must decrease the value of the feedback resistor R_f. However, with this solution that trades the noise performance for a larger range of frequencies, the instrument noise reaches higher levels. In those applications in which the impedance values are relatively small, this is not a problem and in fact is the approach typically implemented in commercial impedance analyzers. However, for large impedance values, such as when addressing the local properties of insulating materials, these noise levels are usually unacceptable and more sophisticated circuits are required for ac CS-AFM, addressing the issue of large bandwidth ac spectroscopy.

A circuit solution consists of using the transimpedance amplifier followed by an amplifier with an increasing gain at higher frequencies than the pole $1/2\pi R_f C_f$, thus obtaining an overall flat frequency response [118–120]. Alternatively, a T-network can be connected in the feedback path of the transimpedance amplifier to compensate the effect of the parasitic capacitance C_f [121]. The drawback of these schemes resides in that accurate manual calibration of the balancing components is necessary to achieve a perfectly flat frequency response of the amplifier.

A dramatic improvement in the bandwidth *and* noise performance can be achieved using an integrator–differentiator scheme (Fig. 12.9b), as implemented in commercial patch-clamp instruments [122]. With the integrator, the noise issue is solved by removing the feedback resistor and operating the current–voltage conversion by means of the simple capacitance, which is a noiseless component in most cases. The differentiator block is then necessary to recover a flat frequency response. The disadvantage of this solution is in that it is not able to manage any dc (or leakage) current, which causes saturation of the integrator a within short measurement time. A periodic discharge, obtained with a switch shunting the feedback capacitor, is then required to reset the amplifier. However, owing to the limited time window, this switch-based solution is not suitable for CS-AFM, which must continuously measure the current during scanning over the sample surface. To obtain a continuous measurement time, the switch must be replaced by a continuously discharging system for the dc current, while measuring the ac current signal through the integrator–differentiator blocks. To this aim, a circuit scheme was recently proposed by Sampietro et al. [123] and successfully demonstrated connected to an atomic force microscope in nanoscale capacitance imaging [51]. The circuit is based on an additional amplifier in the feedback that forces the measured current back to the input node and hence constantly maintains the integrator output at 0 V. Thus, this amplifier offers the possibility to perform wide-bandwidth spectroscopy (from 1 Hz up to a few MHz) with the same noise performance as a traditional transimpedance amplifier with a gigaohm feedback resistor.

Importantly, we note that in all the circuit schemes discussed above, at high frequencies the instrument noise increases proportionally to the frequency and the total input capacitance C_p. This capacitance is the sum of several contributions, namely, the DUT capacitance (100 fF–1 pF in the case of CS-AFM), the operational amplifier input capacitance (a few picofarads) and the stray capacitance of connection cables, which is the major contribution when they are longer than a few centimeters. Thus, to reduce the instrument noise, the input capacitance C_p and hence the length of the DUT–amplifier cable must be reduced as much as possible.

12.4
Conductive Atomic Force Microscopy

The most frequently used CS-AFM technique is C-AFM [10]. As mentioned before, in C-AFM, as shown schematically in Fig. 12.10, the dc electric current is acquired while a a dc bias voltage is applied between the tip and a second electrode, typically the conductive substrate, or alternatively a lateral electrode deposited over the sample. Current detection is normally performed using a transimpedance amplifier

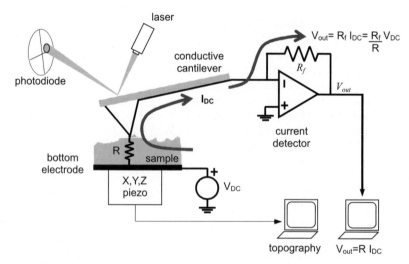

Fig. 12.10. The experimental setup for dc current-sensing AFM (commonly referred to as conductive AFM)

as described in Section 12.3.3, which is available in most commercial atomic force microscopes with different gain values. A logarithmic amplifier [11] and a semiconductor parameter analyzer [117] have also been used in order to measure the current in an extended dynamic range. Another approach consists of directly probing the differential conductance dI/dV [34, 36, 37] by applying a low-frequency voltage signal and measuring the ac current flowing through the sample with a current amplifier and a lock-in detection scheme, exactly as in the NIM configuration described in Sect. 12.5 (see Fig. 12.14).

C-AFM measurements have been performed using two basic procedures, 2D imaging and the point-contact measurement, both in contact mode. In imaging mode, dc conductive images at a fixed voltage are obtained by scanning the tip in the xy plane in contact with the surface under constant deflection feedback. In point-contact measurement, also referred to as spectroscopic mode, the tip is kept in contact with the surface in a fixed position, while the applied tip–substrate voltage bias is swept to measure the current–voltage characteristics $I(V)$.

Over the last 10 years, this method has been used extensively in a whole range of scientific/industrial applications and materials. An extremely large number of C-AFM studies have been reported in the field of semiconductor failure analysis [10–20]. In this context, C-AFM is an invaluable technique to test the reliability and performance of thin oxides on a nanometer scale in semiconductor devices of decreasing dimensions. By means of local current measurements, degradation in the dielectric behavior and changes of oxide thickness can be studied (Fig. 12.11). In addition, the electrical thickness of the oxide has been determined with subnanometer accuracy [13–15] by performing $I(V)$ measurements in the Fowler–Nordheim regime. Similarly, C-AFM has been used to understand the nanoscale conductivity and localized defects of organic semiconductor materials [21–23].

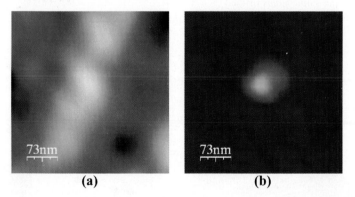

Fig. 12.11. Simultaneous topography (**a**) and dc current (**b**) images on a ultrathin SiO_2 layer (approximately 3 nm thickness and approximately 2 nm roughness) on a gold substrate with 4-V voltage bias applied. The images were acquired after electrically stressing a localized spot in the center of the scan region by obtaining an $I(V)$ curve. The bright spot in **b** corresponds to the stressed spot and clearly shows a large increase of conductance (around 1 nA) compared with the surrounding insulating area. Taking images at lower bias voltage, one observes a decrease in conductivity of the spot, indicating defect-assisted tunneling behavior triggered by high-voltage stress

A second important field of application of C-AFM is in the measurement of electrical conduction through individual molecules. When observing the transport of charge through a single molecule or a few molecules, C-AFM is an ideal technique to avoid information-averaging obtained in macroscale measurements and to provide an unambiguous model of the molecular conduction mechanism [24]. C-AFM has been used to study a wide range of molecular monolayers on a conductive substrate (gold or silicon), such as self-assembled-monolayers, typically alkanethiols [25–28], nitro-based molecules [29,30], carbon nanotubes [31] and single carotene molecules [24]. A characteristic example of a set of current–voltage characteristics obtained on octanethiols is shown in Fig. 12.12.

Another type of C-AFM measurement is the study of electrical conduction of single molecules as a function of the tip position. In this experimental configuration shown in Fig. 12.13a, the molecule is deposited on an insulating substrate (silicon oxide or mica) and in contact with a fixed microelectrode at one end. At the other

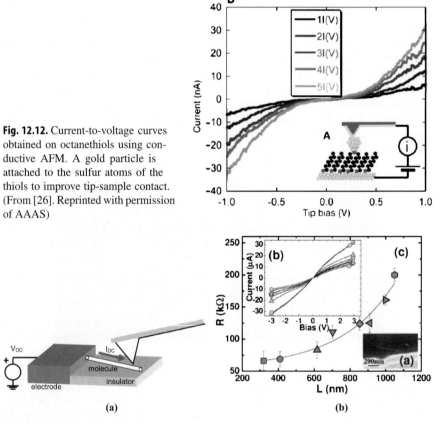

Fig. 12.12. Current-to-voltage curves obtained on octanethiols using conductive AFM. A gold particle is attached to the sulfur atoms of the thiols to improve tip-sample contact. (From [26]. Reprinted with permission of AAAS)

Fig. 12.13. a The conductive AFM experimental setup for measuring the electrical conduction of single molecules as a function of tip position. **b** Conductivity measurement as a function of length on a single-walled carbon nanotube. (Reprinted with permission from [33]. Copyright 2002 by the American Physical Society)

end, the conductive atomic force microscope tip is used as a movable electrode, brought into contact at a selected point along the molecule, and measurements are repeated at different tip–microelectrode distance. Such a configuration has allowed the study of the resistance per length of long molecules, such as carbon nanotubes (Fig. 12.13b) [32–34] and DNA [35] as well as resistance changes as a function of the loading force applied by the tip [36, 37].

More recently, C-AFM has been applied to the emerging field of bioelectronics. Studying the electron transport of biomolecules, such as membrane proteins, and their changes when they are immobilized on the electrode surface is of fundamental importance for the development of biocompatible devices. So far, besides studies on DNA molecules [35, 38, 39], C-AFM has been used for electrical conductivity measurements of proteins such as ferritin [40], metalloproteins [41], photosynthetic complex [42] and bacteriorhodopsin [43].

12.5
Nanoscale Impedance Microscopy

Whereas dc CS-AFM has been widely performed, the issue of nanoscale electrical characterization of ac conduction is yet to be fully addressed. The first impedance measurement at the nanoscale using CS-AFM, commonly referred to as nanoscale impedance microscopy (NIM), was performed by Shao et al. [44]. As in C-AFM, it combines standard contact AFM with the two-terminal current-sensing configuration. Figure 12.14 shows a diagram of the measurement setup. Here, the frequency-dependent current transport is measured by applying an ac voltage between the tip and the substrate, and hence acquiring the ac current flowing through the tip in contact with the surface. From the measured electrical current, the electrical impedance can be extracted as described in Sect. 12.2. The great advantage of this technique is that it allows the study of electron transport over a wide range of frequency set by the measurement instrument.

In performing ac current measurements with the atomic force microscope, one must be aware that the measured impedance of the system includes some parasitic contributions of the experimental setup. The most important of them is the stray capacitance contributions C_{stray} associated with the micrometer-sized tip cone and cantilever, the millimeter-sized chip, the conduction cables and the macroscopic probe assembly, including the holder and cables, which can be of the order of 100 fF–1 pF. The presence of stray capacitance makes the absolute capacitance measurement possible only when the capacitance probed by the apex of the tip is much larger than the stray contribution (Fig. 12.15). For example, this is the case when NIM is used to characterize microdevices by contacting micrometer-sized or nanometer-sized electrode pads. In the remaining situations, i.e., when the apex capacitance is smaller than the stray contributions, the nanoscale capacitance is completely overwhelmed by stray contributions of the probe assembly, which are typically 5 or 6 orders of magnitude larger. Absolute capacitance measurements are not allowed and only measurement of capacitance variations can be attempted. However, since also stray contributions can significantly change during lateral and vertical scanning, probing the tiny variation of local capacitance over stray variations

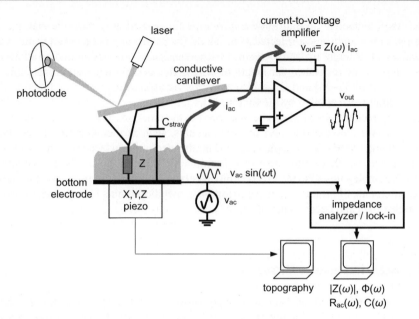

Fig. 12.14. The experimental setup of ac current-sensing AFM used for nanoscale impedance microscopy

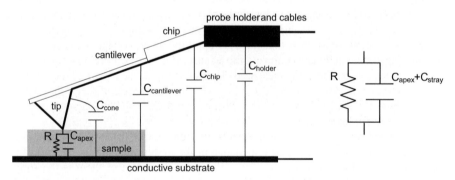

Fig. 12.15. The AFM probe system with the equivalent circuit components corresponding to the local impedance (probed by the nanometer-scale tip apex) and stray capacitance contributions of the probe assembly

is very challenging, requiring instrumentation with extremely high resolution and a sophisticated calibration procedure of stray effects. In particular, this is the case in ac electrical transport measurements through single molecules, dielectric thin films and many nanostructures in which the apex capacitance is typically tens of attofarads.

So far, the majority of experimental studies using NIM have been applied to conductive samples in the presence of dominating stray capacitance, limiting the impedance characterization to the ac resistance. In this case, impedance measurement can be performed by connecting the atomic force microscope to commercial impedance analyzers, which can measure impedance spectra in six or seven decades of frequency up to hundreds of MHz [124]. Using this approach, Shao et al. [44] demonstrated NIM in 2003 probing the ac resistance of polycrystalline ZnO in imaging mode as well as in spectroscopic mode, and obtained impedance spectra of individual grain boundaries at different dc bias voltages (Fig. 12.16). In the same year, Layson et al. [45] studied the ionic transport through solid films, and acquired impedance spectra in fixed positions of a conducting polymer electrolyte film. O'Haire et al. [46] extended the electrochemical application by obtaining impedance images and spectra of Faradaic charge transport through a Nafion electrolyte membrane.

Concerning capacitance measurements, some studies have employed the NIM technique to probe large capacitance values, orders of magnitude larger than stray capacitance ($C_{apex} > C_{stray}$) [47,48]. In this case, the microscope is used as a nanoprobe to precisely locate small features, such as submicron contacts and interconnects of integrated circuits, and perform standard impedance measurements in a specific position. Here, the locality of the capacitance measurement is not of concern, since the apex capacitance is not limited to the tiny area below the tip apex, but is extended to the whole surface of the contact pad and any metal wires connected to it, thus resulting in large and measurable values. In this context, the NIM method is a promising tool for identification of failures and frequency-response analysis of submicron semiconductor technologies.

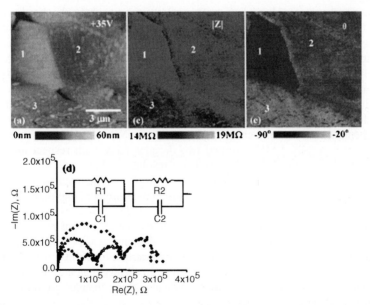

Fig. 12.16. Nanoscale impedance microscopy of polycrystalline ZnO ceramics. Images acquired at 35 V tip–sample dc bias (ac signal of 0.1 V amplitude and 10 kHz frequency): **a** topography; **b** impedance modulus $|Z|$ (log scale) and **c** impedance phase θ, indicating different impedance values in different grains. **d** Impedance spectroscopy in a fixed position at different applied voltage biases (5 V, 3 V, 2 V). (Reprinted with permission from [44]. Copyright 2005 American Institute of Physics)

A different situation is when the tip–sample capacitance is overwhelmed by stray capacitance ($C_{\mathrm{apex}} \ll C_{\mathrm{stray}}$). In this case, the localized capacitance is an issue, since detecting capacitance variations of local capacitance ΔC_{apex} as small as a few attofarads can be a major problem. In addition, a procedure to distinguish the local variations from the changes of stray components $\Delta C_{\mathrm{stray}}$ is required. To this aim, the chief requirement is the high capacitance resolution of the measurement system, at least around 1 aF over a sufficiently wide range of frequency to perform spectroscopic studies. Despite the wide range of frequency, commercial instruments for impedance spectroscopy are typically limited to the femtofarad range [124, 125] and hence are not capable of detecting such small capacitance variations. In order to perform impedance spectroscopy at the nanoscale, currently the best-suited detection approach is measuring the ac current flowing through the tip–sample system using a low noise current-to-voltage amplifier with a lock-in detection scheme, as described in Sect. 12.2. The first application of this method was reported by Lee et al. [49] in 2002 in order to extend the SCM technique to non-voltage-dependent capacitance measurement. They combined a transimpedance amplifier with a lock-in detector, adding a circuit solution to cancel the displacement current through the stray capacitance to further improve the system resolution. In addition, they analyzed the position-dependent artifacts given by stray capacitance in the lateral and vertical directions and proposed a procedure to take them into account. Using a similar setup, Pingree and Hersam [50] showed impedance images on a multiwalled nanotube.

Despite the attofarad resolution achieved, these setups have the disadvantage of a low-frequency bandwidth, up to a few tens of kHz, owing to the inherent trade-off between the noise and the bandwidth of the standard transimpedance amplifier discussed in Sect. 12.3.3. To this aim, the groups of Gomila and Sampietro [51] have recently implemented CS-AFM specifically to extend the capability of NIM to higher frequencies and to address the capacitance locality issue. With use of the lock-in detection scheme with an innovative design of a low-noise wide-bandwidth amplifier (also discussed in Sect. 12.3.3), the bandwidth limitation has been overcome. The instrument extends the frequency range available for NIM, from approximately 1 Hz up to approximately 1 MHz, around two decades more with respect to the state of the art, while achieving high impedance resolution, down to attofarads and the teraohm range in capacitance and resistance with 1-V signal applied. Furthermore, the possibility of simultaneous ac and dc current measurement makes the instrument a versatile tool for combining high-resolution dc and ac CS-AFM.

With use of these refined instrumental setups, the locality of nanoscale capacitance measurements has been experimentally demonstrated [49–52]. In particular, the authors demonstrated on a nanostructured ultrathin film of SiO_2 (Fig. 12.17) that the nanometer changes in the oxide thickness are detected as attofarad changes in capacitance [51]. A theoretical analysis indicates these changes are due largely to the apex capacitance.

A further demonstration that these experimental setups are able to detect small capacitance variations occurring underneath the apex of the probe was achieved by obtaining capacitance–distance curves ($C(Z)$) on a conductive substrate [49, 51, 52]. The $C(Z)$ curve allows the vertical distance below which the apex contribution becomes dominant with respect to stray variations to be determined. To accurately

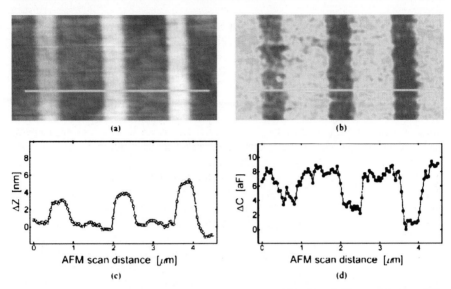

Fig. 12.17. Nanoscale capacitance imaging of a nanostructured thin film of silicon oxide (three SiO$_2$ lines of thickness of about 3, about 4 and about 5 nm over a 5 nm-thick SiO$_2$ layer). **a** AFM topography and **b** capacitance image. **c** The AFM topography profile and **d** the AFM capacitance profile corresponding to the lines indicated in the respective images (1 V amplitude and 130 kHz applied signal; 0 V dc bias; 0.5 s measurement time; 5 μm × 2, 5 μm, 64 × 128 pixels; 0.016 Hz scan frequency; and about 1 h imaging time; data acquired using WSxM software [115]). (Reprinted with permission from [51]. Copyright 2006 Institute of Physics)

assess the tip–substrate distance, the capacitance–distance curve needs to be measured simultaneously with the typical force–distance curve, as sketched in Fig. 12.18. At far distances from tip–substrate contact, a constant capacitance slope is measured owing to the major increase in stray capacitance. At a distance of a few nanometers from the surface, a rapid increase of the capacitance slope is detected owing to the onset of the apex contribution. This vertical distance sets the sample height below which nanoscale capacitance images are dominated by local electrical properties, since the stray gradient is smaller than the local capacitance gradient. Importantly, this distance depends on two parameters: (1) the tip radius and (2) the slope of the stray capacitance. The radius of curvature of the apex is by far the most important, since it determines the effective electrical area of the tip, thus directly affecting the amount of apex capacitance. Apex capacitance variations of 100–200 nm tip radius are on the order of a few tens of attofarads below about 20 nm distance from the conductive substrate. To increase the vertical region of localized capacitance, it is important to minimize the amount of stray capacitance, which gives the constant slope. This can be reduced below 1 aF/nm using an optimized setup with a shielded probe assembly (Fig. 12.19) [51]. To further decrease the parasitic slope, properly shielded probes are necessary, as already mentioned in Sect. 12.3.1.

By using a capacitance–distance curve, Lee et al. [52] for the first time managed to quantify the capacitance given by the aqueous meniscus between the tip and the sample. Furthermore, they demonstrated that this meniscus can be used to increase

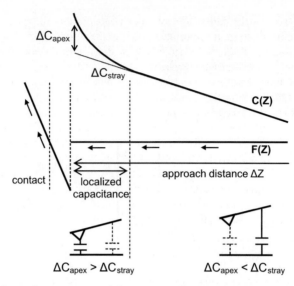

Fig. 12.18. The force–distance curve and the corresponding capacitance–distance curve on a conductive substrate. Below a given tip–substrate distance, which depends on the tip radius and the amount of stray capacitance, the apex capacitance variation becomes dominant over the changes in stray capacitance. In this region, the capacitance measurement is local

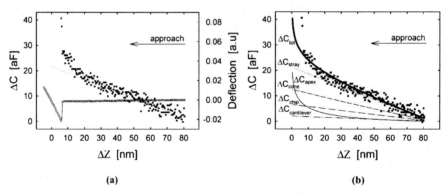

(a)   (b)

Fig. 12.19. a Experimental capacitance–distance (*squares*) and simultaneous force–distance (*circles*) curves on a gold substrate using a 100–200 nm nominal tip radius. The local capacitance contribution is detected at tip–substrate separations smaller than about 15 nm, where it becomes bigger than 0.2 aF/nm stray slope. **b** Theoretical estimation of capacitance and its individual contributions versus tip–substrate separation distance (data acquired using WSxM software [115]). (Reprinted with permission from [51]. Copyright 2006 of Institute of Physics)

the sensitivity of CS-AFM to the local capacitance variations of thin films owing to the high dielectric constant of water.

Thus, by combining high-resolution current sensing with lock-in detection, nanoscale capacitance microscopy allows local capacitance variations to be probed. It is worth mentioning that this method currently requires long acquisition times and a precise protocol for the calibration of stray contributions. In spite of these constraints, it is expected to gain importance, having the unique potential of providing direct information about the dielectric constant or thickness of ultrathin films, such as molecular layers and biological systems.

Finally, we note that there exists a well-known scanning probe technique to measure the capacitance at the nanoscale, referred to as scanning capacitance microscopy (SCM), which typically employs the resonant frequency detection technique. By including the tip–substrate capacitance into the oscillator, tip–sample capacitance variations are directly measured as shifts in the resonant frequency of the circuit with high sensitivity [67–69]. SCM has been typically applied to probe the local depletion capacitance of semiconductors, in which dC/dV measurement are performed and the localized capacitance is not an issue thanks to voltage signal enhancement. However, this method has a number of drawbacks, which make it unsuitable for NIM, including the difficulty of operation, since the circuit has to be properly tuned to resonance, limited frequency bandwidth, the decrease of resolution when the measured impedance has a resistive component and the requirement of large excitation voltage [49, 70]. Another alternative technique has been used to measure impedance, namely, the bridge scheme [126]. In this case, the tip–sample capacitance is one of the bridge components and variations alter the bridge balance, determining a voltage change between the two bridge arms. Brezna et al. [53, 54] have combined this detector with an atomic force microscope in order to perform local capacitance measurements of oxide layers. Although the bridge detector has proven attofarad resolution, here again this technique is not optimal for impedance spectroscopy, owing to a limited bandwidth of frequency (1–20 kHz), high excitation amplitudes (several volts) and difficulty of operation.

12.6
Electrical Noise Microscopy

Electrical noise is a fundamental characteristic of the transport medium, which is expected to gain importance in nanometer-scale systems, where atomic and defect fluctuations can dominate the average current flow. As an example, according to Hooge's law, low-frequency noise increases with shrinking dimensions of the system [127]. Until now electrical noise measurements at the nanoscale have been performed using microfabricated electrodes or by using STM. For instance, with use of a micrometer-sized electrode approach, experimental studies of electrical noise have recently been performed on single carbon nanotubes [128, 129] and quantum dots [130]. Concerning STM noise studies, different types of noise in the STM tunneling current have been observed, including $1/f$ noise in feedback operation mode and thermal noise at zero applied voltage [78, 79], thus leading to a modified scanning tunneling microscope (the scanning noise microscope) working at zero bias and based on the use of thermal current noise for feedback control. From the other side, noise in tunneling current has recently been used to monitor single-atom dynamics [80, 81]. However, as a characterization method of local electrical transport with nanoscale spatial resolution, STM noise spectroscopy is extremely limited by overwhelming noise fluctuations of the tunneling gap. Measurement of the electrical noise spectrum at the nanoscale by means of a scanning tip requires the use of a current-sensing atomic force microscope capable of maintaining the tip–surface contact and as a consequence probing the current fluctuations of the sample.

Despite that, nanoscale electrical noise measurements by means of AFM have been largely neglected up to now. This is mainly due to the very low level of current and therefore of current fluctuations of nanoscale systems, typically well below the resolution of conventional current noise meters. As for local capacitance measurement, nanoscale noise spectroscopy is a difficult task, requiring highly sensitive instrumentation.

So far, nanoscale noise spectroscopy by means of CS-AFM has been unexplored to our knowledge, except for proof-of-principle measurements recently reported by Fumagalli et al. [55]. In order to perform high-sensitivity noise spectroscopy, the current-sensing atomic force microscope was been used in a correlation spectrum analyzer configuration, described in Sect. 12.3. Figure 12.20 shows a diagram of the experimental setup of the modified atomic force microscope. Two low-noise wide-bandwidth amplifiers have been employed, one connected to the atomic force microscope tip and the other to the gold substrate. A bandwidth of approximately 1 MHz was available in the noise experiments. To demonstrate the instrument performance, discrete resistors of known value (1, 10 and 100 MΩ) were connected in series for a gold substrate.

Figure 12.21 shows the current power spectra obtained by the instrument at zero voltage bias and the corresponding theoretical level. According to the Nyquist formula [104], the thermal noise of the resistor R should be measured, which is given by $S_i(f) = \frac{4kT}{R}$, where $S_i(f)$ is current noise power density, k is Boltzmann's constant and T is the absolute temperature of the sample. Measurements and theoretical data are in good agreement over the spectral range from around 10 Hz up to a few kHz. At higher frequency, the correlated instrumental noise due to the voltage noise of the input amplifiers becomes dominant, thus preventing the measurement of thermal noise. These measurements demonstrate that the atomic force microscope in the

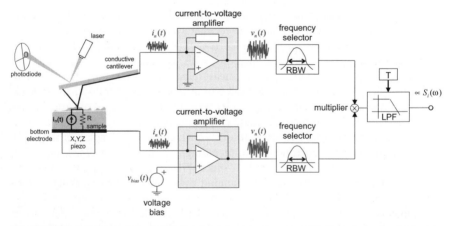

Fig. 12.20. Experimental setup for nanoscale noise spectroscopy. The atomic force microscope is connected to two wide-bandwidth current-to-voltage amplifiers in a correlation spectrum analyzer scheme to perform high-sensitivity noise spectroscopy measurement

Fig. 12.21. Noise spectroscopy using current-sensing AFM of a discrete resistor of known value (1, 10, or 100 MΩ) connected in series to the gold substrate. *Solid lines* indicate the corresponding theoretical level of Nyquist noise, showing good agreement with noise measurements. (Reprinted with permission from [55]. Copyright 2005 American Institute of Physics)

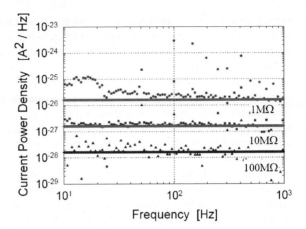

cross-correlation spectrum analyzer scheme can be successfully used as a tool for probing the noise spectral density at the nanoscale.

We remark that the major difficulty in performing noise measurements using CS-AFM resides in the high-level of electromagnetic pickup. In conventional noise measurements, external interference can be properly screened by shielding the whole setup, composed of the DUT and the cables. In the case of AFM, the major source of external interference is the atomic force microscope controller itself. Atomic force microscope control signals are directly picked up at the electrical connection between the nanoscale system of the microscope and the spectrum analyzer. Therefore, although the atomic force microscope and the cables are carefully shielded, some pickup cannot be suppressed.

Furthermore, we notice that noise measurements are extremely sensitive to the tip–surface contact. A discontinuous electrical contact determines current fluctuations, which completely change the noise spectrum. A stable electrical contact has to be maintained during the whole time required by noise spectroscopy (that can be a few minutes in the case of low-frequency measurements).

The noise measurement of Fig. 12.21 also shows another application of noise spectroscopy which can be used as an alternative method for probing the nanoscale resistance. This can be measured using the C-AFM technique, which, however, requires a voltage bias across the sample. Owing to the extremely small tip radius and tip–substrate distance, voltage bias can induce a high electrical field, thus sample heating and unnatural mechanical stress develops on soft samples, such as biological materials. Given the Nyquist formula, thermal noise measurement enables us to obtain the local resistance at zero voltage bias, and is thus a suitable measurement technique for delicate materials.

12.7
Conclusions

CS-AFM is a very versatile technique to probe the electrical transport properties of materials and devices at the nanoscale. By measuring the electric current flowing through a sample, information on the resistance, impedance and electric noise transport properties can be obtained with nanoscale spatial resolution. The level of development of the technique varies substantially depending on the measuring technique considered.

C-AFM used to measure dc currents and to extract resistance or conductivity information is currently a mature technique available in most commercial atomic force microscopes, and does not require of new technical developments at present. It has been particularly applied to the study of thin oxide films, single-molecule conduction mechanisms (both in organic as well as in biological molecules), nanotube characterizations, etc.

On the other hand, NIM used to obtain information on the electrical impedance is still under development and is not available commercially. Especially new advances are required in those setups intended to measure the local impedance properties of the materials, in which the measuring sensitivity or the frequency range has to be greatly increased in order to overcome the unavoidable stray capacitance contributions. Conversely, in those cases in which the measured impedances are smaller than the stray ones, as is very often the case in electronic circuit tests, the technique can be easily applied by coupling the atomic force microscope to commercial impedance analyzers and has already shown its usefulness.

With respect to the electrical noise measurements, they have only been performed with CS-AFM at the level of proof of principle, and hence a full range of opportunities remains in this area.

Although not covered in this chapter, it is worth remarking once more that extracting the material parameters (resistivity, dielectric constant, trap density distribution, etc.) from the transport measurements requires in all cases use of the appropriate theoretical models, as usual in semiconductor device modeling. In the case of CS-AFM, the modeling can be somewhat more difficult owing to the stringent geometry related to the measuring probe, which prevents in most cases the derivation of analytical formulae. Also here, the more advanced models allow interpretation of the dc current measurements, while a lot of work is still necessary in interpreting the ac current and noise measurements.

In summary, CS-AFM is currently a very active field of research and it is expected that in the forthcoming years, when some of the drawbacks existing at present will have been solved, that fascinating information on the electrical transport properties of nanoscale systems will be revealed, thus providing us with new insights into the properties of the nanoworld.

Acknowledgements. The authors gratefully acknowledge Prof. Marco Sampietro (Politecnico di Milano) and Prof. Josep Samitier (Universitat de Barcelona) for assistance in the development of CS-AFM research. This work was supported by SPOT-NOSED European project (IST-2001-388899) and the Nanobiomol Spanish project (NAN2004-09415-C05-01).

References

1. International Technology Roadmap for Semiconductors (2006) ITRS 2006 update. http://www.itrs.net/Links/2006Update/2006UpdateFinal.htm
2. Deen MJ, Pascal F (2006) J Mater Sci Mater Electron 17:549

3. Reed MA, Zhou C, Muller CJ, Burgin TP, Tour JM (1997) Science 278:252
4. Collier CP,Mattersteig G, Wong E, Luo Y, Beverly K, Sampaio J et al (2000) Science 289:1172
5. Joachim C, Gimzewski JK, Aviram A (2000) Nature 408:541
6. Schon JH, Meng H, Bao Z. (2001) Nature 413:713
7. Porath D, Bezryadin A, de Vries S, Dekkar C (2000) Nature 403:635
8. Davis JJ, Morgan DA, Wrathmell CL, Axford DN, Zhao J, Wang N (2005) J Mater Chem 15:2160
9. James DK, Tour JM (2004) Chem Mater 16:4423
10. Murrel MP, Welland ME, O'Shea SJ, Wong TMH, Barnes JR, McKinnon AW et al. (1993) Appl Phys Lett 62:786
11. Sato T, Kasai S, Hasegawa H (2001) Appl Surf Sci 175:181
12. O'Shea SJ, Atta RM, Murrell MP, Welland ME (1995) J Vac Sci Technol B 13:1945
13. Olbrich A, Ebersberger B, Boit C (1998) Appl Phys Lett 73:3114
14. Olbrich A, Ebersberger B, Boit C, Vancea J, Hoffmann H, Altmann H et al. (2001) Appl Phys Lett 78:2934
15. Mang KM, Khang Y, Park YJ, Kuk Y, Lee SM, Williams CC (1996) J Vac Sci Technol B 14:1536
16. Porti M, Nafria M, Aymerich X, Olbrich A, Ebersberger B (2001) Appl Phys Lett 78:4181
17. Porti M, Nafria M, Aymerich X, Olbrich A, Ebersberger B (2002) J Appl Phys 91:2071
18. Boxley CJ, White HS, Gardner CE, Macpherson JV (2003) J Phys Chem B 107:9677
19. Bailon MF, Salinas PFF, Arboleda JPS (2006) IEEE Trans Devices Mater Reliab 6:186
20. Zhanga L, Mitani Y (2006) Appl Phys Lett 88:032906
21. Kelley TW, Granstrom EL, Frisbie CD (1999) Adv Mater 11:261
22. Seshadria K, Frisbieb CD (2001) Appl. Phys Lett 78:993
23. Nakamura M, Yanagisawa H, Kuratani S, Iizuka M, Kudo K (2003) Thin Solid Films 438:360
24. Ramachandran GK, Tomfohr JK, Jun Li, Sankey OF, Zarate X, Primak A, Terazono Y et al (2003) J Phys Chem B 107:6162
25. Wang W, Lee T, Reed MA (2003) Phys Rev B 68:035416
26. Cui XD, Primak A, Zarate X, Tomfohr J, Sankey OF, Moore AL, Moore TA et al (2001) Science 294:571
27. Wold DJ, Frisbie CD (2000) J Am Chem Soc 122:2970
28. Zhao J, Uosaki K (2003) Appl Phys Lett 83:2034
29. Fan F-RF, Yang JP, Cai LT, Price DW, Dirk SM, Kosynkin DV, Yao YX et al (2002) J Am Chem Soc 124:5550
30. Rawlett AM, Hopson TJ, Nagahara LA, Tsui RK, Ramachandran GK, Lindsay SM (2002) Appl Phys Lett 81:3043
31. Paulson S, Helser A, Buongiorno Nardelli M, Taylor RM II, Falvo M et al (2000) Science 290:1742
32. Dai H, Wong E, Lieber C (1996) Science 272:523
33. de Pablo PJ, Gómez-Navarro C, Colchero J, Serena PA,Gómez-Herrero J, Baró AM (2002) Phys Rev Lett 88:36804
34. Gómez-Navarro C, dePablo PJ, Gómez-Herrero J (2006) J Mater Sci Mater Electron 17:475
35. de Pablo PJ, Moreno-Herrero F, Colchero J, Gómez Herrero J, Herrero P, Baró AM et al (2000) Phys Rev Lett 85:4992
36. Gómez-Navarro C, Pedro J, dePablo PJ, Gómez-Herrero J (2004) Adv Mater 16:549
37. Gómez-Navarro C, Saenz JJ, Gómez-Herrero J (2006) Phys Rev Lett 96:76803
38. Cohen H, Nogues C, Naaman R, Porath D (2005) Proc Natl Acad Sci USA 102:11589
39. Cohen H, Nogues C, Ullien D, Daube S, Naaman R, Porath D (2005) Faraday Discuss 1:1
40. Xu D,Watt FD, Harb JN, Davis R (2005) NanoLett. 5:571

41. Zhao J, Davis JJ (2003) Nanotechnology 14:1023
42. Stamouli A; Frenken JWM, Oosterkamp TH, Cogdell RJ, Aartsma TJ (2004) FEBS Lett 560:109
43. Casuso I, Fumagalli L, Padrós E, Gomila G (2007) Appl Phys Lett 91:63111
44. Shao R, Kalinin SV, Bonnell DA (2003) Appl Phys Lett 82:1869
45. Layson A, Gadad S, Teeters D (2003) Electrochim Acta 48:2207
46. O'Haire R, Lee M, Prinx FB (2004) J Appl Phys 95:8382
47. Pingree LSC, Martin EF, Shull KR, Hersam MC (2005) IEEE Trans Nanotechnol 4:255
48. Houzé F, Chrétien P, Schneegans O, Meyer R, Boyer L (2005) Appl Phys Lett 86:123103
49. Lee D, Pelz JP, Bhushan B (2002) Rev Sci Instrum 73:3525
50. Pingree LSC, Hersam MC (2005) Appl Phys Lett 87 233117-1
51. Fumagalli L, Ferrari G, Sampietro M, Casuso I, Martínez E, Samitier J, Gomila G (2006) Nanotechnology 17:4581
52. Lee D, Pelz JP, Bhushan B (2006) Nanotechnology 17:1484
53. Brezna W, Schramboeck M, Lugstein A, Harasek S, Harasek S, Enichlmair H, Bertagnolli E et al (2003) Appl Phys Lett 83:4253
54. Brezna W, Harasek S, Lugstein A, Leitner T, Hoffmann H, Bertagnolli E, Smoliner J (2005) J Appl Phys 97:93701-1
55. Fumagalli L, Casuso I, Ferrari G, Gomila G , Sampietro M, Samitier J (2005) AIP Conf Proc 780:575
56. Martin Y, Abraham DW, Wickramasinghe HK (1988) Appl Phys Lett 52:1103
57. Stern JE, Terris BD, Mamin HJ, Rugar D (1988) Appl Phys Lett 53: 2717
58. Nonnenmacher M, O'Boyle MP, Wickramasinghe HK (1991) Appl Phys Lett 58:2921
59. O'Boyle MP, Hwang TT, Wickramasinghe HK (1999) Appl Phys Lett 74:2641
60. Kalinin SV, Bonnell A (2001) Appl Phys Lett 78:1306
61. Kalinin SV, Jesse S, Shin J, Baddorf AP, Guillorn MA, Geohegan DB (2004) Nanotechnology 15:907
62. Sorokina KL, Tolstikhina AL (2004) Crystallogr Rep 49: 541
63. Colton RJ (2004) J Vac Sci Technol B 22:1609
64. Fujihira M (1999) Annu Rev Mater Sci 29:353
65. Palermo V, Palma M, Samori P (2005) Adv Mater 17:1
66. Kalinin SV, Shao R, Bonnell DA (2005) J Am Ceram Soc 88:1077
67. Williams CC, Hough WP, Rishton SA (1989) Appl Phys Lett 55:203
68. Williams CC (1999) Annu Rev Mater Sci 29:471
69. Tran T, Oliver DR, Thomson DJ, Bridges GE (2001) Rev Sci Instrum 72:2618
70. Lanyi S (2002) Acta Phys Slov 52:55
71. Matey JR, Blanc J (1984) J Appl Phys 57:1437
72. Binnig G, Rohrer H, Gerber C, Weibel E (1982) Phys Rev Lett 49:57
73. Datta S, Tian W, Hong S, Reifenberger R, Henderson JI, Kubiak CP (1997) Phys Rev Lett 79:2530
74. Jager ND, Marso M, Salmeron M, Weber ER, Urban K, Ebert P (2003) Phys Rev B 67: 165307
75. Sakai A, Kurokawa S, Hasegawa Y (1996) J Vac Sci Technol A 14:1219
76. Kurokawa S, Sakai A (1998) J Appl Phys 83:7416
77. Arakawa A, Nishitani R (2001) J Vac Sci Technol B 19:1150
78. Möller R, Esslinger A, Koslowski B (1989) Appl Phys Lett 55:2390
79. Möller R, Esslinger A, Koslowski B (1990) J Vac Sci Technol A 8:590
80. Stroscio JA, Celotta RJ (2004) Science 306:242
81. Stroscio JA, Gavazza F, Crain JN, Calotta RJ, Chaka AM (2006) Science 313:948
82. Binnig G, Rohrer H, Gerber C (1986) Phys Rev Lett 56:930
83. Stevens GC, Baird PJ (2005) IEEE Trans Dielec Electr Insul 12:979

84. Greene ME, Kinser CR, Kramer DE, Pingree LSC, Hersam MC (2004) Microsc Res Tech 64:415
85. Barsoukov E, Macdonald JR (eds) Impedance spectroscopy: theory, experiment and applications, 2nd edn. Wiley, Hobeken
86. Sze SM (1981) Physics of semiconductor devices. Wiley, New York
87. Schroder DK (1988) Semiconductor material and device characterization. Wiley, New York
88. Schroder DK, Park J-E, Tan S-E, Choi BD, Kishino S, Yoshida H (2000) IEEE Trans Electron Devices 47:1653
89. Viscor P, Vedde J (1997)US Patent 5,627,479
90. MacDonald DD (1991) In: Varma R, Selman JR (eds) Techniques for characterization of electrodes and electrochemical process. Wiley, New York, chap 11
91. Katz E, Willner I (2003) Electroanalysis 15:913
92. Landauer R (1998) Nature 392:658
93. Van Kampen NG (2001) Fluct Noise Lett 1:1
94. Jones BK (1993) Adv Electron Phys 87:201
95. Vandamme LKJ (1994) IEEE Trans Electron Devices 41:1936
96. Ralls KS, Skocpol WJ, Jackel LD, Howard RE, Fetter LA, Epworth RW, Tennant DM (1984) Phys Rev Lett 52:228
97. Kandiah K, Deighton MO, Whiting FB (1989) J Appl Phys 66:93
98. Mueller HH, Schulz M (1998), J Appl Phys 83:1734
99. Blanter YaM., Buttiker M (2000) Phys Rep 336:1
100. Gomila G, Reggiani L (2000) Phys Rev B 62:8068
101. Saminadayar L, Gattli DC, Jin Y, Etienne B (1997) Phys Rev Lett 79:2526
102. de Picciotto R, Reznikov M, Heiblum M, Umansky V, Bunin G, Mahalu D (1997) Nature 389:162
103. Xiao M, Martin I, Yablonovitch E, Jiang HW (2004) Nature 430:435
104. van der Ziel A (1970) Noise: sources, characterization, measurement. Prentice Hall, Englewood Cliffs
105. Sampietro M, Fasoli L, Ferrari G (1999) Rev Sci Instrum 70:2520
106. Ferrari G, Sampietro M (2002) Rev Sci Instrum 73:2717
107. Sampietro M, Accomando G, Fasoli LG, Ferrari G, Gatti EC (2000) IEEE Trans Instrum Meas 49:820
108. O'Shea SJ, Atta RM, Welland ME (1995) Rev Sci Instrum 66:2508
109. Lantz MA, O'Shea SJ, Welland ME (1998) Rev Sci Instrum 69:1757
110. Trenkler T, Hantschel T, Stephenson T, De Wolf P, Vandervorst W, Hellemans L et al (2000) J Vac Sci Technol B 18:418
111. Frederix PLTM, Gullo MR, Akiyama T, Tonin A, de Rooij NF, Staufer U et al (2005) Nanotechnology 16:997
112. Menozzi C, Gazzadi GC, Alessandrini A, Facci P (2005) Ultramicroscopy 104:220
113. Pingue P, Piazza V, Baschieri P, Ascoli C, Menozzi C, Alessandrini A, Facci P (2006) Appl Phys Lett 88:043510-1
114. de Pablo PJ, Colchero J, Gómez-Herrero J, Baró AM (1998) Appl Phys Lett 73:3300
115. Horcas I, Fernandez R, Gomez-Rodriguez JM, Colchero J, Gómez-Herrero J, Baró AM (2007) Rev Sci Instrum 78:013705
116. Horowitz P, Hill W (1989) The art of electronics, 2nd edn. Cambridge University Press, Cambridge
117. Blasco X, Nafria M, Aymerich M (2005) Rev Sci Instrum 76:16105
118. Carlà M, Lanzi L, Pallecchi E, Aloisi G (2004) Rev Sci Instrum 75:497
119. Kim D, Koo J (2005) Rev Sci Instrum 76:023703
120. Ciofi C, Crupi F, Pace C, Scandurra G (2006) IEEE Trans Instrum Meas 55:814

121. Howard R (1999) Rev Sci Instrum 70:1860
122. Molecular Devices (2007) Axopatch 200B. http://www.moleculardevices.com
123. Sampietro M, Ferrari G, Natali D (2005) Int Patent WO 2005/062061
124. Agilent Technologies (2007) Impedance Analyser 4294A precision impedance amplifier. http://www.agilent.com
125. Gamry Instruments (2006) Accuracy contour plots. http://www.gamry.com
126. Adams ED (1993) Rev Sci Instrum 64:601
127. Hooge FN(1969) Phys Lett A 29:139
128. Philip G. Collins, M. S. Fuhrer, A. Zettla (2000) Appl Phys Lett 76:894
129. Ishigami M,Chen JH, Williams ED, Tobias D, Chen YF, Fuhrer MS (2006) Appl Phys Lett 88:203116
130. Gustavsson S, Leturcq R, Simovic B, Schleser R, Ihn T, Studerus P, Ensslin K (2006) Phys Rev Lett 96:076605

Subject Index